Cultural Geography

Contents

OXFORD

UNIVERSITY PRESS

Oxford University Press is a department of the University of Oxford.
It furthers the University's objective of excellence in research, scholarship,
and education by publishing worldwide. Oxford is a registered trade mark of
Oxford University Press in the UK and in certain other countries.

Published in Canada by
Oxford University Press
8 Sampson Mews, Suite 204,
Don Mills, Ontario M3C 0H5 Canada

www.oupcanada.com

Copyright © Oxford University Press Canada 2014

The moral rights of the author have been asserted

Database right Oxford University Press (maker)

First Edition published in 2000
Second Edition published in 2006

Library and Archives Canada Cataloguing in Publication

Norton, William, 1944–
Cultural geography : environments, landscapes, identities,
inequalities / William Norton and Margaret Walton-Roberts.
— 3rd ed.

Includes bibliographical references and index.
ISBN 978–0–19–542954–1

1. Cultural geography. 2. Human geography. I. Walton-
Roberts, Margaret, 1968– II. Title.

GF41.N66 2013   304.2   C2013-901109-9

Cover image: © Hemis/Alamy

Frederickson's extract, page 229: By permission of Oxford University Press, USA.

Oxford University Press is committed to our environment.
This book is printed on Forest Stewardship Council® certified paper
and comes from responsible sources.

Printed and bound in Canada

1 2 3 4 — 17 16 15 14

Cultural Geography

Environments, Landscapes,
Identities, Inequalities

Third Edition | William Norton • Margaret Walton-Roberts

OXFORD
UNIVERSITY PRESS

Chapter 8 ❖ Power, Identity, Representation 246

Chapter 9 ❖ Living in Place: The Socio-Spatial Relationship 279

Chapter 10 ❖ Cultural Geography—Continuing and Unfolding 311

Figures, Tables, Boxes, and Photo Essays

Figures

Tables

Boxes

Photo Essays

Preface and Acknowledgments

The intent of this third edition of *Cultural Geography* is unchanged from the previous edition, namely to offer undergraduate students a comprehensive introduction to cultural geography as a subdiscipline of human geography. But the manner in which this intent is addressed and accomplished is much changed and improved, largely through the transition from a single-authored to a joint-authored text. This edition is more sharply focused throughout, less concerned with disciplinary history and methodological details, more nuanced conceptually, and empirically more relevant. Most notably, Chapters 8 and 9 have been completely rethought and rewritten, and now include analyses of current, global case studies.

This third edition is structured as follows. There is a brief introductory statement discussing the subdiscipline of cultural geography. This is followed by two chapters that combine to provide the conceptual background needed to help students understand the contents of later chapters. Chapter 2 covers what is sometimes called traditional cultural geography, with emphasis on ideas about humans and nature and the seminal writings of Carl Sauer, while Chapter 3 focuses on the many ways in which cultural geography has been rethought since the 1970s as Marxist, humanist, feminist, and postmodern theories have largely replaced more traditional approaches. The next six chapters deal, respectively, with environmental issues, the evolution of cultural landscapes, regional landscapes, global identities and inequalities, intersections of identity, and power, place and identity. A new feature in this edition is the inclusion of icons that direct students to relevant Further Readings that are included at each chapter end: ➔➤➤

The publication of this new edition owes much to both the professionalism and friendship of Tamara Capar, who supported the writing process throughout, combining constructive criticism of draft chapters with gentle cajoling as writing deadlines neared. Among the many others at Oxford who deserve our thanks are Peter Chambers, Maria Jelinek, and Caroline Starr. Margaret wishes to thank William Norton and Tamara Capar for their patience, Huzan Dordi for his contribution to Chapter 9, and Ben Roberts for his understanding and selfless support. William wishes to thank Barry Kaye, who continues to take great interest in the contents of this book and to identify new sources of ideas and information; also, thanks again to Pauline, who has been unfailingly supportive of this writing endeavor.

CHAPTER 1

Introducing Cultural Geography

Cultural geography is about understanding the relationship between people and the places they occupy by analyzing cultural identities and cultural landscapes. It derives from a longstanding interest in culture—especially the geographic expression of culture in landscape—as a causal mechanism, and from a more recent concern with cultural politics—especially the social and spatial constitution of cultures. It is concerned with the local and the global, and acknowledges that much of what is evident at the local scale is linked to global matters. Cultural geography incorporates both traditional and newer conceptual bases, and is closely related to other areas of geographic interest, especially social, economic, political, environmental, and physical geography; to other academic disciplines, especially anthropology, history, political science, psychology, and sociology; and to such interdisciplinary concerns as cultural, women's, and ethnic studies.

This introductory chapter aims to provide you with the necessary broad context for tackling the chapters that follow. In this chapter you will find

- ✤ four examples of cultural geographic research, which are intended to provide at least a flavor of what is to come;

- ✤ a discussion of the larger scholarly context of cultural geography, both as an area of social science and as an interest within human geography. This discussion will help you appreciate that cultural geography does not function in a scholarly vacuum, but has origins and links to other academic interests;

- ✤ an introduction to key terms and ideas, together with an explanation of the varying interpretations of these;

- ✤ an outline of six prominent themes in the study of cultural geography along with the conceptual bases for these themes. It is acknowledged that multiple and ever-changing themes to a changing subject matter are welcome and necessary;

- ✤ a summary statement.

If cultural geography is already beginning to appear like a comprehensive, far-reaching, and challenging field of study, then that impression is correct. To better understand this explanation of cultural geography, consider four seemingly very different geographic phenomena that are discussed in this textbook: wheat fields, monuments, gay neighborhoods, and segregated beaches.

What these four have in common is that they are all parts of cultural landscapes created and recreated over time by people as they relate, through their cultural identities, both to local and to more distant places. Initially, as people seek food and shelter, they change and add to local physical landscapes in an incremental and repetitive manner; through time, as cultures and technologies evolve, these processes of **landscape** change become increasingly rapid and complex. ➙➤➤ **1**

Culture, deriving from the verb "to cultivate," suggests the ability to take actions that result in improvement; this can be in the creation of desired landscapes, or of the self. Culture then needs to be seen as a process. This objective of improvement is never wholly achieved, for at any given time there are examples both of intended landscapes and of unintended landscapes that are expressive of change. To paraphrase a popular definition of geography, cultural geography examines what landscapes are located where, why they are where they are, and what the significance of place-to-place variations is. ➙➤➤ **2**

It is not surprising, then, that cultural geography is not simply of academic interest: it is a pragmatic concern. The role of place as it relates to the human condition was stressed by the distinguished cultural geographer Wagner (1990:41) when he asserted that "a theoretically well-grounded, intellectually vigorous, and practically effective social and cultural geography might well assume, in time, a major role in guiding and guarding the evolution of humanity's environments." Certainly, students of cultural geography are exposed to a diverse body of concepts and facts that, together, raise questions and offer insights into the condition of our world and its inhabitants. Indeed, within the larger discipline of human geography, cultural geography is recognized as an increasingly central concern, with Zelinsky (1995:750) claiming that cultural geography is a "scholarly discourse that has shifted from the comfort of placid marginality toward the overheated vortex of ferment and creativity in today's human geography." This shift began about 1970 and continues apace as cultural geography has undergone a transformation from an essentially descriptive, empiricist enterprise to a more conceptually varied search for understanding and meaning of both peoples and places. ➙➤➤ **3** In order to paint as comprehensive a picture as possible, this textbook reflects both the traditional and the newer emphases and, where appropriate, highlights the continuity of the larger cultural geographic enterprise.

Doing Cultural Geography

The first of the four brief discussions designed to whet your appetite for what follows reflects a traditional approach to the study of cultural geography—what you will come to know as the **landscape school**. The primary objectives of this approach are to identify cultural groups, study the relationships these groups maintain with the natural world, delimit the geographic regions they occupy, and describe the landscapes they create.

The second, third, and fourth discussions reflect a more recently developed set of concerns that you will come to know as the **new cultural geography**, although, as this term first appeared during the 1980s, much of the new cultural geography is not so new anymore. In the second discussion the focus is on the complex intertwining of nature and culture, recognizing that natural environments cannot be objectively discussed but rather are necessarily best interpreted as cultural understandings. The third discussion explores recent developments in cultural geography that explore affect, or emotion, in this case considering the geographies of violent affect, or the relationship between affect (emotion, performance) and war. The fourth discussion focuses on matters of cultural identity, especially on how identities are formed with reference both to other groups of people and to places as these places are contested between groups.

Identifying Groups and Describing Landscapes

A traditional concern of cultural geography is to identify groups, usually according to some criterion such as language or religion, and to describe the regional landscape created by these groups in the areas they occupy. Most work of this kind has treated both "culture" and "landscape" as uncontested terms, although as discussed later, both terms are complex and uncertain. One example of a distinctive regional cultural landscape is that part of the American inter-montane west settled by members of the Church of Jesus Christ of Latter-day Saints, popularly known as Mormons. Who are the Mormons, and what is distinctive about the landscape they have created?

The Mormon religion originated in upstate New York in 1830 from the experiences, claims, and achievements of Joseph Smith Jr. According to his own testament, Smith was visited by God and Jesus and told to found a new church. Converts accepted both the principle of prophetic revelation and Smith as a prophet, believed that Jesus had restored his Church through Smith, and accepted the Book of

Mormon—an account of Hebrew immigrants to the western hemisphere around 600 BCE—as a companion to the Bible. Church growth was rapid, and a series of movements through Ohio, Missouri, and Illinois culminated in the 1847 migration to the Salt Lake Valley—understood as the Promised Land. From this center Mormons began a systematic and disciplined occupation of the surrounding valleys in Utah and neighboring states. Two characteristics of the church help explain why church members functioned as a group, creating a distinctive landscape in the areas settled. First, all members were expected to participate in the work of the church, including the work of settling the landscape around Salt Lake City that began in the years after 1847. Second, Mormons are a society of believers practicing co-operative effort and support for other members. The concept

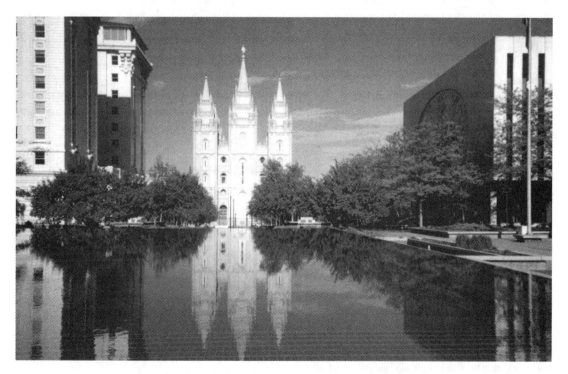

Located in a two-block area in downtown Salt Lake City, the Temple Square complex includes the Temple, Tabernacle, Assembly Hall, two visitors' centers, Church Office building, Joseph Smith Memorial Building Church Administration Building, Lion House, Beehive House, and various monuments and memorials. Located close by is a church museum, Family History Library, historic log cabin, and Mormon handicraft store. The urban landscape speaks clearly of the Mormon presence and of Mormon power. (William Norton)

of community is evident in the favored pattern of Mormon village settlement.

The Mormon village and agricultural landscape contains distinctive features that are absent in surrounding areas. ➔➤➤ 4 In towns, these features include especially wide streets, unpainted barns and granaries adjacent to residences, ward chapels, many I-style homes, and the use of brick for construction. In the surrounding agricultural area, roadside irrigation ditches, an open field landscape, crude unpainted fences, and wooden derricks used for loading hay characterize the landscape. This Mormon village and agricultural landscape is distinctive primarily because members of the Mormon cultural group share characteristics that affect their activities, and it is these activities, related to their cultural identity, that explain how the landscape evolved to become what it is today. Notwithstanding the worldwide reach of the contemporary church, for many Mormons the landscape created in the intermontane west continues to be a fundamental component of their cultural identity.

Nature and Culture: Social Nature

Typically when one thinks about environment, one might separate it from people and imagine it as separate from, rather than part of, human domain. Today the world we live in has become so commodified it is hard to imagine a place that is not influenced by human activity. Think of some of the most pristine environments today, and what you find is something that is fully integrated in human systems of monitoring and control. The Arctic ice caps are subject to scientific monitoring and human-induced change; on the African plains of the Serengeti animals are remotely monitored and tracked. However, societies do not construct nature as they please; there are myriad "social" and "natural" entities conjoined as human/non-human "hybrids" in millions of actor–networks organized at multiple geographical scales. One important example of such networks is captured by the work of Donna Haraway on **cyborgs**. Haraway explains that "a cyborg is a cybernetic organism, a hybrid of machine and organism, a creature of social

reality as well as a creature of fiction" (Haraway 1991). Haraway considers items as, for example, prosthetics, reproductive fertility such as "test tube babies," and the oncomouse, a trademarked mouse that has been genetically altered and specifically bred for use in cancer research experiments, all to be examples of the hybrid, cyborg interactions between human and non-human. Other scholars that have similarly explored the cultural dimensions of science and nature include French sociologist Bruno Latour, who devised the idea of actor–network theory (ANT) to demonstrate how integrated the human and non-human are in the contemporary world. For example, his work on climate change ponders the fact that a single article on climate change "mixes together chemical reactions and political reactions. . . . A single thread links the most esoteric sciences and the most sordid politics, the most distant sky and some factory in the Lyon suburbs, dangers on a global scale and impending local elections or the next board meeting" (Latour 1993:1).

Affect, Emotion, and War

Recent US-led wars in Iraq and Afghanistan have been as much about managing perception and support as they have been about warfare and violence on the ground. War and its public perception have always been managed through the war propaganda machine in order to gain domestic and international support, and recent conflicts reveal even greater intensity in terms of the use of "affective violent geographies." Anderson (2010) explores this issue in relation to the US-led Iraq war. He examines military journals that describe various types of effect- (material) based military strategies (for example, bombing and air raids) but that are also designed to undermine affect-based sources of strength, such as morale and national will to fight. Such military approaches employ various types of psychological warfare and in the process;

a version of morale is assembled that makes morale equivalent to the secret source of an enemy's ability to sustain the activities of war. But it can be targeted "indirectly" by acting over those "manifestations." The result is that potentially all of life is targeted. This is the only way to expose a "morale" or "national will" that is both everywhere and nowhere (Anderson 2010:227).

Modern warfare thus produces material *effects* in terms of death, violence, torture, bombings, and imprisonment, but it also targets *affective* capacities in that methods of warfare can be designed to demolish morale and the national "will to fight," but in doing so it potentially targets everything. Such a "geography of violent affects" (Anderson 2010:230) can also be used to evoke hope on the "home front" and create a sense of security and moral righteousness to sustain warfare; nevertheless it also balances the sense of security with enough fear to generate continued support, especially for unpopular or unjust wars. As Anderson argues, such strategies of violence use material and affective **power** directed at variable objects of self, institutions, and networks. This is one example of how recent forms of cultural geography that use ideas of affect, or emotion, and performance (repetitive actions that consolidate identity and belonging or exclusion—"you are either with us or against us") in politically relevant contexts are increasingly shaping the nature of the world we live in.

Dominant and Other Cultural Identities

Another concern of cultural geography is to describe and understand the tendency for some groups of people, who may or may not be associated with a specific place and who may lack a clear identity, to find themselves in a struggle to maintain a common identity, often in opposition to some other dominant identity. Throughout much of Europe, Roma populations—also known as Gypsies—struggle to maintain their traditional lifestyle and to access places that facilitate that lifestyle.

Of uncertain origin—Gypsies were once thought to be from Egypt (hence the name, Gypsy) but are most likely descended from seventh-century migrants from India—and of uncertain number—estimates range from 4 to 15 million—Roma speak several dialects of a language, Romany, related to Hindi and have no common religion. Nevertheless, they are generally acknowledged as a group at least partly because they prefer to live on the margins of dominant cultures. They lack interest in conventional wage labor, a characteristic that has many implications for lifestyle and for landscapes. They are very mobile, and this contributes to their being viewed and treated as outsiders when they stay in an area for a period of time. Unlike most **ethnic** groups, they are not associated with a particular territory. →»» **5**

Long viewed with suspicion by more settled populations, Roma have experienced much hostility and persecution. Many were forced into slavery in Romania during the nineteenth century, and between 250,000 and 500,000 were killed in the Holocaust. Roma continue to be stigmatized and ridiculed in ways that would be unacceptable if directed toward any other group.

Hostility toward the group is compounded by a distaste for the landscapes Roma occupy. Roma landscapes are typically marginal. They may be waste areas inside cities or may be roadside areas. Regardless, they might appropriately be described as landscapes of exclusion, since they lie beyond the places occupied by the dominant cultural group. Further, these landscapes are different because they are characterized by seeming disorder and are difficult to understand for those who do not belong to the group, especially when compared to the order of a suburban housing estate. Both Roma and their landscapes are in some sense outside of mainstream society and mainstream space, and this contributes to a negative image of both Roma and the places they occupy.

Throughout much of Europe, governments are making it increasingly difficult for Roma to maintain their traditional lifestyle. Istanbul, thought to be the oldest continuous Roma settlement in the world, is the site of an urban regeneration project that will displace the current Roma population. Perhaps more tellingly, popular opposition to Roma is high: a 2008 survey in Italy showed that 68 per cent of Italians wanted all Roma expelled, including the many who are Italian citizens. Physically and psychologically outside the mainstream and increasingly cast as different from majority populations, Roma lifestyles and landscapes face an uncertain future.

What This Book Is About

The four preceding vignettes identify some of the principal concerns of the cultural geography that is discussed in the following chapters. As these vignettes illustrate, cultural geography encompasses a diverse subject matter and diverse conceptual backgrounds. To accommodate these and other concerns, this textbook is organized as follows:

- Chapter 2 introduces the tradition of cultural geography, with emphasis on ways of understanding how humans relate to each other and to nature, and on the introduction in the 1920s by the doyen of cultural geographers Carl Sauer (1889–1975), of the landscape, or Sauerian, school. This tradition, concerned especially with groups and related regional landscapes, continues to be influential, especially in North America, where it has been a dominant concern not just for cultural geographers but also in cultural ecology, cultural archeology, and environmental history.
- Chapter 3 introduces some revisions to the landscape school tradition that derive from a variety of conceptual backgrounds, such as Marxism, feminism, and postmodernism, beginning in the 1970s. These revisions added to, perhaps even reinvented, cultural geography and still contribute to what is sometimes labeled the "new" cultural geography. In the era of new cultural geography, interest in the visible and material landscape as an object of study, evident in the tradition initiated by Sauer, becomes concerned with the role of power and symbolic and social identity. Further, there is recognition that landscape is not the sole object of study, which coincides with a growing interest in studying cultural identities as they are formed and reformed.

- Chapters 4, 5, and 6 are informed by the contents of Chapters 2 and 3—especially 2—being concerned respectively with humans and nature, with landscape evolution, and with regional landscapes. In much of this work there is an underlying assumption that culture is some form of causal mechanism.
- Chapters 7, 8, and 9 are informed by the contents of Chapters 2 and 3—especially 3—being concerned respectively with power, identity, place, and difference. Chapters 8 and 9 pay particular attention to intersections of identity, and the recursive socio-spatial relationship as communicated through place, identity, and landscape change.
- Chapter 10 provides an opportunity to evaluate the content of the text and to make some observations on the current and possible future status of cultural geography, emphasizing that the traditional and newer approaches to the discipline share some fundamental concerns about peoples and places.

Four Terminological Challenges

As is so often the case in social science, key terms tend to be understood differently by different practitioners and in different contexts. The following section highlights four such terms that are particularly important in the study of cultural geography.

Culture

One of the greatest terminological challenges facing cultural geographers is the very term **culture**, which is the subject of considerable debate. One review stated that culture "is one of the two or three most complicated words in the English language" (Williams 1976:87). Other compositional subdisciplines of human geography—such as political and economic geography—do not encounter this definitional problem to nearly the extent that cultural geography does. Quite simply, unlike the terms "politics" and "economy," the term "culture" is contested, creating much uncertainty about an appropriate working definition of culture in geography. While it might be assumed that this uncertainty is a fundamental concern for cultural geographers, in fact the debate about the culture concept is, while important, not central to the study of cultural geography. To anticipate the conclusion reached in a discussion of the debate later in this chapter, multiple meanings of culture are legitimate, indeed valuable.

Cultural and Social

A second challenge comes from confusion about the closely related term "**society**" and about the differences between the words "cultural" and "social" as used in the study of cultural geography. Generally, this debate is rooted in the fact that there have long been uncertainties about these terms in other social sciences, especially anthropology and sociology. But there are also regional differences in preference, with "cultural" being the term traditionally favored by North American geographers, and "social" the term favored by British and other European geographers. As with the uncertainty about the word "culture," this issue is addressed later in this chapter, where it is concluded that while these two terms refer to separate concepts, the debate about their precise differences is of no substantive intellectual merit for the contemporary cultural geographer, being essentially an accident of disciplinary history.

In addition, it is important to acknowledge that cultural, social, political, and economic processes are interrelated. For this reason, the traditional insistence in human geography of teaching and researching these subdisciplines separately has some limitations. Indeed, the phenomenon known as the **cultural turn** can be seen as extending cultural geography into political, economic, and social spheres. The phrase "cultural turn" refers to advances in philosophy and social theory—such as those of post-structuralism, **postmodernism**, and postcolonialism—that together encourage a greater appreciation of the importance of culture in studies of human life. Political geography, for example, is comfortable incorporating cultural as well as social and economic content as needed to explain and understand the changing role of political processes in our world and changes in theoretical understanding of such

changes. Similarly, much contemporary economic geography incorporates cultural as well as social and political processes in order to comprehend economic spatial transformation and organization. Finally, social geography is increasingly combining the cultural perspective into its examination of social change and organization. This textbook recognizes both the importance of the cultural in other traditional subdisciplines of geography and the relevance of social, political, and economic processes in cultural geography. →≫≫ **6**

Central to discussions of the contemporary political and economic worlds is **globalization**, a term that refers to the functional integration of internationally dispersed activities and the ongoing process of change that is highly contingent and therefore in many ways unpredictable. It is also an issue that needs to be at the forefront of cultural analyses of people and place. Both identity and landscape are affected by globalization, not necessarily in the sense that it diminishes the importance of local identity or of place, but rather in that it produces tensions and exchanges between the global and the local, and in the process also creates new forms of spatial articulation and organization.

Nature

The third terminological challenge centers on meanings of the word "**nature**." While culture is indeed a complex term, "nature is perhaps the most complex word in the language" (Williams 1976:184). Given that one of the principal concerns of cultural geography is to analyze and understand relationships between nature and culture, this uncertainty about the meaning of nature poses real difficulties. As the second discussion above also suggests, these complexities are even more prolific now that many argue that nature resides in culture and vice versa. At this stage of the textbook, the meaning of nature is contrasted to culture; thus, nature can be construed as the material world excluding humans, but as we suggest above, more complex readings of nature do not imagine it as a realm divorced from cultural meaning and influence. This meaning is further debated in Chapter 2.

Landscape

The fourth contested term is **landscape.** This term causes difficulties because of uncertainty about its initial meaning and, to a lesser degree, because of the various challenges to the mainstream Sauerian understanding of the term that has been adopted in cultural geography. The classic Sauerian interpretation of landscape refers to the cultural transformation of the natural world, emphasizing visible and material characteristics and close links between land and life—this is the landscape that is lived in. This interpretation will be expanded on later in this chapter and fully detailed in Chapter 2. ➜➤➤ 7

Uncertainties about the meaning of landscape stem mainly from the fact that landscape is the English rendering of a composite German word, *Landschaft. Land* refers to the area used to support a group of people, and *schaft* refers to the molding of a social unity; *landschaft*, then, refers to group activities and experiences that occur in a particular place. Much early German and French cultural geography was concerned with locating and describing groups and their territories, and it is this interpretation of landscape that was introduced in North America by Sauer. But a second meaning of the term also became popular in the English language. This is the pictorial, visual meaning of landscape as scenery, referring to something that is beyond, rather than a part of, ourselves. This is the landscape that is looked at and that we are able to relate to in aesthetic terms. Both meanings of landscape serve to privilege vision.

Providing Context

Cultural geography is one of many areas of academic and practical interest concerned with human activities and human identities, and this section outlines a larger intellectual and disciplinary context to help you understand the particular concerns of cultural geography. In order to appreciate this larger context it is important to understand the **social construction** of knowledge. All knowledge, including theories, develop in specific social, cultural, and historical contexts. The influence of such contexts is such that all knowledge—including knowledge of what is considered to be real—can be considered socially constructed. This important idea is closely associated with some areas of the new cultural geography and resurfaces on several occasions in this textbook. ➜➤➤ 8

Although the scholarly history of cultural geography is lengthy, with speculations about cultural and social matters—especially about human behavior—evident in Greek, Chinese, and Islamic civilizations, it was not until the mid-eighteenth century in Western Europe that such issues received widespread attention. Throughout the extended period from classical Greece to the mid-eighteenth century, there were scholarly advances in both history and the physical sciences, but little discussion of social and cultural issues.

The Rise of the Social Sciences

By about 1750, social science—as opposed to the social sciences—was an established concern of intellectual elites, though it was not as yet the subject of academic study in universities. Although there were various proposals concerning the goals and practice of social science, two basic assumptions prevailed. First, the aim of social science was the discovery of

---

Questions for Critical Thought

————— ✤ —————

As these opening remarks suggest, cultural geography today is an evolving and challenging subdiscipline with a changing, often contested focus. Is this situation to be applauded or deplored? On the one hand, cultural geography is an academic and practical concern that responds both to an increasingly sophisticated body of concepts and to the changing needs of society. Such ability and willingness to respond are indicative of a healthy and vibrant subdiscipline. On the other hand, because such responses are ongoing and may be difficult to place in a larger context, cultural geography is rather like a moving target, never quite as easy to pin down as we might like it to be. New concepts and a changing subject matter—are these circumstances suggestive of a dynamic or an uncertain subdiscipline?

a few general principles that would form the basis for scientific inquiry. **Empiricism** was to prevail in this endeavor; this is an approach that asserts that all factual knowledge is based on experience, with the human mind being no more than a blank tablet (tabula rasa) before encountering the world. Second, a central aim of social science was to promote social progress through revealing truths about ourselves.

Together, these two assumptions reflected the prevailing philosophical ideas of the **Enlightenment** period. The subsequent rise of social science disciplines coincided with, and contributed to, changes that occurred with the onset of the **Industrial Revolution** and the social and economic system of **capitalism**. It is not surprising that this period of dramatic change spurred the establishment of disciplines, each of which functioned to serve society by identifying problems and proposing solutions. Thus, it was during the nineteenth century that the various social science disciplines were differentiated and institutionalized in universities, where their various proponents began to articulate their particular methods and interests. This specialization was a trend that involved a rapidly increasing world of facts, the rise of universities, the appearance of specialized societies and journals, and the desire of groups of scholars to have their discipline placed on an equal footing with other emerging disciplines. All of these developments may be characterized as part of the rise of **modernism**. All involved some claims about the ability of the new social sciences to generate truths about humans in much the same way that the physical sciences were seen to generate truths about nature and in contrast to earlier approaches that focused on religion or **metaphysics**. Certainly, there were arguments against social science's acceptance of a physical scientific approach, arguments that were mostly based on the grounds that it was dehumanizing (that is, neglecting what it means to be human), but the dominant trend was clear.

By about 1900, the principal social science disciplines were firmly established, and since that time additions and changes have been limited. This is important because, as Johnston (1985:5) noted, academic disciplines, once created, "have a defined existence and are invested in by individuals, who wish to protect their capital." Furthermore, while it might seem surprising to suggest that the establishment of academic disciplines can be seen as "a turning-point in human self-discovery," Smith (1997:575) argues that the very presence of social science disciplines is a "visible sign of belief in and the special status of specialist knowledge and techniques about human nature." Indeed, in many ways, the late nineteenth century was the beginning of the modern study of social science. In principle, each of the social sciences can be traced to the recognition of a group of interrelated questions concerning human behavior, a term that is intimately linked with our central concern—culture.

The Rise of Human Geography

What can be said of geography in the light of these general observations about the larger group of social sciences? The study of geography (as opposed to the more specific *human* geography) has a long and distinguished academic pedigree, with substantive contributions made by early Greek, Chinese, and Islamic civilizations, and a consistent history of growth from the fifteenth century onwards in Europe. Throughout, there were central concerns with mapping and written descriptions of lands and peoples. The seventeenth-century scholar Bernhardus Varenius (1622–50)—confronted both with the explosion in European geographic knowledge related to overseas activity and with the need to provide a framework for organizing this knowledge—recognized that geography was a physical and also a human science involving studies of regions and systematic interests. In the late eighteenth century, the philosopher Immanuel Kant (1724–1804) identified geography as essentially a concern with regions.

The growth of geography in the nineteenth century paralleled that of other social sciences. There was an ever-increasing factual base, evident most notably in attempts by two great German geographers, Alexander von Humboldt (1769–1859) and Carl Ritter (1779–1859), to produce comprehensive world geographies, a task that proved to be no longer within the reach of a single scholar. Special interest societies, such as the Geographical Society of Paris (1821) and the American Geographical Society (1851), were founded during this time, and there was a gradual recognition of geography as a university discipline with the creation of individual chairs, for example at the Sorbonne in Paris in 1809 and in Berlin in 1820. An even more important development was the establishment of academic departments of geography, first in Prussian universities in 1874, then in universities in other European countries, and in the United States beginning in 1903. Professional scholarly associations served to further validate the discipline.

This capsule account of the emergence of geography in the nineteenth century raises a key point for understanding the character of the new university discipline. Put simply, it is not appropriate to consider geography only in the context of the social sciences because geography was, of course, much more than human geography. Indeed, human geography is different from the other social sciences because of the explicit and longstanding associations with physical science, specifically physical geography; works by Varenius and Kant clearly acknowledged the contributions of both physical science and human science to the study of geography.

This relatively early integration of physical and human geography distinguished the discipline from the other social sciences that were, by comparison, not as varied and had considerably less intellectual capital invested. But, more importantly, this integration helps account for the late nineteenth-century uncertainty about the identity of geography. As Martin and James (1993:164) noted, "There was no professionally accepted paradigm to serve as a guide to the study of geography." For this reason, the need to answer the question "What is geography?" was paramount. Each of the other social sciences had to grapple with a comparable question, but they were not similarly disadvantaged by the quantity of intellectual baggage that accompanied geography and hence were able to respond to their question largely in the context of late nineteenth-century circumstances, specifically the rise of social science questions related to aspects of human behavior. By about 1900, four definitions of geography were

evident: geography as physical geography, the study of regions, the influence of physical environment on humans, and the study of the human landscape.

These definitions bear witness to the contradiction that geography faced, namely accommodating the legacy of the past with the pressing demands of late nineteenth-century science. Nowhere is this more evident than in the matter of the unity of physical and human geography, with most practitioners questioning whether it was possible and appropriate to include radically different concepts and methods in a single discipline. The consensus answer, institutionally speaking, was yes.

Thus, geography has a long history, such that during the late nineteenth-century period of formal discipline creation, geographers had much to consider when defining their academic field. Furthermore, this history was not one that could be easily accommodated into the emerging structure of disciplines that largely confirmed the acknowledged division between physical and social sciences. The majority of geographers asserted that geography continued to be a unifying discipline dealing with both physical and social facts. This collective "decision" by geographers to maintain a focus on both physical and social science—or, perhaps more correctly, to focus exclusively on neither—is an ongoing source of discussion with debates about the need for and viability of an integrating discipline.

Human Geography since 1900

The Regional Approach

The four definitions of geography noted above grew out of the need to secure an intellectual niche for the new university discipline. By the 1920s it was regional geography that assumed dominant status, a dominance that continued until the mid-1950s. Emphasis was on the delimitation and description of regions, and the underlying philosophy was a form of empiricism that saw the acquisition of geographical knowledge as a way to correct or verify facts about places and peoples. Typically, regional accounts of human-and-land relationships implicitly accepted the logic of **environmental determinism**, which is the argument that the physical environment is a principal cause of human activity. Regional geography thus incorporated both physical and human subject matter, and for many practitioners geography was a bridging science that both straddled and linked physical and human sciences.

Spatial Analysis

By the mid-1950s, some practitioners were expressing concern about human geography's lack of a scientific focus, a shortcoming blamed by some for the discipline's apparent failure to keep pace with advances in other social sciences. The response was to borrow heavily from economics in the introduction and application of what became known as spatial analysis. This approach involved both description and explanation, with particular focus on explaining why things are located where they are. Like some other social science interests, the spatial analysis approach derived from physical science, was objective in character, and had theoretical and quantitative content. The underlying philosophical support came from **positivism**, which was earlier accepted by most other social sciences as well as by the physical sciences. Human geography was late in accepting the logic of positivism, indeed so late that positivism was being rejected by the social science mainstream at the very time it was introduced into human geography. One consequence of this larger shift in conceptual emphasis was that spatial analysis was a major concern in human geography for only a brief period, from the late 1950s until about 1970, and it has never been of major importance in cultural geography; rather it now frames its own subdiscipline of GIS (Geographical Information Systems) and spatial analysis. ➔≫≫ 9

The Landscape School

Although it is fair to characterize human geography from about 1900 until about 1970 as dominated, first, by a regional approach and, second, by spatial analysis, such a capsule account, from the perspective of the cultural geographer, omits the single most important tradition. In the early 1920s Sauer formulated the landscape school (also known as the Sauerian school), and this approach is central to the twentieth-century evolution of cultural geography. It

has proven to be a longstanding approach, still influential today, although it has been somewhat eclipsed by additional concepts and ideas.

The landscape school developed on the premise that culture, operating on physical landscapes through time, was responsible for the transformation of those landscapes into cultural landscapes. It is therefore explicitly opposed to the premise of environmental determinism, although it can be seen as conceptually similar in that it incorporates a cause-and-effect logic. Foundations for this school are evident in the early nineteenth-century work of von Humboldt and Ritter (noted earlier), in the pioneering studies of human impacts on environment conducted by the American geographer George Perkins Marsh (1801–82), and in the work of such geographers as Paul Vidal de la Blache (usually known as Vidal) (1845–1918), Friedrich Ratzel (1844–1904), and Otto Schlüter (1872–1959). Much of the material discussed in the early chapters of

Carl O. Sauer (Reprinted courtesy of the Department of Geography, University of California, Berkeley)

this textbook, especially that relating to human and land relationships, to landscape evolution, and to regional landscapes, is informed primarily by this research tradition.

Marxism and Humanism

The demise of spatial analysis was partly a reflection of changes in other social sciences, especially the move away from objective analyses inspired by physical science toward a range of conceptual concerns that focused explicitly on human beings as subjects, not objects. In short, spatial analysis was criticized for being dehumanizing. In response, since about 1970, human geography has been a willing recipient of, and participant in, a wide range of concerns that, in various ways, place humans at the center of analysis. The two principal philosophical approaches that have embraced this emphasis are Marxism and humanism. Both are discussed more fully later in this textbook, and it is sufficient at this stage to emphasize that, notwithstanding the many differences between them (and, indeed, within each of them), they typically reject any suggestion that humans and human behavior can be studied in the objective terms of physical science.

Marxism and humanism proved to be refreshing additions to the conceptual arsenal available to all human geographers, including cultural geographers, and they continue to stimulate empirical work. Indeed, cultural geographers who were disenchanted with perceived inadequacies of the landscape school were among the first to apply these two methods to the study of human geography.

Feminism and the Cultural Turn

Most recently, since about 1980, at least four additional philosophical movements have added to the conceptual repertoire of the human geographer. That feminism, poststructuralism, postmodernism, and postcolonialism have had a particularly strong impact on cultural geography is a clear indication of the increasing importance of this subdiscipline. As with Marxism and humanism, each of these is explicitly opposed to any physical science analogies. But, more important, they combine to assert the relevance of an understanding of human cultural identity, to highlight the social construction of identity, to acknowledge that **power** relations are organized in various ways, including through the expression of **discourse**, to emphasize the need to break through the conventional gender blindness of human geographic analyses, and most generally to insist on questioning the assumptions embedded in studies carried out by cultural geographers. ➔➤➤ 10

What Is Culture?

> Few scholars using the concept of culture have rigorously defined it. Here geographers have been exemplary rather than exceptional. Most users agree that the term resists simple definition. (Mathewson 1996:97)

The most important yet most confused concept in cultural geography is that of culture. Indeed culture is one of the most difficult concepts to interpret in all of social science. In other words, it is not only geographers who have difficulty with the term, but also anthropologists, sociologists, and others. Many of the difficulties encountered by cultural geographers stem from issues that arise in these other social sciences.

There are several good reasons for the continuing uncertainty about the meaning of this basic term. There was an early association of "culture" with the biological term "cultivation," the tending of natural growth, and a logical extension of this usage to refer also to human growth or **development**. By the mid-eighteenth century, culture was used to refer mainly to human organizations, gradually replacing the closely related term "**civilization**" over the next hundred years. By 1871, culture, or civilization, was defined by the anthropologist Edward Tylor (1832–1917) as "that complex whole which includes knowledge, belief, art, morals, custom, and any other capabilities and habits acquired by man as a member of society" (quoted in Kroeber and Kluckhohn 1952:81). By 1952, two leading anthropologists had combined to distinguish definitions proposed by 110 writers on the basis of 52 discrete concepts employed in those definitions (Kroeber and Kluckhohn 1952).

Culture and Society

Society and culture are closely related concepts but do relate to different phenomena, a point that has been generally accepted since 1958, when two leading figures, the anthropologist Alfred L. Kroeber (1876–1960) and the sociologist Talcott Parsons (1902–79), together produced a statement that aimed to disentangle the terms and propose authoritative definitions. It is worth quoting their opening statements.

> There seems to have been a good deal of confusion among anthropologists and sociologists about the concepts of culture and society (or social system). A lack of consensus—between and within disciplines—has made for semantic confusion as to what data are subsumed under these terms; but, more important, the lack has impeded theoretical advance as to their interrelation.
>
> There are still some anthropologists and sociologists who do not even consider the distinction necessary on the ground that all phenomena of human behavior are sociocultural, with both societal and cultural aspects at the same time. (Kroeber and Parsons 1958:582)

Kroeber and Parsons proceeded to explain the confusion they identified in terms of earlier uses of both terms. Thus, in the formative periods of disciplinary evolution during the late nineteenth and early twentieth centuries, leading anthropologists such as Franz Boas (1858–1942) defined culture as human behavior that was independent of genetic characteristics, while sociologists such as Spencer, Durkheim, and Weber (1864–1920) treated society in much the same way. "For a considerable period," Kroeber and Parsons argue, "this condensed concept of culture-and-society was maintained with differentiation between anthropology and sociology being carried out not conceptually but operationally" (1958:583). They proposed the following definitions, which have been generally accepted within the social sciences (1958:583).

- Culture refers to "transmitted and created content and patterns of values, ideas, and other symbolic-meaningful systems as factors in the shaping of human behavior and the artifacts produced through behavior."
- Society refers to the "specifically relational system of interaction among individuals and collectivities."

 11

Thus, in anthropology it is usual to stress the non-genetic character of culture, referring only to those things that people invent, develop, or pass down through generations. In this sense, culture is extrasomatic—literally, beyond the body. Culture is conventionally seen as having three aspects: the **values** and abstract ideals that members of a human **group** hold, the **norms** and rules that they follow, and the material goods that they create. In sociology, society is understood as the system of interrelationships that connect individuals who share a common culture. This meaning of society is in accord with earlier usage that distinguished society (an association of free individuals) from the term "state" (an association based on power relations). In response to the attractions of physical science modes of explanation, both disciplines during the formative stages of their development incorporated the idea of culture or society as a causal mechanism, although this idea is less important today. Anthropology and sociology offer many other interpretations that do not concern us at this time, though certain ones have carried over into cultural geography and are considered in later chapters as needed.

Culture in Cultural Geography

The meanings that cultural geographers attach to the word "culture" reverberate throughout this textbook, and it is inappropriate to attempt any detailed appraisal in this introductory chapter. Instead, the relevant meanings are discussed later, as needed, while the current discussion offers an overview (see Box 1.1), with details to follow.

Culture and the Landscape School

Sauer introduced the dominant view of cultural geography in the 1920s in a series of writings that effectively set the agenda for the discipline in North

Box 1.1 Cultural Geography, Human Geography, and Geography

It is fair to say that cultural geography, as a compositional subdivision of human geography, is concerned with some particular subject matter within the larger discipline. Note, however, that the flexibility of the term "culture" has permitted two broader interpretations.

First, because culture can be equated with everything produced by humans, as distinct from everything that is a part of nature, there has been a tendency, especially in the United States, to treat cultural geography as synonymous with human geography, such that many courses and textbooks labeled as cultural are more appropriately seen as introductory human geography courses and texts.

Second, for many geographers, cultural geography, when broadly interpreted as the study of humans and land, is geography, both human and physical. This interpretation is supported by an alternative view of culture that includes nature. The editors of the *Companion Encyclopedia of Geography* articulate and endorse this view of geography explicitly: "Geography is, and at root always has been (despite excursions into spatial science and other exotic themes) about the interdependence of people and their environment, and about the evolving intercourse between humans and their earthly, and to a lesser extent celestial, habitat" (Douglas, Huggett, and Robinson 1996:ix).

Identifying the study of human and land relationships as the core of geography is both a hindrance and an aid. It is a hindrance because it opens the cultural geographic door to such a diverse body of material that the subdiscipline almost disappears—it does in fact become all of geography. But it is an enormous aid because it obliges cultural geographers to acknowledge that cultural geography is not only about humans as individuals and as group members but also about the land they live on and with.

Both of these broader interpretations are at least partly a consequence of the uncertain status of the term "culture." Both are based on minor points concerning usage and need not concern us further, but they are instructive in illustrating the rather casual way key terms can be employed. Neither is a substantive issue because there has always been a majority interpretation of cultural geography as one part of human geography, and that interpretation is dominant today. Cultural geography, as defined earlier, focuses on peoples and places with particular reference to cultural identities and cultural landscapes. Human geography focuses on peoples and places more broadly with reference to culture, economics, and politics. Thus, human geography includes such subdisciplines as cultural, economic, and political geography.

America until the first serious reservations were voiced in the 1970s. Sauer regarded three factors as basic to an understanding of landscape: the physical environment, the character of people, and time. Sauer (1924:24) thus defined geography as "the derivation of the cultural area from the natural area." His classic statement of intent was perhaps the following: "Culture is the agent, the natural area is the medium, the cultural landscape the result" (Sauer 1925:46). Although he saw culture as the "shaping force," he viewed physical landscape as being "of fundamental importance, for it supplies the

materials out of which the cultural landscape is formed" (Sauer 1925:46).

For Sauer, then, the importance of culture to the study of geography was a given, such that the principal focus of his research was investigating the impact of culture on landscape, an impact evident in the creation of distinctive material landscapes that permitted the delimitation of cultural regions. But this concern with the material and visible landscape did not imply an exclusion of cultural identity, and practitioners frequently identified groups according to their way of life—what Vidal earlier labeled *genre de vie*—based on such characteristics as language, religion, and ethnic identity. Thus, culture was also viewed as a system of shared values and beliefs. This view, or perhaps more correctly these views, of culture remained largely uncontested in geography until the 1970s.

In this interpretation of culture there is an emphasis on culture as cause, on landscape outcomes, and on identifying characteristics. This view prompted a cultural geography—the landscape or Sauerian school—that functioned as a leading subdivision of human geography with three principal themes: landscapes as the outcomes of human and land relations, landscape change, and regional landscapes. (These three themes are addressed in Chapters 4, 5, and 6 respectively.)

Questioning the Meaning of Culture

Challenges to the Sauerian view of culture, and hence to the landscape school of cultural geography, emanated from several directions, with the strongest being the claim that Sauer *reified* culture—that is, he treated culture as a reality and not as a concept or human construct. Expressed differently, critics raised concern at the use of culture alone as a cause of landscape without also focusing on the individual human decision-making behind it. This criticism was central to a seminal article by Duncan (1980), who argued that Sauer was reflecting the **superorganic** or cultural determinist view made popular in the anthropology of the 1920s and beyond by Kroeber. According to the superorganic perspective, any understanding of the cultural world had to be rooted in culture because it was there that decisions were made. Kroeber, along with the like-minded Robert Lowie (1883–1957), was a contemporary of Sauer at the University of California, Berkeley.

It is useful to see Duncan's argument as a major critique of the landscape school conception of culture and as contributing to the rise of the new cultural geography. Certainly, his argument struck a chord in a 1980s intellectual environment very different from that of the 1920s, such that the changes it sparked are but one component of the larger disenchantment with modernism in social science. The Sauerian tradition was criticized also for its rural

and antiquarian emphasis and for its focus on material landscapes, as cultural geographers proposed a more complex and interpretive view of landscape as cultural construction. →≫ 12

Demands for a new approach to cultural geography came not just from scholars outside the landscape school tradition with an agenda that was both cultural and social in inspiration. The landscape school was also criticized from within, as several major figures called for and proposed new ideas as early as the 1970s. For example, Mikesell (1978:13) asked practitioners to "give more serious thought to how they wish to use the concept of culture," while English and Mayfield (1972:6) bemoaned the "intense preoccupation with the visible material landscape" that "led to an unfortunate neglect of the less obvious, invisible forces which in some cases form cornerstones in the explanation of spatial patterns of human behavior." The important point here is that changes were being initiated by scholars immersed in the landscape school tradition, not only by those looking from outside of the tradition. Furthermore, it is rather misleading to imply that opposition to the Sauerian tradition and calls for new cultural traditions were evident throughout English-language geography. Writing with specific reference to New Zealand, Berg and Kearns (1997:1) noted that "many cultural geographers here, but especially those with interests in 'Maori issues', have always had a political edge to their work" such that "the simple 'old/new' binary constituted in representations of American cultural geography fails to capture the complexity of the situation" in New Zealand.

Culture and the New Cultural Geography

There is no doubt, however, concerning the impact on North American cultural geography made by scholars who objected to the Sauerian view principally because of the perceived tendency to reify culture. Uncertainty about how to correct this perceived flaw of the earlier view contributed to the emergence of several new understandings of culture, a reflection of the diverse theoretical landscape of the times. One particularly distinctive feature of this period was the interest shown in cultural geography by British social geographers; indeed, a 1987 conference entitled "New Directions in Cultural Geography" was organized by the Social Geography Study Group, which subsequently became the Social and Cultural Geography Study Group. Not surprisingly, then, several of the new conceptions of culture had a strong social flavor. The appearance of several varied concepts does not, however, disguise a fundamental similarity. ➔➤➤ 13

For the new cultural geography, culture is not seen as an explanatory variable, nor is culture approached as a cause. Rather, the new cultural geography emphasizes the plurality of cultures, defined as those values that members of different groups share in particular places at particular times, but that may contradict cultural norms of more powerful groups. In brief, culture is seen as a medium or process rather than as an object; we might describe this as a change from an interest in culture to an interest in cultures. In an influential agenda-setting article, Cosgrove and Jackson (1987:99) defined culture as "the medium through which people transform the mundane phenomenon of the material world into a world of significant symbols to which they give meaning and attach value."

The transformation of cultural geography that followed the expression of concerns about the Sauerian view of culture encouraged humanistic studies of behavior and landscape and paved the way for interest in power and identity, in other voices and landscapes, and in imagined landscapes and landscapes as text—the concerns of Chapters 7, 8, and 9 respectively. This transformation also resulted in much questioning about the directions to be taken by cultural geography and the possible fragmentation of the subdiscipline. Regardless, there is little doubt that the new cultural geography was institutionalized by the end of the 1980s, not just because of its reaction to what came before, but also because these changes intersected with more widespread transformation across the social sciences, which came to be known as the "cultural turn." This institutionalization of "new" approaches to cultural geography is confirmed by the publication of cultural geography textbooks that exclude much of the Sauerian

tradition, focusing instead largely on certain aspects of the new cultural geography. ➔➤➤ 14

New Cultural Geography and Cultural Studies

The understandings of culture in the new cultural geography are not simply critical responses to earlier understandings in the Sauerian tradition but are also reactions inspired by developments in the area of **cultural studies**. This is a term that could have been introduced earlier during our discussion of twentieth-century human geography, along with feminism, poststructuralism, postmodernism, and postcolonialism, but it is perhaps most appropriately seen as an umbrella term comprising all of these together with aspects of Marxism and humanism. As an interdisciplinary field, cultural studies embraces numerous concepts, mostly involving a rejection of modernism, and is represented in a number of schools of thought, four of which are noted briefly at this time.

First, a British tradition focused attention on working-class culture, on consequences of the Industrial Revolution, and on such new cultural forms as film and television. Much of the work conducted under the auspices of the Birmingham Centre for Contemporary Cultural Studies is central to this tradition. Second, an American tradition evolved from an anthropological concept of culture that centered on the meaningful ordinary lives of people from a culture other than that of the investigator. Anthropology from this perspective is thus an encounter with otherness. Third, a Marxist-inspired Frankfurt school of critical theory focused on a critique of modernity. A basic inspiration is the conviction that social science must not be the instrument of particular ideologies, notably those associated with dominant groups, including the political state; rather, social science must be emancipatory. Fourth, there are traditions, such as postcolonialism, feminism, and studies of racism, that are linked by their explicit concern with exposing the Eurocentrism that has been and often continues to be associated with the production of knowledge. ➔➤➤ 15

Evaluating the Culture Concept

There are many views of culture within cultural geography, and Box 1.2 both summarizes and expands on the discussion so far by listing some representative definitions.

Clearly, understanding what is meant by the word "culture" is not a simple task and undoubtedly provides both challenge and promise for cultural geography. There are at least three general reasons for this state of affairs. First, there has been much debate concerning the realness of culture, and this has contributed to confusion. The modernist approach has been to regard culture as a thing that is capable of explaining other things, an idea that is now rejected by most cultural geographers and other social scientists. Second, culture has never been easy to define separately from other related words, especially society, but also economy and politics. Third, the dominant current view is to acknowledge that culture has different meanings for different groups in different places and at different times.

Despite these comments, there are three basic statements with which most would agree:

- First, culture refers to our human way of life in the sense that it is what makes one group of people different from some other group.
- Second, culture is the principal means by which humans relate to nature and to each other; we are able to adapt to and change physical environments and are able to co-operate with and sometimes conflict with other groups.
- Third, culture is extrasomatic, distinguishing humans from other species—although increasingly some argue that primates have their own systems of culture, in that different groups will use different hunting techniques and tools depending on their habitat, and these practices are learned and developed within the group.

Human Identity—Who Are We?

So far in this text, the word "culture" has been used to convey some idea of what cultural geographers study, namely the values, norms, behaviors, and material goods of humans. However, the degree of uncertainty about this term means that it has not been possible to convey a clear or comprehensive picture. One way to address this difficulty is to focus on socially constructed characteristics of individual and collective human **identity**—that is, characteristics that are made or acquired rather than inherited. →➤➤**16**

The following discussion addresses ten socially constructed characteristics that cultural geographers have identified: place, language, religion, ethnicity, nationality, community, class, gender, sexuality, and age. A key concern of many social scientists today is the question of how it is possible to understand identity using group labels but at the same time avoid **essentialism**, that is, attributing any essential characteristics to a group. While it is typically necessary to group people, the resulting implication that groups have common traits or behavior patterns may be unwarranted given the often uncertain basis for defining group membership. The most popular alternative to essentialism is **social constructionism**, the idea that identities are formed and continually reformed through interaction with others and

Questions for Critical Thought

———— ✣ ————

There continue to be differences of opinion concerning the claim that the culture concept used by landscape school geographers involved an almost unquestioning acceptance of the singularity of culture, whereas those concepts used by new cultural geographers recognize cultural differences within specific groups. Indeed, it was difference that became the main interest of these new approaches. How do you react to these comments acknowledging uncertainty—indeed disagreement—about the single most important concept used by cultural geographers? Does this present a problem for you as you commence your studies, especially in light of the earlier uncertainty about the larger subdiscipline? Or are you comfortable with this situation? It is useful here to mention similar situations that prevail in other areas of social science. To cite just three examples, there are multiple meanings of society, behavior, and personality, as any study of sociology and psychology textbooks makes clear. Viewed from this perspective, some uncertainty seems both normal and desirable. Do you agree?

Box 1.2 *Some Geographic Interpretations of Culture*

This box details some representative geographic interpretations of the culture concept, organized chronologically. The intent is to provide an indication of the variety of definitions available and to indicate changes in definition through time, but also to suggest that, variety and change notwithstanding, there is an essential consistency.

Sauer (1925:30)
Culture is *"the impress of the works of man upon the area."*

This is one of the classic statements in the seminal 1925 article by Sauer, emphasizing the idea of culture as a cause of landscape.

Sauer (1941a:8)
"Culture is the learned and conventionalized activity of a group that occupies an area."

This less deterministic definition is perhaps closer to much of the empirical work accomplished by landscape school geographers than is the 1925 statement above.

Gritzner (1966:9)
"A culture is a human society bound together by a common complex of culture traits, each trait being anything which to the culturally-bound group has either material form and applicable function, or an expressed value."

With this clear statement reflecting the consensus view and activities of the landscape school, Gritzner stresses that culture is transmitted spatially and temporally through the unique human ability to symbolize.

Spencer and Thomas (1973:6)
"Culture is the sum total of human learned behavior and ways of doing things. Culture is invented, carried on, and slowly modified by people living and working in groups as each group occupies a particular region of the earth and develops its own special and distinctive system of culture."

This textbook definition expands on the idea that culture is learned behavior. In many respects this "was an unorthodox, yet revolutionary textbook for its day," arguing, for example, that "cultural geography might profit from a greater concern with institutional political processes, not simply landscape imprints" (Arreola 2006:670).

Zelinsky (1973:70)
Culture is *"a code or template for ideas and acts."*

This brief definition, taken from a larger and highly influential statement, is a clear assertion of culture as a causal mechanism.

Wagner (1974:11)
"Learned behavior is pretty much what we mean by culture."

Wagner (1975:11)
"The fact is that culture has to be seen as carried in specific, located, purposeful, rule-following and rule-making groups of people communicating and interacting with one another."

Together, the two statements by Wagner reiterate the importance of learning, stress that culture refers to human behavior, and emphasize the importance of communication and interaction between and among group members.

Jackson and Smith (1984:205)
"Culture, in the sense of a system of shared meanings, is dynamic and negotiable, not fixed or immutable. Moreover the emergent qualities of culture often have a spatial character, not merely because proximity can encourage communication and the sharing of individual lifeworlds, but also because, from an interactionist perspective, social groups may actively create a sense of place, investing the material environment with symbolic qualities such that the very fabric of landscape is permeated by, and caught up in, the active social world."

This definition, one of the first from the new cultural geography, incorporates elements of earlier definitions, notably Wagner (1975), but includes some additional ideas, notably a shift from a concern with behavior to a concern with meaning.

Jackson (1989:ix)

Culture is *"a domain, no less than the political and the economic, in which social relations of dominance and subordination are negotiated and resisted, where meanings are not just imposed, but contested."*

This is one of the first statements to explicitly recognize that culture needs to be understood in the context of struggles between different groups of people. In both this and the preceding statement, there is explicit reference to the social world.

Shurmer-Smith and Hannam (1994:5–6)

Culture is *"that negotiated intersubjectivity which allows human beings as individuals to reach a tenuous understanding of one another, to experience each other jointly."*

Rejecting a cultural geography concerned with mapping traits onto landscapes, this definition instead emphasizes ways of being in the environment.

McDowell (1994:148)

"Culture is a set of ideas, customs and beliefs that shape people's actions and their production of material artifacts, including the landscape and the built environment. Culture is socially defined and socially interpreted. Cultural ideas are expressed in the lives of social groups who articulate, express and challenge these sets of ideas and values, which are themselves temporally and spatially specific."

After observing that culture is "a notoriously slippery concept," McDowell (1994:148) suggests that "a broad consensus" concerning the above definition "might not be difficult to achieve."

Jordan, Domosh, and Rowntree (1997:5)

Culture is *"learned collective human behavior, as opposed to instinctive, or inborn, behavior. These learned traits form a way of life held in common by a group of people."*

This statement, taken from a highly influential introductory textbook, maintains the idea of culture as it evolved within the landscape school. This view remains popular with many contemporary cultural geographers, notwithstanding the undoubted achievements of the new cultural geography.

Mitchell (2003:2)

"Culture is decidedly not everything. Indeed . . . Culture needs to be understood as no-thing, which is to say that rather than an object, 'culture' is an on-going, struggled-over set of social relations that give rise to social meaning, to differences within and across social groups and places, and to the exercise of power."

Mitchell is stressing that culture as way of life needs to be understood in political terms as it always serves particular groups and particular aims; thus, culture is both a component and a consequence of power relations.

Summary Comments

Although these selected definitions are but a few of many, they do combine to provide a meaningful overview of the approaches taken by geographers. No one single definition serves the purposes of this textbook, but it is appropriate to suggest that the final three reflect many of the current interests of contemporary cultural geography. However hard geographers and others try to achieve a definitive wording, such a goal is unlikely to be attained. As Jackson (1989:180) acknowledged at the end of a book-length study, "the stuff of culture . . . is elusive."

thus are fluid and negotiable. Not only are identities fluid but they also intersect, so that one of the more important approaches in cultural geography today is that of identity and **intersectionality**, which refers to how different aspects of identity intersect, and how society creates hierarchies of social inclusion and exclusion based on aspects of identity. In some cases people are also strategic about certain aspects of their identity. This is called **strategic essentialism** and can be spatialized; for example, in certain places one's religious identity might be more important to emphasize than one's class identity. Identity, intersectionality, and place are explored in more detail in Chapters 7, 8, and 9.

Place

Cultural geographers of all persuasions, and increasingly other social scientists, are sympathetic to the idea that **place** matters. Identity is not simply a matter of *who* we are, but also *where* we are. Humans in groups occupy areas and thus tend to create material and symbolic landscapes that might be susceptible to delimitation as regions. For most of us, place—for example where we live, where we come from—is an important element in our individual and collective identity.

Language

North American cultural geographers, particularly those working in a landscape school tradition, typically distinguish groups of people according to what are viewed as primary cultural characteristics, notably language and religion. Such variables, which are similar to those used by anthropologists to classify peoples, are particularly appropriate to studies of tribal peoples and to areas of European overseas settlement. For many groups, language is the principal expression of culture and may embody a particular view of the world. In fact, during the nineteenth century, language was one of the principal ways for a population to assert a distinctive national identity. But attaching such cultural significance to language can be dangerous for a minority group whose language is dominated by that of a larger group in the same political unit; in such cases, the loss of language may mean the loss of the larger culture.

Religion

Religion, the second of the two traditional cultural universals, addresses matters of ultimate significance, such as questions about human meaning, about human presence in the world, and about the place of humans in the larger context of god, or gods, and the physical world. As a result, religion plays an important role in any consideration of human relations with nature and with the landscapes associated with particular groups of people. In particular, religion is also playing an increasingly important role in the formation of presumed ethnic identities.

Ethnicity

Ethnicity refers to the perceived distinctiveness of one group relative to others, usually based on a shared history that may be partly mythical and that has political associations. Many groups strive to create ethnic solidarity and enhance ethnic pride by emphasizing origins, history, claims to a homeland, and distinctive material and other cultural traits. In landscape school cultural geography, ethnicity is often used instead of language and/or religion to identify a group.

"Ethnicity" is a problematic term. The assumption that ethnic groups exist typically implies that such groups have their own distinctive culture, which is another instance of reification. Ethnicity is also closely associated—and even confused—with the even more problematic term "**race**." As discussed in Chapter 7, the idea that there are distinct human races is a myth. Unfortunately, "race" is often treated as a synonym for ethnicity, or the impression is created that ethnic differences exist as a direct consequence of supposed racial differences.

Nationality

"Nationality" is also a problematic term. Humans are divided into national groups (however artificial these may be), and also into political units known as nation-states. But these two concepts are rarely congruent, and as a result, contrary to the aspirations of nineteenth-century nationalist thought, national groups are rarely, if ever, neatly packaged into a political unit, and political units rarely, if ever, comprise a single supposed national group. Regardless, the idea of national identity is central to many accounts of the contemporary world.

Community

Unlike the word "culture," community has never held a privileged position in the discourse of social science, although, like culture, it has been granted a great many different meanings. A term that has traditionally had such positive connotations as loyalty, shared concerns, and personal contact, community is today more frequently used as an alternative term for ethnic group.

Class

In the nineteenth century "**class**" was used to refer to people with a common social or economic status. Karl Marx (1818–83) moved beyond classification by making a distinction between those who produce surplus (laborers) and those who appropriate it (capitalists) and concluding that each class has a collective consciousness and organization that places it in a perpetual class struggle. Marx used class consciousness in the sense that others used culture in the late nineteenth century. Today, class is a key variable in the area of political economy, which has contributed some ideas to cultural geography in the guise of the political ecology discussed in Chapter 4. Class is increasingly examined as an element of intersectional identity in contemporary cultural geography.

Gender

Gender refers to attributes that are culturally ascribed to women and men; it is a cultural construct subject to change. Thus, studies of gender emphasize the role played by male power and domination and assumed female subservience in the creation and maintenance of gender roles. It is usual to stress that there is no necessary link between gender and biological sex, as women may have masculine characteristics and men may have feminine characteristics. But not only is there recognition that gender is not an essential characteristic—that is, it relates to culture—but there are also arguments that **sexuality** and sexual preference relate to culture in such a way that they also can be considered a social construction.

Sexuality

"**Heteronormativity**" is the term used to reference the powerful socio-political structures that promote, enforce, and maintain **heterosexual** identities and relationships as the "natural" foundation of sexual-social organization. In this type of social organization the roles, **authority**, and expectations of men and women are socially maintained and mandated along "natural" biological lines. They can be extremely difficult to change because some of the most fundamental dimensions of society depend upon sexual roles remaining stable and intact. These gender roles offer some of the most powerful structuring identities in society. Deviations from sexual "norms" have historically been seen as deviant and/or illegal, and thus those who identify as Lesbian, Gay, Bisexual, and Transsexual (**LGBT**) have been subjected to social, economic, and political exclusion and marginalization. Changing these presumed sexual roles can also affect women and men who choose non-traditional roles in society; for example, "stay-at-home dads" and "working women" have traditionally been seen as deviating from their "normal" roles in society, and their sexuality has often been questioned as a result. Today many countries recognize the rights of LGBT communities, but there are still many places where those who deviate from their assumed biological sex role and place in society are victims of repressive and violent action.

Age

With age comes certain rights and responsibilities. Children and youth, in most countries, can expect an education to varying levels. In democratic nations people of a certain age can vote if they are citizens. We have expectations of people's behavior based on their age. Sometimes those expectations can be limiting, as we construct certain subjectivities around "old" people, and "infantilize" youth. Age becomes an important axis of identity. Since there are powerful assumptions about behavior and age, we are expected to "act our age." Age is also spatialized, in that places and environments are constructed to exclude or include people based on their age, and this also infers ability and interest. There are often assumptions about who will use a space, so we find that some communities and service areas may be devoid of amenities for children (consider airports), or designed in ways that make it difficult for elderly people to easily access space (consider buildings with stairs and narrow hallways that people in wheelchairs or with strollers cannot easily move around in). These limited examples underscore how identities are framed by age, and how the young and the old can be marginalized in spaces that are designed without consideration for the specific needs of different age groups. Cultural geographers are increasingly considering how age informs spatial perspectives, especially research that examines mental maps, for example.

Themes in Cultural Geography

This textbook uses six themes to organize the ideas and work of cultural geographers. They are not formally institutionalized within cultural geography, but are general themes that can be identified when

---

Box 1.3 Some Organizational Frameworks

Although cultural geographers "have not been prone to defend their interests or issue programmatic statements" (Mikesell 1978:1), there have been several efforts to clarify both subject matter and approaches. This box notes, in an abbreviated format, some of the principal efforts that have been made; further details of particular issues are included in this textbook as needed. Students seeking additional information are advised to refer to the sources noted below, to the progress reports on cultural geography and/or social geography published annually in the journal *Progress in Human Geography*, and to the assessments of cultural geography included in Dohrs and Sommers (1967), Mikesell (1967), Miller (1971), Salter (1972), Spencer (1978), Norton (1987), and Ellen (1988).

In their pioneering and highly influential book of readings, Wagner and Mikesell (1962:1) proposed the following five themes as constituting "the core of cultural geography":

- culture
- cultural area
- cultural landscape
- culture history
- cultural ecology

According to Gritzner (1966), cultural geography

- begins with the anthropological concept of culture,

- considers culture traits and groups in terms of development,
- studies culture/nature,
- interprets landscapes, and
- divides the world into culture regions and subregions.

Wagner (1972, 1974, 1975) renounced the five themes outlined with Mikesell in 1962 and offered a revised focus based on institutions and communication.

Rather than specifying themes, Spencer and Thomas (1973) recognized the following four conceptual entities:

- population
- physical environment
- social organization
- technology

In this more process-oriented schema they also proposed six interoperative relationships, namely those involving

- population and environment,
- population and social organization,
- population and technology,
- social organization and technology,
- physical environment and social organization, and
- physical environment and technology.

tracing the development of cultural geography from about the 1920s onwards. Box 1.3 summarizes a selection of earlier attempts designed to impose order on the subdiscipline.

There are at least two general reasons why the absence of a definitive list of themes is not uncommon in social science. First, research themes are not typically identified and then carefully adhered to. Rather, researchers conduct analyses that, although often prompted by earlier work, may be quite different in focus to that earlier work; they may even conduct research that is prompted by new ideas imported from other disciplines. Second, the nature of much social science work is such that commonalities often become apparent only as a body of work evolves, which is a principal reason for the difficulty of classifying new social science work; put simply, new work may require a new classification.

Logically, then, the six themes identified and used in this textbook could not have been easily

Mikesell (1978) identified seven persistent preferences:

- historical orientation
- humans as agents of environmental change
- focus on material culture
- rural bias in North America, non-Western or pre-industrial bias elsewhere
- links to anthropology
- substantive research
- fieldwork

He also identified three recent developments:

- environmental perception
- cultural ecology
- focus on United States

Norton (1989) proposed four themes:

- evolutionary
- ecological
- behavioral
- symbolic

In Murphy and Johnson (2000a), a tribute to Marvin Mikesell (subtitled "enduring and evolving geographic themes"), the editors group papers into three categories:

- constructing cultural spaces
- remaking the environment
- claiming places

Buttimer (2003) outlined five principles:

- one earth inhabited by multiple cultural worlds
- evolving geographical knowledge of nature and culture
- inventions and conventions in texts and images
- insiders and outsiders, mindscapes and landscapes
- self and other, dialogue toward mutual understanding

Strohmayer (2003) argued that in addition to the key concepts of culture and geography, the following should be prominent:

- landscape materiality
- contestation
- identity
- globalization

In Anderson, Domosh, Pile, and Thrift (2003b), with the understanding that cultural geography is a contested and unfolding intellectual terrain, the editors identified five themes:

- culture as distribution of things
- culture as a way of life
- culture as meaning
- culture as doing
- culture as power

Questions for Critical Thought

———— ✧ ————

Do these ten aspects of human identity capture the spirit of that term as you might apply it to yourself or to others, or do you sense that identity is better captured by reference to other characteristics? Some people choose to identify certain characteristics as being important depending on the context (strategic essentialism). Some possibilities are membership of or identification with an environmental group, political party, or religious group. Most cultural geographers today emphasize the social construction of identity through the subjective and the structural, in that personal agency is relevant, but so are the structures of power that contain and direct our practices and options. Can you think of your own examples that illustrate how your identity is predetermined by certain structural factors, and when you have been able to shape your identity as you see fit?

identified, say, twenty years ago, and are unlikely to be appropriate, or at least sufficient, twenty years from now. In some cases the association of ideas and empirical work with a single theme is a necessary simplification designed to suit the purposes of this textbook, which assigns the work of cultural geographers to one of six themes. Other themes could be identified, and in some instances work could be assigned to more than one theme. In short, there is nothing "correct" about the classification; it is simply one way to interpret the literature of cultural geography as it has developed and changed since about 1920. This cautionary statement does not serve to invalidate the themes identified but to make it clear that the classification must not be treated as sacrosanct. Times also change, and different issues become more or less relevant to society.

It is important to note that the six themes, while they are different, are complementary. They do not compete with one another in some search for mythical universal truths. Rather, each one has made and is likely to continue making contributions to the fundamental goals of cultural geography, namely, making sense of the real worlds of both people and place.

Six Themes

The first three themes are introduced in Chapter 2, in the discussion of the tradition of cultural geography, especially through the account of the landscape school. All three continue today as important and changing themes. The second three themes are more closely identified with developments since the 1970s than with the landscape school and are introduced in Chapter 3, in the account of the rethinking of cultural geography. These three are informed by different meanings of the culture concept and, more generally, by a rejection of the larger intellectual environment of the landscape school. Table 1.1 identifies the six themes and provides notations concerning conceptual inspirations, principal figures, and key concepts.

Concluding Comments

- Cultural geography is a contested arena in that there is an abundance of approaches and research interests, with a particular distinction to be made between traditional and newer approaches.
  - › The Sauerian view of culture and the related landscape school, first introduced in the 1920s and linked to the larger modernist project, continues to influence contemporary research—particularly in North America—that examines human and land relationships, landscape evolution, and regions.
  - › The "new" cultural geography that emerged in the 1970s is critical of attempts to reify culture, and instead emphasizes group and landscape inequalities, symbolic as well as material landscapes, and diversity and difference within social systems. ➔➤➤ 17
- From the 1920s to the 1970s a Sauerian view of cultural geography dominated in the United States but was absent in other English-speaking countries and in Europe, possibly because the superorganic interpretation of culture never appealed to either anthropologists or geographers outside the US, who had their own approaches to the study of human geography.

Table 1.1 Cultural Geography—Six Principal Themes

Theme	Conceptual Inspirations	Key Concepts	Principal Figures
Environments, Ethics, Landscapes	landscape school, ecology, anthropology, Marxism, political ecology	humans and nature, nature, culture, ecology, adaptation, human impacts	Barrows, Butzer, Mikesell, Mathewson, Peet, Sauer, Wagner, Zimmerer
Landscape Evolution	landscape school, history, Annales school, world systems	landscape, nature, culture, time, diffusion, acculturation, frontier	Carter, Clark, Conzen, Darby, Hoskins, J.B. Jackson, Kniffen, Lewis, Meinig, Sauer
Regional Landscapes	landscape school, regional geography, new cultural geography, culture worlds	culture, region, homeland, landscape, globalization, world system	Hart, Hartshorne, Hudson, Jordan, Meinig, Sauer, Shortridge, Zelinsky
Power, Identity, Global Landscapes	new cultural geography, Marxism, feminism, postmodernism	power, identity, ethnicity, language, religion, nation, globalization	Blaut, Harvey, P. Jackson, Mitchell, Taylor, Wallerstein
Power, Identity, Representation	new cultural geography, folk and popular culture, sociology, Marxism, feminism, postmodernism	social construction, discourse, difference, identity politics, others, gender, sexuality, resistance	Cresswell, Domosh, Duncan, P. Jackson, Mitchell, Rose, Sibley, Valentine
Living in Place: The socio-spatial relationship	new cultural geography, folk and popular culture, psychology, humanism, postmodernism	discourse, home, place, sense of place, landscape as text, identity, consumption	Carney, Cosgrove, Daniels, Duncan, Jakle, Ley, Tuan

Although incorporating some new content, the first three themes are traditional in that
- their principal conceptual inspiration is the landscape school, and
- they include a view of geography as an integrating physical and human discipline.

The final three themes are more recent in origin. They introduce different content to the ideas and practice of cultural geography and reflect
- generally close contacts between cultural geography and other social sciences,
- a concern with social theory and cultural studies, and
- views of culture that explicitly reject the idea that culture is a causal variable, and instead understand how it is related to issues of power and resistance.

- Today the majority of cultural geography reflects approaches that developed since the cultural turn. These approaches are conceptually distinct from earlier Sauerian approaches, but there remains value in charting the intellectual trajectory of cultural geography for the following reasons:
  › As one part of the larger social scientific endeavor, cultural geography changes as the needs of society change.
  › Differences between older and newer versions of cultural geography are sometimes overstated, and it may be that many of the purported differences result from the understandable tendency of those introducing new ideas to do so through a rejection, rather than development, of earlier ideas.
  › For this textbook the new cultural geography is seen as part of the evolution of the subdiscipline.
  › Accordingly, this text attempts to integrate different ideas and approaches as part of a continuum that reflects the tradition of cultural geography as initiated by Carl Sauer in the 1920s, and capitalizes on the reassertion and reinvigorated subdisciplinary presence in the post-cultural turn period.

Further Reading ➡➤➤

The following are useful sources for further reading on specific issues.

1. In a challenging philosophical text, Sack (2003) develops the idea of the human creation of landscapes that ought to be.

2. Gritzner (2002) outlines the study of geography as "What is Where, Why There, and Why Care?"

3. Throughout this text there is a concern with both concepts and analyses. Concepts shape cultural geographic knowledge while analyses—performed by cultural geographers—test these concepts. The need to consider both concepts and analyses is well established in social science generally and is demonstrated effectively for human geography by Hubbard, Kitchin, Bartley, and Fuller (2002).

4. Francaviglia (1978) provides the seminal description of the visible Mormon landscape.

5. Sibley (1995) discusses Gypsy identity and landscape, and Holloway (2003) explores the racialization of Gypsy identity and the contested construction of Gypsy places.

6. Philo (1988) stresses the need to integrate cultural, social, economic, and political processes in human geographic research. Barnes (1996b) comments on the cultural understood as materialized in the economic. Knox and Agnew (1994), Berry, Conkling, and Ray (1997), Lee and Wills (1997), and Taylor and Flint (2000) are examples of textbooks that include cultural processes in their accounts of political or economic geography. Wheeler (2002) outlines the increasing cultural content of urban geography. Barnes (2003) discusses culture and economy in geography, and Gertler (2003) outlines a cultural economic geography of production.

7. Morin (2003) provides a concise history and geography of landscape studies, while Olwig (2002) provides a counterpoint to scenic understandings of landscape. Examples of alternative understandings of landscape include those by Benediktsson (2007), Mels and Setten (2007), and Qvistrom (2007).

8. Berger and Luckman (1966) discuss the social construction of knowledge.

9. Sopher (1972) outlines a spatial analytic cultural geography.

10. E. Graham (1997) outlines philosophical debates in human geography.

11. Feder and Park (1997:10) discuss the meaning of culture in anthropology, and Giddens (1991:35) discusses the meaning of society in sociology. Kuper (1999) provides a comprehensive account of the culture concept in anthropology.

12. Dodson (2000) provides an interesting commentary on new cultural geography as it developed at least partly as a reaction to the conservative agenda of the landscape school.

13. Cosgrove (1978) and Jackson (1980) are important early statements of the new cultural geography; McDowell (1994) and Kong (1997) offer overviews; Price and Lewis (1993a) and Mitchell (1995) provide critical accounts. R.J. Johnston (1997) is the fifth edition of an influential book, *Geography and Geographers*, which, for the first time, includes a chapter titled "The Cultural Turn."

14. Jackson (1989), Shurmer-Smith and Hannam (1994), Crang (1998), Anderson and Gale (1999), Cook, Crouch, Naylor, and Ryan (2000), Mitchell (2000), Shurmer-Smith (2002), Anderson, Domosh, Pile, and Thrift (2003a), Blunt et al. (2003), and Duncan, Johnson, and Schein (2004) are cultural geography books that choose to exclude much if not all of the Sauerian tradition. A dictionary of key concepts by Atkinson, Jackson, Sibley, and Washbourne (2005) similarly focuses on recent

work. On the other hand, Murphy and Johnson (2000a) edited a tribute to Marvin Mikesell that appropriately includes a wide range of cultural geographic studies. Hoggart (1957), Williams (1958), and Thompson (1968) are important works anticipating the British cultural studies tradition.

15. Geertz (1973) details the anthropological concern with the study of otherness. Said (1978, 1993) identifies some of the biases that would be evident if a specifically European view of the world is adopted.

16. Philo (1989), Parr and Philo (1995), and O'Dwyer (1997) comment on particular identities and behaviors. Moser (2008) discusses personality.

17. Smith (2003) summarizes a survey of perceptions of cultural geography held by cultural geographers. Blaut (1993a) discusses cultural geography as a contested area with a particular distinction made between traditional and newer concerns. For accounts of changing approaches to cultural geography, see Chapters 3, 4, and 5 in Duncan, Johnson, and Schein (2004).

CHAPTER 2

The Tradition of Cultural Geography

This chapter tells the story of the cultural geographic enterprise, especially as it developed in the late nineteenth and early twentieth centuries as part of the larger institutionalization of geography as an academic discipline, and as it was practiced through to about 1970. The next chapter continues the story through to the present with emphasis on the many new concepts introduced and related changes in practice. Of particular concern in this chapter are the geographic perspectives of environmental determinism and the landscape school, and the ideologies through which humans understand their earth-transforming activities.

In this chapter you will find

✤ a discussion of the reasons for viewing humans and nature as separate—what is often called the *dualistic preference*—and for distinguishing between presumed different types of human;

✤ an account of environmental determinism—that is, nature as it impacts on humans—as the principal example of dualistic thinking in the geographic tradition;

✤ a discussion of approaches that emphasize the human use of nature, rather than the impact of nature on humans;

✤ an outline of the landscape school approach to cultural geography, which was introduced in the 1920s and remained largely unchallenged until the 1970s. Possible links to the influential anthropological concept of culture as superorganic are stressed;

✤ a summary of alternative ecological traditions that favor attempts to integrate humans and nature—what is often called the *holistic preference*. Included here are the controversial claims of evolutionary naturalism;

✤ a reaffirmation of the value of understanding contemporary cultural geography in both larger historical and disciplinary contexts.

Three Questions

An overriding concern in this chapter, as the outline of the chapter's contents suggests, is the broad issue of human relationships with nature and related landscapes, a concern that prompts three questions.

First, when we refer to nature, are we limiting the concept to what geographers typically label the physical environment—geology, landforms, climate, weather, flora, and fauna—or do we understand humans as belonging to nature? Expressed simply, are humans seen as *apart from* or *a part of* nature? For many cultural geographers working in the classic landscape, or Sauerian, tradition, the separation of humans and nature was taken for granted. Indeed, the phrase "humans and nature" immediately identifies a separation of the two, a **dualism** that is a traditional component of Western thought. Today, there are several perspectives on this important question.

Second, is human behavior affected, perhaps even determined, by physical environment, or does culture play a determining role? The emphasis on separating humans and nature resulted in what is perhaps an unfortunate tendency, that of privileging one at the expense of the other, and there are numerous examples in this chapter of this tendency toward **reductionism**. Both versions of **determinism**—environmental and cultural—played important roles in the history of cultural geography. Building largely on previous work, many early twentieth-century cultural geographers assumed that human behavior and cultural landscapes resulted primarily from characteristics of the natural world. Major challenges to this view were put forward by French geographers, especially Vidal, and by Sauer. Sauer's work has in turn received criticism for favoring a cultural determinist approach.

Third, what distinguishes humans from other living species, particularly from other primates, and why have some cultures often regarded other cultures—and also some groups within their own culture—as being closer to other living species than they are themselves? Most early twentieth-century cultural geographers simply ignored this question, as it was not seen as necessary to an understanding of cultures and landscapes. More recently, however, there has been much interest in uncovering how the nature/culture dualism informs the production of unequal identities and landscapes.

Overall, responses to these three questions changed markedly as the larger cultural geographic enterprise changed, first with the development in the 1920s of the landscape school and later with the emergence, beginning in the 1970s, of the new cultural geography. Most notably, the various challenges to the landscape school initiated in the 1970s recognized that the earlier cultural geography had ignored some important matters. Consensus views are not yet evident, although one important development is that it is becoming increasingly popular to talk about the social construction of nature (as this term was introduced in Chapter 1), with the broader concern being the way in which we divide our world through

> ## Questions for Critical Thought
>
> —— ✤ ——
>
> Regardless of cultural background, it is likely that the phrase "humans and nature" is part of your taken-for-granted world. For most people socialized in a Judeo-Christian tradition—which is the tradition that has had the greatest impact on the cultural geography considered in this book—it is usual to conceive of the two parts of the phrase as both separate and unequal entities, with nature subordinate to humans. But such a view of the world does not prevail in some other traditions, including Buddhism, Hinduism, and many indigenous cultures, in which humans are more characteristically seen as part of nature, neither separate nor superior. Is one view correct, and the other incorrect? Or is one view in some way better than the other? Or are the differences between cultural traditions best seen as simply different ways of thinking, different ways of structuring the world in which we live? These are challenging questions, and you will be asked to rethink this matter at the end of this chapter when you have some additional bases for formulating a response.

our language as one means of naming, defining, and imposing order on that world. →≫≫ 1

Separating Humans and Nature

Humans have probably always wondered about their place in the world and about their relationship with nature. People have raised these questions in various cultural contexts and in many different ways that tend to reflect prevailing religious attitudes and scientific knowledge. One characteristic feature of Western thought is the presumed separation of humans and nature and the contrasting of one with the other to emphasize difference. The authoritative account of the changing views of humans and nature in the Western world from the classical period through 1800 is by Glacken, who began:

> In the history of Western thought, men have persistently asked three questions concerning the habitable earth and their relationships to it. Is the

earth, which is obviously a fit environment for man and other organic life, a purposefully made creation? Have its climates, its relief, the configuration of its continents influenced the moral and social nature of individuals, and have they had an influence in molding the character and nature of human culture? In his long tenure of the earth, in what manner has man changed it from its hypothetical pristine condition? (Glacken 1967:vii)

In a brief consideration of nineteenth-century thought, Glacken (1985) acknowledged the rise of a fourth question: how have humans interpreted nature in esthetic, subjective, and emotional ways?

Each of these four questions implies a separation of human and natural worlds. Although other traditions asked different questions that often stressed integrated rather than separated human and natural worlds, it is Western thought, as noted earlier, that has been most influential in the evolution of cultural geography. So, why is it that Western ideas about humans and nature have typically seen the two as separate entities?

Historical Perspective

Greek and Christian Thought

The first of Glacken's three questions concerns the idea of a designed earth, a view that reflects the doctrine of **teleology**. In accord with this idea, **Stoicism**, a philosophy that flourished in Athens after about 300 BCE and in Rome some 400 years later, held that because the natural world was clearly suited to human life, then nature must have been created for humans. All things had a purpose, with plants created for animals and animals created for humans. Many of these views, and similar ideas from Judaism, were absorbed into early Christian thought, and they have often been discussed in terms of the Biblical story of the Garden of Eden and the Fall, in which God grants humans dominion over nature: *Be fruitful and multiply, and fill the earth and subdue it; and have dominion over the fish of the sea and over the birds of the air and over every living thing that moves upon the earth* (Genesis 1:28). But as soon as

Adam and Eve ate from the forbidden tree, Adam removed himself from the natural world of the garden, emancipating the human soul from nature. This separation was followed by God again granting humans dominion over all animals after the flood: *Every moving thing that liveth shall be meat for you* (Gen. 9:3). The account of the Fall in the Christian world and the resulting dualism of humans and nature can be contrasted with other **creation myths**, such as those of many indigenous cultures, that often involve seeing humans as a part of the natural world.

The Scientific Revolution

Stoic, Jewish, and Christian traditions of a designed earth and the right of humans to rule over all other living things were widely accepted in Europe for many years. They were influential in the birth of modern science in the sixteenth century, in the philosophical movement of the Enlightenment in the seventeenth and eighteenth centuries, and in the rise of positivism in the nineteenth century. Each of these developments further contributed in some way to the idea that humans and nature were separate. Descartes, Hobbes, Locke, and Hume carried on the scientific method advanced by the empiricist philosopher Francis Bacon (1561–1626), which, broadly speaking, had both a **mechanistic** worldview and a belief in human control of nature as central ideas (see also **Cartesianism**). For Bacon, a key purpose of the scientific endeavor was to emphasize human dominion over nature, a dominion that some saw as partly lost as a consequence of the Fall. Indeed, natural history, comprising both botany and zoology, was studied not only for reasons of scientific curiosity, but also for explicitly practical and utilitarian reasons.

Enlightenment and Nineteenth-Century Thought

The empiricist philosophy that guided the practice of this early science asserted that all factual knowledge is based on experience, thus denying the view that the human mind is provided with a body of ideas before it encounters the world. This empiricism has proven to be the dominant **epistemology** and method of science, including the early social sciences that established the credibility of the discipline in the nineteenth century.

As an outgrowth of the development of science, Enlightenment philosophy carried on the established empiricist tradition. Nature was seen as something that could be explained in terms of a system of universal laws, such as the law of gravity, rather than as a number of unrelated parts understood only by reference to God. Continuing Christian tradition, humans were seen as possessing immortal souls and as dependent on the natural world, a concept known as **materialism**. Thus, human behavior was explicable by reference to laws similar to those being discovered in physical science; this is the concept of **naturalism**. Together, these closely related views (some philosophers argue that they are virtually synonymous) further contributed to the rise of positivist social science in the nineteenth century. The principal opposition to such naturalism came from idealism (see also **realism**). →>>> 2

The positivism that came to the fore in nineteenth-century social science is associated with the writings of the French philosopher Auguste Comte (1798–1857), although the basic ideas were anticipated by the empiricism of Bacon, and this positivism has close links with both materialism and naturalism. Recall that the key assumption of empiricism is that knowledge is dependent on experience. Expressed simply, Comtean positivism claimed that all knowledge, human and natural, is derived from systematic study of the world and the formulation of laws about the world.

Not all scholars accepted that human observers had access to an objective view of nature, unaffected by subjective considerations. Friedrich Nietzsche (1844–1900), for example, insisted that it was not possible to discover nature, only to imagine and interpret nature. Nevertheless, the principal social science disciplines that developed in the nineteenth century—economics, sociology, anthropology, psychology, and, to a lesser degree, geography—became typically empiricist and positivist in emphasis.

The linking of human and natural worlds within one method of science did not lead to any real integration of the two subject matters. Although they were to be studied by the same method, they remained separate.

What It Means to Be Human

Although traditionally most Western thought took it for granted that humans were fundamentally different from other forms of life, ideas about the separation of human and natural worlds spurred interest in determining precisely what it was about humans that made them different.

Aristotle observed that only humans possess a rational or intellectual soul, and Christian theology insisted that to be human was to have an immortal soul. But such requirements did not make it easy to distinguish humans from animals and, from Aristotle onwards, much thought has been given to explaining the distinction. Regardless of detail, it has been usual to see the two as different in some fundamental way and to regard humans as superior. In addition to stressing anatomical distinctions, three specific characteristics distinguishing humans were proposed: speech, the ability to reason, and the possession of a conscience and religious instinct. Descartes added to the debate by proposing that, like animals, humans resembled machines in that many of their actions were instinctual, but that only humans also possessed an intellect. Ideas such as these served to confirm the separation of humans and animals and, more generally, of culture and nature. Animals were awarded low moral status, justifying such activities as hunting, domestication, the eating of meat, extermination of undesirable species, and the displaying of animals in enclosures for scientific, educational, and entertainment purposes. →>>> 3

Different Humans

Ideas related to these sometimes prompted European thinkers, from the Greeks onwards, to relegate certain classes of people—different cultural groups, the insane, the poor, children, women—to the status of lesser animals. Thus, the various attempts to explicitly define *human* had implications not only for the way in which the natural world was treated but also for the way humans treated other humans. Further, it was usual in both Western and other traditions to regard nature as female. There were two contrasting components to this image. On the one hand, especially prior to the birth of modern science, nature was seen as a nurturing mother providing humans with their various needs. On the other hand, there was the wild aspect of nature, suggesting the need for control.

Women are devalued in most societies, a fact that has been interpreted in terms of this association of women with nature, where nature is something less than human, women are closer than men to nature, and therefore women are inferior humans. Why are women seen as closer than men to nature? It may be because it is women who procreate and because of the related domestic social roles that women play. Men, on the other hand, have traditionally been seen to occupy a position closer to culture. An additional distinction that can be made is that traditionally men are concerned with the welfare of the social whole while women are concerned with more particularistic family matters. This is the classic socially constructed distinction made between the male, public domain and the female, private domain. ➡➤➤ **4**

It is difficult to exaggerate the significance of this "transcoding of feminine qualities to Nature, and of naturalness to women" for the cultural geographic enterprise, both traditional and new (Rose 1993:69). Indeed, it might be argued that all of human geography is structured around the separation of humans and nature, especially as this is reinforced by the distinction between urban and rural ways of life that became increasingly apparent in the nineteenth century.

These various ideas concerning human authority over nature, the perceived variable human quality of particular people, and the linking of women with nature are themes that recur frequently throughout this textbook.

Advances in Natural and Social Sciences

During the nineteenth century, developments in the natural and social sciences improved the methods used to distinguish humans from nature. Advances in geology and biology enabled natural scientists to clarify the **evolution** of the human species, while social scientists began to focus explicitly on the concept of culture as an exclusively human attribute. Although it is correct to note that these advances that provided new biological and cultural understandings

did not allow scientists to make absolute distinctions between humans and nature, they did help to clarify some of the ideas previously used in this endeavor. More important, a consideration of these developments in the physical and social sciences reveals that both are relevant in any attempt to define what it means to be human.

Both geology and biology developed dramatically from the late eighteenth century through the nineteenth century. In the case of geology, these developments caused scientists to question seriously the Biblical account of the earth, which is based on a belief in **catastrophism** and which attributes to the earth a maximum age of roughly 6,000 years. In 1774 the French scientist and pioneer of paleontology Count Buffon (1707–88) advanced the concept of **uniformitarianism**, the idea that changes to the earth had occurred over time as a result of such observable physical processes as erosion and deposition, not because of a number of isolated and unobserved catastrophes. This perspective of the earth as having been shaped over time by natural, uniform processes was further developed by two Scottish geologists, James Hutton (1726–97) and Charles Lyell (1797–1875), whose influential texts, produced in 1788 and 1830 respectively, made it evident that the earth was much older than the 6,000 years previously believed.

These advances in geology were not the only developments to alter the established view of natural science at this time. Through developments in the study of plant and animal biology it was becoming clear that not only had the earth evolved slowly into its present form, but animal and plant species had also evolved. The first proposal for biological change was put forward by the French naturalist Jean-Baptiste de Lamarck (1744–1829). Lamarck correctly noted that animals and plants were adapted to their specific environments, but his explanation, known as **Lamarckianism**, as to how such evolution occurred—namely, that organisms develop into more complex forms—was incorrect. It was English scientists Charles Darwin (1809–82) and Alfred Russel Wallace (1823–1913) who, independently, proposed **natural selection** as the solution to the question of plant and animal evolution put forward by Lamarck. Natural selection refers to the process by which nature effectively selects members of a species that are most suited to the environment; hence the most adaptive traits are reproduced and increase while the least adaptive traits decrease. Darwin published his monumental *On the Origin of Species by Means of Natural Selection* in 1859.

Darwinism as a theory is non-teleological in the sense that it explains how an event contributed to some result but does not explain the occurrence of that event. Thus, species variations occur randomly, but those variations chosen by the environment were not designed with some particular result in mind. For many, this absence of teleological explanation means that the basic logic cannot be directly applied to cultural change, an important consideration in assessing attempts to apply Darwinism to the social sciences. The Lamarckian evolutionary process, involving the idea that the organs and behavior of species can change and that new organs and behavior can be transmitted from individual species members to their offspring, may be more appropriately applied in a cultural context.

Building on these advances in natural science, natural and social scientists began to ask and answer questions about the evolution of one species in particular: humans. The natural science answer—still incomplete—was based on biological developments in such areas as brain size and the ability to walk upright. One important approach addressing this issue is sociobiology. The principal social science answer was based on the always uncertain and contested concept of culture and on the assumption that cultural evolution had occurred. An alternative social science answer argues that we have an understanding of nature, but not of ourselves. These three bodies of ideas are now discussed.

Sociobiology

Human beings have a great deal in common both biologically and culturally, and it is critical that cultural geographers not lose sight of this essential human unity. But humans also have much in common with some animal species, such that we can say that humans are a part of nature. At the same time humans possess culture, and this represents a difference from

nature. In short, humans are both natural and cultural beings. It is this recognition that, for many, justifies the argument that most studies of humans are too **anthropocentric** and that much can be learned by treating the human cultural experience as but one part of the larger biological continuum. ➔➤➤ **5**

Sociobiology is a particular expression of the relationship between biological and cultural or social evolution that argues for a reduction of human behavior to biological concepts. In *The Descent of Man*, Darwin studied social organization generally by applying evolutionary theory combined with the biological knowledge he derived from observing both animal and human behavior. This early linking of biology and human subjects was criticized from religious quarters, and also by sociologists such as Durkheim, who argued for the autonomy of sociology and hence for separation of the human and the natural. Versions of this debate have resurfaced in recent years, with a number of sociobiological theories being proposed.

Several sociobiological theories make explicit use of Darwinian concepts to describe a process of natural selection of cultural characteristics that produces, through time, an increasingly complex, increasingly well-adapted, and hence improved, culture. Changes in human culture may be seen as similar to organic mutations, with parallels between biological and human cultural evolution along with the added idea that cultural evolution is a form of learning process leading to ever-improving human circumstances. This is a particular interpretation of behaviorism as developed in psychology. But the most elaborate interpretation of biological evolution in human terms incorporates the idea that cultural change is often intentional and purposive, as in the Lamarckian evolutionary process, rather than accidental, as in Darwinian theory. The temptation of relating biological and cultural evolution has also proven irresistible for some biologists, who accept the idea that the process of cultural evolution is Lamarckian. One particular view is based on a set of theories that assert that human behavior is a product of the co-evolution of human biology and culture.

Not surprisingly, sociobiology has been subject to criticism. Most notably, it represents a major challenge to much contemporary social science because its emphasis on a naturalist perspective puts it out of tune with the prevailing social science view of humans as active subjects. Further, feminists have expressed concern with its male construction of science: "The patriarchal voice of sociobiology is less the effusive sexism that ripples over the whole plane of the text than it is the logic of domination embedded in fashioning the tool of the word" (Haraway 1991:74). Given the generally critical appraisal of sociobiology from the social sciences, as well as related internal disagreements, it seems unlikely that the approach will make a major impact on the contemporary understanding of culture and culture change regardless of the validity of the arguments. Certainly, the principal cultural geographic initiative in this direction—namely, habitat and prospect-refuge theories—has not been widely accepted. At the same time, such ideas should not be ignored, as they reflect a clear concern with the idea that humans are a conventional animal species.

Cultural Evolution

The evolutionary ideas advanced in eighteenth- and nineteenth-century geology and biology were received and applied with great enthusiasm by many nineteenth-century social scientists. Specifically, the thesis that cultures evolved and thus display only limited ties to nature—in other words, that culture changed while nature remained unchanging—was a dominant feature of the early disciplinary experiences of both anthropology and sociology. Although by no means a new idea—indeed it was evident in classical Greece and became especially important during the period of European overseas expansion— the concept of cultural evolution became the subject of numerous theories in the period after about 1860. But despite the number and diversity of evolutionary theories, all of them have three features in common: first, they identify a number of distinct cultural stages; second, they recognize that the probability of remaining in a stage is greater than the probability of returning to an earlier stage; third, they identify a probability of moving through the sequence of stages.

Three great social thinkers of the nineteenth century, Comte, Marx, and Spencer, adopted evolutionist

perspectives, while the anthropologists Edward Burnett Tylor (1832–1917) and Lewis Henry Morgan (1818–81) portrayed culture and cultural change in terms of **unilinear evolution**—a series of stages from savagery to civilization. Together, Tylor and Morgan established the comparative method of inquiry, collecting ethnographic data from different cultures in order to demonstrate the increasing complexity of culture through time. Their study reflected an acceptance of the Enlightenment doctrine of progress, and it was implicitly assumed that European civilization represented the apogee of evolution, while primitive peoples were both biologically and culturally inferior. As one component of evolutionist thought, the anthropological typologies employed new data produced by European overseas expansion to support their theories of unilinear evolution, which postulated that all cultures would, by means of independent invention, pass through a series of increasingly superior stages.

In *Primitive Culture*, first published in 1871, Tylor (1924) proposed three stages of cultural evolution and their characteristics as follows:

- savagery—hunting and gathering, limited technology
- barbarism—agriculture, settled villages and towns
- civilization—writing

More generally, Tylor noted:

On the whole it appears that wherever there are found elaborate arts, abstruse knowledge, complex institutions, these are results of gradual development from an earlier, simpler, and ruder state of life. No stage of civilization comes into existence spontaneously, but grows or is developed out of the stage before it. This is the great principle which every scholar must lay firm hold of if he intends to understand either the world he lives in or the history of the past. (Tylor 1916:20)

Although the three stages are distinguished essentially in terms of technology, the greatest emphasis was placed on language, religion, and myth. Further, Tylor employed the concept of cultural survivals, components of a culture that, for reasons of tradition, persist from one stage to a later stage.

In similar fashion, Morgan (1974), in his 1877 book *Ancient Society*, proposed seven stages of cultural evolution and their characteristics as follows:

- lower savagery—beginnings of human life, gathering of wild fruit
- middle savagery—eating of fish, origins of speech, and use of fire
- upper savagery—use of bow and arrow
- lower barbarism—use of pottery
- middle barbarism—agriculture, irrigation
- upper barbarism—use of iron tools
- civilization—writing

Morgan distinguished the stages in technological terms, although he wrote primarily about the evolution of social and political forms and about a transition from communal to private ownership.

A lack of interest in evolutionism in the early twentieth century ended with a revival prompted by Childe in the 1930s and continued by White in the 1940s and Steward in the 1950s. Childe used evolutionist thought in accounts of the origins of agriculture and subsequent development of an urban way of life. White employed a superorganic concept of culture, specifically a form of technological determinism, to argue that technology determined the availability of energy and the efficiency of its uses, and further, that the structure and successful functioning of culture was the direct result of the amount of energy produced and the manner in which that energy was used. →≫ **6**

Technology, through energy, was therefore the cause of cultural characteristics and change. Steward, a student of both Kroeber and Sauer, was critical of earlier unilinear evolutionist ideas, especially their often grand claims and gross generalizations. He favored instead a more modest **multilinear evolution**. The aim of multilinear evolution was to discover the laws that determined cultural development by identifying parallels in the development of different cultures.

Boas initiated arguments against evolutionism in anthropology with strong support from Kroeber,

Lowie, and Wissler. He based his criticism on the belief that the crucial cause of cultural change was **diffusion** and related culture contact, not independent evolution. He also raised objections to the racism implicit in evolutionism's emphasis on progress, to the lack of objectivity in evolutionary theory, and to the rigidity of the sequences proposed by Tylor and Morgan. Boasian anthropologists emphasized diffusion and contact as these related to the creation of cultural areas. He described human history "as a sort of 'tree of culture', with fantastically complex branching, intertwining and budding off—each branch representing a uniquely different cultural complex, to be understood in terms of its own unique history rather than compared in cultural complexes in other world regions in some grand scheme of 'stages of evolution'" (Pelto 1966:24).

Although the evolution of culture is seen as a central component of the landscape school methodology—it is one of the five themes identified by Wagner and Mikesell (1962) (see Box 1.3)—it is nevertheless an interest that has been ignored by most cultural geographers (but see Newson 1976). It appears that cultural geography's interest in landscape outcomes of cultural **occupance** largely precluded any detailed concern with cultural change, and, further, that such discussion was seen as more properly belonging to the discipline of anthropology. Indeed, the principal evidence of evolutionist thought in human geography is in association with environmental determinism.

Understanding Ourselves

The traditional way of defining *human* was to distinguish between humans and other animals in order to identify some presumed distinctive characteristics, such as having been made in the image of a Christian God, possessing a soul, or being spiritual as well as corporeal. But at the same time, humans have always been intrigued by the possibilities of intermediate species, of creating human life from non-living things, or of creating machines that are akin to humans. The basis for distinguishing humans and other animals lessened with mechanistic logic, which provided explanations of the whole material world—humans and nature alike—and with the advances made by Darwin that showed that human biological evolution was subject to laws similar to those governing the rest of the natural world. To offer a conclusion on this matter is difficult; it may be that a persuasive definition of what it means to be human is impossible. One perspective on this matter is discussed in Box 2.1.

Environmental Determinism

Nineteenth-century thinking about humans and nature incorporated some contradictory elements, but the dominant view was to see the two as separate. Viewing humans and nature as separate did not, however, prevent scholars from holding the view that both were susceptible to study by the same scientific method expressed in the philosophy of naturalism and the associated idea of positivism. For the discipline of geography, the separation of nature and humans was most evident in the explicit distinction typically made between physical geography and human geography, but it was apparent also in the acceptance and application of the idea of environmental determinism, an approach to the study of geography that was closely related to naturalist epistemology.

The doctrine of environmental determinism asserts that the physical environment—nature—determines or at least influences the human world, both culture and cultural landscape (see Figure 2.1a). Thus, all of our human actions are to be understood primarily if not exclusively in the context of the physical world of climate, landforms, soils, and vegetation. This idea was often treated as though it were a law of science rather than an assumption. Recall that concern with physical geographic impacts on humans was one of the three questions identified by Glacken (1967) as characterizing Western thought about humans and nature from Greek times through to about 1800.

This concern is one particular interpretation of the larger philosophical position of determinism—the idea that all events are effects of earlier events. In physical science, determinism was the dominant view for at least 200 years after Newton, although it has now been replaced by the indeterminism of contemporary physics. As noted, the social science disciplines found determinism very appealing at the time of their

Questions for Critical Thought

Determinism implies that events have causes and that, in principle, it is possible to uncover those causes and perhaps even to formulate laws that relate causes and effects. Applied to human behavior, this idea has profound implications concerning free will and moral responsibility for our actions. If the argument that all of our behaviors are predetermined by prior events is accepted, then does this mean that humans cannot be held responsible for their actions? If this is the case, do you agree that this is unacceptable? Are you surprised that social sciences were attracted to determinist logic, especially as applied at a group rather than an individual scale? Moreover, are you concerned to hear that the social sciences used determinist logic without offering any meaningful critique of the larger philosophical context? It might be helpful, when considering environmental determinism as developed and applied within geography, to have preliminary responses to these critical questions in mind.

foundation, as such arguments appeared to give credibility to their specific disciplinary subject matter and to render their enterprise scientific in the nineteenth-century sense of science as empiricist and positivist.

Historical Perspective

Environmental determinist thought has been supported throughout its long history by countless scholars, many of them writing from very different philosophical positions. Indeed, some combined thoughts about environmental influence with thoughts about human impact on the earth, the two being divergent but not contradictory ideas. The following sections trace the history of environmental determinist thought.

Greek and Christian Thought

Early references to environmental determinism in classical Greece are found in the writings of Hippocrates (fifth century BCE), as Glacken (1967:80) explains: "From early times there have been two types of environmental theory, one based on physiology . . . and one on geographical position; both are in the Hippocratic corpus."

The physiological theories derive from the idea that human health varied according to specific locations, and could be affected by such factors as altitude

Box 2.1 *Achieving Effective Knowledge of Ourselves*

According to Gellner (1997), effective knowledge of nature exists, but effective knowledge of humans and their institutions does not. During the long period between the initial domestication of plants and animals and the scientific and industrial revolutions, European culture was based on the production and storage of food, on a rather unchanging technology, and on minimal advances in science. Accordingly, a Malthusian situation prevailed, with resources remaining essentially constant and any increases in population or decreases in food supply creating stress; this is, of course, the first stage of the demographic transition model. Under these circumstances, lack of sufficient food for the least powerful was an ever-present possibility, causing the social hierarchy to function essentially as a queue to the storehouse. The principal concern for individuals and groups was, therefore, not with the issue of total food production, but rather with their ability to exert power over others in order to enhance their social status, or, more simply, to move up in the queue. Accordingly, the prevailing cultural values were those that enhanced personal and group authority and strength, and not those related to technological progress and innovation.

Under these conditions, humans lacked any real knowledge of either nature or themselves. There was necessarily some understanding of parts of the natural world, but the integration of ideas was not encouraged in circumstances where the most powerful assumed ownership of new knowledge and innovation in order to further enhance their status. So how did this situation change? The answer lies in the rise of northwestern Europe beginning in the seventeenth century, which involved a transition from coercive or martial society to productive or commercial society. In Marxist terms, this is a change from a feudal to a bourgeois order. There had been similar transitions at earlier times that

proved unsuccessful in changing the larger cultural circumstances. This particular transition was successful because it occurred in conjunction with a second development, namely the rise of science.

Accordingly, since the seventeenth century, success in European cultures has not been based only on power, but also on a sound knowledge of nature, on how to increase production, on progress. The contemporary world thus has a sound grasp of nature and an ability to change nature, but no equivalent grasp of ourselves or how to change ourselves in desirable ways—a process some call social engineering. There are five possible reasons for what we might regard as our continuing failure to achieve an understanding of ourselves comparable to our understanding of nature. These are

- the inherent complexity of the human world;
- the idea that human behavior is affected by human ideas and interpretations;
- the idea that everything humans do is linked to everything else that humans do;
- the idea that cultural evolution is Lamarckian;
- the role played by chance in human affairs.

In discussing the prospects for advances in understanding ourselves comparable to advances that we have made in understanding nature, Gellner (1997:16) concludes that

we simply do not know whether the social sciences will have their 17th century, whether the breakthrough and the subsequent fall-out will occur. What we do know is that if it did occur, it would, once again, completely change the rules of the game, as radically—perhaps more so—than was the case when technologically based natural science made possible Perpetual Growth and the vision of World-as-Progress.

a) Environmental Determinism

b) Environmental Possibilism

c) The Landscape School

d) Environmental Probabilism

e) Environmental Perception

f) Ecology

Figure 2.1 Geographic Approaches to Human-and-Nature Relationships. This figure summarizes five of the approaches to the study of humans and nature detailed in this chapter. These schematic diagrams necessarily simplify the very real complexity evident in the various debates. It is not appropriate to interpret these approaches as different from one another in any absolute sense; rather, each lies somewhere on a continuum ranging from the dualism most evident in extreme environmental determinist arguments through to the holism evident in some ecological arguments. Source: Adapted from D.N. Jeans, "Changing Formulations of the Man-Environment Relationship in Anglo-American Geography," *Journal of Geography* 73 (1974):37.

and proximity to water. Such theories tended to be raised in discussions of, for example, the best locations for houses. More significant for our purposes are the geographical position theories that derive from comparisons of large areas and concern the character of groups of people rather than the health of individuals. In this category are discussions of environments and related human characteristics, focusing especially on the contrast between the well-off populations of the eastern Mediterranean and the more penurious populations of Europe north of the Mediterranean area. In similar fashion, Aristotle described some eastern Asiatic populations living in warm areas as thoughtful and skillful but lacking in spirit and courage, in contrast to the less intelligent but braver and more warlike northern Europeans. According to Aristotle, the inhabitants of Greece, living in an intermediate environment, combined the better—interestingly, not the worst—qualities of both these populations.

Many other Greek scholars, especially Posidonius (*c*.135 BCE–51 BCE), a Stoicist who also discussed the influence of the position of the sun and stars on both physical geography and humans, continued this tradition of thought. In this sense, early environmental determinist thinking does not necessarily imply a separation of humans and nature, because all things are related. Using more contemporary terminology, much of this early thought was ecological in character in that all aspects of the physical and human worlds were seen as linked; humans grow in an environment and are affected by that environment.

Strabo (*c*.64 BCE–20 CE), a member of the later Stoic school, adopted a broader cultural perspective in discussing the geography of the known world. Although environmentalist overtones are evident in much of his writing, especially concerning the physical geographic reasons for the rise of Rome, there is also recognition of the human ability to overcome environmental challenges. The writings of Strabo highlight the important general point that many scholars who contributed environmental determinist arguments also recognized other human-and-nature relationships.

After the demise of the Roman Empire in the fifth century, many of the Greek and Roman contributions were unavailable to European scholars, and the most studied and influential book for about the next thousand years was the Bible. The Biblical design argument was most influential and contributed to the number of accounts of human use of the earth; original environmental determinist ideas were less common, although there were contributions from such scholars as Albertus Magnus (*c.*1200–80) and St Thomas Aquinas (*c.*1225–74). Accounts of humans and nature played an important role in Islamic scholarship, with ibn Khaldun (1332–1406), one of the greatest philosophers of the period, emphasizing the impact of environment on social organization and economic life.

The Scientific Revolution

In Europe, the onset of the scientific revolution and the beginning of significant overseas exploration combined to produce additional examples of the presumed impact of physical environment, especially climate, on humans. The new era of sea travel meant that the basic methodology of cause and effect, reinforced by advances in science, could be applied to an increasing number of locations in the expanding world known to Europeans. One twentieth-century environmental determinist was fond of quoting Bacon, who said: "True knowledge involves the study of Causes" (Taylor 1928:3). Certainly, this way of thinking flowered in many European works during this period. The French philosopher Jean Bodin (*c.*1529–96) built on earlier work to argue that cultural differences were a result of astrological, latitudinal, and local factors. One conclusion he reached was that people in temperate climates were both more talented than those in the colder north and more energetic than those in the warmer south. However, even such an enthusiastic proponent of environmental determinism as Bodin acknowledged as well the role of other factors, such as the consequences of cultural contacts related to war and migration.

Enlightenment and Nineteenth-Century Thought

Montesquieu (1689–1755) restated many of the earlier positions on environmental determinism using new information that was becoming available because of European overseas movement. He used climate to explain cultural differences—he posited that people in cold areas were strong and courageous while those in warm areas were more suspicious and cunning—and to explain cultural persistence, as in the presumed unchanging cultures of Asia. By this time, environmental determinism was much more than simply a way of explaining culture: it also justified cultural relativism. Thus, Montesquieu, though he was opposed to slavery, understood that it could develop where a population was so indolent that slavery was the only means available by which that population could be used as labor. Again, as with Bodin, Montesquieu did not adhere to environmental determinism at the expense of recognizing other aspects of the human world; he included in his writings discussions of human impacts on environment.

The idea of natural controls is also evident in the writings of Thomas Malthus (1766–1834), who argued that it was not possible for a genuinely happy society to evolve because of the universal circumstance that population numbers increased more rapidly than did the supply of food, which was limited by nature. It was in his famous 1798 essay that Malthus first referred to the "struggle for existence."

One of the most explicit statements of environmental determinism during the Enlightenment came from the French philosopher Victor Cousin (1792–1867):

> Yes, gentlemen, give me the map of a country, its configuration, its climate, its waters, its winds, and all its physical geography; give me its natural productions, its flora, its zoology, and I pledge myself to tell you, *a priori*, what the man of that country will be, and what part that country will play in history, not by accident but of necessity; not in one epoch, but in all epochs; and, moreover, the idea which it is destined to represent! (quoted in Febvre 1925:10)

The second half of the nineteenth century provided many examples of the apparent relevance of the new ideas about biological evolution for human beings. Darwinian theory served to confirm the largely non-scientific arguments of earlier writers, with the result that many versions of environmental determinism—such as that proposed by the English historian H.T. Buckle (1821–62), which explained the onset of the scientific age in Europe in terms of climate, soil, and configuration—began to assume increased prominence. One proposal for a positivist social science incorporating environmental determinism was especially important, namely that associated with the evolutionary perspective most fully developed by Spencer. As noted earlier, the second half of the nineteenth century was the heyday of evolutionary theories, with major contributions also from the anthropologists Morgan and Tylor. But it was Spencer, in his articulation of what is usually labeled **social Darwinism**, who argued for an analogy between animal organisms and human groups with regard to the struggle to survive in a physical environment. The logic of social Darwinism is straightforward, hypothesizing that cultural groups evolved in accordance with their ability to adjust to physical environments. Accordingly, Spencer, for example, saw history as a steady move from warmer to colder areas. ➔≫ 7

During the nineteenth century, then, environmental determinism continued to be popular, and in fact became a component of other conceptual schemes, such as cultural evolution and social Darwinism. But the century culminated in the adoption of the approach in a quite traditional, straightforward form by the newly institutionalized discipline of geography. This adoption was notwithstanding the fact that several highly influential mid-nineteenth-century geographers had expressed alternative views. Environmental determinism appealed to geographers for a number of reasons, including the following:

- it had a long scholarly tradition;
- it offered a way to account for ever increasing evidence of cultural differences in different physical environments;
- it belonged to the body of mechanistic science, which continued to be popular;
- it was supported by Darwinian theory and related ideas that applied to cultures.

Geography as the Study of Physical Causes

The Impact of Ratzel

The German geographer Friedrich Ratzel (1844–1904) accepted many of Spencer's ideas concerning the similarities between human societies and animal organisms, strongly emphasizing that humans were subservient to nature. Indeed, Ratzel is often seen as the originator of modern environmental determinism, proposing the approach as a specifically geographic concern in 1882 in the first volume of *Anthropogéographie*. As restated by Ratzel, environmental determinism proved an immensely popular approach to the study of humans and nature even though various alternatives emphasizing other relationships, including a unity of humans and nature, were available at the time. It is clear that Ratzel regarded this concept only as a generalization, not as a scientific law, and in the second volume of *Anthropogéographie* he substantially modified the view. But the early Ratzel appears to have been the more influential, probably because it was the early Ratzel whose views mirrored the needs of geography at the time, namely a subservience of human to physical geography and an apparent scientific emphasis. ➔≫ 8

In addition to the various reasons for the attraction of environmental determinism already noted, it bears mentioning that one compelling reason for other geographers' positive reception of Ratzel was the fact that his approach gave credibility to the geographic enterprise, stressing, as it did, the role of physical geography in human affairs. At a time when the various social sciences were competing to generate explanations, the practitioners of a discipline typically seized upon any approach that highlighted their discipline at the expense of others. A related reason for the use of environmental determinist logic was its implicit incorporation in the dominant regional approach to the study of geography.

The writings of Ratzel played a critical role in what was to become cultural geography, such that it is difficult to exaggerate the importance of environmental determinism during the early years of the newly institutionalized geographic discipline. Thus, the leading British geographer at this time, Halford J. Mackinder

(1861–1947), accepted the view that human geography was derivative of physical geography. Similarly, in the United States, the environmental determinist view was articulated by the founder of the Association of American Geographers, William Morris Davis (1850–1934), and, more generally, became an integral component of the new university discipline. But the principal examples of environmental determinist thought are contained in the works of two American geographers, Ellen Semple (1863–1932) and Ellsworth Huntington (1876–1947).

Semple and Huntington

Semple visited Germany in the 1890s to study with Ratzel and subsequently introduced his understanding of physical influences on humans to American geographers. In *Influences of Geographic Environment*, Semple famously began:

Man is a product of the earth's surface. This means not merely that he is a child of the earth, dust of her dust; but that the earth has mothered him, fed him, set him tasks, directed his thoughts, confronted him with difficulties that have strengthened his body and sharpened his wits, given him his problems of navigation and irrigation, and at the same time whispered hints for their solution. She has entered into his bone and tissue, into his mind and soul. (Semple 1911:1)

These opening sentences set the tone for a book that contains determinist explanations throughout. Many of these explanations support the longstanding view of climate as a principal cause of human character, human activities, and human landscapes. Semple's work also reflects the prevailing thought about racial differences between groups of people. The following brief quotation is not atypical of the thrust of the 1911 book: "The northern peoples of Europe are energetic, provident, serious, thoughtful rather than emotional, cautious rather than impulsive. The southerners of the sub-tropical Mediterranean basin are easy-going, improvident except under pressing necessity, gay, emotional, imaginative, all qualities which among the negroes of the equatorial belt degenerate into grave racial faults" (Semple 1911:620).

In similar fashion, Huntington accepted the basic logic of environmental determinism but used the idea to focus on the supposed links between climate and the growth and decline of civilization. The basic theme was that areas of stimulating climate encouraged the growth of civilization, whereas the monotonous tropical climate was especially inimical to such growth:

A prolonged study of past and present climatic variations suggests that the location of some of the most stimulating conditions varies from century to century, and that when the great countries of antiquity rose to eminence they enjoyed a climatic stimulus comparable with that existing today where the leading nations now dwell. In other words, wherever civilization has risen to a high level, the climate appears to have possessed the qualities that today are most stimulating. (Huntington 1915:4)

But Huntington did not simply produce general arguments; he also provided details of climatic change, ways of measuring civilization, maps of civilized areas, and implications of climatic controls for humans. As an example of his measures of civilization, Huntington (1945:6) noted "intermediate peoples, such as the Chinese, occupy intermediate places, with more cars than the pygmies, less than the Bulgarians." His concern with the present human condition is well stated as follows: "The climate of many countries seems to be one of the great reasons why idleness, dishonesty, immorality, stupidity, and weakness of will prevail. If we can conquer climate, the whole world will become stronger and nobler" (Huntington 1915:294).

Appraisal

Certainly, the single most disturbing aspect of environmental determinism as an approach for the newly institutionalized discipline of geography is that it quickly proved to be embarrassingly simplistic and racist in its readings of culture and climate. Nevertheless, although environmental determinism may well be an essentially discredited idea, it continues to reverberate in many discussions about humans and nature.

Generally speaking, those who applied the approach assumed the argument to be correct and sought examples rather than approaching geographic facts in a more objective and critical-minded manner. Thus, the attraction of environmental determinism was well expressed by Griffith Taylor (1880–1963): "As young people we were thrilled with the idea that there was a pattern anywhere, so we were enthusiasts for determinism" (Taylor, in Spate 1952:425). Indeed, Taylor made an important contribution in arguing for a modified determinism: "Man is able to accelerate, slow or stop the progress of a country's development. But he should not, if he is wise, depart from the directions as indicated by the natural environment. He is like the traffic-controller in a large city, who alters the rate but not the direction of progress; and perhaps the phrase 'Stop and Go Determinism' expresses succinctly the writer's geographic philosophy" (Taylor 1951:479). ➞➤➤ **9**

Although advocates such as Semple, Huntington, and Taylor acknowledged the need for cautious applications and also recognized the role played by humans, it is fair to say that in many geographic studies of the late nineteenth and early twentieth centuries, culture, technology, and economics were subservient to physical environment. Further, a substantial body of literature, including much earlier geographic literature arguing for the unity of humans and nature, was ignored.

One useful way to think about environmental influence is in the context of human technology. Expressed very simply, it is possible to distinguish between two types of nature, benign and difficult,

and between two types of technology, high and low. The outcomes of different pairings of humans and nature are identified in Table 2.1. The table suggests that the variable "determining" the outcome of a particular relationship between humans and nature is the human variable of technology, not nature. This is because when the human variable changes from one level of technology to another, the outcome, the nature of the relationship, changes. When nature changes, from one type to another, the outcome is unaffected.

Human Use of Nature

Recall that many scholars from classical Greece onwards combined thoughts about environmental influence with thoughts about human impact on the earth, the two ideas being divergent but not contradictory. Indeed, the two ideas share a conceptual focus in that both imply the presumed separation of humans and nature. Nevertheless, it is not usual to find scholars before the eighteenth century addressing the question of human impacts in the same detailed philosophical terms as they discussed physical influences. For this reason, and because the larger philosophical context is covered in the earlier discussion of environmental determinism, this discussion of writings about human use of the earth is relatively brief.

Historical Perspective

Greek and Christian Thought
The idea that humans were able to change the physical environments in which they lived is a logical

Table 2.1 Human-and-Nature Relationships

Relationship	Outcome	Example
High technology/ benign nature	Humans not limited by nature—various activities	Temperate areas today
High technology/ difficult nature	Humans not limited by nature—various activities	Hot, dry deserts today
Low technology/ benign nature	Intimate human–nature relationship—foraging	Temperate areas 12,000 years ago
Low technology/ difficult nature	Intimate human–nature relationship—foraging	Hot, dry deserts 12,000 years ago

consequence of human life itself. Scholars in classical Greece typically justified this on philosophical grounds by recognizing that humans, being capable of reasoning, were different from nature and as such were orderers, or organizers, of the natural world. This rationale was an important component of the designed earth argument: God created the earth for humans, who were then granted authority and dominion over it to change or indeed complete it as they saw fit. As Europeans became aware of environments outside of Europe, they viewed these new lands as simply awaiting their finishing touch.

These classical Greek and Judeo-Christian ideas about the inherent responsibility of humans to organize, change, and complete nature provided ample justification for the effects of such human activities as irrigation, drainage, mining, and animal and plant domestication. As there was no real evidence of undesirable changes resulting from these activities, they were seen in a positive light. Indeed, many of the scholars who commented on the role of physical influences—such as Montesquieu and Sebastian Münster (1489–1552), whose 1544 book *Cosmographia Universalis* was the authoritative volume on world geography for about the next 100 years—commented also on the positive contributions humans were making in the sphere of environmental change. This broad interpretation of human impacts prevailed until the nineteenth century.

Questions for Critical Thought

❖

The preceding account of environmental determinism hopefully makes it clear that the approach was extremely popular among geographers from the late nineteenth century until the mid-twentieth century. This is the case despite the fact that it seems appropriate to view the approach as fundamentally flawed, except perhaps when employed in very broad terms to help uncover general relationships, such as that between global physical geography and population density and distribution. Intriguingly, the bestselling Pulitzer Prize-winning book *Guns, Germs, and Steel: A Short History of Everybody for the Last 13,000 Years,* a thoughtful and provocative discussion of the global distribution of wealth and power, has been subject to criticism by some geographers because of the perceived use of environmentalist determinist logic. For example, Judkins, Smith, and Keys (2008:22) suggest that the book is fundamentally flawed because of the imposition of "a modernist scientific determinism to the question of human history, society formation, political action and conquest." Expressed simply, the key criticism of the work is that factors other than physical environment are excluded from consideration. *Guns, Germs, and Steel* is the best known of several influential books addressing global issues that might be similarly criticized (other examples are by Landes, 1998, and by Sachs, Mellinger, and Gallup, 2001).

Consider the following four observations and ask yourself if you agree with them. First, interpreted as geography, environmental determinism neglects much earlier work, defines the discipline in terms of one specific causal relation, neglecting alternative forms of interpretation, and typically excludes subject matter that does not accord with the approach. Second, it is overly simplistic, as perhaps are all versions of determinism when applied to human behavior; the complexities of human behavior are ignored in the desire to identify a cause-and-effect relationship. Third, although the physical environment is an important consideration in much decision-making—for example, there are obvious links between behavior and diurnal and seasonal cycles and between behavior and such dramatic physical events as floods and earthquakes—it remains the case that all such behavior still needs to be understood in terms of human environmental contexts, such as those of culture, society, politics, and economics. Fourth, as is readily apparent, all human landscapes change through time notwithstanding the fact that physical environments are relatively unchanging. After considering these four questions, are you inclined to conclude that, at best, environmental determinism is only superficially attractive given that it does not stand up to rigorous scrutiny?

The Scientific Revolution

Ideas about human impacts were central to the thoughts of the first scientific philosophers, Bacon and Descartes, and to the tradition that they initiated in that there was new confidence in the ability of scientific knowledge to control physical environments. This confidence combined with prevailing religious understanding of human dominion over nature contributed in the seventeenth century to the general sense that humans were both improving themselves and improving nature through their impacts on nature.

Enlightenment and Nineteenth-Century Thought

It was in the eighteenth century, beginning with the work of the French naturalist Buffon, that the favorable view of human impact on nature assumed real importance. Buffon acknowledged that humans needed to change environments in order to enhance their civilization and stressed the improvements that had been achieved through orderings of nature. He wrote extensively about animal and plant domestication, the removal of forests, changes to soils, and the creation of cultural landscapes in general. Central to his arguments were distinctions between humans and animals, which focused on human intelligence, creativity, and adaptability.

The first suggestions by European scholars that humans might be abusing rather than using, or indeed improving, physical environments surfaced between the eleventh and fourteenth centuries, a period of agricultural expansion that involved the transformation of forest and marsh, the establishment of settlements, and increased mining activity. But it was the American geographer Marsh who made the first real advances in this area with the 1864 publication of *Man and Nature*. He portrayed humans as essentially a destructive power and noted their hostile influences on environments: "Man is everywhere a disturbing agent. Wherever he plants his foot, the harmonies of nature are turned to discords. The proportions and accommodations which ensured the stability of existing arrangements are overthrown" (Marsh 1965:34). In addition, he presented considerable evidence to demonstrate the frequency, magnitude, and deleterious effects of human activity, including detailed examples concerning plants, animals, forests, water bodies, and coastal areas. Overall, Marsh provided a compendium of factual information that focused on the theme of human activities and their detrimental consequences for nature.

Cultural Geographic Interpretations

Possibilism

Ratzel published the first volume of *Anthropogéographie* in 1882, and the determinist arguments contained therein were accepted and further developed by Semple and others. The second volume, published in 1891, advocated a different approach to human geography. Rather than focusing on the influences of physical environment on humans, Ratzel emphasized the role of humans in their cultural groupings. As we have seen, this approach, like that of environmental determinism, had a long scholarly precedent, and it was also favored by the contemporary German geographer Alfred Kirchhoff (1838–1907).

It was this approach, advocated in the second volume of *Anthropogéographie*, that was further developed by the eminent French geographer Vidal. In this tradition, the physical environment was regarded not as a determinant of human activities but as a factor that set limits on the range of possible human options in an environment, and it is this emphasis on environmental possibilities that prompted the label "possibilism" (see Figure 2.1b). More generally, possibilism is one component of *la tradition Vidalienne*, a tradition that also includes concerns with local regions (*pays*), with immediate physical surroundings (*milieu*), and with the culture or way of life of a group of people (*genre de vie*). For Vidal and those who followed him, such as Lucien Gallois (1857–1941), Emmanuel de Martonne (1873–1955), and especially Jean Brunhes (1869–1930), *genre de vie* was the key to understanding which of the various possibilities offered by the environment was selected for use by a human group. Vidal was also a key influence in the development of the *Annales* school of history in France, which included such early members as Lucien Febvre (1878–1956) and Marc Bloch (1886–1944), all of whom consistently studied the relevance of environment in human history.

Vidal and his followers—often referred to as Vidalians—recognized close relationships between humans and environment, explicitly rejecting the idea of environmental determinism. The approach continued the dualism evident in much of the previous thought, although the emphasis on the separation of humans and nature is less explicit than is the case with environmental determinism. Further, possibilism arose in direct opposition to the concept that social phenomena are explicable only in terms of other social phenomena, an approach associated with the French sociologist Durkheim. For Vidal, neither determinism—environmental or social— was acceptable. ➡️≫**10**

For a fascinating account of the practical relevance of these questions, see Box 2.2.

Variations on a Theme
Although the basic logic of possibilism was generally accepted by many cultural geographers, it remained

the subject of debate during the first half of the twentieth century, and a small number of geographers proposed modifications. In addition to the stop-and-go version of determinism proposed by Taylor, Spate argued that neither environmental determinism nor possibilism was adequate; he coined the term "probabilism" as an appropriate middle road (Spate 1952:422). If one accepts the basic logic of possibilism, then this variation makes sense. Probabilism simply recognizes that the various possibilities have varying probabilities of occurrence (see Figure 2.1d), a logic that was indeed implicit in possibilism.

At the same time, there was also a continuing concern, rooted in the tradition initiated by Marsh, with human impacts. Sauer was one of the principal instigators of a seminal volume of readings, the motivation for which was well stated by Thomas:

> Every human group has had to evaluate the potential of the area it inhabits and to organize

---

Box 2.2 A Historic Encounter

Geography's concern with the relative merits of different ways of conceiving relationships between humans and nature is not, of course, only an academic issue. A fascinating early example of the practical significance of the differing viewpoints of environmental determinism and possibilism involves the debate about the prospects of settling the arid Australian interior after the First World War. Taylor, a proponent of stop-and-go determinism, was teaching geography in Australia at the time and was an important figure in the debate. Openly critical of settlement proposals, Taylor referred to "Nature's Plan," identifying "optima," meaning climatic conditions ideally suited to settlement, and "limits," meaning the minimum climatic conditions suited to settlement; he published a textbook that was banned by the Western Australian education authorities because of its pessimistic views on settling the interior. On the other side of the debate were those who favored the expansion of settlement, including a number of politicians and the Canadian explorer and ethnologist Vilhjalmur

Stefansson (1879–1962); he was well known in Canada for his accounts of time spent in the Arctic, in which he expressed his belief in the potential of that area for human habitation and development. Stefansson visited Australia in 1924.

The very different perspectives of Taylor, the determinist, and Stefansson, the possibilist, prompted very different assessments of the Australian interior, assessments that were widely reported in the press. Stefansson optimistically referred to the prospects of the area and, without any real justification, compared it favorably to the American West. His expansionist views were precisely what the Australian politicians and public wanted to hear. Taylor subsequently described Stefansson as a "dangerous anthropogeographer"; he became an isolated figure and left Australia in 1928 (Powell 1980:181).

Who was right? Certainly, Taylor underestimated the ability of humans to adapt to difficult environments, but, just as certainly, Stefansson greatly exaggerated the potential of the area.

its life about its environment in terms of available techniques and the values accepted as desirable. The identification, use and care of resources is in the end a problem of human values and behavior. Cultural differences in techniques and values, and hence in utilization of the physical biological environment and its conversion into a human habitat, have distinguished one human group from another. The effects of man on the earth are geographically varied and are historically cumulative. (Thomas 1956:xxxvi)

Another variation on the possibilist approach involved a more explicit acknowledgment that the way groups of people perceive, and therefore behave in, environments varies according to their particular group characteristics (see Figure 2.1e). Differences between real and perceived environments are emphasized in this approach, which became popular during the 1960s.

Controlling Nature and Controlling Others

"Human use of nature": the very phrase conveys the impression that nature is something to be manipulated according to human needs and wants. Returning to the Christian ideas of the separation of humans and nature and of humans being granted dominion over all other creatures, and recognizing the long tradition of subsequent religious and scholarly thought, it becomes evident that one feature of the human experience, especially in the Western world, has been the quest for ever greater control over the natural world. Further, this ambition to control has been extended to include human control of other humans, notably those who are perceived, for whatever reason, as perhaps less than human. These are, then, old ideas that have long been acknowledged, but they are also ideas that are subject to various interpretations.

As noted, in the Jewish, Greek, and early Christian worlds, humans were generally enthusiastic to establish **hegemony** over nature. Hierarchies were established with Aristotle, for example, speculating about human nature and developing a sliding scale with human males at the top and human females second, followed by ranks of other living

beings. The principal Christian view was to see humans as masters of animals, with wild animals there to be hunted and domesticated animals there to perform labor. Thus the experiences of hunting and of animal and plant domestication served to confirm human domination of the natural world.

By the end of the eighteenth century, the growth of science had given rise to a rational rather than a religious justification for human domination of the natural world and for identifying different groups of humans. **Slavery**, forced labor, **feudalism**, and the isolation of the mentally ill and of lepers were all instances of humans identifying differences within the human species and exercising control over those seen to be inferior. If to be human was to have some qualities defined by those in positions of power, then it was inevitable that some humans would be regarded as in some way less than other humans. Indeed, just as there had been those who had claimed that plants could be domesticated or removed and that animals needed to be domesticated, hunted, or displayed in zoos, so there were those who claimed that humans needed to possess culture—European culture, that is—or else be exterminated, enslaved, or forcibly moved. ➡➤➤ 11

An especially important phase in the development of these ideas occurred during the period of European overseas expansion, when several European powers effectively carved up and divided a large portion of the globe among themselves. During this period the idea that humans, specifically European males, could also dominate others became paramount. In some respects, this European achievement was seen as the culmination of the Enlightenment, involving a triumph of progress and reason. The domination of others was considered legitimate because Europeans considered themselves superior; indeed, the conquerors were replacing the perceived impoverishment of others with the values of justice and reason. The "White Man's Burden" was to civilize the savage. This logic was reinforced by the evolutionary biology of Darwinian theory, and especially by social Darwinism.

It was not unusual for groups that were relatively powerful, such as the English establishment, to view groups that were relatively powerless—such as the Irish, Africans, infants, women, the poor, the mentally ill—as inferior, and even to describe them in animal terms. Once seen in animal terms, people were liable to be treated in analogous ways—by being domesticated, for example, or even hunted. Another way to express this general point is to note that it became possible to measure human progress, advances in civilization, in terms of the domination of nature and of others.

The Landscape School

Initially, the possibilist approach was closely identified with, and indeed essentially restricted to, a group of French geographers. But two closely related concerns were initiated outside of France during the early part of the twentieth century, both asserting the primacy of culture over environment. In 1906 the German geographer Schlüter built on the ideas of Kirchhoff and proposed the study of a cultural landscape, *Kulturlandschaft*, which, he suggested, was a landscape created by a cultural group from a previous physical landscape, *Naturlandschaft*. It was this idea of landscape transformation that attracted Sauer and that provided impetus for the highly influential landscape school. Expressed simply, Sauer proposed that geography be concerned with the development of the cultural landscape from the natural landscape. The emphasis was thus on the evolution of the visible material landscape.

It is helpful to interpret the approaches of Schlüter and Sauer as a further move away from an environmental determinist argument in that, with their explicit incorporation of culture, they allowed for an ongoing process of environmental change and **adaptation** rather than an acceptance of certain environmental options. Thus, in a textbook inspired by the landscape school, Carter (1968:562) noted that it is "human will that is decisive, not the physical environment. The human will is channeled in its action by a fabric of social customs, attitudes, and laws that is tough, resistant to change, and persistent through time."

In 1925, Sauer wrote the article "The Morphology of Landscape," which is usually viewed as the seminal statement of the landscape school. The substantial impact of this one piece on most work in cultural geography prior to about 1970 and on much work since that date is undeniable, and yet there are a number of intriguing questions surrounding the article:

- Was it an attempt to delimit a subfield of human geography—which is what the article effectively did—or an attempt to set a course for all of human geography?
- Was it influenced more by earlier European work, such as that of Ratzel in volume 2 of his *Anthropogéographie*, of Schlüter, and of Vidal, or more by the anthropological arguments about the meaning of culture as elaborated especially by Kroeber?
- Did Sauer himself, in his considerable body of empirical work, adhere closely to the methodological principles outlined in the "The Morphology of Landscape"?

The discussion of these questions as part of this larger account of the landscape school enriches understanding of the identity of the cultural geography that built on the ideas of Sauer. →≫12

Origins

At the time when Sauer was expressing his ideas about landscapes, the larger discipline of geography had an uncertain identity. Following the first establishment of geography departments in Prussian universities in 1874, there were various claims and counterclaims about the identity, methodology, and subject matter of the newly institutionalized discipline. The debate took place against a complex intellectual backdrop. Geography, both physical and human, was a long-established area of both intellectual and more general concern, with traditional interests in exploration, mapping, and written descriptions of places and peoples. Scholars writing about geography before 1874 contributed to many important debates, such as those about humans and nature, about populations and resources, about human differences, and about changes in ways of life. Further, the late nineteenth century was a period during which two seemingly contradictory trends were evident: there was an increasing separation of physical and human sciences as the social sciences became institutionalized, and yet there was much sympathy in both areas of endeavor for cause-and-effect analyses. Geography occupied a distinctive position in the organization of knowledge at this time because it was both a physical and a social science, and most geographers who achieved leadership roles found that it was incumbent on them to define the discipline of geography.

Principal statements about the identity of geography included the declaration by Ferdinand von Richthofen (1833–1905) in favor of a geography centered on the regional concept, and the two ideas promulgated by Ratzel, namely geography as environmental determinism, in volume 1 of *Anthropogéographie* (1882), and geography as the study of human landscapes, in volume 2 of *Anthropogéographie* (1891). Vidal followed the later Ratzel in advocating the study of human landscapes. For most American and British geographers at this time the central concern was the study of earth and life relationships, but they saw two additional tasks as particularly important: determining the influence of physical environment on human geography and human history, and delimiting and analyzing regions. Certainly, both the physical cause and the regional approaches to geographic study were widely accepted in the English-speaking world. The formative physical cause work was accomplished especially by Semple and Huntington, while the formative regional statements were contributed by Mackinder and others in Britain and by Davis in the United States. It was into this somewhat confused and quite uncertain scholarly landscape that Sauer ventured in the 1920s.

Sauer taught at the University of Michigan from 1915 to 1923, after which he moved to the University of California at Berkeley. ➔➤➤ **13** He remained at Berkeley until 1957, when he began an active 23-year period of retirement. It was during his early years at Berkeley that Sauer made his contribution to the debate about the identity of geography. In the period from 1924 to 1931, Sauer succeeded both in articulating a subject matter, namely landscapes as created by humans in their capacity as members of cultural groups, and in outlining an approach to that subject matter, an approach that emphasized relations

between cultures and the physical world, landscape evolution, and region delimitation. The discipline's interest in physical and cultural landscapes was anticipated in earlier writings; the evolutionary concern was similar to earlier ideas of Vidal and Schlüter; the regional concern was in accord with the dominant view of the discipline. Sauer's methodological achievements were to treat geography as one of the social sciences, to synthesize and expand upon earlier geographic works, to express this synthesis in clear and forceful language, and to influence the activities of the many younger geographers who were attracted to work with him.

The considerable literature about Sauer and the landscape school includes a variety of interpretations about precisely what Sauer and those who followed him believed. Varied interpretations are perhaps inevitable when a scholar writes often and at length, contributes to knowledge over an extended period, and proves to be highly influential. ➔➤➤ **14**

But it is important to note that two evaluations of Sauer's philosophical stance arrived at quite different conclusions. First, Entrikin (1984:387) stated, "Culture history became his model of social science, and its underlying methodological and philosophical basis was drawn from natural science rather than from historiography." This view is supported by Leighly, Sauer's closest associate at Berkeley, who identified geomorphology and plant ecology as principal influences and noted that, on one occasion, Sauer described himself as an earth scientist. Second, a rather different view, argued by Speth (1987), identifies a historicist perspective as Sauer's principal inspiration, an opinion that has been accepted in broad terms by humanist geographers who have traced their roots back to the landscape school. This disagreement is, of course, but one particular version of a larger philosophical debate. Notwithstanding these two different interpretations, six basic ideas promulgated by Sauer can be identified. ➔➤➤ **15**

Six Key Ideas

1. *First, Sauer was adamant in his rejection of environmental determinism at a time when both Semple and Huntington were major figures in*

American geography and when much regional geography was sympathetic to the approach. For Sauer, accepting environmental determinism involved needless generalization. This generalization resulted in a loss of objectivity, since advocating a particular viewpoint typically produced results that were inferior in value to results that were achieved without some initial assumptions having been made. Thus, facts were the first priority, to be followed—not preceded—by interpretation, with interpretation proceeding without either stressing or ignoring physical geography. Sauer's rejection of environmental determinism was based on flaws he perceived in the argument and not on any downplaying of the importance of physical environment. He consistently referred to possibilism, as discussed by Vidal and Febvre, in sympathetic terms: "Within the wide limits of the physical equipment of area lie many possible choices for man, as Vidal never grew weary of pointing out" (Sauer 1925:46).

2. *Second, rather than either of the two preferred approaches to geography, namely environmental determinism and regional geography, Sauer favored the study of the facts of landscape as the subject matter of the discipline:*

> It then becomes the task of geography to grasp the content, individuality, and relation of areas, in which man comes in for his due attention as part of the area, but only in so far as he is areally significant by his presence and works. This is a unitary and attainable objective. The landscape is constituted by a definite body of observational facts that may be studied as to their association and origin. . . . A phrase that has been much used in German literature, unknown to me as to origin, characterizes the purpose perfectly: "the development of the cultural out of the natural landscape." (Sauer 1927:186–7)

Although a regional emphasis was implicit in this landscape methodology, the term "landscape" was thought to be a clearer alternative to either of the then favored terms, "area" and "region."

3. *Third, Sauer was also critical of the regional approach to geography that centered on the question of "where":* "The geographer is hardly likely to prosper by giving his main attention to all kinds of distributional studies" (Sauer 1927:185). Sauer expressed concern that focusing on the distribution of any and all phenomena would result in geographers studying some subject matter in which they lacked competence. He did, however, acknowledge the importance of regions as areas that exhibited a characteristic expression as a result of cultural occupance.

4. *Fourth, Sauer saw a focus on landscape as constituting both physical geography (the physical landscape) and the impacts of humans on physical geography (the cultural landscape).* Sauer consistently recognized the unity of physical and cultural components of landscape.

5. *Fifth, Sauer stressed that cultural impacts on landscapes prompted landscape change, and also that occupying cultures underwent change:* "An additional method is therefore of necessity introduced, the specifically historical method, by which available historical data are used, often directly in the field, in the reconstruction of former conditions of settlement, land utilization and communication" (Sauer 1931:623). According to Sauer, neither cultural landscape nor culture was fixed, and he made reference to a succession of landscapes with a succession of cultures. This evolutionary emphasis was similar to the concept of sequent occupance, which related changing landscapes to changing cultural occupances (Whittlesey 1929). Sauer outlined the evolution of both natural and cultural landscapes, the latter being summarized as follows: humans, through their culture, transform the natural landscape (see Figure 2.2). Cultural landscape, not culture, is the object of analysis. Recall from Chapter 1 that, although Sauer (1925:46) referred to culture as "the agent" and the "shaping force," he referred to the physical landscape as "fundamental."

6. *Sixth, as part of the opposition to environmental determinism, Sauer regularly stressed that*

culture functioned as a cause of landscape. The anthropological impetus for this claim, known as the superorganic concept of culture, is discussed in Box 2.3, and the extent to which Sauer and his followers embraced this culture concept is debated in Box 2.4.

These six methodological statements combined to encourage three related themes that have been central to most subsequent landscape school research (Sauer 1941a):

- Cultural landscapes result from occupance of a physical landscape; hence there is an ecological component to cultural geography.
- Cultural landscapes change through time, as indeed do the cultures occupying the landscape; hence there is an evolutionary component to cultural geography.
- Cultural activities result in the creation of a relatively distinct cultural landscape; hence there is a regional or cultural area component to cultural geography.

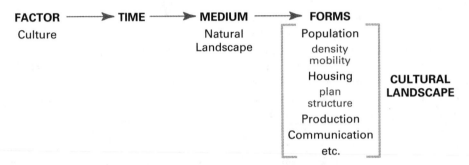

Figure 2.2 The Morphology of Landscape. This is the classic diagram included in the seminal 1925 article by Sauer, "The Morphology of Landscape." It is, in fact, the second of two very similar diagrams, the first outlining the established research area of the morphology of the natural landscape, which Sauer (1925:41) described as "the proper introduction to the full chorologic enquiry which is our goal." Thus: "The natural landscape is being subjected to transformation at the hands of man, the last and for us the most important morphologic factor. By his cultures he makes use of the natural forms, in many cases alters them, in some destroys them." (Sauer 1925:45) Although "cultural morphology might be called human ecology" (Sauer 1925:45), the term "morphology" was favored over the alternative term "ecology" at least partly because, following the assertion by Barrows (1923) to the effect that geography was human ecology, ecology was identified with the idea of environmental determinism. Referring to the diagram, Sauer (1925:46) wrote:

> The cultural landscape is the geographic area. . . . Its forms are all the works of man that characterize the landscape. Under this definition we are not concerned in geography with the energy, customs, or beliefs of man but with man's record upon the landscape.... The cultural landscape is fashioned out of a natural landscape by a cultural group. Culture is the agent, the natural area is the medium, the cultural landscape the result. Under the influence of a given culture, itself changing though time, the landscape undergoes development, passing through phases, and probably reaching ultimately the end of its cycle of development. With the introduction of a different, that is, alien culture, a rejuvenation of the culture landscape sets in, or a new landscape is superimposed on remnants of an older one.

This diagram, and the related commentary, provide a succinct summary of the six key ideas of the landscape school, namely that
- physical landscape is not a determining factor;
- cultural landscape is the object of study;
- studying cultural landscapes involves more than a concern with distributions;
- physical and cultural geography cannot be separated;
- both cultures and cultural landscapes change through time; and
- culture can be understood as a cause of landscape.

Source: C.O. Sauer, "The Morphology of Landscape," *University of California Publications in Geography* 2 (1925):37. (Reprinted courtesy of the Department of Geography, University of California, Berkeley)

Box 2.3 The Superorganic Concept in Anthropology

Until the mid-twentieth century, the most influential way of thinking about humans as members of cultural groups was the superorganic interpretation of culture.

Recall that this essentially mechanistic and determinist way of thinking was developed within anthropology as one component of cultural evolution logic, and that there were similar developments in sociology, with both Comte and Spencer conceiving of society as an integrated entity, comparable to a physical system and entirely determining the behavior of the people within it. The superorganic concept flows from this idea, seeing societies as wholes created either by growth from within or by absorption of groups from without. Similarly, Durkheim argued for the separation of sociology from both biology and psychology, using the term "social organism" to refer to the social as a unique phenomenon, a process separate from the qualities of individual members of a society and one that cannot be explained by reference to psychology and biology. Both the social organism of Durkheim and the superorganic of Spencer see the social or cultural as a force constraining individual behavior.

There are close conceptual parallels between what was happening in geography, namely the approach of environmental determinism, and what was happening in sociology and anthropology, namely the approaches of social and cultural determinism. All these approaches were versions of larger determinist logic, all were mechanistic, all were reductionist, and all were privileging one part of the world to the essential exclusion of other parts. The superorganic concept is especially important to cultural geographers because of the argument that Sauer's own acceptance of the approach was such that it became a central part of the landscape school.

As used by Kroeber, the term "superorganic" referred to the non-organic human product of societies, cultural institutions, modes of production, and levels of technology. Although the term is borrowed from Spencer, it was similar to Spencer's only in the sense that it referred to non-biological aspects of societies. Kroeber's superorganic concept made a distinction between social processes and biological or organic processes, emphasizing that in contrast to biological evolution or changes in organic structure, human societies did not use the principle of heredity to transmit new adaptations to other members of the species. Further, "the distinction between animal and man which counts is not that of the physical and mental, which is one of relative degree, but that of the organic and social, which is one of kind. . . . in civilization man has something that no animal has" (Kroeber 1917:169). Although humans may be biological organisms, they are social animals, and it is through language and social and cultural institutions that they learn the values of the civilization in which they live:

> All civilization in a sense exists only in the mind. Gunpowder, textile arts, machinery, laws, telephones are not themselves transmitted from man to man or from generation to generation, at least not permanently. It is the perception, the knowledge and understanding of them, their ideas in the Platonic sense, which are passed along. Everything social can have existence only through mentality. Of course civilization is not mental action itself; it is carried by men, without being in them (Kroeber 1917:189).

Advocates of the superorganic opposed explanations that focused on the natural world and employed environmental determinism, and also opposed explanations that focused on individuals and employed psychological logic. According to the argument against psychological explanations, the length of time an individual was involved in culture precluded the possibility of a paramount role for individuals as initiators of human activities.

It is possible to exaggerate the specific meaning of this concept. Although many subsequent commentators stressed the explicit cultural determinism evident in the superorganic, others take a rather

Continued

different view. In a major overview of anthropological theory, Harris (1968:342) concluded that Kroeber was not a cultural determinist. Indeed, Kroeber (1917:205) himself noted that to infer "that all the degree and quality of accomplishment by the individual is the result of his moulding by the society that encompasses him, is assumption, extreme at that, and at variance with observation. . . . [N]o culture is wholly intelligible without reference to the non-cultural or so called environmental factors with which it is in relation and which condition it." This is what Mikesell (1969:231) called a "cautious philosophy roughly comparable to the geographic concept of possibilism."

Box 2.4 A Superorganic Concept of Culture in the Landscape School?

In light of some later commentaries on the work of the landscape school, it is important to consider whether or not Sauer was proposing an approach that incorporated a cultural determinist or superorganic perspective. There are three obvious reasons why this might not be the case.

First, Sauer was highly critical of environmental determinism because he saw it as "based on a belief that a single natural law can explain the social order" (Sauer 1927:173); it is reasonable to assume that Sauer would not replace an environmentally centered law with a culturally centered law that shared a similar failing. Second, Sauer consistently acknowledged the importance of physical landscape in all of his methodological writings as well as in many of his empirical studies. Third, the European sources that Sauer cited favorably were largely concerned with questions about the interactions between humans and nature from a possibilist or similar perspective. Indeed, there was a debate between Vidal and Durkheim on precisely this issue, with Durkheim objecting to the inclusion of physical landscapes in attempts to explain human behavior.

On this basis, it seems reasonable to conclude that Sauer was not any version of determinist, environmental, or cultural. But there is an alternative argument that has support from at least two areas. First, Sauer did refer to culture using terms such as "cause." In this context, Figure 2.2, the classic landscape school diagram, has culture acting on physical landscapes to produce cultural landscapes. Second, there is no doubt that Sauer was intellectually close to both Kroeber and Lowie, who were in the anthropology department in the same university as Sauer. Leighly (1976:339–40) emphasized these close ties: "Basic to Sauer's course in world regional geography was the concept of 'cultures', now used in the ethnological sense, which he learned from the anthropologists at Berkeley." Sauer (1974:192) also acknowledged this association and influence: "Anthropologists were our tutors in understanding cultural diversity and change. Robert Lowie in particular introduced us to the work of such geographers as Edward Hahn and Ratzel as founders of an anthropogeography that I had not known." The intellectual association between Sauer and both Kroeber and Lowie is significant in that both anthropologists were sympathetic toward a superorganic concept of culture.

Was Sauer an explicit superorganicist, proposing explanations of cultural landscape exclusively in terms of culture as cause? A balanced answer seems to be no, given the considerable weight that Sauer accords to physical landscapes. The aim of cultural geography, as he defined it, was to "explain the facts of the culture area, by whatever causes have contributed thereto" (Sauer 1931:623). The close links between the landscape school and *la tradition Vidalienne* suggest a shared concern with broadminded interpretations of the human-and-land relationship that were incompatible with either the superorganic of Kroeber or indeed the similar social organism concept proposed by Durkheim. For both

Continued

Kroeber and Durkheim, the major concern, respectively, was with cultural and social causes of human behavior, but in cultural geography both culture and society were neglected, relatively speaking, in favor of visible material landscapes as objects of study. Arguments about possibly excessive influences from Kroeber also ignore the fact that Sauer was intellectually close to Herbert E. Bolton, a Latin American historian at Berkeley. More generally, a scholarly association does not necessarily result in indiscriminate acceptance of ideas. ➤➤ **16**

On the other hand, there is little doubt that several later landscape school geographers either explicitly or implicitly accepted the superorganic concept. Duncan (1980) details compelling examples. Thus Zelinsky (1973:70) claimed: "The power wielded over the minds of its participants by a cultural system is difficult to exaggerate." In similar fashion, Carter (1968:562) stated: "It is human will that is decisive, not the physical environment. The human will is channeled in its action by a fabric of social customs, attitudes, and laws that is tough, resistant to change, and persistent through time." More generally, many cultural geographers referred to Kroeber and other anthropologists with superorganicist inclinations in their discussions of the concept of culture. Although there is no unequivocal final word on debates of this type, it is perhaps worth noting that, in a retrospective comment, Sauer (1974:191) reflected: "The morphology of landscape was an early attempt to say what the common enterprise was in the European tradition."

Impact

The impact of Sauer's early writings was both immediate and long-lasting. In 1934, a major overview of geography as a social science quoted the 1925 article at length and included discussion of the cultural landscape in regional analysis: "Wherever man enters the scene he immediately alters the natural landscape, not in a haphazard way but according to the culture system which he brings with him, his house groupings, tools, and ways of satisfying needs" (Bowman 1934:149–50).

Geography as the Study of Landscape
The landscape school was one of five attempts made to define geography between about 1900 and 1960; the other four were environmental determinism associated with Ratzel, possibilism associated with Vidal, regional geography associated with Hartshorne, and spatial analysis associated with Schaefer. Cultural geography has often served as a synonym for human geography, and in fact Sauer was perhaps attempting to determine an agenda for this larger discipline rather than generating a subdiscipline. And although landscape school cultural geography never really dominated the larger field of human geography, it has proven to be an especially enduring set of ideas and practices, initiating a compositional subdiscipline of human geography that has made major contributions from the 1920s through to the present. ➤➤ **17**

The prevailing consensus is that the legacy of Sauer is a concern with cultural landscapes, although this is an association that is critically discussed on several grounds. For instance, along with the concept of landscape, Sauer and his followers focused on links with anthropology, on regional studies in Latin America, and on ecological analyses. Further, many geographers identified with Sauer—Denevan and Peet, for example—centered their studies on topics other than landscape. At the same time, it is worth noting that some other cultural geographers not directly associated with Sauer—such as Hart, Jakle, Lewis, and Meinig—stressed the landscape concept. Among the geographers directly associated with Sauer who did focus on landscape are Kniffen, Parsons, Salter, and Zelinsky.

Revisions of the Approach
Prior to the first indications of a major revision of the agenda of cultural geography in the 1970s, the landscape school concepts introduced by Sauer were subject to four revisions, both by Sauer and by associated scholars.

The first revision was a decreasing emphasis on the evolution of cultural landscapes from physical

Questions for Critical Thought

———— ✤ ————

Given your understanding of geography by the 1920s, do you believe that Sauer's ideas were innovative and merited adoption by others? Interestingly, the reaction from other American geographers at the time was mixed. Some of Sauer's contemporaries followed his lead and practiced the cultural geographic study of landscape, but many others, including most of the older generation of geographers, were largely unmoved, preferring instead to concentrate on regional geographic studies. Further, Sauer's ideas had little impact outside of North America, with most European cultural geographers preferring what is best described as a possibilist emphasis in their studies. Regardless, it is difficult to exaggerate the importance of Sauer's contribution to the development of cultural geography as this was pursued in North America, and much of the content of chapters 4, 5, and 6 derives from the methodological principles enunciated by Sauer.

landscapes, which was owing at least in part to the considerable amount of specialized knowledge required to analyze physical landscapes. The second revision to landscape school concepts was a diminishing emphasis on maintaining a regional focus, as implied in the concept of sequent occupance, and an increasing emphasis on visible landscape. The third revision was the close integration of historical and cultural analyses, prompted by concern with landscape evolution. The importance of time for Sauer and the landscape school may have been influenced by the work of anthropologists, who saw time as continuous, a view to which Sauer had fully subscribed by 1941. The fourth revision concerned the decreasing emphasis on culture as cause—an implicit rejection of the superorganic and its unproven assumptions.

Such transformations were relatively minor and did not radically change the character of cultural geography as a concern with the evolution and description of visible material landscapes—landscapes as repositories of **material culture** (see also **non-material culture**). Meaningful criticisms

of the historical landscape tradition first appeared in the 1960s and centered on the perceived emphasis on visible, material landscapes, and the related neglect of values and beliefs. Such criticisms appeared notwithstanding Sauer's explicit proposal to the effect that "human geography, then, unlike psychology and history, is a science that has nothing to do with individuals but only with human institutions, or cultures" (Sauer 1941a:7). ➜≫≫ **18**

Toward Holistic Emphases

Rationale

Consider the following four statements:

- "It is wrong to say that the western tradition has emphasized the contrast between man and nature without adding that it has also emphasized the union of the two. The contrasting viewpoints arise both because man is unique and because he shares life and mortality with the rest of living creation" (Glacken 1967:550).
- "The old dichotomy between man and Nature, the view that environment is an antagonist that must be conquered, or to which one must passively submit can only lead to disaster or stagnation" (Tatham 1951:162).
- "May I remind you that much faulty thinking on the part of historians, philosophers and others is due to the fact that they always will set man against nature as if they were two distinct categories. Man versus Nature instead of man and his environment being regarded as a single complex" (Taylor, in Spate 1952:425).
- "The danger is in setting up a false duality and so involving ourselves in insoluble or unnecessary questions of the chicken-and-egg order" (Spate 1952:419).

These four quotations introduce an approach to the study of humans and nature that aims to think in terms of unity rather than separation, in terms of **holism** rather than duality. There is a long scholarly tradition of attempting to view humans as a part of nature, although, unlike the theme of separation, this tradition has never been dominant. This is not surprising given the prevalence of separation ideas in the major philosophical traditions as well as in Christianity. Many of the scholars associated with a holistic approach, although they argued for a consideration of humans and nature together, found it necessary to use these terms separately, thereby implicitly acknowledging a division between the two in accord with prevailing thought. Indeed, the historical perspective presented below is similar in overall emphasis to that contained in the earlier section on human use of nature.

Historical Perspective

Greek and Christian Thought

The themes of interdependence among all things and of the unity of all living things have been a part of human thought since at least the time of classical Greece. Both Herodotus (484–c.425 BCE) and Plato observed that all life functioned as a unity and served to maintain a stable universe. (This is the tradition of a mother earth, or Gaia.) A holistic theme is evident also in the work of Posidonius, referred to earlier in the discussion of environmental determinism. Philo (c.20 BCE–c.50 CE), known especially for achieving a broad synthesis of Greek and Jewish thought, viewed humans as a part of nature in the sense that all were created by God. In his view, the ranks of different species in nature were determined by the order of their creation, with the superior humans created last. Indeed, there are two somewhat different Christian creation stories in Genesis, with that referred to earlier in this chapter having humans created first and all other creatures subsequently.

It can be suggested, then, that from the onset of the Christian tradition, humanity is both apart from and a part of the community of all living things. St Augustine (354–430), the greatest of the Latin church fathers, similarly saw nature as a continuum and ranked living things above those without life, and living things with intelligence—humans—above living things without intelligence—animals. Again, the Jewish philosopher Maimonides (1135–1204) observed that humans needed to understand their place in nature rather than believing that nature was created for humans. Related to these ideas about humans being one part of a hierarchy, other scholars, such as Henry More (1614–87), noted that nature was unthinkable without humans.

Thus, in addition to the many arguments concerning a separation of humans and nature are numerous arguments, again often based on religion, that emphasized a harmonious relationship between or integration of humans and nature. Such ideas continued to be stated until the late eighteenth century, by which time the first of a group of scholars, often with explicitly geographic concerns, provided a new set of arguments about the unity of humans and nature.

Geographical Thought

From the Renaissance to 1800, a handful of scholars took a holistic approach to the study of geography. Johann Reinhold Forster (1729–98), who traveled around the world in the 1770s, studied the world as a coherent whole, while Anton Friedrich Büsching (1724–93) wrote a geography of Europe that incorporated the idea that all nature, affected by humans or otherwise, was one. Two of the leading scientists of the late eighteenth century, the geologist Hutton and the biologist Lamarck, both conceived of humans as a part of nature.

But it was the nineteenth-century geographers von Humboldt and Ritter who made the greatest contributions to the holistic theme. Both were scholars of tremendous importance and influence in the first half of the nineteenth century, achieving prominent positions in science and society. "Never before or since have geographers enjoyed positions of such prestige, not only among scholars but also among educated people all around the world" (Martin and James 1993:112–13). Their contributions to geography and to the advance of knowledge in general are considerable. Our concern is with their shared commitment to a view of the earth as an organic whole, and to the idea that all things on the earth's surface are related—a concept known as *zusammenhang*, meaning literally "hanging together."

It is important to observe that such a commitment did not necessarily exclude the idea that there could be some environmental impacts on humans: writing in 1808, von Humboldt acknowledged that climate and configuration could affect agriculture, trade, and communications. Yet for both von Humboldt and Ritter, the interrelationship of humans and nature was clearly apparent and could not be ignored. To quote von Humboldt (in Tatham 1951:44): "The earth and its inhabitants stand in the closest reciprocal relations, and one cannot be truly presented in all its relationships without the other. Hence history and geography must always remain inseparable. Land affects the inhabitants and the inhabitants the land." Similarly, Ritter (in Dickinson 1969:37) stated: "My aim has not been merely to collect and arrange a larger mass of materials than any predecessor, but to trace all the general laws which underlie all the diversity of nature, to show their connection with every fact taken singly, and to indicate on a purely historical field the perfect unity and harmony which exist in the apparent diversity and caprice which prevail on the globe, and which seem most marked in the mutual relations of nature and man." Ritter, unlike von Humboldt, expressed a teleological philosophy, believing that the earth was designed to be the home of humans. ➔≫≫ **19**

Several other nineteenth-century geographers held views similar to those of von Humboldt and Ritter in that they clearly favored a holistic emphasis, especially rejecting any suggestion of environmental influences. Thus, the French geographer Élisée Réclus (1830–1905) emphasized the processes by which cultures used environments to satisfy their needs, and the Russian geographer Peter Kropotkin (1842–1921) saw nature as a dynamic whole that included humans and that was always changing. ➔≫≫ **20**

Geographers were not the only scholars reacting to the traditional conceptual separation of humans and nature. In addition to the idea of a human-and-nature separation, nature has been seen as external in that it is beyond the realm of humans, and as universal in that there is both external nature and human nature. A universal view implies that humans are as natural as are the external components of nature. Although Marx can be read in various ways, for some he introduced a different way of thinking that rejected the external/universal dichotomy and that emphasized the continuity between humans and nature, with

humans viewed as part of nature. Some geographers have recognized this contribution of Marxist thought referring to inner actions within nature rather than to interactions between humans and nature.

Ecological Emphases

Perhaps the most developed contribution to the idea that humans and nature are best treated in a holistic fashion, focusing on the unity of the two rather than on their separation, is contained within the broad theme of ecological thought (see Figure 2.1f). Indeed, much of the work previously noted, especially that by such geographers as Marsh, von Humboldt, and Ritter, is often interpreted today as being within the ecological tradition (a tradition discussed more fully in Chapter 4).

The word "ecology" is derived from two Greek terms: *oikos*, meaning "place to live" or "house," and *logos*, meaning "study of." Literally then, ecology is the study of organisms in their homes. The key impetus for the development of ecology was the third chapter of *The Origin of Species*, which centered on the various adaptations and interrelationships of organisms, implicitly including humans, and environment. Haeckel first used the term in 1869, although it was Darwin, in *The Descent of Man* (published in 1871), who explicitly included humans. The broad theme of ecology proved to be of compelling interest in both physical and social science. Thus, along with the specifically physical approaches, such as plant ecology and animal ecology, a variety of human and cultural ecologies appeared in the various social sciences. ➔≫≫ **21**

In principle, these various ecological approaches rejected the dominant late nineteenth-century concern with the separation of humans and nature. In particular, these approaches repudiated any idea of human dependence on nature or human independence from nature. Instead, they encouraged studies of humans and nature without privileging one at the expense of the other, thereby avoiding a reductionist perspective. Ecology is based on the premise that things cannot be studied except in context, such that relations with other things are critical. In practice, however, many specific applications of ecology maintained a separation, at least in part because much of the terminology of the approach originated in physical science. As a result, several ecological approaches—including the approach incorporated in the landscape school as well as the **cultural ecology** initiated by the anthropologist Steward, which was probably the most influential ecological

approach—incorporated a view of humans as members of cultural groups dominating nature. It seems reasonable to suggest that humans are both constrained and enabled by nature, but also affected by their culture. See Box 2.5 for a discussion of the evolutionary naturalist argument.

Box 2.5 *Evolutionary Naturalism*

There is little doubt that much contemporary cultural geography is comfortable with the various critiques of earlier ways of thinking about humans and nature, but there is another interpretation of the scholarly tradition in social science. For Hutcheon (1996), the problem lies not in physical science's seduction of social science, but rather in the failure of social science to fully articulate a position based on the twin scientific pillars of evolution and naturalism. According to her interpretation, the social sciences have largely failed to take advantage of the scientific approach, with the various attempts to do this being steps in the right direction that have never been adequately pursued. Recall that naturalism implies the universality of cause and effect, while evolution is seen to apply to all things. Evolutionary naturalism, as Hutcheon describes it, is a way of thinking based on "the premise that human beings are continuous with all of nature, and that all of nature is continuously evolving" (Hutcheon 1996:vii). This is an important argument that merits consideration in our discussion of humans and nature.

According to the evolutionary naturalist argument, two currents of philosophical thought have been established in the Western world since about the sixth century BCE. The dominant tradition is based on a metaphysical dualism, a separation of the physical and the spiritual, or the natural and the supernatural. This implies that there are two categories of experience—namely secular and sacred—and two ways of knowing—namely experiential and mystical. Application of the scientific method, of causal logic, has been essentially restricted to the physical realm. The minority tradition, described as evolutionary naturalism, sees both humans and the universe as natural, and therefore encourages

the study of human behavior and human institutions using procedures similar to those applied in physical science. Roots of this minority tradition are evident in Buddhist thought as well as in anti-teleological Greek Epicurean philosophy.

It was Descartes who provided a developed logical base for the dominant dualist perspective stressing the separation of science from religion, and it was Hobbes, Locke, and Hume who, although sharing the concern with a scientific approach, countered dualism with the argument that humans could also be studied scientifically. Hutcheon (1996) traced the history of social scientists' efforts to employ the methods of physical science, specifically in the form of evolutionary naturalism, and argued that such efforts have always been countered by approaches explicitly accepting that humans need to be studied differently from the physical world. Principal examples of social scientists who have included elements of the evolutionary naturalist perspective in their work include Spencer, Durkheim, George Herbert Mead (1863–1931), and B.F. Skinner (1904–90). Hutcheon (1996) also identified a series of links between contemporary physical and social science, evident especially in the sociobiological emphasis outlined earlier in this chapter.

In a related interest, the cultural geographer Wagner (1996:1) introduced the admittedly speculative claim that "human beings are innately programmed to persistently and skillfully cultivate attention, acceptance, respect, esteem, and trust from their fellows," an idea that forms the basis for a materialist and evolutionary selectionist perspective on human behavior. This idea is in close accord with the basic argument of evolutionary naturalism since both theories involve an explicit rejection of the separation of humans and nature.

Concluding Comments

- This chapter first introduced a number of philosophical concepts and associated scholars in the attempt to clarify our understanding of the meanings of the terms "humans" and "nature," and the character of relationships between the two.
- This chapter also details several variations of the presumed relationship between humans and nature, which are relevant to an understanding of the human world, versions of which continue to be evident in contemporary cultural geography.
- Views that argue for cause-and-effect relationships remain attractive for many cultural geographers. The idea of physical geography as cause, while no longer a widely accepted argument, remains important in some specific contexts.
- The view that culture is a principal determinant of human activities also continues to be an important component of some cultural geography.

- However, much contemporary cultural geography downplays cause-and-effect approaches in general, whether physical or cultural, favoring instead approaches that advocate some holistic perspective. The most notable such approaches are those that are ecological in emphasis.
- One transition in the cultural geographic approach to the study of humans is, by now, very clear. There has been a substantial redirection of scholarly work concerned with human behavior and cultural change from the search for causes, whether the causes are physical or cultural, to the search for meaning. Although both endeavors have always been present, the search for causes achieved paramount status during the period from the seventeenth century until about the 1970s; since then, the search for meaning has become increasingly central.
- It is significant that the search for causes was especially influential during the formative period of the social sciences in the late nineteenth century, and this emphasis was therefore able to exercise a significant impact on the development of cultural geography at a key stage.

Questions for Critical Thought

—— ✿ ——

Have you revised your understanding of humans and nature in light of the material introduced in this chapter? Certainly, it is difficult not to be impressed by both the quantity and quality of the cultural geographic and other work on this topic. Indeed, you will encounter additional material relevant to humans and nature throughout this book. At this stage, you may wish to conceive of nature in two rather different ways: first, there is nature without humans present, wilderness; and second, there is nature with humans present, the cultural landscape. In both cases nature has meaning to us as it is a part of our perceived world. Does it seem sensible to suggest that humans can be understood as both apart from and a part of nature? We are apart from nature in the sense that we have culture and are able to do one thing and not another according to our judgment. But we are also a part of nature in that it is not possible to conceive of humans without there being a natural world that sustains us. To raise an additional question: during the late twentieth century a series of developments, notably postmodernism, poststructuralism, and feminism, raised further concerns about the claims that there is a real and knowable world that humans can objectively study. Rather these approaches suggest that knowledge is socially produced, and its production is shaped by the positionality of the spectator. Such arguments are powerful, since the maxim "knowledge is power" (Foucault) suggests that those who dominate knowledge production at any point in time are also those who hold and reproduce power, and this includes the power to determine our relationship to and use of nature. Can you think of any examples that illustrate this idea?

Further Reading ➡️➤➤➤

The following are useful sources for further reading on specific issues.

1. The history of human-and-nature relationships is discussed in Teich, Porter, and Gustafsson (1997) and in Coates (1998).

2. Bhaskar (1989) and Sayer (1992) discuss merits and demerits of the realist position.

3. Flew (1978) explains the difficulty of defining what it means to be human.

4. Ortner (1972), Ortner and Whitehead (1981), Fitzsimmons (1989), and Johnson (1996) review possible links between women and nature.

5. Langton (1979) outlines parallels between biological and human cultural evolution. Cavalli-Sforza and Feldman (1981) interpret biological evolution in human terms. Durham (1976:89) claims that "both biological and cultural attributes of human beings result to a large degree from the selective retention of traits that enhance the inclusive fitness of individuals in their environments." Rossi and O'Higgins (1980) consider various ways of thinking about humans as members of cultural groups.

6. Childe (1936), White (1949), and Steward (1955) are examples of evolutionist writing. Wright (1983), Sanderson (1990), and Trigger (1998) outline evolutionist arguments.

7. Because geography followed social Darwinism rather than Marxism, Peet (1985) claims that the discipline was prevented from achieving a meaningful approach to humans and nature.

8. Olwig (1996a:86) argues that the concept of nature was "tainted by its use in the context of a nature/culture dichotomy in which nature was seen to have some form of determinant relation to culture—a relation that had to be, and largely was, erased from academic geographic discourse." See also Olwig (2003).

9. For more on "stop and go determinism" see Taylor (1937:459). Rostlund (1956:23) claims that environmental determinism was "not disproved, only disapproved."

10. Berdoulay (1978) and Andrews (1984) discuss the debate between Vidal and Durkheim, with Berdoulay noting that neither environmental nor social determinism was acceptable to Vidal.

11. Anderson (1995, 1997) outlines the human interest in displaying animals in zoos. Whatmore and Thorne (1998) discuss geographies of wildlife.

12. Mathewson and Kenzer (2003) provide a valuable volume of readings about the work and legacy of Sauer.

13. Leighly (1976) and Macpherson (1987) focus on the specific Berkeley context for the growth of geography and related disciplines.

14. Mikesell (1968, 1969), Hooson (1981), Entrikin (1984), Solot (1986), Macpherson (1987), Martin (1987a), Bowen (1996), and Mathewson (1996) comment on the methodological writings of Sauer.

15. Sauer (1924, 1925, 1927, 1931) outlines the basic arguments of the landscape school.

16. Aschmann (1987:137) identifies links between Sauer's methodological and empirical writings.

17. Platt (1962:39) and Parsons (1979:13) consider links between Sauer and the anthropologists Kroeber and Lowie. Duncan (1993, 1998), Price and Lewis (1993a, 1993b), and Penn and Lukermann (2003) discuss the superorganic in landscape school cultural geography. Leaf (1979) and Heyer (1982) discuss the superorganic more generally.

18. Rowntree (1996:130–1) identifies the diverse character of landscape school cultural geography. Brookfield (1964) and English and Mayfield (1972) note the possible exclusion of human values and beliefs in landscape school cultural geography.

19. Tuan (1997:6) examines the acknowledgment by Humboldt that physical geography could affect human geography. Minca (2007) offers an insightful discussion of Humboldt's understanding of landscape.

20. Olwig (1980), Bowen (1981), and Breitbart (1981) discuss nineteenth-century geographers sympathetic to an integrated humans-and-nature perspective.

21. Young (1974:8) notes: "Human ecology may be defined (1) from a bio-ecological standpoint as the study of man as the ecological dominant in plant and animal communities and systems; (2) from a bio-ecological standpoint as simply another animal affecting and being affected by his environment; and (3) as a human being, somehow different from animal life in general, interacting with physical and modified environments in a distinctive and creative way."

CHAPTER 3

Rethinking Cultural Geography

The principal aim of this chapter is to continue telling the story of cultural geography, especially as it has been affected by and taken part in the several transformations that the larger discipline of human geography experienced from about 1970 onwards. To help place the transformations in context, recall that landscape school cultural geography, as introduced in the 1920s, derived from a clear statement about the evolution of landscape, tended to see culture as a causal mechanism, emphasized historical and usually rural landscapes, included both material and symbolic aspects of culture, involved field and archival research, and, notwithstanding the explicit 1925 and subsequent statements, was essentially empirical and not theoretical. This agenda has been interpreted by some critics as inherently conservative, and the distinction with those approaches introduced from about 1970 onwards is striking, with new interests in places, landscape interpretation, symbolism, human identity, difference, and inequality. The perspectives and approaches introduced through the newer bodies of thought have clearly had considerable impact on the concepts and practice of all areas of cultural geography, and indeed on all areas of social science. All are complex, interrelated, and even contested bodies of thought that are continually evolving. Accordingly, the discussions in this chapter do not attempt to reflect the full richness and diversity of these ideas. Rather, what is attempted here is a series of capsule accounts with emphasis on material that is relevant to the cultural geographic studies discussed in the following chapters, especially Chapters 7, 8, and 9. In this chapter you will find

❖ a discussion of the rise of spatial analysis and its limited links with cultural geography;

❖ a discussion of geographic interest in Marxist social theory;

❖ a description of the rise of several humanist approaches;

❖ an outline of five principal psychological approaches to the study of behavior;

❖ an outline of two different geographic approaches to studies of behavior in the context of what is described as the model of humans;

❖ an outline of feminist approaches, with emphasis on the oppression or subordination of women and other oppressions. The challenges that feminist thought poses for the foundations of the geographic enterprise are identified;

❖ a discussion of the cultural turn, with reference to a variety of innovative conceptual contributions including poststructuralism, postmodernism, and postcolonialism;

    ✢ a discussion of some implications of the cultural turn, including its critique of the previously accepted mode of representation, a concern with human difference, and a concern with the social construction of knowledge;

    ✢ a comment on the introduction of culture into traditional social geography;

    ✢ an indication as to how these developments inform the contents of later chapters.

Spatial Analysis

Regional geography did not place any priority on studying culture and can be baldly characterized as empiricist, descriptive, atemporal, and implicitly if not explicitly environmental determinist. ➔➤➤**1** The growing discontent with this approach was most effectively articulated in a landmark article that argued for a more scientific approach to geographic subject matter—for a **nomothetic** rather than an **idiographic** emphasis (Schaefer 1953). This discontent resulted in a major revision of the geographic enterprise with the onset of what can be described as the quantitative revolution and ushered in the approach known as spatial analysis, an approach that dominated geography from about 1955 to about 1970. Table 3.1 lists basic differences between regional and spatial approaches.

Cultural geography intersected with spatial analysis through diffusion, which was evident in the Berkeley School, and, following the work of Hägerstrand (1967), created opportunities for computer modeling and **simulation** of diffusion processes across space. While some geographers adopted this diffusion process with vigor (for example, in terms of the diffusion of different building types), criticisms of the apolitical nature of diffusion modeling demanded a reconnection with more detailed analysis of cultural processes on the ground, particularly with reference to how power might influence the diffusion of certain practices (Blaut 1977). Certainly diffusion took hold in areas such as the study of disease, but it was less influential in cultural geography. Why was it that, unlike most other subdisciplines of geography, cultural geography was essentially unaffected by spatial analysis? The general answer is that

Table 3.1 Regional Geography and Spatial Analysis

Regional Geography 1900 to Late 1950s	Spatial Analysis Late 1950s to 1970
is idiographic	is nomothetic
asks what? and where?	asks what?, where?, and why?
aims to describe	aims both to describe and to explain
is atemporal	is atemporal
focuses on facts	includes both factual and conceptual content
is essentially a classification	uses theories, hypotheses, and quantitative methods
is implicitly environmental deterministic	rejects environmental determinism
is linked to empiricist philosophy	is linked to positivist philosophy
declined about 1955 because it was seen as not scientific	declined about 1970 because it was seen as dehumanizing

The above is a basic summary of some differences between the traditional regional and the spatial analytic approaches as practiced in twentieth-century cultural geography. The dates are approximate.

cultural geography was separate from the dominant tradition of geography that was challenged by spatial analysis, namely regional geography. The cultural geographic interest in evolution and in the study of process through time was an especially important difference between cultural and regional geography. Challenges to the regional concept were aimed explicitly at regional geography's concern with regions as the core of geography, not at cultural geography's concern with regions as outcomes of evolutionary processes. The demise of the region as the core geographic concept was of major significance to such interests as economic geography, which lacked a substantive conceptual basis other than that provided by regional geography; it was of considerably less significance to cultural geography because of its firm foundations within the landscape school. Further, because spatial analysis maintained the atemporal emphasis of the regional school, cultural geographers saw little that spatial analysis could offer them. ➜➤➤ **2**

Geographers who advocated spatial analysis directed their critical attention especially to the areas of economic and physical geography, which seemed to offer the greatest potential for practicing theory construction and quantitative methods in a manner guided by a positivist philosophy, specifically **logical positivism**. The deductive procedures involved in theory creation were certainly unattractive to cultural geographers, who were often intent on analyzing the particular and not the general. For a research tradition concerned primarily with questions of history and culture, a positivist perspective was necessarily of limited appeal, arguing as it did for objective research and for research strategies adopted essentially from the physical sciences. (This despite the fact that the superorganic content of the landscape school could be interpreted as reflecting the "behaviorist claim that habit should be construed not as thought but as activity" [Duncan 1980:194–5].)

In one respect the failure of spatial analysis to affect cultural geography is surprising, because the discipline of history was undergoing some methodological debate and developing an increasingly positivist concern at about the same time. Further, both the French *Annales* school of historical method and the mostly North American new economic history embraced some theoretical and quantitative content. Although these developments in history, which were pursued by some cultural geographers, are comparable to spatial analysis, they did not lead cultural geography any closer to adopting a spatial analytic approach. Interestingly, toward the end of the 1960s, when spatial analysis was being challenged by other approaches, the deficiency of excluding historical work was recognized and a process-to-form methodology proposed. Process-to-form ideas involve identifying causal processes and linking these directly to spatial outcomes. Although this methodology is explicitly historical, it was not welcomed by most cultural historical geographers.

Overall, the failure of spatial analysis to affect cultural geography, while unsurprising, is perhaps regrettable. Although spatial analysis was quickly challenged by other approaches, it benefited most other subdisciplines of geography by introducing new ideas and initiating new methods of analysis. However, beginning about 1970 many human geographers saw spatial analysis as a dehumanization of the discipline and, finding inspiration instead in Marxist concepts, initiated new studies concerned with inequalities and subjectivity.

Marxisms

Versions of Marxism were introduced to human geography around the late 1960s under the guise of **radical geography**, and prior to the rise of humanist geography and the cultural turn of the 1980s. Indeed, in some respects Marxism was one of the approaches that humanistic geography and the cultural turn reacted against due to the emphasis Marxism placed on grand theory—the claim that there is necessarily one single answer to a problem—and also because of the essentialist character of the Marxist approach. Marxism has contributed some important concepts to cultural geography and is linked to some aspects of postmodernism. Most notably, Marxism informs ideas about human identity, **difference**, power, inequality, and commodification. ➜➤➤ **3**

Marxism, as developed by Marx and Friedrich Engels (1820–95), is both a body of social theory and a political doctrine. During the twentieth century the political doctrine was implemented in modified form in a number of countries, notably the former Soviet Union and China. The concern in the present context is with the social theory Marxism introduced to geography.

Marxist Social Theory

It is often suggested that in developing a social theory Marx followed Hegel and employed a **dialectic** as an alternative to formal logic, although it seems more appropriate to suggest that the dialectic used by Marx was ontological, referring to the fact that societies are organic wholes that evolve only to eventually collapse as a consequence of their internal contradictions. The three components of Marxist social theory are as follows.

The first principal component of Marxist social theory involves the identification of principal types of human society in a historical context. Examples of societies are slavery, feudalism, capitalism, and **socialism** (all four terms are included in the Glossary). Marx discussed each society as a **mode of production**, comprising both **forces of production**—that is, the raw materials, implements, and workers that produce goods—and the **relations of production**—that is, the economic structures of society. It is the forces of production that produce goods, while the relations of production determine the way in which the production process is organized. The most important relation concerns matters of ownership and control. Two insights emerge from this component of Marxist social theory: there is a historical transition from one mode of production to another, and social classes play a key role in social formation and social change. Further, society can be differentiated into **infrastructure** (or **base**)—another term for relations of production—and **superstructure**, the legal and political system and forms of consciousness. These terms are helpful because infrastructure can be interpreted as a determinant of superstructure, meaning that human thinking results from material conditions. This is effectively a form of economic determinism.

The second principal component of Marxist social theory is the transition from one type of society to another explained in terms of two related processes, namely technological change and class struggle. Marx emphasized the emergence of new classes and the fact that an existing dominant class was able to limit the opportunity for further social change. Because a dominant class typically favors maintaining things as they are, Marx recognized the need for class struggle as the means for replacing one mode of production with another mode that comprised a different set of classes or that lacked classes altogether.

The third principal part of this theory is its analysis of nineteenth-century capitalism, which anticipated a transition to socialism. Marx was especially critical of capitalism because the dominant class (owners) was concerned primarily with profits—or what he called the ceaseless accumulation of capital. Further, he saw owners as exploiting the dominated class (workers). Although such exploitation of one class by another was by no means new, Marx recognized that what was previously open in modes of production such as slavery and feudalism was disguised in the capitalist mode of production. Marx envisaged capitalism as the last of the class-based societies. Much of the political doctrine of Marxism is concerned with the various mechanisms by which classless socialist societies would eventually replace capitalism.

The Importance of Division of Labor

Marxist social theory is concerned with the material basis of society and aims to understand society and social change by referring to historical changes in social relations. This approach is termed **historical materialism**. Marx conceived of history in materialistic terms, and, as we have seen, a central feature of Marxist social theory concerns the development of productive forces, social relations of production, and the formation of classes.

For Marx, the first human historical act is the production of means to satisfy material needs, and fulfillment of these needs leads, as population numbers increase, to other needs. The first form of social relation, the family, is required to fulfill the initial material needs; when the family proves unable to satisfy these new needs, a new social formation develops. In this sense, it is appropriate to identify each mode of production with a particular social formation that emerges to satisfy the material needs of society. This is what Marx intended when he said that society has a materialist basis. Thus, in tribal societies there is an elementary division of labor based on age and gender, but with the development of agriculture, civilization, and industry the division becomes increasingly complex. In the industrial world of the nineteenth century, the capitalist mode of production involved a separation and conflict of industrial and agricultural interests, a separation and conflict of capitalist and worker interests, and a separation and conflict of individual and community interests. The implications of these divisions of labor, from the elementary forms found in the family to the much more complex forms found in capitalism, are enormous.

Thus, Marxist social theory maintains that there are unequal distributions of labor, of the products of labor, and of private property. Further, the distinction between individual and community interests leads to the creation of states that serve the community interest. When human relations are reduced to commercial relations—a form of commodification—the result is an estrangement or alienation of social activity. Marx was concerned both with the material conditions of workers and with the way in which their lives assumed an alien form.

Versions of Marxism

Marx was a prolific writer whose works included numerous uncertainties and contradictions as his ideas developed and transformed through time. Accordingly, there have been numerous interpretations of both Marx's social theory and his political doctrine. Certainly, there is no one correct reading of Marx. Two principal versions of Western Marxism that appeared during the twentieth century are outlined here.

The Frankfurt School of critical theory emphasized the humanist aspects of Marxist thought and added elements of social psychology. From this perspective, Marx was a voluntarist, seeing human history as reflecting human intentional activity. This school criticized bourgeois society for being dominated by a form of technological rationality, and social science for its prevailing scientific orientation.

A second version of Marxism rejected the humanist aspects and focused on its structuralist character. A popular approach during the 1960s, structuralism emphasized that understanding a social system required an appreciation of the structural relations between the component parts. From this perspective, human problems were essentially structural in character, as they were rooted in some specific economic system, and humans were not considered as autonomous active subjects as they were in the Frankfurt School. There were, of course, variations in the degree to which human agency was denied; for example, structural Marxists following Althusser were more conscious of geographical and historical specificity, and incorporated greater aspects of human agency in their explanations. ➤➤➤ **4** In addition to these versions, there have also been attempts to read Marx in terms of rational choice theory in economics—this is known as analytic Marxism—while Marxist theory has also been influential in some feminist thought and in cultural studies generally. Regardless of the specific version, Marxism is always concerned with analyzing inequality, oppression, and subordination, and with identifying some means to overcome these. The basic solution for Marx was, of course, the creation of a socialist society involving the abolition of class, of private ownership of the means of production, and of commodity production.

Marxisms in Geography

As the account of Marxist social theory has shown, there is no single Marxist tradition to which geographers could turn, and because of this contested situation, any Marxist approach adopted in geography has been challenged not only by those geographers who are generally unsympathetic to the Marxist approach but also by other Marxist geographers who favor an alternative version. The most serious debate has been between those geographers inclined to the humanist-oriented Frankfurt School of thought and those inclined to structural Marxism. The disagreement between these two is one part of the larger "structure and agency" debate—that is, the dispute about the ability of human beings to function under the constraints imposed on their behavior by social structures. ➡➤➤ 5

There are two important claims relevant to understanding the contribution of Marxism to cultural geography. First, it has been asserted that "Marxism and cultural geography share important basic presuppositions concerning the significance of culture, but in different ways and for different reasons both have failed to sustain those presuppositions in their practice and have not developed a dialogue with each other" (Cosgrove 1983:1). Cosgrove's attempt to correct this failing recognized a common focus on the historical aspect of human and land relations and stressed conceptual similarities such as a shared Marxist and Vidalian view of the relationship. Second, it has been argued that a Marxist cultural geography corrects perceived deficiencies in the discipline, such as cultural geography's failure to recognize that culture has a political component and its failure to recognize that cultures are divided into classes. Further, according to Blaut (1980), a Marxist view is one that avoids two of cultural geography's humanist tendencies: the tendency to set individuals

and culture in opposition, and the tendency to neglect external constraints from society at large. The most substantive recent work along these lines sees cultural geography as necessarily involving a conception of culture that focuses on questions of power and authority (Mitchell 2000). There has also been resurgence in Marxist-informed scholarship in response to the "crisis of capitalism" that appears to have surfaced during the end of the first decade of the twenty-first century.

Humanisms

While Marxism was influential in geography in the late 1960s and early 1970s, there was a corresponding rise in resistance to the overly structuralist approach it offered. For many cultural geographers, outside of launching a proletariat revolution, Marxism dismissed the importance of human agency. This reduced model of human agency also related to a substantive philosophical debate concerning whether or not the study of humans can be modeled on approaches developed in physical science. **Humanism** views the idea of a physical science approach to studying humans as untenable: to treat humans objectively would be to treat them essentially as objects and not as thinking, feeling individuals. This is a basic point. For many cultural (and other) geographers, the fundamental problem with Marxism and spatial analysis was its inability to accommodate individual human characteristics, and this is one reason that various humanist approaches became popular beginning around 1970. ➡➤➤ 6

Three key aspects of humanist philosophy are relevant. The first is the view that humans are ontologically and epistemologically irreducible. This means that claims about knowledge cannot be derived from physical science. This humanist tenet involves emphasizing the uniqueness of such human phenomena as creativity and understanding: "human beings present a unique reality, irreducible to the order of animal behavior" (Wertz 1998:53). The second relevant aspect of humanist philosophy is its focus on human experiences and symbolic expression rather than abstract principles, and its acknowledgment of many different truths. The third aspect is humanism's respect for individual freedom and dignity.

In responding to the ideas of humanist philosophers, especially Dilthey and Edmund Husserl (1859–1938), many scientists argued that the subject matters of physical and human science necessitated different concepts and methods. The most obvious justification for this separation was the fact that humans are free to act as they choose. But a further justification was based on the fact that human phenomena are internal, not external, to the experience of the scientist. This is a critical observation, and Wertz (1998:49) expressed it this way: "Because physical phenomena and their laws of interconnection are not directly given, they must be explained by constructed models, theoretically derived hypotheses, and experiments revealing functional relationships among isolated factors. Human phenomena, because they are immediately given to knowledge and internally related to each other, need not be hypothesized but are to be described and understood in their meaningful connections." ➔➤➤ **7** Humanists demand a model of humans that is quite different from the one that results from the use of physical science methods. In particular the issue of methodology differentiated humanistic geography from other approaches in that values and emotion were inserted into the debate. Humanistic geographers were encouraged to use ethnographic and other participatory types of research methods to prevent the objectification of the research group in order to capture and reflect the nature of human subjectivity.

Concepts and Procedures

In geography, it was primarily cultural and historical geographers who introduced humanist ideas, which offered a powerful critique of spatial analysis. Box 3.1 outlines the three principal humanist interests advocated by cultural geographers—phenomenology, existentialism, and idealism.

Most work in humanist geography is associated with phenomenology and stresses human subjectivity and free will, the relevance of participant rather than observer research, and the need for a hermeneutic or interpretive focus. For Bunkse (1996:361), the key humanist element in cultural landscape creation is imagination—"the unique human ability to step outside the immediate context of life, to, as it were, absent ourselves from the here and now and to enter into other realities or to create entirely new realities." Perhaps surprisingly, the attraction of humanist ideas in the 1970s for certain geographers, including many cultural geographers who were disillusioned with the positivist spatial analytic tradition and Marxism, did not involve close links with psychology. Rather, as noted in Box 3.1, humanist geographers turned directly to some of the philosophical sources. Regardless, the interpretation of humanism made by psychologists and by cultural geographers was fundamentally similar.

Despite some questions concerning the preferred individual social scale of analysis and also a specific focus, humanist approaches have generated some important concepts. ➔➤➤ **9** Most generally, it is customary to talk about the **taken-for-granted world**, sometimes called the lifeworld, of everyday living and thinking. This is the intersubjective world of lived experiences and shared meanings, including previously hidden aspects of our personal geographic environment. Intersubjectivity refers to the shared basis of experience in everyday life, to the understanding that individuals have of each other, and to their similar perceptions of the world. Our human experience of this intersubjective or taken-for-granted world comprises the facts that the phenomenologist explores. Studying the taken-for-granted world involves recognizing the centrality of mundane, everyday experiences, and one task of the humanist cultural geographer is to grasp the dynamism of this world.

Other humanist concepts are used to discuss parts of the surface of the earth as these are understood, perceived, lived in, and experienced. Thus, the need to focus on **place** and not space became the key feature of early humanist geography and continues to be a paramount concern; indeed, much humanist geography is devoted to illustrating and clarifying this term. The meaning of a place is inseparable from the consciousness of those who inhabit it, such that place is not so much a location as a setting, not a thing but a relationship. Closely tied to this focus on

Box 3.1 *Phenomenology, Existentialism, and Idealism*

Much of the support for the humanist movement in cultural geography as it emerged in the early 1970s originated in opposition to the model of humans embraced by positivist spatial analysis and structural Marxism. Humanism contends that nature can be explained but that humans, social life, and individual behavior need to be understood through a human-centered approach. Necessarily, then, humanists consider the scientific method inappropriate for research into human behavior, because humans have intentions; this is the fundamental principle of subjectivity. Further, humanist thought maintains that social phenomena are not entirely external to the researcher, and this has led to the development of a hermeneutic tradition, which aims to reveal expressions of the inner life of humans by *verstehen* (see also **hermeneutics**). Ley and Samuels (1978:9) describe this "principal aim of modern humanism in geography" as "the reconciliation of social science and man, to accommodate understanding and wisdom, objectivity and subjectivity, and materialism and idealism."

The three humanist approaches referred to in this box are well established in the larger philosophical literature. Although it has been customary to use this tripartite classification in geography, there is much overlap and integration of the three approaches given the common interest in subjectivity and individuals.

Phenomenology, as introduced especially by Tuan and Relph, has prompted numerous variants. For example, although it was initially articulated as a study of individuals, phenomenology has been adapted to the social scale by Alfred Schütz. A key concern of the perspective is the contention that there is not an objective world independent of human experience, and hence there is a focus on the individual lived world of experience and the understanding of meaning and value. The implications for cultural geography are varied, but phenomenology certainly presents cultural geographers the major task of reconstructing individual worlds.

Existentialism was proposed especially by Samuels, although "the boundary between existentialists

place are three related ideas. The first, **topophilia**, refers to love of place, to the affective ties between humans and environment. As people live in and become comfortable with a particular place, they develop a **sense of place** that serves as a basic emotional underpinning for everyday life. The idea of **placelessness**, referring to locations that lack meaning, has not proven very useful, possibly because any location that can be identified necessarily has meaning to someone.

Unlike positivists, who generally adhere to a scientific method, humanists do not employ a particular method in their studies, as there is general agreement that humanism cannot be defined as a set of procedures or techniques. Humanist geographers are generally hostile to statistical techniques and opposed to formal reasoning; they tend to prefer more creative forms of representation to explore the phenomena they study. Some use conversational language to capture the nuances of everyday experience,

while others adopt a more self-conscious literary style to uncover profound experiences in everyday life. Probably the most popular methods used by humanist geographers is participant observation, in which the researcher interacts and engages with the subject of research, and in some cases the researcher may openly empathize with the researched. This method, which had become important in both anthropology and sociology by the 1970s, was quickly adopted by humanist geographers, who favored its explicit recognition that people and their lives do matter. Overall, humanists use qualitative methods that involve researcher observation of and involvement in everyday life, which they regard as central to an understanding of humans and human landscapes.

One problem with much humanist geography is that it typically prioritizes individuals at the expense of groups. Such an emphasis is not in accord with much current work that sees a concern with individuals as a reflection of Western, middle-class and phenomenologists cannot be drawn precisely. . . . existentialists concern themselves with the question of the nature of 'being' and understanding human existence" (Entrikin 1976:621). With existentialism as with phenomenology there is concern with individuals and their relationship both with the world of things and the world of others. There is little evidence to suggest that this philosophy is emerging as central to cultural geography.

Idealism as advocated by Guelke is one component of a larger idealist alternative to naturalism. The Guelke version derives specifically from historical idealism as proposed by the historian Collingwood. Idealism insists that phenomena are only significant when they are a part of human consciousness; to comprehend the world, it becomes necessary to rethink the thoughts behind actions in order to discover what decision makers believed rather than why they believed it. Although this approach was cogently advocated, and although there have been several applications, most notably

by Guelke (especially 1982), idealism conceived in this manner has not proven to be especially attractive to other cultural geographers.

Each of these three approaches stresses human subjectivity and free will. Each shares the view that a spatial analytic geography is dehumanizing because it treats humans as objects, not subjects. But most contemporary cultural geographers do not accept the claim that the appropriate scale of analysis is that of the individual, as they recognize that the behavior of the individual is subject to many cultural constraints. Jackson and Smith (1984:21), for example, noted that a problem of humanist philosophies was that of "building an effective bridge between individual cognition, perception, and behavior, on the one hand, and an appreciation of man's place in society, on the other." The tendency of humanism to focus on individuals, to explain behavior in terms of individual mental processes, has led to its being criticized as psychological reductionism. ➔➤➤ **8**

values. Indeed, notwithstanding many arguments to the contrary, one leading humanist geographer has argued that no real humanist tradition in geography has developed, and that even some ten years after being introduced, the approach consisted of "little more than a few expressions of possibilities" (Relph 1981:134). If this claim is correct, humanist geography can be unfavorably compared with corresponding interests in sociology and psychology, in which scholars have succeeded in developing ideas on the basis of substantive research activities. Regardless, humanism has contributed some important concepts and is closely allied with some aspects of postmodernism.

Behavioral Geographies

Following landscape school principles, the approach traditionally preferred in cultural geography was to see individual behavior as subsumed within a larger

cultural context, so that cultures were understood to be the best means of explaining landscape change—this is one component of the superorganic concept of culture. But by about 1970 two revisions to this landscape school interest were evident. The first involved the rise of a positivist spatial approach to the study of behavior, and the second involved a more humanist and individually based approach.

In formulating these two behavioral geographies, geographers turned to psychology, focusing in particular on those psychological concepts that appeared to relate directly to their traditional concerns with space and environment. ➔➤➤ **10** Thus, geographers largely ignored the many difficult issues that psychologists confronted. Of course, the way these concepts were interpreted depended on which particular geographic approach was used, and it is for this reason that two different behavioral geographies, each incorporating a different model of humans, emerged. In both cases it is helpful to consider the underlying psychological theories prior to outlining the behavioral geographies.

Understanding Human Behavior

Knowledge and learning are divided into disciplines that change through time, such that the contemporary social sciences are social constructions with boundaries that are neither necessary nor unchanging. Nowhere is this more evident than in attempts to understand human behavior. Two points are relevant here. First, although long an area of scholarly interest, it is only in the twentieth century that the study of human behavior has become identified particularly with the disciplines of psychology and sociology; prior to this time, such work was conducted primarily by philosophers and historians. Second, the current division of interest between psychology and sociology may be unfortunate given the shared interest in human behavior. The division reflects certain trends in the way disciplines were formed in the nineteenth century, especially the desire of Comte, and later of Durkheim, to establish sociology as an independent science of society. Hence, while any full understanding of human behavior requires consideration of both individuals and collectivities, most theoretical formulations are derived from disciplinary origins and are limited to one or the other. Box 3.2 provides a brief overview of this question of social scale.

During the nineteenth century, psychologists introduced two approaches to the study of the human mind, in the original Greek tradition of psychology. Some argued that the mind could best be understood through **structuralism**, analyzing its parts or structures, with the principal research method being that of introspection. Others argued that the mind was to be understood in terms of ongoing thought processes that are responsible for human learning; this approach, known as **functionalism**, was influenced by Darwinian ideas of evolution and adaptation.

From these bases, twentieth-century psychology introduced many varied ideas about how to understand and explain behavior. It is not possible to do full justice to the diversity of psychological interest in human behavior, but the principal schools of

thought are readily identifiable, and those to which geographers have turned for inspiration are relatively few. The approaches of psychoanalysis, behaviorism, Gestalt psychology, humanism, and cognition are outlined in Box 3.3, which shows that one of the major sources of disagreement among the different approaches was how to define the subject matter of psychology. Was psychology the study of the human mind, of human behavior, or of both of these? As the contents of Box 3.3 suggest, behavioral concepts developed in psychology are substantial and varied, and there is an abiding tension as psychologists have often desired the objectivity of the natural sciences but have been concerned that they do not and cannot possess such objectivity because some of what they study, such as mental states, seems necessarily subjective and incapable of measurement.

Two Models of Humans

Space and Behavior

In the late 1960s, many social scientists considered human geography to be one of the social sciences, an understanding that was in accord with the claim that all social sciences study social life or "patterns of conduct that are common to groups of people," such that it is "not their subject matter but their approach to it that differentiates the various social sciences" (Blau and Moore 1970:1). Indeed, geography was routinely included in every "list of the social sciences" with each discipline having "its own subject matter, its own part of the study of man" (Senn 1971:60). Also, during the 1960s heyday of spatial analysis, many human geographers viewed their discipline as a social science concerned with some aspect of behavior, with behavior often used as an umbrella term to refer to human activities in general. Thus, Ginsburg (1970:293) noted that human geography was concerned with "questions of human behavior to the same degree, though not necessarily in the same way, that the other social sciences are." Similarly, a leading cultural geographer asserted: "Regardless of the tradition they follow or the methodology employed in the definition of their role, most human geographers have no hesitation in identifying themselves as social scientists" (Mikesell 1969:227).

This conjunction of circumstances sparked a geographic interest in the study of behavior that subscribed to a model of humans with close links to spatial analysis and some tenuous links with behaviorism and cognition as discussed in Box 3.3. This behavioral geography was unconcerned with human creativity and understanding, and the model of humans its exponents employed saw humans as responding to particular stimuli but paid only minimal attention to questions of human freedom and dignity. In addition to this rather narrow view of humans as rational decision makers, which was borrowed from economics and used typically in economic geographic analyses, behavioral geography held a more flexible view of humans as optimizers. This idea was in accord with some work in economics that viewed humans as boundedly rational animals—that is, as following simple step-by-step rules given constraints of limited

Box 3.2 The Question of Social Scale

Individuals or groups? This is a recurring issue in social science.

Because individuals live in societies, and because societies comprise individuals, there is no clear distinction between the two, and it can be claimed that they must be considered together. Indeed, it was only in the nineteenth century that social scientists established disciplines distinguished on this rather flimsy basis. Certainly, psychology as the study of individuals, and both sociology and anthropology as the study of groups, suffer from poorly defined key terms.

Even today, the distinction between individual and group is less than clear, with university courses on social psychology often included in both psychology and sociology departments. Psychological social psychology typically studies social relations from the individual outwards by reference to such individual characteristics as attitudes, and the approach is often derived from naturalism. Sociological social psychology typically employs group identities such as class and community, and is more receptive to methods related to the idealist perspective.

What is the appropriate social scale for analyzing the kind of human behavior that is of interest to cultural geographers? There is no simple answer to this question, since different approaches favor different scales. Indeed, the question itself is misleading, as a group can range from two individuals to the global population.

Of course, Sauerian cultural geography focused explicitly on culture: "Human geography, then, unlike psychology and history, is a science that has nothing to do with individuals but only with human institutions, or cultures" (Sauer 1941a:7). Further, most contemporary cultural geography accepts the need for a group scale. For example, Marxist approaches contend that humans are more than individuals, and that there is a need to consider the intersubjectivity of social life. From this perspective, the question becomes one of who does something—an individual or a class, an institution or a culture? The Marxist answer is that individuals need to be defined within the wider context, the social relations and the mode of production that structures their life and livelihood. On the other hand, as discussed in Box 3.1, humanist geographers, because of their concern for subjectivity and free will, employ philosophies that typically encourage an individual scale of analysis.

Not surprisingly, then, cultural geographers today display a willingness to study people at the full range of social scales. Most of those who continue to work in an essentially Sauerian tradition focus on particular cultural groups in particular settings, as do those whose inspiration is from Marxist or feminist thought. The many cultural geographers whose work is informed by postmodern and related thought show less commitment to a predetermined social scale, preferring instead to employ the scale that seems most appropriate for the topic at hand, although it is fair to say there is a particular concern with cultures understood as fluid, changing identities. Only humanist geographers are, on the whole, committed to the individual scale. On balance, it is the group scale that is most popular in the practice of cultural geography today.

Box 3.3 *Psychological Approaches to the Study of Behavior*

Psychoanalysis as introduced and developed by Sigmund Freud (1856–1939) is a deterministic approach claiming that human behavior is beyond the control of the individual such that the role of free will is minimal. Proponents of this approach believe that because it is the presence of a dynamic unconscious that explains why people behave as they do, the task of a psychoanalyst is to uncover the forces that operate in the unconscious. But this unconscious cannot be accessed directly, and there is accordingly an emphasis on uncovering this hidden world through other means, such as the analysis of dreams. Despite the controversial nature of these ideas within psychology, the complexity of the concepts, and the implicit determinism, versions of psychoanalysis are attractive to some cultural geographers (Kingsbury 2004). ➔➔➔ **11**

Behaviorism, as introduced by John B. Watson (1878–1958) in 1913, claimed that the subject matter of psychology was behavior, not the human mind, and that behavior was to be studied by means of objective procedures as practiced in the physical sciences. Radical behaviorism was developed by Skinner, who introduced the concepts of antecedent conditions, operants, and consequences, with the key idea being the law of effect—that is,

that consequences influence behavior. Few cultural geographers are attracted to these ideas (Norton 2003). ➔➔➔ **12**

Gestalt psychology developed in the early twentieth century from the idea that human behavior is determined by a psychophysical field analogous to a gravitational or electromagnetic field. Thus, human perception of a particular object is explained by reference to the larger field within which the specific object is contained. The human mind and perception are one whole and not a number of parts. This way of analyzing conscious experience is different from those proposed by both psychoanalysis and behaviorism. The emphasis is on how the world is perceived rather than on how individuals behave because it is perception that determines behavior. Thus, behavior can be understood by uncovering the perceived world of individuals rather than by studying the objective world. The principal method of analysis used by Gestalt psychologists is a phenomenological approach involving the study of individual perception and subjectivity. Some geographers have been attracted to these ideas because of the emphasis on perception and subjective environments, the concern with individuals as members of groups, and the explicit acknowledgment

---

knowledge and time. A classic study of farming practices in Sweden further established that humans adopted strategies designed to find a satisfactory outcome, and this led to the introduction of the term "**satisficing behavior**" (Wolpert 1964). Many of the studies conducted in this tradition studied human behavior and inferred cognitive characteristics such as **perception**. Major geographic analyses informed by these ideas included work on natural hazards.

Place and Behavior

Most cultural geography concerned with human behavior favored a different model of humans, one that incorporated humanist concerns such as subjectivity. It is for this reason that geographers were attracted to such psychological concepts as mental

or **cognitive maps** and perceived environments that involve identifying a psychological environment, or what might be called the subjective frame of reference. This focus on subjective environments involves the recognition that the physical world can affect individuals only in so far as it is perceived or experienced. Thus, in much geographic work, objective reality is not seen as a cause of behavior; rather, it is objective reality, as perceived or assigned meaning by the individual, that is judged important.

While several geographers were devoting themselves to the substantive body of work centering on perceived environments, one geographer adopted the transcendental phenomenology of Husserl. According to this view, humans have misunderstood the character of the world by believing that a of the role played by external, non-psychological environments.

Humanism rejects the scientific tradition of conducting research that is replicable and verifiable by other researchers, seeking instead an understanding of the world as it is appreciated by the individual or group being studied. It asserts that behavior results from subjective understandings—perceptions—of the objective world, and all versions of humanist psychology maintain a central concern with human experience and personal growth. Hall and Lindzey (1978:229) describe the position of humanism in relation to psychoanalysis and behaviorism thus: "Humanistic psychology opposes what it regards as the bleak pessimism and despair evident in the psychoanalytic view of humans on the one hand, and the robot conception of humans portrayed in behaviorism on the other hand."

Cognition is an approach arguing that mental processes, especially thinking but also perception, memory, attention, pattern recognition, and problem solving, are principal causes of behavior. Cognitive psychology claims that the presence of mental factors means that any attempt to explain all human behavior in terms of physical laws is necessarily inadequate, and that while it may be the case that behavior is shaped by consequences, it is still the case that cognition must be studied in its own right. With its emphasis on information processing, it is related to the cognitive science approach, which aims to explain cognitive processes in terms of the means by which information is handled by the mind, usually proposing that the mind works in a series of logical steps rather like a computer. Some cognitive psychologists have borrowed concepts from such areas as computer science, communications science, and neuroscience. Additionally, cognitive psychology has a related geographic interest in the computational process modeling of spatial cognition and behavior. ➝»» 13

All too clearly, psychologists do not agree on the best approach to study of human behavior. Each of the approaches noted has some merit, and it is not possible to definitively state that one is in some way inherently superior to the others. A balanced response to the above overview may be to suggest that different approaches are suited to different specific problems.

scientific, mathematical, and objective view of the world is also the world itself. Rejecting such a view, Husserl claimed that it was not appropriate to interpret events in our lifeworld scientifically; what was more important was to attempt to understand the general structures of meaning of the lifeworld. To do this, Husserl explained, it is necessary to suspend our preconceptions by putting aside any suggestion that the real world is naturally ordered. With these ideas as a basis, Hufferd (1980:19) argued that "in order to explain why an object (human or not) acts or reacts in whatever fashion, it is necessary to understand it in relation to the perspective resulting in its action." According to this argument, any attempt to understand human behavior needs to proceed as follows. First, the researcher must record the accounts offered by numerous participants as precisely as possible until major themes become evident and no additional new themes are forthcoming. Second, the researcher must list the frequently mentioned topics, classify them, and compare the different interpretations. Third, the researcher must summarize the topics and identify key features so that the perception held by the group can be reconstructed as authentically as possible. Arguably, this reconstruction forms the basis for an understanding of human activities.

One or Two Geographies of Behavior?

Behavioral geography as it developed from the spatial analytic tradition was not closely associated with work in cultural geography, unlike the humanist-flavored behavioral geography promoted by cultural geographers. But this simple distinction in principle, specifically involving different inspirations and different models of humans, was much less clear in practice as geographers struggled to introduce behavioral analyses into the discipline. A classic early study intentionally and effectively integrated positivist and humanist geography to explain and understand American inner-city life with specific reference to an inner-city area of Philadelphia with a high concentration of African-American residents. For some contemporary cultural geographers, such an integration of approaches is necessarily flawed, although Ley later insisted that objective measurement and subjective interpretation were compatible:

> For while one may conduct a multilocale ethnography of more than one place, an upper limit is quickly reached. In moving from two sites to 20 some standardization of accounts—usually measurement—becomes necessary. I see nothing but intellectual gain from incorporating the harvest of an intensive interpretation of one or a few sites with an extensive analysis of a large number. (Ley 1998:79)

Overall, however, geographers undertaking studies of behavior favored either a positivist or a humanist perspective, and thus were making a decision about naturalism, specifically about what might be called the model of humans. They either considered it appropriate to view humans as objects and to work toward the development of laws governing behavior, or else they preferred to view humans as subjects, active participants in their worlds. In retrospect, it is clear that most behavioral work favored the humanist perspective. Thus, research focused on the idea that humans are subjects, emphasizing presumed links between perception, behavior, and the creation of landscape, and often building on the assumption that perception caused behavior, which in turn caused landscape.

The Inference Problem

It is not satisfactory to approach an understanding of the cultural landscape solely from the rather naive proposition that perception causes behavior

causes landscape. All too clearly, factors other than perception influence behavior, and factors other than behavior influence landscape. Nevertheless, much useful and innovative work has centered on the role played by perception. But should cultural geographers organize their research around the logic that perception causes behavior, or should they infer perceived environments on the basis of actual behavior? Both of these research frameworks involve difficulties.

On the one hand, using **image** to explain behavior is difficult in that it can be argued that perceived environments cannot even be objectively measured. Further, a close correspondence between perception and behavior is often difficult to demonstrate. On the other hand, inferring images from behavior also poses logical problems, as it is one example of the inference problem—that is, assuming the identity of causes on the basis of an analysis of effects.

Questions for Critical Thought

Although several geographers writing around 1970 anticipated that a behavioral approach might prove a panacea for perceived ills of the discipline, this has not been the case. But why not? One reason might be that two quite different versions of behavioral geography developed, based on quite different conceptions of what it means to be human. Viewed retrospectively, it seems clear that in both cases the underlying philosophical impetus, positivism in one instance and humanism in the other, proved unacceptable to the majority of geographers. Further, Walmsley and Lewis (1984:4) identified the confused character of behavioral geographies by referring to such features as a focus on individual decision-making units, the relevance of both acted-out and mental behavior, an emphasis on the world as it is rather than as it should be under some theoretical constraints, an interest in economic location topics, and use of models. This list is suggestive of the mixed inspirations and varied practice of behavioral geographies. With these complexities in mind, identify a topic that you think could be suitably pursued using a behavioral approach.

On balance, the second concern is the more problematic. Inference is an uncertain form of discovery, for there are no guarantees that the inferred cause is indeed a true cause. To demonstrate the dangers of inference in this context, consider the following simplified circumstance. Farmer A perceives that either crop x or crop y will produce a satisfactory income, while farmer B perceives that either crop y or crop z will produce a satisfactory income; both choose to cultivate crop y, and therefore their two landscape decisions are identical, despite being based on different perceptions of reality.

Feminist Thought

Providing a singular definition for **feminism** as it applies to geography is problematic, because intellectually and politically "universalizing feminism is history" (Pratt 2008:69). In its place there has emerged recognition that all forms of universal knowledge can enact violence upon the excluded, be they women or sexual, supposed racial, or other minorities. In this regard, various approaches to feminism, as employed by geographers, have increasingly sought to translate meaning between different contexts, and in the process produce what Katz (2001) has called countertopographies, lines of connection that reveal how seemingly disparate groups face challenges that are actually the product of similar structural processes, be it capitalism, **patriarchy**, heteronormativity, or other forms of socially and politically enforced hierarchy. Such structures of exclusion can also intersect to reproduce and maintain power, privilege, and status for some, while exacting the cost of such inequality upon others. In revealing such connections, the goal of some feminists' work is to forge political alliances to expose, overcome, or ameliorate the resulting systemic oppressions. Some examples of this approach in geography include the work of Geraldine Pratt on Canada's live-in care program, and how Canadian immigration policy exploits global uneven development to solve its own deficient domestic system of child care provision by importing women from the developing world to provide low-cost, but high-quality, care labor. While feminists have resisted the grand narrative of Marxism,

and the universalist tendency that accompanies it, there are clearly similarities in terms of challenging structures of exploitation and inequality (indeed this relationship has been referred to as "the unhappy marriage of feminism and Marxism" [Hartmann 1979]). Likewise, with humanist geography there is a similarity in the choice of methodologies that take seriously the agency of the subject, but also a rejection of the often masculine and heteronormative assumptions embedded in early humanistic approaches. This discussion provides an appropriate basis for consideration of the social construction of **gender** and of landscapes of sexuality that come later in the text. ➡➤➤ **14**

Three Versions

The basic tradition of feminist thought, dating back to the late eighteenth century and known as the first wave or liberal feminism, aimed to obtain equal rights and opportunities for women. Note that from some later perspectives, such a seemingly meritorious approach can be seen as fundamentally flawed because, in aspiring to equal status with men, liberal feminism supports the view of the male model as the norm or benchmark that all others must aspire to, and not to reach such status infers some kind of deficiency. Such benchmarking can be enforced through supposed racial, sexual, and gendered tropes.

The feminisms that appeared during the 1960s and continued into the 1980s are often described as the second wave of feminist thought. These tended to assume that it was possible to identify a specific cause of the oppression of women. Suggested causes included men's ability to control women's fertility and the capitalist requirement of a pliable workforce. Versions of cultural feminism, with a focus on technologies and ideologies developed by men and the idea of essential relationships, belong in this second-wave category. These more radical feminisms often have the general intent of moving beyond the redistribution of rights and resources pursued by liberal feminists in order to initiate changes to the structure of society. Further, many of these radical feminisms contend that the oppression of women is so deeply rooted in both psychic and cultural processes that much more than superficial change is necessary.

During the 1980s, the third wave of feminist thought appeared. This involved a critique of the earlier assumption that it was appropriate to attempt to universalize the experiences of women (usually White and middle class) in the more developed world. Feminists thus developed a fuller concern for the multiple oppressions that structure our gendered identities. These multiple oppressions are based on social class, skin color, income, religion, age, culture, geographic location, among other differences. Third-wave feminism also involves an expanding interest in sexual difference as well as in gender, an interest stimulated by psychoanalytic theory. There is an increasing realization that to insist on equality is to deny the importance of sexual difference. Finally, it has involved an engagement with aspects of poststructuralism and postmodernism that includes a concern with the social construction of gender. Most current work in feminist geography is identified with this third wave. ➝➤➤ **15**

Note that for most versions of contemporary feminist theory, sex is a category based on biological differences between the female and male (but see the discussion of sex and gender in Chapter 8), while gender, although based upon the natural category of sex, is a social construction that is performed in a repetitive manner over time, enforced within heteronormative frameworks, and refers to what society deems to be appropriate behavior for females or males within specific contexts, times, and places.

Locating Feminism in Geography

Unlike many of the other conceptual terms that have entered geographical discourse in the past twenty years (Marxism, structuralism, postmodernism), feminism provides neither a methodology nor a theory for human geography, although methodological and theoretical differences have been derived from it. Rather, feminism as it was originally introduced to geography had a far more radical goal—it provided a political basis for a critique of the practices of geography, practices that had allowed for the invisibility of women as both practitioners in the field and objects of enquiry. (Domosh 1996:411)

In light of this assessment, it is unsurprising that, despite a considerable body of conceptual, empirical, and methodological literature by geographers and others, feminism does not merit any significant explicit mention in many standard reviews of geographic theory and method. The contrast with the ways that Marxism and postmodernism were received by the discipline is striking. It does appear that geography is less than certain about how to accommodate a body of thought that not only includes new perspectives and approaches and implies new subject matter but also questions how the discipline of geography is masculinist in its historical and intellectual development. Cloke, Philo, and Sadler (1991:xi) (three male geographers) explained their decision not to include a chapter on feminism in their account of approaches to human geography as follows: because "the issues involved are more important than this 'ghettoising' might imply," because of the belief that feminism is something "more than just 'another' approach," and because "such discussion is most ably pursued by women."

In addition to this uncertainty, there are debates about feminisms and feminist geographies within the feminist geography literature itself. Some have identified areas of shared feminist and geographic interest and argued for integration, while others express concern about the fundamental masculinism of geography that might effectively prohibit meaningful integration. The term "masculinist" refers to work that claims to be exhaustive and universal but that ignores its own naturalized exclusions, be they women or others, and that accordingly is concerned only with the position of men, and naturalizes power hierarchies already in place. Another important debate arose in the writing of a feminist geography textbook aimed at introductory-level students (Women and Geography Study Group of the Royal Geographical Society with the Institute of British Geographers 1997). This book was written by a collective that made the decision ("not an easy one") to exclude men from the writing team (Rose et al. 1997:3). ➝➤➤ **16**

Feminist Challenges to Geography

The concerns of feminist geography have changed through time in general accord with changes in the larger feminist context. In the early 1970s and in accord with the logic of liberal feminism, there was concern on the part of some feminists about women in geography, especially since the discipline of geography, like many other disciplines, was dominated by men. There was also an attempt to incorporate gender as a variable into geographic studies, along with more usual variables such as culture and ethnicity.

By the 1980s, studies of oppression had begun to consider oppression both within the discipline of geography and within the larger world, and many of these studies adopted a Marxist orientation. The early concern about the lack of women in the discipline had expanded to include concerns about the essential masculinism of geography, especially because of geography's traditional link with exploration and its continuing involvement with fieldwork. Most recently, a more radical tradition with links to postmodernism has emerged. This tradition attempts to construct theory in response to the idea that geography is a masculinist discipline that claims to speak for everyone but that speaks only for White, middle-class, heterosexual males.

Probably the single most important concept in feminist geography is that of gender, along with the derivative ideas of **gender roles**, **gender relations**, and patriarchy. Gender roles are sets of behaviors deemed to be socially appropriate or inappropriate established on the basis of sex, and in line with dominant views of social relations within specific societies. Few, if any, of these behaviors rely on inborn sexual differences. Most important, gender roles vary culturally and are, accordingly, best thought of as learned differences in behavior, or as the theorist Judith Butler (1990) argues, are embedded and naturalized through a process of **performativity**, whereby there is no essential gender role; rather a series of learned behavioral repetitions are enacted (Butler 1990). There is an inherent geography or spatiality to this, as McDowell shows through her work on gender in the workplace (McDowell 1997b). The concept of gender relations recognizes that gender roles are not necessarily accepted; rather, gender roles are enforced, sometimes contested, and change through time and across space. If imaged as a performance, then gender norms, as we know them, can also be destabilized. These characteristics of gender roles refer to the idea that gender involves unequal power relations between women and men, specifically the relationship of patriarchy, whereby men dominate women. These arguments about sexual identity are explored further in Chapter 8. ➜❯❯ **17**

It is clear that feminism poses several challenges— perhaps better thought of as invitations—for geographers to consider. Since about the 1960s, forms of feminism have vigorously challenged virtually all of the belief systems and institutions of dominant patriarchal cultures, including geography. Feminism has questioned contexts for knowledge as well as basic concepts such as reason and logic. For geography this has involved questioning the legitimacy of approaches such as positivism, humanism, and Marxism on the basis of what is perceived as their universalistic claims often uncritically based upon sexist biases. Indeed, feminists were among the first to recognize the importance of the social construction of knowledge, meaning that those who produce knowledge do so from specific positions, and this positionality determines the research problem, the data to be used, and the methods to be employed, rather than some untainted, neutral objective scientific rationality.

Feminists, along with postmodernists, use a number of concepts that, together, highlight the need to assess precisely what is happening during the research process. Most fundamentally, there is a need to acknowledge the importance of **situated knowledge**—that is, to recognize the discourse within which knowledge is produced. Similarly, there is a need for **reflexivity**, meaning self-reflection at all stages of research, in order to understand how particular characteristics of the researcher may influence the research activity, and to disclose such partialities. A third related concept is that of **gaze**, referring to how those in positions of power have the ability to transform subjects into objects through their gaze, and it also constructs the differentiation

Questions for Critical Thought

———— ✛ ————

It is perhaps inevitable that the discussion of feminism in geography is much more than a mere academic discussion. As with Marxism, feminism has a powerful political dimension, and it appears that cultural and other geographers are unsure how to accommodate the combination of academic and political challenges. Since women typically conduct most feminist geography, is it possible this may ghettoize feminist geography, which is a problem because, after all, gender includes men's and women's roles and social relations. For example, the issue of sexual violence demands critical analysis of masculinity. Think of another example where you might study men in order to advance a feminist research project.

of active self (the male self-possessed observer) and the passive **other** (the subject/object who has been made available for their consumption). In geography "the gaze" has been critically invoked in relation to the nature of exploration, landscape observation, and fieldwork. Feminists have advocated for positions that return the gaze, that is, resist and reverse the power of the masculine gaze to objectify the subject. ➝➤➤ 18

The Cultural Turn

A Context

The "cultural turn" refers to the period post-1980s that saw an intellectual shift across the humanities and social sciences placing increased importance on culture while downplaying other considerations such as economy and politics. An important aspect of this cultural turn is the emphasis on language, or "linguistic turn," a term that is sometimes used as an alternative to cultural turn. Specifically, there is increasing realization that language is important, since it enables the communication of meanings, but that it is also constraining because of the limitations resulting from the meanings we attach to words; we do not know the world for what it is but rather only

as it is mediated through language and other symbolic systems. Expressed simply, the linguistic turn challenges the ability of language to function as a neutral conveyor of information. In short, language cannot "serve as a perfectly transparent medium of representation" (Duncan and Ley 1993:5).

This discussion considers the cultural turn from four interwoven perspectives, namely those of cultural studies, poststructuralism, postmodernism, and postcolonialism. This is a less than perfect way to categorize material that is by definition fluid and dynamic, and it is appropriate to stress that the very spirit of the cultural turn is unsympathetic to classification and the imposition of order. This is an idea that resurfaces on several occasions in this discussion, but it is also one that proves difficult to adhere to at times. ➝➤➤ 19

Another point of information is in order. Although it may not be difficult to understand that there could be a cultural turn in some other disciplines, it may sound confusing to claim that the turn is affecting cultural geography—how can there be a cultural turn in cultural geography? The short answer is that the meanings of culture involved in the cultural turn are different from the meaning incorporated within Sauerian cultural geography—refer back to the various definitions of culture included in Box 1.2, and recall the deterministic implications of the classic Sauerian definition in contrast to the understanding of culture as a negotiated intersubjectivity implied in some of the more recent definitions. To illustrate this shift some people refer to cultural geography in the post–cultural turn phase as "new" cultural geography. Needless to say, the intellectual influence of the cultural turn deems that most cultural geography taught today has been informed by, if it is not already, the "new" cultural geography variety.

Cultural Studies

Cultural studies is perhaps most appropriately seen as an umbrella term incorporating a wide range of philosophical and social theoretical ideas, including feminism, poststructuralism, postmodernism, postcolonialism, and aspects of Marxism and humanism. Central to cultural studies is the claim that established disciplines are unable to cope with the complex circumstances of the cultural world such that the advantages of an interdisciplinary approach are stressed.

Box 3.4 provides an overview of some of the more important authors and ideas that are emerging as influential in cultural geography. The information in this box is designed to promote a general understanding of the cultural turn.

Cultural studies literature typically rejects modernism in addition to the Enlightenment concerns with an empiricist epistemology and with a search for universal truths. Together, the varied bodies of thought that are labeled "cultural studies" constitute a key feature of the cultural turn in the social sciences and humanities generally. Principal components of this work include several British studies of working-class culture (for example, Thompson 1968), the concept of culture introduced by Geertz (1973), and the Frankfurt School of Marxist critical theory. Also important in the current context are several advances in social theory associated with such writers as Gramsci, Derrida, Foucault, Baudrillard, and Deleuze, as these are discussed in Box 3.4, along with several traditions—such as feminism, postcolonialism, subaltern studies, and studies of racism—that are linked by their concern with exposing the Eurocentrism associated with the production of knowledge in the Western world since the Enlightenment. ➔➤➤**20**

Understandably, different bodies of thought that are gathered under the cultural studies umbrella emphasize different matters: feminism, for example, is critical of much of the gender blindness of traditional Marxism, while Marxism is a major source of inspiration for many postmodern social theorists. But together these works stress the importance of culture as a dynamic and primary force that is not necessarily always predetermined by political, economic, and social forces. Further, all see culture as a site of negotiation and conflict in societies where power is exercised through multiple and variable channels that are splintered especially in terms of ethnic, religious, sexual, gender, regional, and class identities.

Questions for Critical Thought

❖

Cultural geography clearly welcomes the importation of ideas from elsewhere, including from the cultural studies tradition, but there may be a real difficulty in such movement because cultural studies is seen by many of its practitioners to be a postdisciplinary and critical practice. Thus, from a cultural studies perspective, does it make sense that ideas be exported to disciplines and subdisciplines and then employed within precisely those disciplinary and subdisciplinary frameworks that cultural studies challenges? It might be suggested that cultural geographers seem to want the best of both worlds—both the critical theories developed in cultural studies and the traditional subdiscipline of cultural geography in which to apply these critical theories. But for some practitioners of cultural studies the two are irreconcilable, not least because disciplines and subdisciplines are seen as something other than neutral when it comes to discussions of such topics as racism and **sexism**. Cultural studies does not take the existence of disciplines for granted, whereas most new cultural geography implicitly accepts the existence of a body of knowledge labeled cultural geography. Simply put, there is a very real challenge involved in attempting to employ theories derived from an avowedly postdisciplinary tradition in a disciplinary context. Indeed, for many in cultural studies, it is a challenge that should not be taken up.

Poststructuralism

Structuralism is one of several bodies of thought that view culture as a system of ideas based on advances in linguistic theory that see a sound and the object it represents as entirely arbitrary. In other words, structuralism's chief concern is with relationships rather than with those things that create and maintain relationships. For example, a structuralist interpretation of gift-giving in primitive societies focuses on the act of giving rather than on what is given, or by whom or to whom it is given. Structuralism aims to remove the difficulties of cross-cultural comparisons by showing that elements of all cultures are the product of a

Box 3.4 *Cultural Studies—Scholars and Ideas*

Although many of the scholars and ideas associated with cultural studies are also closely identified with feminism, postmodernism, poststructuralism, and postcolonialism, they are discussed together in this box for reasons of convenience. Specifically, because much of the empirical material in subsequent chapters (especially 7, 8, and 9) relies on the ideas presented in this box, it aims to serve as a reference informing many of those studies.

Antonio Gramsci (1891–1937) was a Marxist theorist and political activist whose most influential conceptual contribution was to elaborate the idea of hegemony. According to Gramsci, any class that achieved a position of economic importance and desired to attain power in a larger society first needed to achieve a degree of cultural and intellectual hegemony. This hegemony allowed the class to express its worldview, to structure social and other institutions so that they accorded with the class's aims, and most generally to create a context sympathetic to the class in power. To maintain hegemonic control the class in power needs to elicit the willing acquiescence of those it dominates in order to maintain and reproduce its position of authority in society.

Gramsci also introduced the term "subaltern" to refer to those socially subordinate groups that lacked both the unity and the organization of more dominant groups that are able to exercise authority and control. The term has been popularized by Indian Marxists and generally refers to those groups considered to be in some way socially inferior to other groups.

The work of Michel Foucault (1926–84) both crosses and challenges disciplinary boundaries and is especially difficult to classify. One of his important contributions concerns the closely related ideas of discourse and **episteme**. Foucault saw subjectivity as constructed within and through discourses, with discourse defined as a system, comparable to a language, that enables the world to be made intelligible. Discourses are important because they serve to legitimize a particular view of the world that then becomes part of the taken-for-granted lexicon. Discourses define our world for us, as Foucault (1980:52) explains: "The exercise of power perpetually creates knowledge and, conversely, knowledge constantly induces effects of power." Thus, power is practiced through discourse, and discourses consolidate power.

For example, it can be argued that the discipline of geography evolved within the late nineteenth-century discourses of **imperialism**, **colonialism**, and **racism**; these discourses produced certain forms of knowledge that in turn consolidated the "taken-for-granted" normalization of colonial rule under the guise of "the white man's burden." However, Foucault stressed the necessarily partial and situated character of any particular discourse, and this is where the related idea of episteme comes in; it refers to the worldviews or structures of thought that a society holds at a particular time, including the multiple displacements and discontinuities of discourses that occur over that time, or as Foucault (1991:56) states: "an analysis of transformations in their specificity." As an example, consider the shift in current political discourse from a stated concern for the rights of *citizens* to that of *taxpayers*. What work does this discursive shift do?

Jean Baudrillard (1929–2007) is a particularly controversial postmodern writer. As with many

common single mental process. This idea has proven attractive to some anthropologists, but despite the interest it appears that many of the arguments are highly impressionistic. Two of the principal insights of structuralism—both antihumanist—are important, in modified form, in **poststructuralism**.

The first insight, following the linguistic philosopher Ferdinand de Saussure (1857–1913), is that language is a medium for understanding social organization. Thus, meaning is not conveyed through language but rather produced within language. Simply put, the facts do not speak for themselves. The second insight, following the Marxist philosopher Louis Althusser (1918–90), concerns the interpretation of subjectivity and claims that notions of individuality are seen as ideological other twentieth-century theorists, Baudrillard challenged Marxism both for being deterministic and for reducing culture to a secondary status at the expense, particularly, of the political and the economic. Baudrillard sees contemporary society as a **consumer culture** based on consumption and not on production. He also introduced the idea of the simulacrum, arguing that the contemporary world is dominated by signs, images, and representations, obliterating the real and rendering any search for truth and objectivity fruitless.

The French philosopher Jacques Derrida (1930–2004) has demonstrated a continuing concern with the taken-for-granted character of much philosophical thought and, according to some postmodernists, has effectively shown that philosophy has no privileged access to meaning and truth. Derrida is best known for the introduction of deconstruction and the related idea of text. Essentially a school of philosophy and literary criticism, deconstruction has had a significant impact on many of the traditional social science and humanities disciplines. However, it is by no means universally accepted, especially by philosophers, at least in part because it is such a radical set of ideas. Most generally, it has been used to challenge some of the various structures of the dominant European cultural tradition.

Gilles Deleuze (1925–95), a philosopher, and his occasional collaborator Félix Guattari (1930–92), a political theorist and psychoanalyst, are best known for their attempts to integrate the ideas of Marx and Freud. One of their joint efforts has been described as a "vast, chaotic rag-bag of a book" (in Honderich 1995:183), and although they have been welcomed

into cultural geography, especially by Shurmer-Smith and Hannam (1994), the long-term significance of their work remains to be determined.

Edward Said (1935–2003), a literary and cultural theorist, contributed an influential body of ideas—**Orientalism**—contending that the Orient is an invention of those who study it from outside, and that North America and Europe employ ideas from within their cultures, ideas such as freedom and individualism, to facilitate their conquest and domination of other regions. Thus, the Orient is a necessary European image of the other, a construct of a dominant European discourse. For Said, culture is best seen as a hegemonic environment with specific prevailing modes of thought. The contemporary importance of these ideas is evident in the claim that the United States sees the Arabic world in dehumanized terms, a tendency that is exacerbated by terrorist activities and by the ongoing conflict between Palestinians and Israelis.

Much of the work that is considered part of the cultural studies tradition, including feminism and studies of racism, qualifies as critical theory—a term that refers to the conviction that both the social sciences and the humanities need to be emancipatory and not ideological instruments. Critical theories are fundamentally different, epistemologically, from theories in physical science in that they are reflective and not objectifying, and do not suggest detached objectivity is a central condition of knowledge production. These critical theories have origins in the Frankfurt School of critical theory, especially in the work of the German philosopher and sociologist Jürgen Habermas (b. 1929).

constructs. Thus, humans cannot act autonomously; rather, their actions are constrained by various structures.

"Poststructuralism" is a broad term. Its central ideas reflect an expansion of, rather than a break with, structuralism. Many of the most important concepts, including those modified from structuralism, are often included as one component of the

cultural studies tradition and are, as noted, also identified with postmodernism. →≫ 21

Postmodernisms

Box 3.5 provides a basis for outlining the key arguments and content of postmodernism by offering distinctions between four terms, namely "modernity," "modernism," "postmodernity," and "postmodernism." According to many (but by no means all) observers, the contemporary world is increasingly displaying the characteristics of postmodernism.

Although this discussion offers a series of generalizations about postmodernism, it is not intended to convey the impression that there is some clearly agreed-upon set of ideas. Postmodernism is a highly

Box 3.5 Modernity and Postmodernity— Modernism and Postmodernism

Postmodernity refers to a new phase of history and a new cultural system, while postmodernism is a narrower term referring to a set of ideas, a scholarly movement that opposes the intellectual logic of modernism. Much academic literature describes both the historical phase and the scholarly movement using the one term postmodernism; however, it is certainly useful to distinguish the two in principle, although this proves to be rather more difficult in practice. By extension, modernity and modernism can be similarly distinguished from each other (Duncan 1996).

Modernity, the modern period, was initiated by the eighteenth- and nineteenth-century processes of industrialization, urbanization, secularization, and nation-state creation. Durkheim identified the increasing division of labor as the key to the onset of the modern period, while Tönnies identified the transition from Gemeinschaft to Gesellschaft. In broad terms, the contemporary world continues to display the basic characteristics of modernity. However, many observers recognize an emerging phase of postmodernity that is gradually replacing modernity.

The principal characteristics of postmodernity are those of a postindustrial economy accompanied by globalization processes, a loss of national identities, and a corresponding rise of more local ethnic identities. Harvey (1989) analyzed postmodernity as a condition of society. For many geographers, one appeal of the concept of postmodernity is the interest shown in spatial difference as compared to the emphasis on progress through time. Jameson (1991) described postmodernity, using Marxist terms, as a mode of production, a period of late (taken-for-granted) capitalism that is characterized by multinational capitalism and by the commodification of culture.

Modernism, as a scholarly movement, is associated with the Enlightenment ideals of progress and rationality, with an empiricist epistemology, and with advances in scientific knowledge. It was in this intellectual context that the various academic disciplines were established in universities, a process of institutionalization described in Chapter 1.

Postmodernism is a scholarly movement based on a loss of faith in the values and ideas that support the modernist project. Postmodernism emphasizes a blurring of disciplinary boundaries and the rise of new interdisciplinary movements and, in some instances, even argues for a postdisciplinary academic context.

Conflating these ideas, as many scholars do, it can be argued that such a significant set of changes is occurring today that the modern Western world is in a state of crisis, reflected in such circumstances as the general weakening of Western political traditions, the lessening of authority of some social institutions, and the resurgence of religious fundamentalism.

Note that, unfortunately, the terms "modernity" and "modernism" are inherently flawed because what is modern at one period of time is not necessarily modern at some later period. It is for this reason that the rather clumsy terms "postmodernity" and "postmodernism" can be described as following, subsequent to, and even in opposition to modernity and modernism respectively. Note also that the terms "postmodernity" and "postmodernism" are problematic because they appear to preclude the possibility of a coexistence of modernity with postmodernity and of modernism with postmodernism. The use of the prefix "post," meaning after or behind, is usually interpreted to mean that one circumstance is replaced by another.

contested arena with numerous competing ideas that together constitute a real challenge to established theories and understandings of social science.

Postmodernism effectively rejects all other attempts at social theorizing because of the emphasis placed on establishing foundations for knowledge. Social theories developed within modernism (Marxism, for example) are seen as privileging science and the related foundational bodies of knowledge that discuss matters in terms of truth and falsity. Postmodernism, on the other hand, is antifoundational, opposing attempts at grand, all-embracing theories and favoring instead more local and particularistic studies that emphasize difference.

Postmodernism also questions—even denies—the ability of science and theory to establish truth, and rejects the claim that any such truths, were it possible to attain them, could be liberating. These are challenging ideas based on the argument that all knowledge, including truth claims, is socially created. Postmodernism proposes that the myths and ideologies of modernism serve only to promote ethnocentrism, specifically Eurocentrism. This postmodernist way of looking at human identity encourages studies of disadvantaged groups such as women, the working class, and ethnic minorities—groups that were previously marginalized or excluded because of their lack of power. In addition to encouraging studies of the disadvantaged and oppressed, postmodernism asserts that the voices of such groups are as important as the voices of so-called authorities. Authors and authority, as identified in modernism, are not privileged in postmodernist theory. For a postmodernist, individuals are repositories of cultural systems and are therefore unable to speak for themselves; rather, they speak for the cultural system. It should be clear from these comments that any adoption of postmodernist claims has some significant implications for cultural geography.

Postmodernism emphasizes the idea of social construction. Postmodernists share the feminist concern with gender, class, and ethnic identity, recognizing that these are social constructions rather than established realities and, accordingly, that they can be changed. Postmodernists further acknowledge, with feminists, that academic disciplines such as geography are social constructions that exist only because they were created (and can therefore be dismantled) by society. Note that this general idea, without the accompanying philosophical argument, was central to the account of the discipline of geography included in Chapter 1. Also, like feminists, postmodernists stress the importance of positionality and the situated character of knowledge. The claim that language is at the heart of all knowledge, in the sense that social constructions are reflected in and often maintained through language, means that it is necessary to uncover the limitations that language places on our thinking and understanding. For this reason, a key feature of the postmodern agenda is the philosophy and method of **deconstruction** (see Box 3.4). Deconstruction seeks to uncover what is not said in a **text** and what cannot be said because of the constraints imposed by language. Further, deconstruction regards all aspects of a text as necessarily intertextual in the sense that meaning is produced from one text to another rather than being produced between the world and any given text. More generally, this **intertextuality** refers to the idea that all things are related. Any reading of a text, meaning also any view of reality, derives its meaning from other texts. As applied in the context of social science generally, a key purpose of deconstruction, which postmodernists prefer to theory construction, is to reveal the interests served by theories. The risk of deconstruction, though, is that it can become nihilistic; with potentially no end to the deconstruction process, and as an intellectual and political project it risks becoming vacuous. Feminist geographers have taken some guidance in dealing with this from Donna Haraway's (1991) ideas of situated knowledge, where knowledge production is explicitly recognized as always embodied and partial; this is expanded upon below. ➔≫ **22**

Postcolonialism

Closely related to postmodernism is **postcolonialism**, which is concerned with theories of identity, the role played by power, and the production of culture, especially in the less developed formerly colonized world. Postcolonial geographic writing explicitly reflects an appreciation of the consequences of colonial activity as it has caused or contributed to circumstances of human and landscape inequality. The impacts of colonialism are profound, and a postcolonial perspective endeavors to uncover issues surrounding the loss of identities and places that characterize the lives of those affected.

It is clear that the larger academic discipline of geography, along with the other social sciences, was not born innocently, but rather was a product of—and was complicit in—larger nationalist, colonial, imperial, racist, and masculinist ventures. The close association between Western ideas about exploration and discovery on the one hand and geography on the other hand cannot be denied. In much the same way that feminism challenges geography to rethink some fundamental assumptions about what is knowledge, so postcolonialism provides a compelling critique of the foundations of the geographic enterprise. Most notably, postcolonialism challenges the Eurocentric perspective of much geography; this challenge was evident in the work of Said, who was especially critical of the efforts made by colonizers to fix the identities of the colonized (see Box 3.4). Much postcolonial work stresses the Western tendency to see the world in dualistic terms, such as self and other, civilized and savage, and seeks to understand how these ideas came into being and how they can be dismantled. ➡️➤➤ **23**

The Mode of Representation

The task of cultural geographers is to represent the world to others, but this raises the question of what is an appropriate form for this **representation**. Prior to the emergence of the cultural studies tradition, the answer to this question was taken for granted. Today, as much of the content of this chapter has implied, it is debated, such that there is what might be called a crisis of representation.

The answer that was previously taken for granted is that any form of representation can be used that claims both to be universally valid and to result in an accurate understanding of the world. In this form of representation, "the task of writing is the mechanical one of bolting words together in the right order so that the final construction represents the thought or object modelled" (Barnes and Duncan 1992:2). Both the Sauerian approach and the occasional attempts to employ a positivist approach are seen to belong in this category. Most important, the Sauerian approach can be seen as an attempt, usually based on observation, to accurately describe and classify the subject matter of cultural geography. Much of the material included in Chapters 5 and 6 reflects this mode of representation.

But there are other modes of representation that reject the suggestion that it is possible to produce universally valid and accurate representations of the world on the grounds that any representation is actually an interpretation. This idea is central in humanist studies and explicit in several of the more influential modes of representation introduced in the context of the new cultural geography. Both feminism and postmodernism, for example, claim that accurate representations are not possible. More broadly, the hermeneutic method explicitly acknowledges the role of interpretation.

As discussed by Barnes and Duncan (1992:2–4), three critical implications arise from accepting the argument that it is not possible to produce accurate representations of the world. The first two of these implications were touched on in the accounts of feminism and postmodernism earlier in this chapter.

First, if there is no pre-interpreted reality for writing to reflect, such that it is not possible for writing to mirror the world, then what does writing reflect? The general answer that has been proposed is that writing reflects earlier texts; this is the idea of intertextuality. From a postmodern perspective, this also implies that all truths are inside, not outside, texts.

Second, all writing reflects a particular and necessarily local vantage point and is inevitably marked by its origins. Bondi stated the feminist argument clearly:

According to feminist philosophers, dominant conceptions of knowledge are "gendered." This claim is elaborated in various ways but, of central importance, is the notion that western intellectual traditions operate through interrelated dualisms, such as reason and emotion, rationality and irrationality, objectivity and subjectivity, general and particular, abstract and concrete, mind and body, culture and nature, form and matter. In each case, the terms are defined as mutually exclusive opposites but are not equally valued: the first occupies a superior position and is positively valued; the second is subordinate and negative. And intertwined within this system of hierarchical dichotomies is a distinction between masculine and feminine. It is through this intertwining that dominant knowledge systems are "gendered": the superior terms in the dualisms are associated with masculinity, the subordinate terms with femininity. (Bondi 1997:245–6)

More generally, it is possible to claim that because all texts are a reflection of a particular and not a universal viewpoint, it becomes critical to understand the position of the author of the text and also the intent of the text. The ideas of situated knowledge and positionality assert that authors are unable to speak for themselves; rather, they speak for the cultural system of which they are a part. Thus, all knowledge is partial. Some postmodernists move beyond these claims to assert that no one text is more privileged or in any sense a better interpretation than any other; this is clearly a claim that is especially open to debate. Ley (1998:80) noted that to abandon authority "leads to the abandonment of responsibility, to a tentativeness that in its reluctance to achieve any closure embraces, sometimes celebrates, ambiguity and indeterminacy." Further, it may be that the tendency for any representation to be **ethnocentric** need not necessarily imply that political projects ought to be excluded from academic work.

The third implication is that there is a need to employ a variety of writing strategies, tropes, and especially rhetorical devices to help convey meaning. In attempts to mirror the world, the principal writing strategies were those of objectivism, often involving a narrative. Metaphors are increasingly employed to describe the unknown using the language of the known. ➙➤➤ 24

Conducting Research

In addition to the challenges raised by Marxism and feminism, the cultural turn has introduced new research norms to the traditional study of cultural geography. Together, these three philosophical movements have taken the subdiscipline in new directions. Two principal developments are noted here.

First, Marxism, feminism, and the cultural studies tradition have encouraged studies of human identity, human **difference**, and the politics related to these matters. Many contemporary cultural geographic studies are characterized by a persistent questioning of concepts, classifications, and categories, a situation best described as one of **constructionism**. Identity, for example, is no longer taken to be something that is pre-given but, rather, is acknowledged to be socially constructed and therefore open to change. This has led cultural geographers to develop an increasing sensitivity to culture, especially concerning relations between dominant groups and other groups, and also an increasing awareness of the politics of difference. It is for this reason that the role played by power relations and the way power is practiced through discourse are central features of many of the approaches discussed in this chapter. There is also much evidence that contemporary cultural geography is including, even welcoming, the voices of those previously excluded.

A second development sparked by these three movements is that any discussion of ideas and facts is now interpreted with a recognition of the specific discourse being employed; this is the idea of **contextualism**. Cultural geography now demonstrates an increased awareness of and sensitivity to how knowledge is constructed, as well as by whom and for what purpose. As a result, much contemporary cultural geography, especially that informed by feminism and postmodernism, explicitly acknowledges the positionality and situatedness of the author.

But these developments, especially those associated with the cultural turn, are not without their critics. Indeed, Duncan (1994) characterizes the debate between Sauerians and new cultural geographers during the 1980s and early 1990s as a "civil war" within cultural geography. Further, concerning the effect that the cultural turn has had on social geography, Gregson (1993:527) observes that "on the one hand social geographers move to deconstruct the socially constructed categories which form the basis for social differentiation (in so doing, frequently criticizing these categories as essentialist), on the other they find themselves having to recognize that these categories are the very ones through which people make sense of themselves and of social life."

There have also been more general expressions of concern about the seemingly ceaseless introduction of new ideas into geography, a process that might be explained in terms of the commodification of knowledge, with competition resulting in flexible

specialization. Berry (1995:95) refers to a "combination of self-defeating trendiness and the calculating acceptance of a mainstream agenda by academics more concerned about creating successful niches than about principle," and suggested that postmodernism in particular is fundamentally flawed because a geography without a rational and science-based core cannot accommodate specialties.

The current philosophical and theoretical background that underpins cultural geographic research is multifaceted, complex, and dynamic, perhaps even uncertain and unstable. It has certainly been responsible for a wealth of empirical analyses.

Studying Society

It is clear that cultural geographers are now actively involved in working with, and in some instances even advancing, social theory. This is interesting because, traditionally, social theory was almost exclusively the property of the discipline of sociology and, accordingly, lacked spatial content. This circumstance changed beginning in the 1980s, when Giddens (1984:368) was able to claim that human geography was a social science, based on the fact that "there are no logical or methodological differences between human geography and sociology." Further, Giddens (1987:9) asserted that the "social sciences developed as a family of disciplines such that the different 'areas' of human behavior that are covered by the various social sciences form an intellectual division of labor which can be justified in only a very general way." Geographers, too, stressed the social theoretic recognition of the importance of space, with Massey (1985:12) asserting that "space is a social construct—yes. But social relations are also constructed over space, and that makes a difference." And Agnew and Duncan (1989:1) identified a "recent revival of interest in a social theory that takes place and space seriously." The inclusion of place or space in social theory, previously almost unheard of, is now quite usual. In this regard, while geography was involved in the cultural turn, many other disciplines within the social sciences, such as economics, were experiencing their own spatial turn (Warf and Arias 2008). →≫≫ **25**

Questions for Critical Thought

—◈—

As you read through the account of post-modernism, are you inclined to think that there is an intriguing ambiguity in the post-modern emphasis on social construction on the one hand and the continuing concern with academic disciplines and disciplinary status on the other hand? Particularly puzzling is that, as noted earlier, cultural studies sometimes aspire to disciplinary status. Thus, Benko and Strohmayer (1997:xiii) noted that "more than any other disciplines, with the possible exception of Cultural Studies, it was Human Geography that came away from this encounter with a renewed sense of mission, vindication, even of pride." The importance of disciplines is evident in this assertion, with the identity of those disciplines reinforced by the use of capitals. Somewhat ironically, others have referred to "the delineation both of a tradition and a future trajectory for cultural studies as the Academy's most upwardly mobile discipline" (Carter, Donald, and Squires 1995:vii). As a student, do you think it is important to have a disciplinary identification?

Social Theory and Space

The first substantial work in this area was the **structuration theory** of Giddens. The key feature of this theory is its focus on the links between human agents and the social structures—both local social systems and larger social structures—within which they function. Structuration theory also explicitly recognizes links between social relations and spatial structures, with space viewed as a medium through which social relations are produced and reproduced rather than as a mere backdrop against which social relations unfold. The term **locale** is used to refer to the settings in which everyday social interactions occur.

Although several geographers have endorsed and used structuration theory, others have questioned its value, and there is little evidence that it will play a major role in contemporary cultural geography. Sympathetic accounts by human geographers tend to be critical of geography for viewing place as inert when it ought to be viewed as a historically contingent process, as a transformation of space and nature that cannot be detached from the transformation of society. But cultural geographers might well resist any suggestion that they have viewed place as inert. The many regional studies by Meinig discussed in Chapter 6 may not be conceptually sophisticated by structuration standards, but they are explicit analyses of related spatial and social change. ➙➤➤ **26**

Another body of work to which geographers have been attracted is concerned with what might be called the production of space, a phrase that refers to the fact that the spaces within which social life occurs are socially produced. In this approach to social theory, a rational and scientific interpretation of space is replaced with an interpretation that sees space as experienced or imagined.

But the most important advances in contemporary social theory for the cultural geographer are those that strive to build on ideas that are explicitly opposed to the Enlightenment concern with an empiricist epistemology and with the search for universal truths. Accordingly, cultural geographers have shown much interest in the works of some critical social philosophers. These theoretical contributions were considered earlier in the context of the discussion of cultural studies and related interests as a source of concepts for cultural geographic analyses of both identity and difference.

Social Geographies

But what of the subdiscipline of social geography? Until the 1980s, social geography could be regarded as the British counterpart to North American Sauerian cultural geography. The different experiences are explained in terms of different intellectual origins. Thus, although cultural geography was stimulated by leading European geographers such as Humboldt, Ratzel, and Schlüter, it was really initiated by Sauer and never fully incorporated into geography curricula outside of North America. The British social geographic tradition has origins in French sociology, in British human geography, and in the Chicago school of human ecology. ➙➤➤ **27**

The term "social geography" was used by French sociologists of the Le Play school in the 1880s and by Réclus in the 1890s, and a detailed proposal was made by Hoke (1907). However, none of these developments had a significant impact on geography, and it was not until the 1930s that social geography emerged as a subdiscipline following a programmatic statement by Roxby (1930), the publication of social geography books by members of the French school, and the completion of some of the sociological human ecological work that emphasized social values. The result was a British social geography essentially concerned with the regional distribution of population, occupation, and religion. There was a focus on identifying different regions according to various social phenomena, and Pahl (1965:81) described the subdiscipline as being concerned with the "processes and patterns involved in an understanding of socially defined populations in their spatial setting."

By the 1960s, this British social geography was an uncertain field because of the varied intellectual stimuli. During the 1960s and 1970s three transformations occurred: a behavioral perspective was introduced; quantitative methods were used; and a focus on social problems, often stimulated by Marxist thought, developed. But social geography continued as an uncertain subdiscipline with much relevant material regarded as belonging to other compositional subdivisions such as population geography and urban geography.

At the same time, in North America cultural geography had a much clearer identity, with a single dominant school of thought, the Sauerian school, focusing on ecology, evolution, and regions. Certainly, British social geographers, like sociologists of the time, tended to see culture as the concern of the anthropologist, not the geographer. A competing North American movement promoted a social geography built on the ideas of symbolic interactionism (see Box 3.6), with social geography defined as the study of human spatial behavior and related human landscapes from the point of view of society as the sum of a population's symbolic interactions. Proponents of the school favored the label "social geography" over "cultural geography" because of its implied emphasis on communication processes through which symbolic interactions take place and its related claim that individuals are constrained in their behavior by the character of the groups to which they belong. The cultural geographer Wagner (1972, 1974) likewise saw distinct landscapes and societies evolving in response to communication processes often generated by social institutions. Neither of these interests has prompted substantial research activity. →≫ 28

Since the 1980s the status of social geography has been especially uncertain. Reflecting this uncertainty, Dennis and Clout (1980:1) began their textbook with the following concern: "Of all the varieties of geography, 'social geography' is probably the most elusive and poorly defined." Similarly, Cater and Jones (1989:viii) began their textbook with a discussion of "social geography's identity crisis." A few years earlier, Eyles and Smith (1978:55) concluded their review of the status of social geography on a cautionary note: "Social geography's greatest achievement may yet turn out to be the acceleration of its own destruction." On the other hand, Hamnett (1996:3) is more certain about the identity of social geography: "The approach I take is that social geography is primarily concerned with the study of the geography of social structures, social activities and social groups across a wide range of human societies."

Converging Cultural and Social Geographies

The differences between social and cultural geography were emphasized during the 1960s, when cultural geography remained outside of, and social geography was involved in, spatial analysis. But a convergence of the two interests was initiated in the 1970s when both subdisciplines found inspiration in humanism and Marxism. According to some observers, this convergence has proceeded apace, especially since the mid-1980s (see Figure 3.1).

Nevertheless, it is misleading to suggest that the two have in some way become one. Indeed, the message delivered in simplified form in Figure 3.1 is readily apparent given the material presented in Chapters 1 and 2: specifically, the extent to which cultural and social geography have experienced a genuine integration is open to much debate. This fact is evidenced by both the form and content of the various Progress Reports written for the journal *Progress in Human Geography*. Certainly, the content of these reports indicates that the viability of a meaningful integration of cultural and social geography is questionable. For example, Gregson (1995:139) noted

Questions for Critical Thought

Does this debate about what social geography is and is not seem of marginal relevance? Perhaps in some respects the details are somewhat marginal, but at the same time, the larger debate highlights a fundamental issue because it is about what is at the core of the subdiscipline, not about what is at the margins. The issue probably arises because of a complicated intersection of circumstances, including the quite separate European social and North American cultural geographic traditions and the labeling of the post-1980 developments as a cultural, not a social turn. Recall from Chapter 1 that the distinction between these two terms has never been especially clear, and it seems likely that confusion and uncertainty will continue for some time. From your perspective, what do you see as the difference between cultural and social geography?

Box 3.6 Symbolic Interactionism

Symbolic interactionism refers to a group of related social psychological theories, inspired by the ideas of the social philosopher George Herbert Mead and subsequently developed especially by sociologist Herbert Blumer, that have close links to the humanist philosophy of pragmatism. There are three fundamental premises:

- Individuals learn the meanings of things primarily through interactions with others, meaning that we learn how to define the world through our experiences of social interaction.
- Although we derive meanings of things from social interaction, we do not simply accept those meanings; they serve only as a basis for our understanding.
- Our behavior results from our interpretations of the meanings presented to us.

In addition to these three fundamental premises, symbolic interactionists agree on an additional seven assumptions:

1. Humans are unique in their ability to use language.
2. Other unique human traits arise because of language.
3. Human development is dependent on social interaction.
4. Humans are purposive and rational animals.
5. Humans seek rewards and avoid costs.
6. Humans are conscious actors.
7. Humans are active.

All major versions of symbolic interactionism accept the three fundamental premises and the seven additional assumptions, although there is often some disagreement concerning additional conceptual content.

A human ecology built on these ideas is different from other versions advocated by sociologists.

Specifically, symbolic interactionism stresses that humans are different from other animals in some very important ways, whereas many of the other versions of ecology, explicitly derived as they are from physical science, emphasize similarities between humans and other animals rather than differences. The most important difference identified by symbolic interactionists concerns the efficiency of human language and the implications for what it means to be human, given that effective communication is the key to complex behavior. Other important differences include the interaction with others that characterizes human life, the purposiveness of human behavior, and the high degree of awareness that humans possess. Ericksen (1980) opted to employ a symbolic interactionist interpretation of human ecology because of its emphasis on interaction, which is central to any ecological approach.

Although the human ecology proposed by Ericksen (1980) has not been adopted by cultural geographers, the symbolic interactionist perspective has been discussed and applied. Most of the relevant work is associated with the cultural geography that criticized the perceived superorganic concept of culture employed by Sauerians. As Duncan (1978:269) noted: "The self is largely a product of the opinions and actions of others as these are expressed in interaction with the developing self." The idea that spatial and social behavior are entwined was accepted in a prominent social geography textbook, which had as its central concern "human spatial behavior and the derived geographical patterns from the point of view of society: the summation of a population's symbolic interactions" (Jakle, Brunn, and Roseman 1976:7). Note also that the definition of culture offered by Jackson and Smith (1984:205) and included in Box 1.2 is based on this approach. ➡➤➤ **29**

"the ascendance of cultural, as opposed to social, theory in social geography (and in human geography more generally)," and wondered whether a discussion of progress in social geography was really an "obituary." Hamnett (1996:4), on the other hand, claimed that the "revival of the 'new' cultural geography in Anglo-America has been very influential in the revival of social geography." Finally, Peach

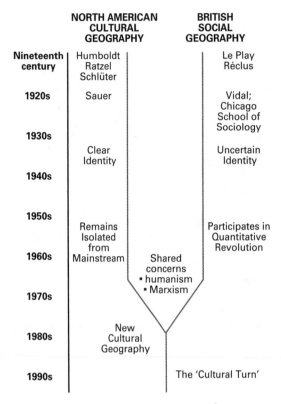

Figure 3.1 Cultural Geography and Social Geography—Origins, Concerns, and Convergence. This schematic diagram highlights the separate origins, different concerns, and the convergence (beginning around 1970) of cultural and social geography traditions.

(1999:286) maintained that work in social geography continued "to have more in common with the spatial side of sociology than with the cultural side of geography." →》》 **30**

Concluding Comments

- As stated in Chapter 1, the new cultural geography can be seen as a welcome addition to, rather than a replacement for, the Sauerian approach.
  - › Although the cultural turn in particular is most correctly seen as a set of new developments, the Sauerian antecedents of some aspects of these new developments are evident.

- › Further, there is no clearly definable new cultural geography because the term does not refer to a singular approach but rather to several varied conceptual advances that, while they have some common content, also demonstrate substantive differences. This is not surprising—the Sauerian approach itself was not and is not a single monolithic approach.
- Because of the relative newness of much of the conceptual material, it is not always apparent what is and what is not important; the consensus judgments made today are unlikely to be identical to those made tomorrow. And as we can see from recent events, the financial turmoil within the global economy has reinvigorated Marxist explanations of the inherent instability within capitalism.
  - › Indeed, several geographers have expressed concern about the seemingly indiscriminate, and perhaps even self-indulgent, borrowing of ideas from other disciplines, and also about the short lifespan of many of the concepts borrowed. More generally, other geographers worry that "human geography is to be subsumed within a broadly defined social science" with a resultant loss of disciplinary identity (Unwin 1992:210).
- Many of the recent conceptual advances in cultural geography are more difficult to understand and apply than is the Sauerian approach.
  - › This is because they are often associated with complex and disputed philosophical and/or social scientific origins, with the result that there are typically multiple versions of each of the newer conceptual advances. It is also because many of these advances are linked to postmodernism, a term that refers to a varied corpus of ideas but that characteristically insists upon the need to take into account non-traditional sources and the need to question established wisdom. Finally, it is because the cultural turn revolves around writing styles and vocabulary that makes it seem exclusionary to some and pretentious to others.

- Following the review of spatial analysis and the behavioral geography that it prompted, this chapter introduced a series of conceptual advances that are linked by a shared suspicion—in some cases an outright rejection—of science and the scientific method. More particularly, and with explicit reference to cultural geography, there is a general concern about the Sauerian tradition, which is now considered by many to view culture and cultures in an inappropriate manner. A crucial addition to our conceptual arsenal is Foucault's idea that power is exercised through the production of knowledge and truth in the context of a particular discourse, meaning that knowledge and truth are uncertain, even changeable. Certainly, a popular view of culture emphasizes that culture is, as noted in Box 1.1, "a domain, no less than the political and the economic, in which social relations of dominance and subordination are negotiated and resisted, where meanings are not just imposed, but contested" (Jackson 1989:ix). Of course, such an interpretation is not universally accepted, as evidenced by the body of ongoing cultural geographic literature, a literature that continues to be informed by a Sauerian or modified Sauerian view. Instead what is evident in cultural geography today is an amalgam of different "post"-prefixed isms (Barnes, 1996a) exhibiting some similar tendencies in their effort to contest central elements of Enlightenment thought.

Further Reading ➡️➤➤

The following are useful sources for further reading on specific issues.

1. Hubbard, Kitchin, Bartley, and Fuller (2002) provide a history of geographic thought that includes clear accounts of the transformation reviewed in this chapter. The edited text by Shurmer-Smith (2002) includes accounts of transformations from 1970 onwards.

2. Sopher (1972) suggests possible links between cultural geography and positivism.

3. There are accessible discussions of Marxism in geography in Cloke, Philo, and Sadler (1991), Unwin (1992), Peet (1996a), R.J. Johnston (1997), Robinson (1998), Blunt and Wills (2000), and Mitchell (2004). Quani (1982) provides a book-length treatment. As discussed by Peet (1977a, 1977b), Marxist social theory was introduced into human geography as a part of an emerging radical geography during the 1960s—the radical journal *Antipode* began publishing in 1969—with the first major Marxist-inspired work being the book *Social Justice and the City* (Harvey 1973).

4. Habermas (1972) outlines the Frankfurt school of critical Marxism, and Althusser (1969) outlines structural Marxism.

5. Peet and Thrift (1989) explain the structure and agency debate.

6. Pioneering contributions in humanist geography are those by Relph (1970) and Tuan (1971), both of whom identify phenomenology as a legitimate approach. There have been partially successful attempts to base humanist geography on a critical encounter with historical materialism (see Cosgrove 1978, 1983).

7. Rose (1981) discusses some ideas of Dilthey as they relate to humanist geography.

8. Tuan (for example, 1972, 1975) and Pickles (1985) outline phenomenology and geography; Samuels (1978, 1979, 1981) outlines existentialism and geography; Guelke (1971, 1975) outlines idealism and geography. For critical commentaries on idealism, see Watts and Watts (1978) and Harrison and Livingstone (1979).

There have been partially successful attempts to base humanist geography on existential philosophy (see Samuels 1978) and on idealist philosophy (see, for example, Guelke 1974).

9. Tuan (1974), Relph (1976), Ley (1977), and Seamon (1979) discuss some humanist concepts introduced by geographers.

10. Gold (1980:7–15) and Martin (1991:6–11) provide brief overviews of psychological approaches to the study of behavior. Golledge (1969), Spencer and Blades (1986), and Kitchin, Blades, and Golledge (1997) discuss the varied relations between geography and psychology.

11. Philo and Parr (2003) provide a useful introduction to what they describe as "psychoanalytic geographies."

12. Watson (1913) is the pioneering statement on behaviorism and psychology, while Skinner (1969) details the most popular version of behaviorism, namely radical behaviorism. Homans (1987) and Lamal (1991) provide guidelines for group-based behavior analyses.

13. Neisser (1967, 1976) outlines cognitive psychology. Smith, Pellegrino, and Golledge (1982) discuss the geographic modeling of spatial cognition and behavior; Golledge (1987) outlines environmental cognition, and Golledge (2003) reflects on cognitive behavioral research.

14. Overviews that provide relatively accessible introductions to feminist geography include Hanson (1992), Rose (1993—generally acknowledged as a classic text), Domosh (1996), Monk (1996), McDowell (1997a), Women and Geography Study Group of the Royal Geographical Society with the Institute of British Geographers (1997), Robinson (1998), Domosh and Seager (2001), Moss (2002), Pratt (2004), and Sharp (2004). Gregson et al. (1997:67–8) discuss a feminist geographic analysis of the term "patriarchy."

15. McEwan (2003) outlines debates between Western and other feminisms.

16. Hanson (1992) discusses areas of shared geographic and feminist interest; Bondi and Domosh (1992) and Longhurst (1994) explain the masculinism inherent in most geography. Recognizing that there are multiple histories of an academic discipline, Monk (2004) examines the careers of women geographers in the United States prior to about 1970. Moss and Falconer Al-Hindi (2008) explore the range and complexity of contemporary feminist thought in contemporary geography with an edited collection that strives to be open and fluid to different approaches and ways of understanding and enacting feminist praxis within the discipline.

17. Dowling and McGuirk (1998) review some of the principal themes reflected in research on gendered geographies.

18. England (1994) and Rose (1997, 2001) provide an understanding of feminism and the situated character of knowledge.

19. Duncan (1980), Ley (1981), Berry (1995:95), and Dunn (1997) outline some concerns about cultural geography borrowing ideas from other disciplines. Chambers (1994:121–2) discusses cultural studies and disciplinary identities.

20. There is little doubt that the various stylistic tactics employed by some of the cultural geographers who are deeply involved in postmodernism are not always well received. Walker (1997:172–3) expresses it this way: "I become easily irritated with the posturing of the postmodernist and the mannered style of discourse that glibly condemns the linear, logical, and evidentiary essay in favor of fragments of literary allusion and freely tossed Lacanian word salads, which leave a faint and convoluted trail of simulacrumbs for the poor reader to follow." Gregson (1993, 1995) and Badcock (1996) discuss some concerns about writing styles associated with the cultural turn.

21. Challenging commentaries on poststructuralist geographies are provided by Doel (1999, 2004).

22. Rosenau (1992) looks at differences within postmodernism; Berg (1993) assesses the claims that even the differences between modernism and postmodernism are based on false logic. Accessible accounts of postmodernism by geographers include those of Dear (1988), Webb (1990–1), Cloke, Philo, and Sadler (1991), Unwin (1992), Duncan (1996), Edwards (1996), R.J. Johnston (1997), Robinson (1998), Dear and Flusty (2000), and Minca (2001).

23. Nash (2004) reviews postcolonial geographies in an attempt to answer the question, "Can the spatial discipline of geography move from its positioning of colonial complicity towards producing postcolonial spatial narratives?" Other discussions of postcolonial geography include Sidaway (2000), Nash (2002), and Ryan (2004).

24. Barnes and Curry (1983), Duncan and Ley (1993), and Duncan and Sharp (1993) examine modes of representation in cultural geography.

25. Soja (1989:Chapter 1) looks at space and social theory; Gregory (1994) used the work of Lefebvre (1991) to facilitate discussion of the historical production of modern capitalist spatiality; other instances of the incorporation of space into social theory include the ideas of habitus (Bourdieu 1977) and of normalizing enclosures (Foucault 1970).

26. Gregory (1981), Pred (1984, 1985), and Gregson (1987) discuss geography and structuration theory.

27. Discussions of cultural and social geographies are provided by Jackson (2000) and Peach (2002).

28. Watson (1951) and Dunbar (1977) chronicle the history of social geography.

29. Jakle, Brunn, and Roseman (1976) and Dennis and Clout (1980) are examples of social geography textbooks that employ a symbolic interactionist focus.

30. Hamnett (2003) worries that contemporary human geography as it is involved in the cultural turn is becoming increasingly detached from social issues and concerns. Clifford and Valentine (2003) include a series of chapters designed to help students conduct geographic research. Much of the content is directly relevant to work in cultural geography. Pain and Bailey (2004) summarize recent trends in British social and cultural geography, noting that there are increasing efforts to wed theoretical and action research.

CHAPTER 4

Environments, Ethics, Landscapes

Cultural geography, and for some even the larger discipline of geography, can be broadly interpreted as the study of humans and environment (which can include, for example, the urban, rural, natural, and industrial environments). This chapter develops this interest both conceptually and practically.

Of the various approaches to the study of humans and environment discussed in Chapter 2, it is several versions of the ecological approach that have attracted increasing interest in recent years. There is an increasing concern with human impacts on the environment, a concern that has given rise to an abundance of new literature in physical and social sciences. There is also a concern based on recent advances in biology that explicitly acknowledges the variability rather than the stability of nature. Further, and not surprisingly, developments in social theory have prompted more sophisticated versions of ecological thought, such as ecofeminism and political ecology. This chapter considers both traditional and these newer ecological interests. The sequence of material consists of

- a discussion of approaches to the study of ecology, stressing the subject's nineteenth-century origins in physical science, the interests in it shown by other social sciences, and the use made of its complex systems concepts;

- a discussion of principal geographic interpretations of human ecology, namely those of Barrows and Sauer;

- discussions of more recent ideas of the production of nature through capitalist processes of commodity production, of the "new ecology" that acknowledges the ever-changing character of the natural world, of environmentalism, of ecofeminism, and of political ecology;

- an introduction to ideas of human–nature interaction in terms of actor–network theory and animal geographies;

- a discussion of environmental ethics that includes accounts of various attitudes and values that relate to our behavior in environment;

- an acknowledgment of the changing character of ecological approaches and a consideration of the prospects for ecological approaches as used by cultural geographers.

Ecology: A Unifying Science?

Two Misconceptions

It is helpful to begin by dispelling two popular misconceptions. First, it is a mistake to equate ecology with concern for environmental **conservation** and preservation; ecology was not introduced in response to concerns about human use of land. Indeed, in nineteenth-century thought there were numerous interpretations of the human-and-nature relationship, including two diametrically opposed views of nature: from one perspective, nature was there to be exploited for the benefit of humans, while from another perspective, nature needed to be conserved in order to minimize the damage being inflicted by humans. Ecology came to the fore in this contradictory intellectual environment; indeed, some scholars regarded ecology as a means of facilitating the continuing exploitation of nature by pointing out ways in which the resultant damage could be minimized.

Second, notwithstanding some attempts in this direction, it is a mistake to think of ecology as an academic discipline. The origins of ecology as a scientific approach can be traced to the third chapter of Darwin's *The Origin of Species*, which considered the various adaptations of organic beings to environment, although the more general context was biologists' attempt to understand how the distribution of species is related to environment. The first use of the term "ecology" is generally attributed to Haeckel in the 1860s, and by the end of the nineteenth century ecological approaches were well established in both plant and animal science. There has never been a single discipline of ecology; rather, ecological concepts, notably the concept of a plant or animal community, were applied to a number of existing disciplines: as Bowler comments (1992:365), "many scientists could do 'ecology' while retaining their primary disciplinary loyalty elsewhere." This multidisciplinary aspect of ecology is fundamental.

Origins

Ecology has its specific origins in physical science, though there are versions of human ecology in several

of the social sciences. The term "ecology" is derived from two Greek words, *oikos*, meaning "place to live" or "house", and *logos*, meaning "study of".

Accordingly, the central concern of ecology as a science is with the ways in which living things interact with each other and with their environments or homes. Ecology is, then, best thought of as an approach to the study of relationships between organisms and their environments and not as a particular body of subject matter, and it is for this reason that there are ecological approaches in many of the physical and social sciences.

The first significant advances in understanding the term "ecology" were made by Clements, an American botanist, who developed the idea that a plant community was a distinctive superorganism that had a life of its own. This idea, which derived its intellectual inspiration from the work of Spencer, had parallels with the superorganic concept of culture. Thus, Clements thought of a plant community as something more than the sum of the individual species and as responding to laws that operated only at the community level. The Spencerian idea of social progress was reflected in the concept of climax vegetation, which Clements viewed as the logical outcome of a process of evolution. Clements's *Research Methods in Ecology* (1905) was the first textbook to describe this new methodology of ecology as applied to plants.

By the early 1900s, ecology was an established approach in biology, botany, and marine science. The British Ecological Society was founded in 1913, and the Ecological Society of America in 1915. Six principal related concepts were used in early ecological analyses in physical science:

- *Community* refers to a group of related organisms in a particular area.
- *Competition* refers to the struggle between member species of a community.
- *Invasion* refers to the ability of some species to take over areas occupied by another species.
- *Succession* refers to the sequence of change as a community moves toward the finished or climax situation.

- *Dominance* refers to the fact that one or more species in a community is normally able to seek out and occupy the most favorable environment.
- *Segregation* refers to the spatial patterning of species.

Ecologies in Social Science

Although ecology began as a study of nature without humans—or, certainly, without privileging humans—versions of human ecology soon appeared. The first direct reference to humans in an ecological framework was by Darwin in 1871 (*The Descent of Man*), but it was Spencer who was the principal intellectual inspiration for the early advances in human ecology, and it was plant ecology that provided the basic concepts. Recall from Chapter 2 that Spencer, in his articulation of what is usually labeled social Darwinism, argued for an analogy between animal organisms and human groups based on the common struggle to survive in a physical environment. The logic of this social Darwinism—a form of naturalism—is straightforward, with cultural groups evolving in accord with their ability to adjust to particular physical environments. Indeed, it was Spencer who introduced the oft-quoted phrase, "survival of the fittest". These ideas are related to views of biological evolutionism as an inexorable form of social progress and, of course, to the conception of society as a superorganic entity.

"Human ecology means different things to scholars in different disciplines" (Porter 1987:414). Although, as this remark suggests, it is evident that there is no consensus in the social sciences as to what constitutes an ecological approach, it is equally evident that much useful research has been conducted under the umbrella of ecology. Conceptual plurality is accompanied by substantial and quite varied research achievements.

The following sections discuss ecological approaches in sociology, anthropology, history, and psychology, especially as these aid in understanding the subsequent account of human ecological approaches in geography.

Sociology

Sociology was the first of the social science disciplines to develop a coherent human ecological approach. This development took place at the University of Chicago, where several members of the geography department (including both Goode and Barrows, discussed below) were also expressing interest in human ecology. The key sociological figure was Park, one of the leaders of early American sociology. In conjunction with others—notably Burgess, McKenzie, and Wirth—Park successfully articulated and employed a human ecological approach to the analysis of urban areas.

For Park and his colleagues, usually described as the Chicago school of sociology, human ecology centered on the distribution of and interactions between humans and human institutions. Their approach involved mapping these distributions and explaining them by reference to the various concepts developed within plant ecology. →»»1

Park treated the concept of an animal or plant community as virtually synonymous with society. He viewed humans as competing for space, a competition reflected in differential urban land values and therefore varied land uses. Segregation—the presence of urban areas usually distinguished by land use or ethnic identity—was one outcome of competition. Park used the terms "invasion" and "succession" to refer to land-use changes through time, and "dominance" to refer to the ability of particular land uses to locate in the most desirable area. Within the city, this most desirable area was termed the "central zone", and the dominant land uses were those belonging to the financial and retailing sectors. The concepts as they were applied, especially in the Chicago context, allowed sociologists to suggest a concentric zone model of intraurban land uses.

Park's accomplishment was a human ecology explicitly modeled on plant and animal ecology—a human science derived from physical science with humans seen as organic creatures affected by the laws of the organic world. But these conceptual advances were stimulated not only by physical science ecologies but also by personal experiences. "I expect that I have actually covered more ground tramping about in cities in different parts of the world, than any other living man," Park explained (1952:5). "Out of all this I gained, among other things, a conception of the city, the community, and the region, not as a geographical phenomenon merely, but as a kind of social organism."

This is precisely the type of approach that geographers might have been expected to adopt, but their interests typically lay elsewhere—in environmental determinism, in the regional approach, and, of course, in the emerging landscape school. Indeed, sociologists placed great emphasis on the intellectual separation of sociology and geography, stressing especially that geography was an idiographic science whereas sociology was nomothetic. For Park and his colleagues, the role of geography was limited to supplying facts that the human ecologist could explain. Park also regarded geography as a discipline concerned with humans and their physical environments, whereas sociology was concerned with humans and their social environments. ➔➤➤ 2

The early human ecological approach in sociology was successful. The conceptual and empirical initiatives, mostly in midwestern American cities, sparked considerable interest. This was an interest that geographers did not share until the rise of a spatial analytic approach in the 1960s, by which time the approach was largely out of favor in sociology.

A principal criticism of human ecology from within sociology concerned the tendency to see and describe human groups in physical science terms. Hence, a major revision occurred with the introduction of the idea that various symbolic associations and sentimental attachments could be identified with city areas, and that these variables were able to counter the ecological processes (Ericksen 1980). Rather than treating the ecological process as dominant, proponents of human ecology emphasized humans themselves as leaders, not followers, of the ecological process. The conceptual inspiration for these ideas is the approach to sociology known as symbolic interactionism (see Box 3.6).

Anthropology

Although the sociological interest in human ecology had little impact on the work of cultural geographers, the anthropological interest is another matter, a fact that reflects the close ties between these two disciplines since at least the 1920s. Anthropological human ecology relies on the same physical science foundations as does the sociological counterpart, but anthropologists have developed their contributions more fully.

The first explicit outline of an ecological approach in anthropology came in the 1950s, and Steward (1955:30) is generally credited with introducing the term used to describe it: "In order to distinguish the present purpose and method from those implied in the concepts of biological, human, and social ecology, the term cultural ecology is used."

For Steward, the various human ecologies had failed as a result of their proponents' insistence on attempting to define each one as a new subdiscipline. Steward believed that the ecological approach was better seen as precisely that: an approach, a means to an end and not an end in itself. The aim of the approach was to explain particular cultural features, not to derive general laws. Steward included culture as a superorganic factor but drew a distinction between the behaviors of humans and the behaviors of other organisms. This was not, however, the extreme superorganic concept, as he (1955:36) also referred to the "fruitless assumption that culture comes from culture".

Steward rejected the concepts of community, competition, invasion, succession, dominance, and segregation (as introduced by physical scientists and as adopted by Park and others) on the grounds that they were inappropriate for humans. Rather, he saw culture as the key to understanding human relationships with nature. He acknowledged, for example, that competition might be present, but suggested that it could be interpreted only in cultural terms and needed to be approached through an understanding of culture and not as some independent process. Further, he saw human culture as extending beyond the range of local communities.

Steward's rejection of physical science concepts and emphasis on culture explain the choice of name for the approach—cultural ecology. The role of physical environment was limited in Steward's approach, which contained the implicit assumptions that "cultural and natural areas are generally coterminous because the culture represents an adjustment to the particular environment," and that "different patterns may exist in any natural area and that unlike cultures may exist in similar environments" (Steward 1955:35). Steward's approach was one of environmental adaptation that took into account the character of the culture. These ideas are comparable to the possibilism introduced by Vidal.

➜➤➤ 3 Three fundamental procedures are evident in Steward's cultural ecology, namely analysis of technology-and-environment relations, analysis of behavior patterns affecting environment, and analysis of relations between these behavior patterns and other aspects of culture. Numerous other anthropologists continued Steward's pioneering work, often in amended form, but there have also been substantive criticisms and developments of the approach. Five are noted:

- First, the superorganic implications have been criticized, particularly in light of some of the larger changes in social theory.
- Second, there have been arguments for a unified science of ecology instead of a specific cultural ecology, thus reintroducing the idea that human and nonhuman behavior could be studied in similar ways.
- Third, proponents of cultural materialism proposed, as an alternative to cultural evolution, a theory of sociocultural evolution based on the doctrine of natural selection and focusing on the demographic, economic, technological, and environmental causes of cultural evolution (see Box 4.1).
- Fourth, Barth (1956, 1969), refining the cultural ecology approach for analyses of ethnic group distributions, introduced the concept of an ecological **niche**.
- Fifth, some other ecological anthropologists argued for "numerous unifying concepts and principles" (Hardesty 1977:vii), including those of adaptation, human energetics, and the ecological niche.

This fifth development is pursued in greater detail in the account of adaptation later in this chapter.

History

In history the introduction of ecological approaches was related to the belated appreciation that human behavior and environment are closely related—as recently as 1984, Worster (1984:2) found it necessary to propose "the development of an ecological perspective in history", while according to Crosby (1995:1180), historians of earlier generations "could not see what they were not ready to see". What they did not see was the significance of the ecological analyses conducted by geographers such as von Humboldt and Marsh; nor did they see the relevance of population growth and technological change as factors basic to historical analyses. For most historians, the significance of environment was merely as backdrop to—and not as a meaningful player in—human affairs. Even at the global scale, both Toynbee and Spengler discussed civilizations with only minimal reference to environment.

In the 1930s and 1940s, two contributions concerned with settlement of the Great Plains focused explicitly on physical environment and culture (Webb 1931; Malin 1947). Both treated Great Plains settlement in the context of human-and-land relations through time, with natural resources such as soil and water discussed in the context of technology and culture. The fact that these important American studies largely failed to generate wider interest was a reflection of historical scholarship at the time. ➜➤➤ 4

Contemporary ecological, now more usually called environmental, history is inspired by the geographic work of Réclus, Ratzel, Semple, and Huntington, by the anthropological work of Kroeber, Wissler, Steward, Rappaport, and Harris among others, and by the historical work of Webb, Malin, and the German historian Wittfogel. Following Wittfogel, and the argument that the "fundamental relation underlying all social arrangements . . . is the one between humans and nature," Worster (1984:4) provided a powerful argument for including more nature in the study of history. Environmental concerns are now a thriving approach in history with two rather different thrusts, namely concerns with natural history and with the history of ideas about nature.

Box 4.1 Cultural Materialism

We need to understand cultural materialism as it emerged from the work of the anthropologist Marvin Harris but moved into cultural geography through readings of the work of Raymond Williams (1981). For Harris cultural materialism is "based on the simple premise that human social life is a response to the practical problems of earthly existence" and "opposes strategies that deny the legitimacy or the feasibility of scientific accounts of human behavior" (1979:ix). At the outset, Harris's cultural materialism distinguishes between emic and etic accounts, with the former allowing cultural insiders to determine the legitimacy of any analysis and the latter referring to accounts provided by an observer external to the cultural group. While humanists favor an emic perspective, cultural materialists following Harris emphasize an etic perspective: "The starting point of all sociocultural analysis for cultural materialists is simply the existence of an etic human population in etic time and space" (Harris 1979:47). Analyses of etic human populations recognize an infrastructure that includes modes of production (such as technologies of food production) and modes of reproduction (such as technologies of population change), a structure that organizes production and reproduction at both domestic and political social scales, and a superstructure that includes activities such as rituals and science.

- With these points as a basic framework, cultural materialism asserts that "etic behavioral modes of production and reproduction probabilistically determine the etic behavioral domestic and political economy, which in turn probabilistically determine the behavioral and mental emic superstructures" (Harris 1979:55–6). This is the key principle of infrastructural determinism, claiming the priority of etic and behavioral conditions and processes over their emic and mental equivalents, and claiming the priority of infrastructural conditions. This is not a form of monocausal determinism, as infrastructure comprises a host of demographic, technological, economic, and environmental variables.

- Cultural geographers become more engaged with cultural materialism through the work of Raymond Williams. Williams (1981) understood culture as an ordinary part of everyday life that is materialist because it is "a process of active making" and therefore formed "the practical activity of *producing* meanings" (Barnett 2009:136). By extending materialism to include culture as a *productive* process, Williams overcame the constraints of Marxist analysis, which had tended to divide culture and economy along the base/superstructure model; rather, Williams theorized the nature of the links between the previously divided cultural, economic, and political realms, while still conceptualizing cultural practices within wider power structures.

Psychology

In psychology, an ecological approach was initiated by Lewin (1944:22–3) as one part of a concern with the role of non-psychological inputs into human behavior: "Any type of group life occurs in a setting of certain limitations to what is and what is not possible, what might or might not happen. The non-psychological factors of climate, of communication, of the law of the country or the organization are a frequent part of these 'outside limitations'." Lewin (1951) thus referred to behavior as a function of individuals and their psychological environment, which together make up the life space, and of the larger non-psychological environment. Together, these three components constitute what is called a field (see Figure 4.1). ➔➔➔ 5

This approach is not clearly related to comparable developments in other social sciences. While the basic principle—namely, that humans are parts of environments and not objects in and separate from environments—is similar, the psychological contribution is very clearly a part of the parent discipline.

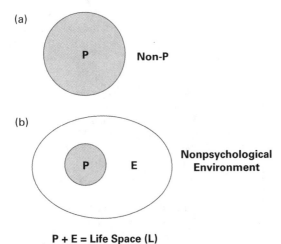

P + E = Life Space (L)

Figure 4.1 Lewin Field Theory. In 4.1a, the individual person, P, is formally separated from the rest of the world, non-P. The reason Lewin represents this distinction in graphic form and not simply in writing is that a spatial representation is susceptible to mathematical manipulation. The shape and size of the figure enclosing P are irrelevant; all that matters is that there is a bounded area completely surrounded by a larger area.

In 4.1b, a second step is taken in the representation of psychological reality. In this case, a second bounded figure is drawn around the bounded figure that encloses P. The area between the boundary of P and the boundary of the second figure is E, the psychological environment. The total area of both P and E is labeled the life space. Finally, the area outside the boundary of the second figure is the nonpsychological environment that includes physical, cultural, and social facts.

Perhaps the principal sharing of ideas concerns the psychological interest in the way that humans adapt to their surroundings, an interest that includes consideration of culture as a "group's adaptation to the recurrent problems it faces in interaction with its environmental setting" (Berry 1984:87).

Human Ecology and Geography

"Geography as Human Ecology"
Although geography might be regarded as the most convivial disciplinary home for human ecology, given the traditional geographic concerns with both physical and human worlds, geographic ecological approaches have a checkered history, possibly because they are not associated with a specific

seminal statement and do not have a coherent body of empirical research. Most of the suggestions that geography adopt a human ecological perspective have come from geographers sympathetic to an integrated physical and human geography. →≫ 6

The first use of the term "human ecology" in the context of geography was in 1907 by Goode, a geographer at the University of Chicago. Following Clements and other plant scientists, Goode envisaged human ecology as an organizing concept for the discipline of geography. Such an idea was popular at the time, with several ecologically inspired papers presented at meetings of geographers, while Huntington became the second president of the Ecological Society of America. The culmination of this interest in a human ecology was the seminal methodological statement by Barrows (1923), "Geography as Human Ecology", initially delivered as a presidential address to the Association of American Geographers in 1922. It is useful to think of this statement as another in the series of attempts to claim a particular identity for the discipline of geography.

The specific aim of Barrows's statement was to define geography as the study of mutual relations between humans and the natural environment and thus as human ecology. More generally, Barrows intended to carve a niche for geography separate from other academic disciplines: "Geography finds in human ecology, then, a field cultivated but little by any or all of the other natural and social sciences. Thus limited in scope it has a unity otherwise lacking, and a point of view unique among the sciences which deal with humanity" (Barrows 1923:7). He emphasized human adjustment to, as opposed to influence of, physical environment, and distinguished geography from geology and other physical sciences on the grounds that it was not concerned with the origins and distribution of physical features—landforms, climates, vegetation types, soil types, and so on—but only with human adjustment to those features. Barrows also made an explicit distinction between geography and history, suggesting that the historian began "with what our remote ancestors saw" while the geographer began "with what we ourselves see" (Barrows 1923:6). In sum, Barrows argued that geography as a discipline needed to concentrate on human ecology interpreted as human adjustment to physical environment.

Overall, geographic contemporaries of Barrows had good reason to be somewhat perplexed at the thrust of this statement. Given the established concerns of geography at the time, some objected to the explicit exclusion of the traditional geographic content of physical geography, while others were concerned at the lack of reference to the traditional regional focus, and still others were dismayed at the rejection of environmental determinism. Overall the statement was not well received, and it is not surprising that it created little positive impact on the discipline of geography; in retrospect, it can be seen as an idiosyncratic attempt to forge an identity for the discipline. In terms of its immediate impact, Barrows's statement inspired an ecological account of the geography of Chicago (Goode 1926), and two university-level textbooks, *Geography: An Introduction to Human Ecology* (White and Renner 1936) and *Human Geography: An Ecological Study of Society* by the same authors (White and Renner 1948). Both textbooks saw geography as the study of human society in relation to the earth. Overall, however, most subsequent statements about geography as human ecology were not sympathetic to the Barrows interpretation.

Human Ecology and the Landscape School

It is appropriate to think of the Sauerian tradition as containing three cultural geographic emphases—human-and-land relationships or ecology, landscape evolution, and regional landscapes. The ecological interest is evident in the concern with human-and-land relationships (see Figure 2.2) and was regularly stressed. For example, Sauer (1924:33) noted: "Land is passing out of economic use in this country more rapidly than new land is being occupied. Timber devastation, soil erosion, overgrazing, soil depletion, failure of irrigation, dry farming, and drainage, here and there causing a pitiless revaluation of areas. Land is increasingly becoming a restricted economic good." Again, Sauer (1927:192) claimed: "To what extent is man as a terrestrial agent, that is by his areal expressions of culture, living harmoniously in nature (symbiotically), and to what extent

is he setting narrowing limits for future generations by living beyond the means of the sites that he occupies? Man . . . appears, periodically, to effect his own ruin. . . . He has certainly been an engine of unparalleled destruction." From the outset Sauer expressed concerns about human impacts, and several of his students completed major ecological analyses, including some studies of vegetation and the exemplary cultural geographic studies by Wagner (1958a) and Mikesell (1961). ➔➤➤ **7**

The ecological content of the landscape school is not as readily acknowledged as are other aspects of the tradition, at least partly because Sauer rarely used the terms "ecology" or "human ecology". This is perhaps because Sauer took exception to the statement by Barrows on the grounds that it involved an inappropriate limitation on the discipline of geography, rendering it simply the study of humans in relation to natural environment. Like the majority of

---

Questions for Critical Thought

❖

Why was the landscape school statement by Sauer so influential while the human ecology statement by Barrows had minimal impact? Here are three suggestions. First, Sauer not only outlined his position but he also restated and clarified that position several times, conducted related analyses, and attracted students to work with him. Barrows on the other hand progressed little beyond the initial statement. Second, Sauer built on a rich European tradition of geographic thought and explicitly incorporated physical geography, unlike Barrows, who was seen as proposing something that lacked relevant intellectual foundations in the discipline. Third, Sauer's position was both clear and forceful, while Barrows's ideas were interpreted in two contradictory ways, with some seeing his position as a diluted form of environmental determinism and others seeing it as excluding physical geography. Are you inclined to agree with these three suggestions (and they are, of course, only suggestions)? Are you able to think of other possible reasons that may contribute to making one type of approach in geography more influential than another?

geographers at the time, Sauer was less than enthused with the human ecology proposed by Barrows.

Other Contributions

Several other geographers issued appeals for an explicit concern with human ecology; four are noted here. ➔➤➤ 8 First, Thornthwaite (1940:343), a student of Sauer, claimed that "human ecology must transcend all of the present academic disciplines, and . . . the development of a science of human ecology must involve cooperation of geography, sociology, demography, anthropology, social psychology, economics, and many of the natural sciences as well." Such a unified view of human ecology, a co-operative enterprise, bears little resemblance to either the earlier statement by Barrows or the Sauerian cultural tradition. Second, Morgan and Moss (1965) argued for an ecological approach that focused on the community concept, drawing parallels between biological and human communities. This account acknowledged that any analogy between animal organisms and human groups, in the Spencerian or social Darwinian tradition, was subject to criticism, but argued the case for viewing human communities in two ways: "1) Ecologically, as aggregates of different species focusing on man, including all the animals and plants which depend on man and all those on which man in turn depends. 2) Socially, as aggregates of one species, consisting of individuals with a variety of interests and functions which, joined together, make possible one social and economic unit" (Morgan and Moss 1965:348). In both cases, the key feature is the uniting of individual community members. Third, in a different vein, Chorley (1973) aimed to show how human ecology provided a unifying link between physical and human geography. Fourth, Norwine and Anderson (1980:vii) built on the idea of the environment as a factor "impacting man's destiny". None of these four statements has had meaningful impact on the discipline of geography.

Summary

Ecology is an approach applied in several physical science disciplines, and human ecology is an approach applied in several social science disciplines. Young

(1983:2) described human ecology as "a strange nonfield, a hybrid attempt to understand the ecology of one species, the being *Homo sapiens*". At first thought, it seems reasonable to suppose that if ecological analyses including humans are to have one logical disciplinary home, then that home should be in geography. But this has not proven to be the case. Put simply, geography, despite extending several invitations, has not proven to be an especially congenial abode for human ecology. Notably, the views of such mid-nineteenth-century geographers as von Humboldt and Ritter centered on human-and-land relations, and these relations were also a concern in both Vidalian and Sauerian traditions, but ecology *per se* was not stressed. It is interesting to ponder how different the history of geography might have been if one of the leading late nineteenth-century geographers had explicitly highlighted an ecological focus rather than the study of regions or environmental determinism.

From Barrows onwards, human ecology in geography has been characterized more by proposals of method than by demonstrations of method. Two observations highlight this circumstance. First, reviewing human ecology in geography between 1954 and 1978, Porter (1978:15) characterized the advances as "a decade of progress in a quarter century". Second, Zimmerer (1996a:161) referred to ecological concepts as being both "persistently foundational and yet doggedly problematic in

---

Questions for Critical Thought

——— ✦ ———

Most cultural geographic work with ecological content has been inspired by the landscape school. For Mathewson (1999:268), ecology's place in many of the cultural landscape studies conducted by landscape school geographers is "implicit and pervasive", and indeed there does not seem "to be any end in sight for this tradition, despite the emergence of new varieties of cultural geography whose practitioners jejunely and routinely issue uninformed dismissals of what has preceded them". Can you think of some examples either illustrating or contradicting Mathewson's point?

their application" in human geography. Overall, twentieth-century geographers identified both a physical science–human science division in their discipline and a division between ecological approaches and other dominant approaches, first regional and later spatial analytic; the failure of geography to center on an ecological approach is at least partly a consequence of these internal uncertainties about geography's disciplinary identity. ➔≫≫ **9**

Ecology and Systems Analysis

As evident from the preceding overview, ecology is a unifying approach that has not been widely adopted. For most interested scholars, especially those approaching problems from a specific disciplinary perspective, the challenges of adopting an ecological approach are simply too great, given that a unified science needs to integrate both physical and human phenomena and be genuinely interdisciplinary. It has not even proven possible to integrate the various human ecologies and, indeed, the evidence points to increasing diversity of approaches rather than to integration. But there is one principal exception to these statements: the application of systems analysis in an ecological context. Many ecological analyses in various disciplines turned to systems concepts to support their work, such that these concepts represent the closest that researchers have come to a genuine integration of ecological approaches.

Rather surprisingly, the term **ecosystem** was not introduced until 1935, when it was coined by the English botanist Tansley. Sometimes called an ecological system, an ecosystem is any self-sustaining collection—an interacting system—of living organisms and their environment. Although this is a simple definition, ecosystems are both complex and dynamic.

➔≫≫ **10** An ecosystem has four principal properties. First, it is monistic, meaning that it integrates living things and their environment, allowing for recognition of the interactions between elements. In a general sense, this property is in accord with the ambitious aims of geographers such as von Humboldt and Ritter, who advocated a holistic view of humans and nature. Second, an ecosystem has

a structure that can, in principle, be investigated. Third, it functions, in that there are interactions occurring within the system. Fourth, an ecosystem is similar in principle to any other system, such as a hot water system or the cardiovascular system.

In addition, three features of ecosystems make them an attractive basis for study. First, it is generally recognized that ecosystems exist at a multitude of scales, from the world to a single pond. Second, the ecosystem is seen as emphasizing the complexity of interrelationships between all member organisms. Third, it is commonly acknowledged that an ecosystem is in a dynamic balance, with the environment constantly adapting in response to changing conditions.

These four properties and three features of an ecosystem, along with the difficulties of conducting ecological analyses using traditional methods, prompted use of the procedures of systems analysis as a means of identifying key variables and regulating factors. Systems analysis is a scientific approach to complex phenomena that proceeds through five stages (see Figure 4.2):

- In the first stage, known as *systems measurement*, objectives are outlined and data collected.
- The second stage, *data analysis*, involves calculating relationships between variables in order to determine the key variables given the objectives of the study.
- *Systems modeling* is the third stage, during which models are built to serve as a theoretical interpretation of the system.
- The fourth stage, *systems simulation*, involves manipulating the theoretical models in order to assess consequences of specific changes.
- Finally, during the stage of *systems optimization*, the aim is to develop strategies for achieving desired objectives.

The use of modeling and simulation is often essential given the size and complexity of many of the systems analyzed. As an example, consider attempts to study the effects of an increase in humans' carbon dioxide emissions on our global ecosystem. Similarly, one of the reasons many

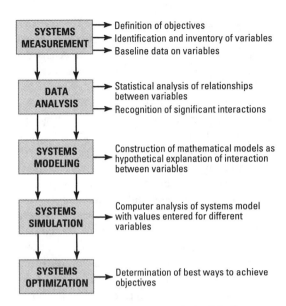

Figure 4.2 Stages in a Systems Analysis. This diagram outlines five stages through which a typical systems analysis of an ecosystem might proceed. Details of each stage necessarily vary according to the specific aims of the analysis; for example, the fifth stage is appropriate only if the aim is to solve a particular problem. The procedure can be modified if the objective is to identify the consequences of a particular ecosystem change, such as the location of an industrial plant. Source: *Global Ecology in Human Perspective* by Southwick (1996). Fig. Stages in a System Analysis, p. 45. By permission of Oxford University Press, USA.

of this kind have been conducted from within the physical sciences, with humans seen as but one component of the system. A typical systems analysis with a social science emphasis employs the ecosystem concept in broad terms, rarely following the stages outlined in Figure 4.2. However, there are a number of favorable overviews and applications of the approach by both geographers and anthropologists. Certainly, the attraction of a systems framework for the cultural geographer seems clear, with the explicit recognition of human-and-land relations and of humans as a part of nature. Nevertheless, following an outline of the approach, Foote and Greer-Wootten (1968:89) noted that they had "presented 'old wine in new bottles'", with the vintage being Sauer in the 1920s.

Further attempts to define the ecosystem as a geographical model have not won many converts. Stoddart (1967:538) proposed the idea of a geosystem as a replacement for an ecosystem, arguing that systems analysis "at last provides geography with a unifying methodology, and using it, geography no longer stands apart from the mainstream of scientific progress"—an appeal that has gone largely unanswered. Such a failure may reflect the fact that geographers do not seek a unifying methodology, especially given the diverse subject matter of the discipline.

attempts to predict future population growth have been so spectacularly unsuccessful is that the complexity of factors involved has rarely been adequately acknowledged. In many cases, futures are predicted on the premise that relevant current trends will continue without significant change, which is rarely a reasonable assumption. In brief, systems analysis, which does at least allow for the complexity of a system to be recognized, is often the only feasible way to gain some understanding of ecosystems and their changing character.

In principle, analysis of an ecosystem that includes humans focuses on human modification of the environment, whether accidental or deliberate, and on the consequences of modification. Typically, human involvement in an ecosystem leads to species destruction and loss of equilibrium. Most analyses

Questions for Critical Thought

Although there is, at least on the surface, a compelling logic to the systems approach, it has not been widely employed by cultural geographers. Why not? Here are two possible answers. First, analyses of complex systems using sophisticated analytical skills typically require input from teams of scientists, and team projects are not part of the tradition of cultural geography. Second, many cultural geographers are somewhat uncomfortable with seeing humans as merely one component of a system, preferring instead to see humans as a distinctive and perhaps separate component. Can you think of one example where cultural geographers would play a central role in an interdisciplinary team?

Another area recently emerging that is relevant to systems ecology is complex systems. This is inherently an interdisciplinary approach to contemporary problems since multiple factors relevant to any system have to be modeled and understood, and therefore social science, natural science, and mathematics subjects are all called upon in order to understand the nature and operation of complex systems. In 1997 the American National Research Council publication *Rediscovering Geography* suggested "research on complex systems also will place geography centrally within an emerging scientific synthesis" (141). However, this approach has become more aligned with computer modeling of mostly natural, rather than cultural, processes.

Adaptation

A useful and yet much debated concept used in cultural ecological analyses, including those with a systems analysis focus, is adaptation. Defined loosely as changes in behavior that are designed to improve quality of life, or rather more formally as the strategies adopted to achieve ecological success, the meaning of the concept is clouded, especially by the recurring issue of environmental as opposed to cultural control and by the choice of social scale of analysis. Not surprisingly, adaptation is a concept introduced, interpreted, and applied in a number of disciplines, including psychology, anthropology, and cultural geography. ➔≫≫ **11**

In Psychology
In psychology, there is concern with the way in which individuals adapt, and an ecocultural psychology centers on adaptation to the recurrent challenges posed by environments, with the argument being that individual behavior must be considered in some complete ecological and cultural context. From this perspective, adaptation is interpreted in terms of the reduction of dissonance within a system. Ecological balance or harmony is promoted through three types of adaptation.

The first and most usual form of adaptation is adjustment, which involves making changes in behavior to reduce conflict with environment. The second type, reaction, involves a form of retaliation against the environment to force the environment to adjust. The third type of adaptation, withdrawal, involves conflict reduction through human removal from the source of conflict. Recognition of these three versions of adaptation is comparable to earlier work in psychology that recognized movement toward, against, or away from a stimulus. This perspective sees all adaptations as individual solutions to particular environmental circumstances.

This psychological concern with individual behavior highlights questions about the social scale at which adaptation occurs, with anthropologists and cultural geographers usually favoring a group scale, although acknowledging the need for a variety of scales. Both individual and group behaviors can be seen as adaptations, although the details of the processes are different according to social scale. What may be a sound adaptive response for an individual may not be so for the group, and further, sound adaptive responses for both individual and group may not be so for the environment. Group adaptation may be interpreted as the state of management of physical resources.

In Anthropology
In anthropology, adaptation is closely linked to cultural evolution, with three levels distinguished: behavioral, physiological, and genetic. Anthropologists are primarily concerned with the behavioral level. According to Hardesty (1977:23), behavior "is the most rapid response that an organism can make and, if based upon learning rather than genetic inheritance, is also the most flexible". From this perspective, two types of adaptive behavior may be distinguished, namely idiosyncratic (that is, unique individual responses as studied by psychologists) and cultural (that is, shared responses as studied by anthropologists and cultural geographers). Group cultural adaptation occurs because of changes in technology, organization, and ideology, all of which are aspects of culture. How do these changes promote adaptation? The usual response is that they provide solutions to the problems posed by environment. But changes in these cultural variables may also help by improving the effectiveness

of solutions, providing adaptability, and providing awareness of environmental problems.

In Cultural Geography

In cultural geography, some cultural ecological work has modified the adaptation concept as it has been traditionally employed. Denevan (1983) and Butzer (1989b) argued that a version of cultural adaptation could be a key explanatory procedure for the cultural geographer, with human survival dependent on an available supply of adaptive responses, the appropriate scale of analysis being the ecological population, and the research aim being to account for the sources of variation and the processes of selection. There has also been an increasing recognition of the importance of ethnicity and of social and political power. Gade (1999) provided a series of analyses in the Andean context, including an intriguing review of rats and humans in Guayaquil, while Knapp (1991) favored a concern with adaptation as human behavior at the individual scale in a study of pre-European wetland agriculture in Ecuador, with emphasis on coping behaviors such as decision making and problem solving.

Box 4.2 offers a particular geographic interpretation based on the idea that all human behavior is social and therefore learned, such that adaptation is appropriately viewed as one part of the matrix of a society. According to this argument, "explanations about variations in adaptive processes and about adaptive innovations . . . are to be looked for in variations of other elements of sociological systems, and not in changes of either ecological or biological systems" (Mogey 1971:81). Regardless of specific interpretation, the central question addressed using this concept concerns why humans behave as they do in the environments they occupy. More generally, Butzer described the concerns of cultural ecology as follows:

> It focuses upon how people live, doing what, how well, for how long, and with what environmental and social constraints. It emphasizes that human behavior has a cognitive dimension and is dependent upon information flow, values, and goals.

Finally, cultural ecologists recognize that actions are conceived and taken by individuals, but that such actions must be examined and approved by the community, in the light of tradition and the prevailing patterns of institutions and power, before decisions can be implemented. (Butzer 1989b:192–3)

Note that, in this tradition, culture is not superorganic; rather, there is a concern with cognition, and individuals are viewed as members of groups. Conceptually, there are closer links with symbolic interactionism (Box 3.6) than with cultural materialism (Box 4.1).

Rethinking Ecological Approaches

This section begins by acknowledging that nature may not be quite so natural after all, but rather may be more correctly interpreted as a social construction, meaning that when we talk about nature we are saying as much about the role human systems play in the production of nature as we are about the thing we are describing. Neil Smith (1984) has been most influential in this critical rethinking of nature in geography. His approach views nature as an artifact of capitalism, in that nature is widely transformed into a commodity and this plays a central role in uneven development. Smith also argues that the idea of nature is processed through a capitalist lens in terms of its value to society, such that "nature" is increasingly translated into some kind of monetary value. ➔➤➤ **12** This critical approach aims at revealing how nature is subject to the dominant societal structures in place at different times. Smith is influenced by a Marxist approach, but other approaches such as feminism and poststructuralism have contributed to the debate over the construction of "nature".

Nature has been widely symbolized in different times and places as female; for example, nature as a nurturing mother providing humans with their needs, and nature as wildness that needs to be tamed or controlled. The latter image became

Box 4.2 *Adaptation as Social Adjustment*

It may be surprising to discover that both cultural geographers and anthropologists continue to debate the relative importance of physical environment and humans to an understanding of the human-and-land relationship in the specific context of adaptation. While the extreme of environmental determinism is no longer seriously argued, the complexity of the relationship and variations from place to place encourage analyses that consider a variety of particular situations. As previously noted, the historians Malin and Webb emphasized human adaptation to fixed and stable environments, rarely recognizing that the human mind was a crucial variable. Speth (1967), meanwhile, in an analysis of the intermontane Mormon landscape, argued that an adaptation focus lay midway between the two extremes of environmental and cultural determinism. Similarly, Porter (1965:419–20) concluded a study of subsistence activities in Kenya by noting: "A model which seeks to describe culture as adaptive, through subsistence, to environmental potential cannot ignore man himself as a causative agent of environmental change." Certainly, discussion of adaptation has renewed debate about determinism, with Mogey (1971), in a penetrating but infrequently referenced argument, proposing that all human behavior is social.

In order to maintain themselves, all societies have social mechanisms that enable them to adapt to environment, mechanisms that include both techniques and objectives and that are learned through time. There are different sets of social mechanisms in specific societies, and these are subject to ongoing change. With these claims in mind, Mogey (1971:80) explicitly argued against any form of environmental determinism, noting that even major natural disasters do not lead to social change with many societies adapted to specific disasters. From this perspective, all adaptation is social. Community was favored as the appropriate social unit for analysis because it is at this social scale that adaptation occurs, with different communities employing different adaptive strategies. The specific differences among communities are based on two factors, technology and values. Following this argument, Mogey advanced six propositions:

- Community is the unit of social structure through which environments affect human behavior. It is this process that is labeled adaptation.
- Adaptive innovations require modification of the value system and of the techniques of the community if they are to be acceptable.
- In equalitarian communities, rates for the appearance of adaptive innovations are low.
- The mere presence of a status system that supports a set of leaders does not itself guarantee the ready acceptance of innovations.
- Adaptive innovations are frequent and acceptable in those communities with bureaucratic organizations, because they have values based on achievement and a status hierarchy that demands performance. However,
- Bureaucratic organizational systems have their own internal rigidities, which may impede their performance in the process of adaptation. Hence, more work on the interrelations of value systems and techniques in political, religious, and family systems is needed.

These are propositions, not conclusions, and do not imply a version of sociological determinism. As noted, they have not been widely tested or adopted by geographers in either cultural or social traditions, but there is an emerging literature on socio-cultural adaptation to climate change (for example, Adger 2003). Climate change experts indicate that adaptation does not occur simply because it is needed. Certainly, many of the institutions of human society are constructed precisely because they facilitate cultural continuity—religion is a prime example—and such institutions are a "prison de longue durée" (Braudel, quoted in Tarrow 1992:179).

dominant during the scientific revolution: "As Western culture became increasingly mechanised in the 1600s, the female earth and virgin earth spirit were subdued by the machine" (Merchant 1980:2). Such a mechanistic image encourages human use of the earth—the manipulation of nature—and can be contrasted with an alternative social construction of nature, an ecological image that sees nature as a whole and humans as a part of that whole. These two worldviews are founded on very different ontological and epistemological premises; the mechanistic view is seen as both masculinist and exploitative of nature as female, imposing some kind of order and containment upon "wild" nature, while the ecological worldview acknowledges the need to break from these social constructions that involve human domination of nature and male domination of female.

Arguments such as these have contributed to a substantial and growing set of ideas on what might be called the invention and reinvention of nature. These ideas center on the way nature is constructed in terms of prevailing ideas about race (or, more correctly stated, the myth of race), sexuality, gender, **nation**, family, and class, and the dominations and inequalities that are contained within these (McClintock 1995). Such arguments require a re-evaluation of much earlier thought, especially concerning the meanings and material consequences of representing nature as female. This deconstruction of nature as a gendered discourse is part of a larger feminist interest in the character and power of the knowledge that passes for science, stressing that such knowledge—including the idea that there is a difference between man (not humans) and nature—was constructed by men at a time when the domination of nature by men was considered desirable.

The key idea behind many of these approaches is that nature is a social construction subject to the power dynamics at play at different historical periods and can play out differently in spatial terms. This is an important argument. According to earlier naturalist scientific views, it is possible to achieve an understanding of nature that is independent of the individual observer; however, more recent views stress the impossibility of any such objective

interpretation, arguing that individuals are capable of constructing many different worlds, with the supposedly objective nature understood by science being but one of these. From a humanist perspective, it is necessary to interpret individual lifeworlds, the worlds of our direct experiences, rather than to observe nature objectively. ➔≫≫ 13

In addition to the intellectual challenges these approaches offer to an epistemology of nature as previously constructed, many have stressed the need for an approach that is sympathetic to concerns expressed by environmentalists and others about negative human impacts on the environment. In a discussion of the North American landscape, Wilson (1992:13) noted that "the whole idea of nature as something separate from human experience is a lie. Humans and nature construct one another. Ignoring that fact obscures the one way out of the current environmental crisis—a living within and alongside of nature without dominating it." From Wilson's perspective, what humans need is a new culture of nature based on a revised set of power relations where humans do not simply rule and dominate.

If nature is best seen as a part of culture, then we might also argue that humans are a part of nature, an idea put forward in several recent arguments. First, there is the concept, building on earlier ecological ideas, of a global ecosystem. This is a holistic idea in that it refers to all things, organic and inorganic, and to the web of relationships between all parts. For many geographers and others, this concept is proving to be an effective stimulus for what is at present the relatively new science of global ecology. Second, the controversial Gaia hypothesis, introduced by Lovelock in 1979, asserts that the environment and all life combine to form a single unified system, and have done so for about the past 3.8 billion years. Further, it claims that life itself has a profound impact on the environment, especially on the composition and temperature of the atmosphere, so that life maintains an environment that is favorable to life. This hypothesis opposes Darwinism in that it can be interpreted as having a teleological component, implying some action that is purposefully directed toward a specific goal.

In addition to these two holistic arguments, there are also views that strive to reconcile the need for integrating humans and nature and the need to acknowledge the distinctiveness of humans. According to this logic, any analysis of humans and of human landscapes must be **ecocentric**, focusing on the larger environmental context, including humans. But at the same time, analyses need to be anthropocentric, centering on humans because humans are different from other parts of the ecosystem, including other animals, in that they have culture. It is in this sense that humans might be seen as both a part of and apart from the rest of nature. In this vein geographers have more recently been influenced by work that attempts to overcome the nature/culture binary (where the two realms are seen as separate and distinct spheres with culture, or humans, dominating nature), but without diminishing nature and the non-human world to the point that it becomes a totally inactive object that is shaped solely by human actions. Such efforts can be classified under the broad rubric of **hybridity**, as outlined in Box 4.3. ➔➤➤ **14**

New Ecology

A basic idea behind much traditional ecological thought, including the systems approach, is that nature strives to be in balance, and undergoes a process of evolution as it moves to a state of balance. The rise of the new ecology can be understood in the larger context of reactions against the intellectual traditions within which this initial ecological approach arose, especially the scientific tradition of mechanistic analysis and the Enlightenment tradition of continuing progress. Proponents of the new ecology maintain that the volatility of the natural world challenges the conventional ecological wisdom that depicts nature steadily progressing toward equilibrium. As a result, a central feature of new ecology is a non-equilibrium view of nature. Originating in biology, this new ecology is especially critical of systems analysis, which is seen as implying that nature can be understood in mechanical terms. ➔➤➤ **15**

Zimmerer (1994:109), explaining the new ecology, noted that "this new perspective calls attention to the instability, disequilibria, and chaotic fluctuations that characterize many environmental

systems as it challenges the primordial assumption of systems ecology, namely that nature tends towards equilibrium and homeostasis." Meanwhile Blumler (1996:32), referring with modest humor to the concepts of vegetational succession and to an eventual climax vegetation as expounded by Clements, remarked: "I have my doubts, given the highly dualistic nature of feminist writings, but it does seem possible that Clements' succession model represents a case of unconscious male psychosexual wish fulfillment: vegetation proceeds from small and flaccid to hard woody and upthrusting, becoming ever bigger and better until the ideal climax is reached."

Although discussions of new ecology focus on issues with implications for ecological analyses, such as the origins of agriculture, there is, as Cronon points out, a larger issue involved in the idea that nature is inherently unstable, namely our human image of nature:

> Recent scholarship has demonstrated that the natural world is far more dynamic, far more changeable, and far more entangled with human history than popular beliefs about the "balance of nature" have typically acknowledged. Many popular ideas about the environment are premised on the conviction that nature is a stable, holistic, homeostatic community capable of preserving its natural balance more or less indefinitely if only humans can avoid "disturbing" it. This is in fact a deeply problematic assumption. (Cronon 1995:24)

It is especially problematic because abandoning our belief in an essentially unchanging nature is psychologically disturbing. How are we to judge our actions without such a benchmark? We have a double difficulty here. First, nature is socially constructed. Thus, our descriptions of it reflect not just nature itself but also ourselves, our values, and our assumptions, implying that we cannot in any objective sense "know" the natural world. Second, nature changes regardless of our human activities.

What does all this mean for us when we ask ethical questions of ourselves about how we should treat nature? There are not, of course, any clear responses

Box 4.3 Hybridity

Hybridity generally refers to the mixing of two things; in the case of the nature/culture binary, the idea of hybridity infers that our human experience of nature is, and always has been, mediated through cultural lenses; this was a central feature of the classic account by Glacken (1967). For example, the image of nature implied by environmental determinism was mediated through religious (specifically Christian) and philosophical (specifically empiricist and naturalist) beliefs and values. In a production of nature account, capitalism structures the view of nature, and in some feminist accounts, patriarchy, heteronormativity, and racism determine the meaning of nature. Only recently have some scholars, in response to ideas of the construction of nature, turned to the concept of hybridity in order to correct the asymmetry that has traditionally accompanied the nature/culture binary (Whatmore 1999; Castree 2005). Part of this approach seeks an explicit interest in overcoming the division between nature/culture, and this division or binary has been influenced by various lines of academic inquiry. For geographers two of the more prominent intellectual influences with regard to the idea of hybridity can be seen in the work of the American feminist scholar Donna Haraway (1991) and the French science studies scholar Bruno Latour (1993).

Haraway is an immensely influential feminist scholar trained in science and philosophy. In 1985 she published an essay known as the *Cyborg Manifesto*. By using the term "cyborg" Haraway invokes the hybrid of machine and organism both natural and artificial, thereby challenging essentialist ideas of identity that inform such things as "the women's movement", and instead promotes awareness of the constant process of circuitry and feedback that exists between seemingly distinct realms (for example, humans–technology, nature–culture, animal–human). Haraway, by arguing that "we are all cyborgs," advocates the blurring of essentialist identities and challenges hierarchies that produce binaries, thereby encouraging the formation of new political alliances and affiliations. Haraway's book *Primate Visions* (1989) further illustrates the fragility of binary categories, critiquing the idea of scientific neutrality by demonstrating, in great detail, how the myth of race and gender inform scientific findings in primate research.

The challenge to scientific neutrality and its replacement with ideas of hybridity has also been productively articulated by Bruno Latour (1993). Latour argues that the separation of nature/culture is the product of the rise of modernity (the Modern constitution), and yet when we examine

Questions for Critical Thought

———— ⚛ ————

Although the implications of new ecology are as yet barely touched upon, do you think that the equal emphasis being placed on both nature and culture means that the basic argument may prove attractive to some cultural geographers? This is an interesting question to consider because, with much physical science emphasizing nature and much social science emphasizing culture, each area of interest knows relatively little about the other. Cultural geographers, in classic geographic fashion, have always

incorporated both concerns. Blumler (1996:26) wrote: "I am more inclined to the Sauerian viewpoint which although strongly opposed to environmental determinism, nevertheless recognizes that nature and culture were equal players, interacting with each other in non-linear fashion. . . . In this sense, the Sauerian approach is consonant with the new ecological paradigm." Are you inclined to agree with Blumler's assessment, or do you feel that he is claiming too much for the Sauerian approach?

how people around the world actually live their lives, these neat divisions do not exist. Instead what exists everywhere is exactly the hybrids that Haraway recounts:

> Now hybrids, monsters—what Donna Haraway calls "cyborgs" and "tricksters" (Haraway, 1991) whose explanation it abandons—are just about everything; they compose not only our own collectives but all the others, illegitimately called premodern. At the very moment when the twin Enlightenments of Marxism seemed to have explained everything, at the very moment when the failure of their total explanation leads the postmoderns to founder in the despair of self-criticism, we discover that the explanations have not yet begun, and that this has always been the case; that we have never been modern, or critical. (Latour 1993:47)

In place of the obsession with purist binaries, Latour encourages studies that engage with empirical research but do not exclude the "residuals" that exist in the middle ground between the poles at the purists' ends of explanation (science or culture but not the two together). For example, instead of understanding the work of science through only the findings of disciplined research, we should instead assess how the daily interactions of scientists in the laboratory shape their work and contribute to the construction of scientific facts. This approach was used by Latour and Woolgar (1979) in their landmark study *Laboratory Life*, and led to the development of an approach known as **actor–network theory** (ANT). ANT refers to the complexity of networks that form in any realm of human activity and consist of heterogeneous elements and associations, including non-human agents granted agency to shape the nature and outcome of networks (for example, animals, viruses, computers). Again this approach poses a challenge to the binaries embedded in the modernist discourse.

There are many emerging currents of interest that align with this hybrid nature/culture approach, and, according to Whatmore (2009:494), three of the most important are the growing interest in animal geographies that aligns with the non-human or more-than-human presence in networks; a critical assessment of accounts of "wilderness" and pristine environments that can work to undermine the rights of indigenous communities; and the growing incorporation of the technological and natural in such areas as prosthesis, cloning, and transgenic developments.

to such a question, but the double difficulty noted does provide a useful backdrop for the remainder of this chapter.

Environmentalism

A classic argument proposes that current environmental problems are rooted in Jewish and Christian traditions because God was seen as giving the earth to humans for them to use as they chose, even if choosing meant destroying. Some commentators reject this argument, claiming there is a substantive difference between the Biblical concept of dominion and the concept of environmental domination.

The subsequent debate around this issue includes discussions of religions other than Judaism and Christianity to suggest that, even where religions traditionally see humans as a part of nature, there is evidence of ruthless exploitation of nature.

Major challenges to the Western tradition that saw humans as different from and superior to nature generally and to other animals in particular came from several and different directions. Malthus envisioned disastrous social consequences because of excessive and apparently uncontrollable population growth, relative to increases in the **resource** base, and John Stuart Mill (1806–73), echoing his concerns, warned that increases in population and in resource consumption could not simply continue. Henry David Thoreau (1817–62) viewed all creatures as a part of the community of nature, while John Muir (1838–1914), the founder of the Sierra Club, urged preservation of selected natural environments. These four thinkers provided pioneering accounts that stressed the fragility of our human use of earth.

Although it was Darwin who provided scientific support for the idea that humans were a part of nature, geographers have long been concerned with deleterious human impacts on their environment, and the contributions made by such major scholars as von Humboldt, Ritter, and Marsh are especially noteworthy. The Sauerian interest in such matters was a principal spur to a 1955 conference that laid the groundwork for the landmark volume *Man's Role in Changing the Face of the Earth* (Thomas, Sauer, Bates, and Mumford 1956). Aldo Leopold (1887–1948) urged society to adopt a new land ethic aimed at protecting the environment by rejecting the idea of land as mere property, and it is frequently claimed that this argument represents the true philosophical beginnings of the contemporary environmental movement. Certainly Leopold (1949) outlined a non-anthropocentric, holistic environmental ethic. But widespread popular interest in the environment dates from the 1960s, and especially from the 1962 publication of *Silent Spring* (Carson 1962)—a powerfully critical account of the effects of pesticides on human life. It was this work that ushered in the age of environmentalism, loosely defined as a concern with changing human attitudes and behavior regarding environment.

→≫≫ **16** The interweaving of the physical and human—cultural, political, and economic—aspects of this topic is undeniable. "It is now taken for granted that the global environmental crisis and a renewed concern with global demography (the return of the Malthusian specter) are inseparable from the terrifying map of global economic inequality" (Peet and Watts 1996:1). Certainly, current concern about environment cannot be detached from concern about people, as the United Nations–sponsored conferences on Environment and Development in Rio de Janeiro in 1992, on Population and Development in Cairo in 1994, and on Women in Beijing in 1995 made

Questions for Critical Thought

It is undeniable that the typical Western attitude toward human impacts on environment is changing today. Indeed, there is a useful distinction to be made between an earlier anthropocentric view and a more current, but by no means dominant, ecocentric view. This transition can be seen as from a dominant emphasis on material well-being toward greater emphasis on overall quality of life. But is a continued transition both logical and inevitable? Pinkerton (1997) described the ecocentric view as one of enviromanticism, politically useful but practically questionable. What do you think of this claim? Is it too simplistic, or does it capture an important idea? Certainly, contemporary environmental movements are complex social and political forces, ranging from those favoring preservation of landscapes to those with more radical ideas, such as dismantling capitalist society and economy. Although it is not difficult to see how environmental movements can translate into formal political movements, as has happened in many more developed countries, is it more difficult to see how such movements might generate real changes in both our understanding of, and our behavior toward, environmental change?

abundantly clear. Today, some commentators portray concern for the environment as being in opposition to progress, science, technology, and population growth, an attitude that might be explained by reference to the close association between the expansion of European empires and related environmental impacts.

Ecofeminism

Ecofeminism refers to links between the feminist approaches and environmental and **deep ecology** approaches. An explicit statement of interest is as follows:

> According to ecological feminists ("ecofeminists") important connections exist between the treatment of women, people of color, and the underclass on one hand and how one treats

the nonhuman natural environment on the other. Ecological feminists claim that any feminism, environmentalism, or environmental ethic that fails to take these connections seriously is grossly inadequate. (Warren 1997:3)

These ideas first appeared in the mid-1970s through an integration of some feminist ideas and the green movement—from feminism comes the idea that humanity is gendered in such a way that women are subordinate and exploited, and from the green movement comes the idea that humans are damaging the natural world.

→ ≫ **17** There are two fundamental versions of ecofeminism. Cultural ecofeminism focuses on the links traditionally made between women and nature, and the related circumstance that men oppress both. Cultural ecofeminism is thus a response to the feminist claim that Western culture has identified women and nature as belonging together and has devalued both. Human nature is seen as grounded in human biology in that sex/gender relations have resulted in different power bases. As such, cultural ecofeminists see the world as dominated by technologies and ideologies developed and controlled by men, and they seek to elevate and liberate women and nature, usually by means of direct political action. Relations between women and the environment are seen as essential.

Social ecofeminism rejects the biological determinism implied in the previous position, focusing instead on nature as a social and political construct and not as a natural construct. Thus, social ecofeminism advocates liberating women by overcoming the constraints imposed by marriage, the nuclear family, patriarchal religion, and capitalism. There is a rejection of the essentialist idea that the linking of women and nature results from biology in favor of the idea that the relations between women and the environment are constructed. Because they are constructed, these relations are not universal; rather, they vary from place to place and through time.

Implications

The fundamental implication of ecofeminism is best described in the larger context of feminism in general. Since the 1960s, the various forms of feminism

have effectively called into question virtually all of the established and cherished belief systems and institutions of dominant patriarchal cultures. The implications of ecofeminism for the larger ecological perspective are enormous. Many ecofeminists regard ecocentric views such as deep ecology as inadequate correctives, though there are others who see ecofeminism as a union of deep ecology and feminism. Ecofeminism asserts

- that the dominant development paradigm sees the earth as a resource to be exploited and developed for human benefit,
- that there are compelling connections between the treatment of nature and the treatment of women, and
- that there is a larger domination of the world by the values of a Western cultural tradition.

Accordingly, ecofeminists strive to move beyond the various dualisms, such as nature–culture and female–male, in order to create a new consciousness of people and the earth in harmony. One way to demonstrate some of these matters, and also to clarify the difference between cultural and social versions of ecofeminism, is to review the example of the Chipko movement, the most often discussed of the many struggles that symbolize the relationship between women and environment.

The Chipko Movement

Chipko, meaning "embrace" or "hug" in Hindi, is a grassroots, non-violent movement initiated in 1974 by women in an area of northern India who expressed concern about the removal of broad-leafed indigenous trees. Since then it has spread remarkably and is credited with saving perhaps 4,634 square miles (12,002 square kilometers) of trees. Why the concern? Because these trees are an essential resource for the rural poor, who depend on the trees for fuel, food, fodder, building materials, and household utensils, and also for limiting soil erosion. Why was it women who initiated the movement? One reason, certainly, is that women are more dependent than men on trees, being the ones who collect fuel and fodder and perform household tasks, and so it is women who would be the principal sufferers if trees were removed. Meanwhile, replacing the indigenous trees with pine or eucalyptus (as the government planned to do) would benefit men by providing them with employment. But a cultural ecofeminist would answer this question differently, focusing on the idea that ancient Indian cultures worshipped tree goddesses and that movements similar to the Chipko movement were in place over 300 years ago. On the other hand, a social ecofeminist might answer that it is because women are marginalized members of society and have consequently created a link with nature that they are the ones to initiate the movement.

➤➤➤ **18** This example can be interpreted in a way that supports the cultural ecofeminist essentialist claim that women are closer to nature and are, therefore, necessarily more committed to caring for the environment. But it can be interpreted rather differently following the social ecofeminist constructionist argument that women in the less developed world view their relations with nature differently from women in the more developed world.

To summarize this discussion of ecofeminism is not easy because generalizations are difficult, but the following statements reflect a moderate ecofeminist position:

- Nature does not need to be ruled over by humans.
- Humans need to recognize their intimate relationship with nature.
- From a religious perspective, God is neither male nor anthropocentric.
- Mutual interdependency replaces the traditional hierarchies of domination that have God above humans above nature, or perhaps have God above men above women above children above animals above plants.
- Prevailing patterns of interdependency based on sex, gender, skin color, class, culture, income, religion, age, and geographic location need to be reconstructed to create more equitable situations.

Political Ecology

Political ecology "combines the concerns of cultural ecology and a broadly defined political economy. Together this encompasses the constantly shifting dialectic between society and land-based resources, and also within classes and groups within society itself" (Blaikie and Brookfield 1987:17). Political ecological analyses recognize and incorporate various social and spatial scales and stress the need to understand a problem in a broad cultural, social, political, and economic context. Thus a political ecologist may conduct an analysis of a local area and of a small group but may also try to understand the problem at other scales; further, while the problem may be specifically economic, the political ecologist will incorporate other human and physical considerations in the analysis, including understanding how cultural practices structure access to and the use of natural resources.

The Approach in Context

Suggestions that political and economic processes need to be considered in human or cultural ecological analyses first appeared in the 1970s. Arguments

---

Questions for Critical Thought

— ✢ —

Much of the content of this section so far is based on the idea—not the fact—that nature is not so natural after all, but is better interpreted as a social construction. For postmodern philosophers, this view is one part of a larger argument that insists on a lack of givens in most areas of inquiry. A contrary argument about nature asserts that "social construction is necessary but not sufficient for our being. Some values on earth are not species-specific to homo sapiens" (Rolston 1997:62). The key point for our purposes is that philosophers continue to debate the logic of constructionist claims, and some have sought to move beyond merely deconstructing arguments, and rather try to build new approaches that attempt to construct new understandings of networks and affiliations. Considering the range of approaches informing the work of cultural geographers that have been reviewed so far in this chapter, which ones are you most convinced by?

focused on a socialist human ecology as the best means of practice, since it is concerned with environmental histories, human adjustments, system stability, and the fate of folk ecologies. Certain analyses of population, food supplies, and famine in parts of the developing world incorporated some of the characteristics of this emerging approach. But political ecology as an essentially Marxist-inspired analysis of the human use of resources and its related impacts on environment is usually linked to two important works, by Blaikie (1985) and by Blaikie and Brookfield (1987). ➝➤➤ **19**

Blaikie, addressing the problem of soil erosion in the less developed world, rejected the conventional explanation that referred to such factors as overpopulation, inappropriate use of resources, environmental difficulties, and collapse of markets. Instead, he emphasized social causes, specifically land managers, or households that were using local resources and that were obliged to extract surpluses thus leading to land degradation. Blaikie identified a chain of circumstances leading to soil erosion: first, the land managers have direct contact with the land; second, land managers relate to each other, to other land users, and to other groups in society; third, they are linked also to the state and to the world economy. The attractiveness of this political ecology needs to be understood within four larger contexts, in addition to those provided by the ecological traditions discussed in this chapter.

First, there is the context of political economy. For geographers, political economy is not easy to define, as it is really an umbrella term covering a variety of radical emphases mostly inspired by Marxism. It can be summarized as the idea that the political and the economic are irrevocably entwined, such that any study of human activities must incorporate both. Central to the political economic approach is the idea that much human activity is bound up with social and political struggles and with competing claims to such things as ownership and right of access. Political ecology is a radical perspective on the political economic approach, emphasizing social justice, equity issues, and unequal power relations between the people and groups that affect environments. There are

certainly similarities here with Smith's (1984) work on the production of nature.

Second, there is the larger context of environmental and social concern that has been evident since about the late 1970s, especially in the context of the less developed world. Unlike the original environmental and social concern that was raised in the early 1960s, this later concern was bound up with larger questions of economic and cultural globalization and with increased awareness that humans were affecting a global environment. Political ecology is particularly critical of environmental policies that do not consider the impacts of global capitalism.

Third, political ecology needs to be considered in the context of its links with structuration theory as outlined by the sociologist Giddens (for example, 1984) (as discussed in Chapter 3). Structuration is a complex set of ideas that seeks to integrate the work of a wide variety of thinkers, including Durkheim, Freud, Marx, and Weber. A key feature of this theory is its focus on links between human agents and the social structures within which they function, with individuals viewed as operating within both local social systems and larger social structures; indeed, there is a substantial and continuing debate in social science—the structure and agency debate—centered around this issue. Structuration theory views individuals as agents operating within both the contexts of local social systems (sometimes called locales) and the larger social structures of which they are a part. The most important characteristic of structuration is its identification of the dualities associated with social structure and human agency. Thus, social structure enables human behavior, while at the same time behavior can influence and reconstitute culture. Further, the rules of any social structure are both *constraining* and *enabling*: they are constraining because they limit the actions available to individuals, but they are also enabling as they do not determine behavior. Certainly, both political ecology and structuration theory consider the ways in which global, national, or regional processes connect over time with local-scale processes.

Fourth, political ecology can be considered in the context of a set of approaches, known as *liberation ecologies*, that explicitly incorporate ideas of entitlement, social justice, and livelihood. In a discussion of these Marxist-inspired liberation ecologies, Peet and Watts (1996) incorporated aspects of development theory, various poststructural critiques and related discourse theory, and social movements and related political forms; these additional conceptual insights were noted in Chapter 3.

→ ≫≫ 20

Political ecology is a multifaceted approach that shares conceptual and empirical concerns with several other interests. As with some of the other topics raised in this chapter—for example, the new ecology and ecofeminism—political ecology is an approach that continues to evolve as the underlying theoretical concepts are developed and clarified. This approach has proved attractive to many cultural geographers who acknowledge that culture is best understood in a political context, with some groups exercising power over other groups. This section now reviews three examples of political ecological research to demonstrate how the concern with multiple cultural, political, economic, and environmental factors work together in the explanation of empirical cases.

Peasant–Herder Conflicts

Since the early 1970s, large numbers of Fulani pastoralists in West Africa have moved south from semiarid environments in the Sahel, especially from Mali and Burkina Faso, into the northern savanna region of the Ivory Coast. Their migration has been welcomed by the Ivory Coast government because of the contribution the Fulani make to beef production, but one unfortunate consequence has been a series of conflicts between the incoming pastoralists and the local Senufo peasant farmers that they have encountered. The principal reason for conflict is the damage inflicted by Fulani cattle on crops grown by the Senufo, for which the Senufo have not been compensated. There are similar conflicts in other West African countries.

Traditional geographic explanations for such movements and conflicts rely on the circumstances of a declining resource base relative to population, while a political ecological focus stresses the need to consider human activities as responses in context, and looks at the chain of causality leading to conflict. As Bassett (1988:469) explains, "It is at the intersection of Ivorian political economy and the human ecology of agricultural systems in the savanna region that one can begin to identify the key processes and decision-making conditions behind the current conflict." Peasant farmers are aggrieved at the consequences of pastoral intrusions into their cropping areas because of the impact of uncompensated crop damage on their already marginal standard of living. But although it might be suggested that this is the cause of conflict, it is not seen as a sufficient cause of the peasant uprising that included the murder of about eighty herders in 1986 and that is continuing today. Table 4.1 indicates the key determinants

Table 4.1 Determinants of Peasant–Herder Conflicts in Northern Ivory Coast

Ultimate Causes	Proximate Causes	Stressors	Counter-risks

Ivorian development model:
• surplus appropriation by
  - foreign agribusiness
  - the state
• livestock development policies
	Low incomes and beef consumption
	Insecure land rights
		Uncompensated
		crop damage
		
		Political campaigns
			Compensation
			
			Fulani expansion

Savanna ecology	Intersection of Senufo agriculture and Fulani semi-transhumant pastoralism	Theft of village cattle	Crop and cattle surveillance
Fulani immigration			Corralling animals at night

Source: T.J. Bassett, "The Political Ecology of Peasant–Herder Conflicts in the Northern Ivory Coast", *Annals*, Association of American Geographers 78 (1988):456. (Reprinted by permission of Blackwell Publishing)

of conflict, distinguishing between ultimate causes and proximate causes, and identifying stressors and counter-risks.

 21

As with any ecological approach, political ecology thus acknowledges complexity and interrelationships, looking beyond key factors that may be evident to understand an issue in context. In the case of peasant–herder conflicts in the Ivory Coast, three ultimate causes are noted, one of which is the Ivorian development model, which a traditional cultural ecological approach would not seriously consider. There are also more specific proximate causes that relate to the particular circumstances of power relations between the groups involved. Note that poverty, as expressed in terms of low incomes, is only a proximate and not an ultimate cause. It is necessary to know how poverty comes about, rather than simply to use it as an explanatory factor.

Banana Exports and Local Food Production

Historically, colonial powers encouraged banana production in colonies with the appropriate physical environment; thus, bananas were imported by France from Martinique and Guadeloupe, by Spain from the Canaries and the Azores, and by Britain from the Windward Islands in the eastern Caribbean. Indeed, for the islands of the eastern Caribbean, bananas have been the principal export crop since the early 1900s, with most production on small, usually less than 5-acre (2-hectare) plots and by farmers who were often owner–occupiers.

In many islands of the eastern Caribbean during the early 1990s, there was evidence of increasing banana production accompanied by decreasing production of local foodstuffs and increasing food imports. The usual explanation for such circumstances assumes that it is the increase in export crop production that is the cause of the other changes. But in a political ecological analysis of this situation in the eastern Caribbean island of St Vincent, Grossman (1993:348) noted that such explanations do not "specify the precise means by which banana production is supposed to interfere with local food production", nor do they "analyze the significance of the role of the state and capital in influencing these relationships".

Grossman (1993) cited, as an important political economic factor, the fact that marketing policies and state intervention result in banana production being advantaged over local food production, especially because of the guaranteed market and fixed prices for bananas; as an important local factor, he noted the reduced possibility of theft of bananas because of technological innovations in harvesting and packing that make it much more difficult for thieves to steal bananas at night and sell them in local markets the following day. He also considered the question of potential conflicts in the productive sphere by reference to any seasonal conflicts of time allocation, to spatial patterns of land use, and to issues of intercropping. Grossman concluded that such conflicts were not the principal cause of local food production decline. Rather, the explanation lies in the different political economic contexts of banana and local food production. The banana market is attractive for reasons related to the state and to foreign capital.

Interestingly, the value of this approach is further evidenced by changes in world trade regulations. In accord with protocols agreed to in 1957, European countries were able to afford preferential treatment to their banana-producing areas. In 1993, the European Union (EU) amalgamated the various protocols into a single preferential trading structure. This system of protectionism for specific banana-producing areas has, however, been condemned by the World Trade Organization (WTO) following a challenge initiated by the United States and the banana-producing countries of Ecuador, Mexico, Guatemala, and Honduras. This global economic change, only one aspect of the movement toward free trade, is a major blow to those banana-producing countries such as St Vincent that previously received preferential treatment. As another example of the complex relations between banana production and larger policies, the European Union issued a statement of standard in 1994 requiring bananas to be at least 1 inch (25.4 millimeters) wide, 5.5 inches (14 centimeters) long, and without abnormal curvature. The bananas grown in the islands of the eastern Caribbean do not conform to such standards. Many of these details were resolved by the WTO in 2001.

Violent Environments

Some commentators from Malthus onwards have predicted that continuing population growth and related resource depletion will produce a human future fraught with problems. This **limits-to-growth thesis** was resurrected in the 1960s and has continued to dominate both scholarly and public opinion since then. The alternative **cornucopian thesis** sees our human future in rosier terms, with less human crowding, less pollution, and fewer ecological problems. Contemporary neo-Malthusian claims of impending anarchy related to the growing incidence of regional violence, such as those made by Kaplan (2000), have been criticized for their geopolitical underpinnings and their proposals for dealing with the perceived problem through a Western policy of containment and exclusion, which would effectively cut off parts of the less developed world. But a more important set of criticisms are those that emanate from a political ecological perspective. →≫ **22**

In broad terms, political ecologists reject the argument that environmental scarcity and related problems of economic activity and migration are the cause of weakened states and subsequent violence. Rather, they see violence as "a site-specific phenomenon rooted in local histories and social relations yet connected to larger processes of material transformation and power relations" (Peluso and Watts 2001:5). In other words, the problems reside in the political dimension of culture. Any account of the political ecology of violence begins not just with politics but with the fundamental geographic circumstance of human-and-nature relationships. Some cultural geographers in this field have been inspired by the work of anthropologist Michael Taussig (1987), who charted how terror and violence formed part of the colonial imaginary and the imposition of capitalist systems of resource extraction in peasant societies in South America. Taussig's work is inspired by cultural techniques of postmodern critique, but offers a politicized reading of terror and healing with relevance to contemporary situations.

This debate about environmental security and violence is important, with Marxist thinkers stressing the historical and social appropriation of nature and favoring a political ecology approach, liberal thinkers being inclined to focus on environmental scarcity and favoring a limits-to-growth argument, and conservative thinkers emphasizing human ingenuity and technological advances to overcome challenges and favoring a cornucopian view.

Environmental Ethics

How are we to determine what is ethically right when we consider our treatment of the natural world? Human attitudes toward environments and non-human animals have undergone significant changes since the dominant early nineteenth-century view that supported both the idea that it was right and proper for humans to use environments as they saw fit and also the idea that humans were different from all other animal species. Today, there is considerable philosophical debate concerning the proper assessment of humans as part of nature. This discussion involves consideration of a fundamental question: should we be behaving in some environmentally friendly manner because we need to for our own sake, or rather because it is ethically right for us to do so in terms of the natural world itself? The following sections outline seven different perspectives on this issue. →≫ **23**

Religion and Environment

The debate about the relationship between Christian ideas and human impacts is but one aspect of the larger question of religion and environment. Most religions identify what is important to humans beyond everyday concerns and offer rules of behavior that are said to derive from some spiritual source. Inevitably, then, religion has something to say about human activities as these affect other humans and the environment that humans occupy. Overall, most religions provide guidelines for human activities and for an understanding of nature.

Many traditional religions were not as separated from the rest of human life as Christian religion has been. In traditional Hawaiian culture, god, humans, and nature were related to one another and formed a single interacting community, with all species—not just humans—considered sentient. Thus the entire world was seen as alive, capable of knowing and acting, in the sense that Western culture sees only humans as alive. Such beliefs allowed for a genuine kinship that is lacking in the Christian tradition; they also made for a quite different perception of the world. These comments apply generally to a great many of the traditional belief systems that were replaced or substantially weakened as Europeans spread overseas, but they apply also to several of the major world religions practiced today. ➝➤➤**24**

In Islam, respect for the natural world is founded on the principle that Allah created all things. Islamic religion also acknowledges that humans are not the sole community living in the world. Use of the natural world to support humans is considered appropriate, but needless destruction is not sanctioned. Both Hinduism and Buddhism traditionally viewed humans as but one part of nature such that abuse of the natural world is also abuse of people. However, most observers suggest that the materialistic orientations of Western culture have affected these traditions with a corresponding decrease in concern for the natural world.

Extending the Principle of Equality

Environmental ethics breaks new ground from a philosophical perspective. All ethics is concerned with seeking an appropriate respect for life, but only environmental ethics asks whether non-human life should be included. Should we extend the principle of equality beyond our own species? If we do extend the principle of equality to include all sentient life, should we also extend it to insentient life? Note that asking these questions would not be necessary in some traditional cultures, as the discussion of religion and environment acknowledged.

Referring to the principle of equal consideration of interests, which allows us to claim that other humans should not be exploited because, for example, they have lesser intelligence or a different culture, it might be argued that it is similarly wrong for humans to exploit members of other species. Following this logic it is, for example, wrong to eat meat, wrong to use animals in experiments, and wrong to display animals in zoos. These are very complex ethical issues that are at

the heart of our human relationship with parts of the natural world. ➝➤➤**25**

What are some of the basic arguments against keeping animals in zoos? Most obviously, the zoo environment is an alien one that restricts the animals' freedom of movement, limits their ability to seek out food, and, more generally, prevents them from behaving in ways that are natural to them. Further, it has been argued that animals often suffer in zoos, where many die young, suffer injury, or develop deformities. There is much evidence of animals engaging in disturbed, trance-like behavior: for example, elephants swaying back and forth, polar bears swimming in circles, parrots grooming themselves until they bleed, and gorillas regurgitating and then re-ingesting food. In response to ethical concerns, there has been considerable debate about the wisdom of keeping animals in zoos and whales and dolphins in aquariums, and several zoos have chosen to close their elephant exhibits in recent years.

But there is another rather more difficult argument against zoos that focuses on the question of why we wish to display animals in zoos. In addition to conventional justifications concerning human entertainment, education, scientific analyses, and species preservation, it can be argued that they serve to reinforce our human sense of superiority. The fact that we are able to confine animals means that they are there for our use and pleasure. It is in this sense that zoos can be viewed as a reflection of an outmoded environmental ethic. The evolution of zoos can be read as a colonial obsession with charting and itemizing the new world, an obsession with representing the world as exhibition, and the supposed racial and gendered constructions that inform the human–animal interface during the modern period (Anderson 1995). Changes in the role and structure of zoos in society then say a great deal more about changes in the human condition.

Transforming Culture

One ecocentric perspective favors a deep ecology, or what is often called *sustainable development*. Naess (1989) proposed that humans need to develop a new worldview that sees nature as having value in its own right, that regards all humans as connected, that recognizes a need to work with rather than against nature, and that views ecosystem preservation as a primary goal. In addition, such a deep ecology regards humans as part of nature, asserts that every life form has in principle a right to live, is concerned with the feelings of all living things, and is concerned about resources for all living species. Thus, deep ecology is both an environmental ethic and a spiritual basis for rethinking what it means to be human. Several of the basic concepts employed by Naess are derived from Hinduism.

In a related fashion, but pursuing an original argument, one writer has proposed that we need to move away from the oppression and emphasis on wealth that are characteristic of modern society through a reintroduction of the specific basic qualities that we once possessed, namely a connection with nature, a sense of belonging, and an egalitarian community (see Box 4.4).

The "Golden Rule"

Another way to approach the question of environmental ethics is to think in terms of how we wish to be treated. We might approach this matter by questioning two of the frequently stated bases for developing an appropriate ethic, namely the related claims that humans are permanently damaging the earth and that we need to learn how to behave as stewards for the earth we are threatening. In one sense, both of these propositions reflect an exaggerated view of our human role because humans are but one of millions of species and have, geologically speaking, appeared on the earth only recently. But to think in such grand terms is, of course, misleading.

To help develop an appropriate ethic, we need to think in more local terms, believing that we are capable of such devastating impacts as eliminating both the natural world upon which we depend and our very species. Following this logic, Gould (1990) proposed that an appropriate environmental ethic is one that can convince people to view clean air, clean water, reforestation, and so forth as being essential to the well-being of our environment and ourselves here and now, rather than one that stresses long-term

> ### Questions for Critical Thought
> ⸎
>
> Is it rather disturbing that so much popular and governmental attention is paid to the question of human impacts on environment when the problems of human impacts on other humans remain so distressing? It is commonplace today to claim that concern for environment is a fair measure of a society, but do we similarly assess ourselves according to our treatment of others? The answer seems uncertain. To paraphrase Wagner (1990:41), a principal goal of cultural geography should be to develop a theoretically well-grounded, intellectually vigorous, and practically effective discipline that might well assume, in time, a major role in guiding and guarding our attitudes toward and treatment of ourselves and others. In Chapter 7 you will read about genocide and ethnic cleansing, and in Chapter 8 about many other circumstances of humans viewing other humans as somehow inferior. Do you know as much about human abuses of other humans as you do about human abuses of the environment? Are our priorities right? Should the focus on human or environmental welfare be constructed in this either/or fashion?

consequences of neglecting such issues. The ethic in question is simply that of treating others as we wish to be treated by others.

Concluding Comments

- As one means of thinking about human–human and human–nature relations, ecology is not central to any one social science, but rather of interest to geographers, sociologists, anthropologists, psychologists, historians, and others.
- Geography might appear to be the most natural home of ecology because of the physical and human content and the longstanding idea that the discipline can serve as some sort of bridge between physical and human sciences, but this has not proven to be the case, and the geographic use of the ecological approach has been erratic. There was limited acceptance of the pioneering call from Barrows.

Box 4.4 Social Evolution and the Environmental Crisis

For Earley (1997), as for many other commentators, it is the continuation of modernity—the modern period—that poses the greatest threat to human survival, in the sense that our current environmental crisis results from the way our world is structured and our related view of nature and our place in nature. To move away from this modernity requires us to change our social structures and institutions. These points are clarified by means of a model of social evolution (Earley 1997:2–4). At the outset of social evolution, humans enjoyed three basic qualities:

- a connection with nature
- a sense of belonging and richness of experience
- an egalitarian community

Through time, populations grew, resources were limited, and conflict occurred, resulting in three emergent qualities that provided for more conscious choice, increased freedom from environment, additional population growth and resource exploitation, and a new understanding of ourselves and of nature. These are

- technology;
- reflexive consciousness; and
- social structure.

The difference between these two sets of qualities is striking. The initial ground qualities—what we were—reflect vitality and organic wholeness, while the later emergent qualities—what we have

become—reflect power and differentiated organization. Put simply, it is the emergent qualities, especially the resultant economic system emphasizing material growth, that are responsible for our alienation from nature, from others, and, indeed, from ourselves. The result is that we

- have gained technology, but destabilized the environment;
- understand more, but empathize less; and
- have developed social structure, but lost equality and community.

According to Earley, it is only in recent years that it has been possible to fully grasp the implications of this process of social evolution and to appreciate that it is insufficient and, of course, impossible, to return to the ground qualities. What humans need is a rediscovery of the ground qualities and an integration of ground and emergent qualities, resulting in

- ecological technology;
- integrated mind and heart; and
- a social structure that promotes equality and community.

In summary, Earley (1997:4) states: "We must now consciously choose to regain our wholeness and vitality in conjunction with our complexity and autonomy, as individuals and communities, as organizations, as a world society, and as a living planet."

- Perhaps the most successful uses of ecology in geography are the current linkage between ecology and environmental concerns, the development of a political ecological focus, and the uses of ecology in analyses of population change as it relates to technology. Geographers have adopted and applied ecology as developed in other disciplines, especially in sociology and anthropology.

› For example, the Chicago school of sociology was a major inspiration for urban geography in the 1960s, despite such limitations as the omissions of a cross-cultural perspective and any reference to symbolism.
› Cultural geographers have also turned to anthropological cultural ecological analyses. Other approaches that focus on the

idea of hybridity have offered geographers a way to move beyond the modernist binary of nature–culture, and a great deal of research is emerging that engages with these ideas.

› More recently, geographers have been attracted to the new ecology in biology and to political economy as developed in politics and economics especially.

• Current evidence strongly suggests that a focus on the human–environment interface is an increasingly popular and even more diverse tradition today than previously. There appears to be every indication that cultural geographers will continue their fascination with emerging approaches in this area, and that conceptual revisions and additions will be accompanied by increasing numbers of empirical analyses.

Further Reading ➙ »»

The following are useful sources for further reading on specific issues.

1. In sociology, some initial human ecological ideas are outlined by Park (1915), with the first textbook account authored by Park and Burgess (1921).

2. The traditional sociological view was that geography was principally concerned with factual information about humans and their physical environments rather than with explanations of human distributions. Thus, Quinn (1950:339) suggests that geographers were concerned with direct human-and-land relations whereas ecologists were concerned with interrelationships. Hawley (1950:72) identifies a similar distinction, but also stresses that geography, specifically regional geography, was atemporal while human ecology was not. Schnore (1961) conveys a more positive view in an enlightened interpretation of then current human geography in ecological terms—work in economic, urban, and population geography is cited for ecological content. Hawley (1950, 1968) builds on the human ecological work of Park to propose closer links with biology and a de-emphasis of humans, while Firey (1945, 1947) argues, with specific reference to Boston, for the importance of symbolic variables rather than physical science–based human ecological variables in any account of urban areas. The geographer Entrikin (1980) discusses human ecology in early sociology.

3. Kroeber (1928) notes a neglect of ecological considerations in then current ethnological studies, but chooses not to pursue the theme any further. Steward (1936, 1938) relates group characteristics to ecological circumstances. Bennett (1976:48) and Vayda and Rappaport (1968:492) outline criticisms of the cultural ecology proposed by Steward. The plethora of ecological approaches in anthropology prompted Meggers (1954) to discuss the limits set on cultural evolution by environments, even to the extent of arguing for cases of determinism, and resulted in Netting (1977) suggesting that any theoretical or methodological overview of the approaches was not practicable. Headland (1997) proposes that anthropology adopt a historical ecology and identifies a number of myths that such an approach could address. Using a cultural ecology framework, the anthropologist Geertz (1963) studies land-use change in Indonesia in generic systems terms, while Rappaport (1963) analyzes human involvement in island ecosystems. A key assertion evident in much of the geographic and anthropological work is that human populations can be treated in much the same way as any other population. Rappaport (1963:70) notes that the "study of man, the culture bearer, cannot be separated from the study of man, a species among other species."

4. Crosby (1995:1186) chooses "for the sake of convenience" to identify the work of Sauer and the 1955 symposium, which produced the seminal volume *Man's Role in Changing the Face of the Earth* (Thomas, Sauer, Bates, and Mumford 1956), as the scientific debut of the ecological approach in history. Reviews of and proposals for ecological history are provided by Worster (1988, 1990), Merchant (1990), White (1990), and Cronon (1992). Some examples of ecological studies by historians include those by Cronon (1983) on environmental changes caused by different cultural groups in New England; by Crosby (1986) on the biological expansion of Europe; by Merchant (1989) on changing ideas of nature, gender, and science in New England; and by Arnold (1996) on environment, culture, and European expansion. A succinct review of the American West environment is by Jacobs (1997), while Cunfer (2005) provides a geographically informed study of agriculture in the Great Plains environment. Griffiths (1997) stresses links between use of environment and political power in a historical context. Somewhat separately, explanations of the rise of the Western world focus on the relevance of physical geography as it relates to human activities (Landes 1998; Diamond 1997). For a historical geographer's perspective on environmental history, see Baker (2003). Articles on the study of past environments by historical geographers and environmental historians are included in a Special Issue of the *Geographical Review* (April 1999).

5. The ecological psychology outlined by Lewin is more fully developed into a substantive research area by Barker (1968), with the term "behavior settings" used to describe the larger context within which behavior occurs. The relative significance of the individual and the environment to an understanding of behavior is related to the character of the environment, with the individual being more important if the environment is relatively stable and the environment being more important if it is varied

and changing. A related advance is that of environmental psychology (Ittelson, Proshansky, Rivlin, and Winkel 1974).

6. Stoddart (1965, 1966) discusses geography and ecology in general terms, and Martin (1987b) details the ecological tradition in geography. Schnore (1961:209) describes the article by Barrows as "little more than a piece of intellectual history", while Koelsch (1969) notes Barrows's impact on historical geography.

7. Sauer's interest in human ecology is noted by Leighly (1987).

8. Clarkson (1970) and Grossman (1977) outline geographic versions of human ecology. Boulding (1950) outlines some links between economics and ecology. Hawley (1998) comments on developments in human ecology.

9. Head (2007) provides a thoughtful commentary on cultural ecology, noting, among other points, the considerable work accomplished by geographers and the tendency to view humans as somehow different from nature.

10. Chapman (1977), Huggett (1980), and Wilson (1981) provide broad-based geographic accounts of the systems approach. The most substantial applications of a systems approach by geographers are studies in the natural hazards tradition (see Kates 1971), while there are also studies of subsistence (Nietschmann 1973) and of civilizations as ecosystems (Butzer 1980).

11. Berry (1984) outlines ecocultural psychology; Berry (1997) uses the approach in a study of immigration and subsequent acculturation; Bennett (1993) considers adaptation and social scale in psychology; Cronk, Chagnon, and Irons (2000) review adaptation in anthropology. Porter (1965) is an early statement on adaptation and cultural geography. Concerning terminology employed in adaptation studies, Goldschmidt (1965:402) notes: "Our investigation is a study in cultural adaptation, in ecological analysis, in the character of economic

influence on culture and behavior in social micro-evolution—depending upon which of the currently fashionable terminologies one prefers."

12. Neil Smith (1984) made several important contributions to geography, but one of his most important is his analysis of the production of space and nature within capitalist systems. His insights with regard to capitalism's production of space are also central to the field of political ecology.

13. There are many geographic and other commentaries on nature as socially constructed and on the related idea of hybridity; these include Gerber (1997), Coates (1998), Proctor (1998), Whatmore (1999, 2002), Eden (2001), Franklin (2002), Huckle (2002), and Castree (2003). Von Maltzahn (1994) focuses on the social construction of nature from a humanistic perspective; Evernden (1992) outlines the need to rethink our culture of nature by refocusing on the wildness that is inherent in nature; Eder (1996) considers the need to understand why humans continue to abuse the environment, proposing that the real culprit is our continuing taken-for-granted idea that nature is there to be dominated; Oelschlaeger (1991) focuses on the cultural idea of wilderness from a historical perspective, beginning with the experience of pre-agricultural societies to show that there has long been an idea of wilderness as other. Wilson (1999) discusses the place of nature in the context of La Sierra, Colorado. On a rather different theme, Salerno (2003) discusses human estrangement from nature. Castree and Braun (2001) present a collection of work that advanced Smith's (1984) critical engagements with the social production of nature by asking who benefits from these engagements with and production of nature. White and Wilbert (2009) offer an edited collection of multidisciplinary research engaging with the post-social idea of nature and the realm of complex networks that incorporate the natural, social, and technological environmental

challenges of the twenty-first century. Wolch and Emel (1995) consider the claim that social theories are becoming increasingly anthropocentric. Botkin (1990) notes some psychological difficulties involved in the idea that nature is constantly changing.

14. Lovelock (1982) details the Gaia concept. Murdoch (1997) outlines actor–network theory. Wolch, Emel, and Wilbert (2003) describe animal geographies.

15. Current geographic interests in ecology are outlined by Fitzsimmons (2004) and Robbins (2004).

16. The origins of Western environmentalism are discussed by Grove (1992); Paehlke (1995) provides an encyclopedia of conservation and environmentalism; de Steiguer (1997) discusses the age of environmentalism; Inglehart (1990) identifies changing attitudes to environment. The life and work of George Perkins Marsh is covered by Lowenthal (2000).

17. According to Plummer (1993), nature, women, and other subordinated groups have been manipulated as a result especially of the two processes of backgrounding (that is, denying value) and instrumentalism (that is, service without recognition). Shiva (1988), Buckingham-Hatfield (2000), Warren (2000), and Domosh and Seager (2001) examine women and environment.

18. The Chipko movement is discussed by Dwivedi (1990), Merchant (1995), Mellor (1997), Rose, Kinnaird, Morris, and Nash (1997), and Warren (1997).

19. Walker (2007) questions whether political ecology is sufficiently political. Neuman (2005) provides a clear introduction to the approach intended for the undergraduate reader. Wisner (1978) develops an early argument for a socialist human ecology (see also Bennett 1976; Ellen 1982). Dickens (1996) notes links between the division of labor and our understanding of

nature. Kjekshus (1977), Porter (1979), Watts (1983), and Grossman (1984) are early studies in the political ecology tradition. More recent studies include those by Zimmerer (1996b) on peasant life in the Peruvian Andes, Zimmerer and Young (1998) on the complex character of environmental change in the less developed world, and Simmons (2004) on land conflict in the eastern Brazilian Amazon. Also see Batterbury (2001), Bebbington and Batterbury (2001), Zimmerer and Bassett (2003), Forsyth (2003), Paulson and Gezon (2005), and the various articles in a Special Issue of *Economic Geography* (vol. 69, no. 4, 1993).

20. Fairhead and Peach (1996:10–15), Mayer (1996), Bryant (1997, 2001), Keil, Bell, Penz, and Fawcett (1998), Low and Gleeson (1998), and O'Connor (1998) provide accounts of the larger intellectual context supporting political ecology. Walker (2005) provides a critical review. Perramond (2007) questions the use of an adaptation framework.

21. Adriansen (2008) provides an insightful analysis of the Fulani in a part of Senegal, emphasizing the importance of specific knowledge systems and noting that mobility is not important to Fulani ethnic identity. With reference to the inland Niger delta of Mali, Benjaminsen and Ba (2009) highlight the role played by national policies, especially laws favoring arable activity, and related corruption.

22. Kaplan (1996, 2000) offers interesting global arguments for adopting a limits-to-growth perspective. Simon and Kahn (1984) present a forceful argument in favor of the cornucopian thesis, and Lomborg (2001) gives a thoughtful and provocative global overview. With reference to the rural American West, Walker (2003) discusses a political ecological approach to environmental conflicts in advanced capitalist societies.

23. Des Jardins (2001) provides an introduction to environmental philosophy. Smith (2004) discusses ethics and the human environment.

24. Links between the Christian tradition and environmental abuse are noted by White (1967); Gore (1992) considers problems with this argument; Kinsley (1994) reflects on aspects of both points of view. Some accounts of religion and environment are Spring and Spring (1974), Gottlieb (1996), Chapple and Tucker (2000), Redekop (2000), and Jamieson (2001). Hunter and Toney (2005) examine the distinctiveness of contemporary Mormon environmental perspectives as contrasted with the larger United States population.

25. Geographic discussions about humans and animals include Anderson (2000) and Philo and Wilbert (2000), while Singer (1993) and Jamieson (1985, 1997) raise philosophical questions relating to our human treatment of animals, including animals in zoos.

CHAPTER 5

Landscape Evolution

This chapter discusses studies of cultural landscape evolution as well as related work in historical geography. The idea that informs most of this work is the recognition that humans, often through their cultural identities, transform physical landscapes through time to create human landscapes. Thus, it is usual to identify close links between the characteristics of the culture and the landscapes resulting from cultural occupance. This chapter contains

- ✤ a short explanation of the need to consider changes through time in analyses of landscapes, with emphasis on the ideas of Sauer;

- ✤ a discussion of the landscape consequences of cultural change as these relate to processes of diffusion, with reference to three distinct approaches, concerned respectively with culture traits, spatial processes, and political economy;

- ✤ a discussion of the landscape consequences of cultural change as these relate to contact between different groups, with emphasis on the movement of Europeans overseas and resultant encounters between Aboriginal and European cultures. There is also reference to our changing understandings of the character of these encounters;

- ✤ a review of the tradition of historical geography, with reference to a number of approaches and to examples of empirical work and with emphasis on the shaping or making of landscape. This review examines approaches to the analysis of change through time, frontier studies, evolutionary regional landscape studies closely associated with the Sauer tradition, an essentially British local-history approach to the study of landscape change, and the tradition of reading landscape as exemplified in much of the work published in the magazine *Landscape*;

- ✤ an account of imagined landscapes, subjective environments, or images that must be considered in almost any discussion of landscape evolution as they may be significantly different from the more objective or material environment. The three regional examples detailed are those of colonial Western Australia, the American Great Plains, and prairie Canada.

- ✤ Throughout there are close links with the discussions in Chapter 6 of cultural regions, which can be considered as dynamic outcomes of evolutionary processes. Several evolutionary regional concepts are also discussed in Chapter 6.

Providing Context

The evolution of the cultural landscape. For many cultural geographers, this one phrase identifies the central concern of the landscape school tradition as initiated by Sauer in the 1920s. As evident from discussions in this chapter, many contemporary cultural geographers continue to work in the landscape school tradition while referring to a more specific evolutionary concern. Indeed, a historical orientation was the first of the seven "persistent preferences" of cultural geography noted by Mikesell (1978:4), while Williams (1983:3) notes that in 1936 Sauer described historical geography as "the apple of my eye". Similarly, Donkin (1997:248) writes: "A fascination with 'origins' is both primitive and widely shared—the product, perhaps, of 'the essential time bond of culture rather than its looser place bond', some rather uncomfortable words of John Leighly, which I've often pondered." It bears repeating that this emphasis on time and change was in marked contrast to the regional geographic approach (as favored by Hartshorne and others) that dominated North American geography between about 1900 and the mid-1950s.

Intriguingly, it is not just cultural and historical geographers who argue the need for evolutionary studies of landscape. Looking back on a long and distinguished career, the theoretical and quantitative human geographer Morrill (2002:34) observed: "I define geography as the study of the evolution of landscape—the physical and human forces that shape the earth's surface." Turning his focus to the human forces, Morrill (2002:34) continued: "The elements of study of geography as a social science are the *behavior* of persons, households, groups, institutions, and social systems that define and change the landscape. Geography as a social science tries to explain this behavior and analyze processes of change. . . ." These statements about human geography are generally in close accord with the principles of the cultural geography subdiscipline introduced by Sauer. Although Sauer's early methodological statements emphasized the behavior of cultures as the cause of landscape change, much subsequent empirical work used the other social scales noted by Morrill.

Writing about the physical landscape, Sauer (1925:36) noted that it was to be understood in terms of both space and time, since it was continually changing in response to geomorphologic and climatic processes. He also noted that the presence of humans and the consequent introduction of cultural processes resulted in landscapes that were no longer only physical but that also reflected cultural occupance. Physical geographers studied the physical processes generating change, while cultural geographers studied the cultural processes generating change. Although Sauer himself apparently showed little interest in such methodological statements after they were written, their impact on others seems undeniable. Mikesell (1978:3) asserted that "Sauer acted as a catalyst for cultural geography, as well as an initiator of specific trends."

One specific challenge laid down by Sauer, then, was to study the evolution of cultural landscapes. This was a challenge because the orthodox view of geography in the 1920s was that it was an atemporal regional discipline; any historical geographic studies were limited to the study of regions in past times. What Sauer proposed was, within the North American tradition, unorthodox because the study of change was a challenge to the hegemony of regional geography. But the response to his proposal was both appreciative and considerable, resulting not only in the rise of a cultural geography studying landscape evolution but also in the rise of a relatively distinct historical geography separate from the tradition of studying the geography of past times. That this historical geography of changing landscapes was accepted into mainstream North American geography was evidenced by the inclusion of a chapter on historical geography in the seminal 1954 volume *American Geography: Inventory and Prospect*. Written by one of Sauer's students, this chapter spells out a historical geography concerned with both changing landscapes and past times (Clark 1954). Thus, in addition to studies of cultural landscape evolution, it is appropriate to see Sauer as a leading figure in the rise of a more general historical geography. This relationship can be simply expressed in the words of historical geographer Cole Harris: "Put most simply, good geographical synthesis is usually built around the relationship between culture and environment, and in the consideration of this relationship lie many of the major issues of our time" (1971:169).

Cultural Diffusion

Given Carter's (1978:56) claim that "in the broadest sense diffusion is the master process in human culture change," it is not surprising that diffusion studies have been a central feature of work in cultural geography since the beginnings of the landscape school. It is notable, however, that the basic thrust of these studies has undergone two significant changes, with the result that it is appropriate to identify three principal approaches in geographic diffusion research. Interestingly, these three approaches mirror the larger twentieth-century history of human geography. The first, the traditional approach, which focuses on the spread of a specific **cultural trait**, is identified with the landscape school. The second, the spatial analytic approach, which is designed to uncover empirical regularities in the diffusion process, evolved from the traditional approach in the 1960s, but is best seen as a part of the theoretical and quantitative movement in human geography at that time. The third, the political economy approach, which is concerned with links between diffusion, culture, and power, began as a reaction against the perceived dehumanization of the spatial analytic approach. Regardless of the specific approach, in most studies the concern is with the diffusion of an **innovation**. ➡️»» 1

The Spread of Culture Traits

The traditional approach to studying cultural diffusion has close ties with early twentieth-century anthropology, especially with the Boasian school, and is an explicit component of the landscape school of cultural geography. Accordingly, the concern is with diffusion that is largely controlled by culture but that does not exclude physical environment. For the landscape school, culture history is the core of human geography, and diffusion is the basis for understanding cultural origins, cultural landscape evolution, and the creation of cultural regions. Not

surprisingly, then, there is concern with those elements of material culture that are visible in landscape. The pioneering exponent of this type of study was Kniffen, who employed this focus largely to demonstrate certain generalizations involving questions of regionalization and diffusion: "The material forms constituting the landscape are the geographer's basic lore. The cultural geographer deals primarily with the occupance pattern, the marks of man's living on the land. He finds his data, his evidence, in buildings, fields, towns, communication systems and concomitant features" (Kniffen 1974:254). Most analyses in this tradition are empiricist and inductive, and focus on a single culture trait—agricultural fairs, covered bridges, and house types, for example—identifying its origins, its routes, and its present distributions. ➡️»» 2

The Example of Covered Bridges

Culture trait studies were justified on the grounds that an understanding of the origins and diffusion of a particular trait aids understanding of spatial variations in cultural landscapes. A study of covered bridges in the United States stressed the feature's European antecedents, with covered wooden bridges especially evident in the Rhine Valley region of Switzerland. The original reason for covering is uncertain, although it does provide shelter and a location for market stalls, offers defensive advantages, and significantly extends the longevity of a wooden bridge. Once covering became common, the practice probably continued for reasons of tradition. In North America, covered bridges became popular after about 1810, with the area from New England to New Jersey proposed as the **cultural hearth**. Kniffen (1951b:123) stated: "The covered wooden bridge tended to spread freely among those possessing kindred patterns of thought and action." More specifically, the feature spread unevenly west, reaching southeastern Ohio by 1820 and southern Indiana by 1840, and attaining a maximum expansion in the eastern region by 1850. After this date, the number of covered bridges within the eastern region increased substantially, including many in the Canadian province of New Brunswick. The principal area in the western United States extends from southwestern Washington to central

California, a largely humid and wooded area west of the Sierra Nevada Mountains. There are isolated instances of covered bridges in Utah and Arizona, but these are seen as special cases, and not as parts of larger regional patterns.

Kniffen explained the diffusion of covered bridges first in terms of physical geographic considerations, specifically the availability of suitable streams and the necessary timber for building, second in terms of cultural considerations, with covered bridges seen as just one component of the larger New England culture, and third in terms of technological considerations, with new iron structures appearing after about 1860. This analysis of covered bridges included several generalizations about cultural differences in North America: for example, Kniffen suggested that within Canada, Quebec was more receptive than Ontario to cultural innovations from the United States.

Linking Trait Studies to Culture

Culture trait research typically focused on describing and mapping the diffusion of a single landscape feature, with emphasis on particular cultural identities and descriptions of related impacts on landscape. But although these analyses tended to study single traits, they often viewed such traits as representative of the larger culture, a tendency that is most apparent in studies of agricultural diffusion.

For example, an analysis of the diffusion of cigar tobacco production in the northern United States, published in 1973, demonstrated links between the diffusion of this cultural trait and certain cultural— and, in particular, religious—identities. The author of the study, Raitz, found a twofold significance of cultural identity: first, he uncovered a relationship between culture (notably religion) and the interest in cultivating a crop that is labor-intensive;

Of the nineteen covered bridges built in Madison County, Iowa, only five remain. Built in 1880, Holliwell Bridge is the longest covered bridge in Iowa at 122 feet (37 meters), and remains in its original site over the Middle River southeast of Winterset. Renovated in 1995 at a cost of $225,000, the bridge is one of several featured in the movie *The Bridges of Madison County*. The movie—adapted from the bestseller by Robert James Waller—tells the story of a four-day affair between an Iowa housewife and a traveling *National Geographic* photographer who is in Madison County to photograph the picturesque covered bridges. Today the covered bridges of Madison County are a popular tourist attraction. (William Norton)

second, Raitz established that networks of local groups promoted the cultivation of cigar tobacco by spreading necessary knowledge. Beginning in southeastern Pennsylvania, cultivation diffused to the Connecticut Valley of Massachusetts and Connecticut in 1833; a single migrant then diffused cultivation from the Connecticut Valley to southwestern Ohio in 1838, with German immigrants later reinforcing the tradition. Diffusion to southern Wisconsin occurred independently between 1846 and 1843 from three sources, namely Pennsylvania, Connecticut, and Ohio. Finally, there were later diffusions west and north from southern Wisconsin toward northern Wisconsin in 1860 and 1903, and also central Minnesota in 1919. The introduction of cultivation was associated in Pennsylvania with Germans, especially with Old Order Amish; in the Connecticut Valley with Polish Catholics; in Ohio with German Lutherans, Baptists, and Old Order Brethren; and in Wisconsin and Minnesota with Norwegian Lutherans. In all major production areas today, there continues to be an association between cigar tobacco production and a specific cultural group.

A second example of this development in culture trait research involves an ambitious study by Sauer (1952), *Agricultural Origins and Dispersals*, in which he expanded earlier local studies of plant diffusion in Mexico into a thesis about agricultural origins and diffusion in a global context. Central to this argument was his claim that cultural changes such as the development of agriculture occurred in areas that lacked environmental stress. In other words, according to Sauer, necessity did not prompt invention; rather, it was a sedentary lifestyle and food surpluses that provided the necessary leisure time for such activities as agricultural experimentation. It was with this argument in mind that Sauer proposed two areas of agricultural origin—Southeast Asia and the lands around the Caribbean, both areas offering food from water and land. Sauer postulated that vegetative planting cultures developed in and diffused outward from these hearths. Although most subsequent work on the origin of agriculture does not provide compelling support for these ideas, they remain as a stimulating contribution to an important

debate, and the 1952 book is widely acknowledged as a classic. ➜➤➤ 3

Migration and Diffusion

Cultural geographers have long been interested in the movement of cultural groups and how the cultural diffusion associated with this movement affects the related landscape. The distinction between migration (the movement of people) and diffusion (the movement of a culture trait that may or may not involve the movement of people) is rarely made in these geographic analyses, a reflection of the fact that the central interest is the effects rather than the process of diffusion. Salter (1971a:3–4) summarized the links between migration and diffusion thus: "The cultural geographer views man's mobility with a tripartite perspective: the catalyst for movement, the effect of movement on trait or people in motion, and the consequences of such movement."

A study of Amish populations conducted in the 1970s addressed aspects of all three of these concerns (Crowley 1978). The Amish are members of a strict Protestant Anabaptist sect that, led by Jakob Ammann, broke from the Mennonite Church in the 1690s. Amish populations migrated throughout much of Western and central Europe by the early nineteenth century, largely as a consequence of religious persecution. About 500 Amish moved to North America between 1717 and 1750, with a further 1,500 moving between 1817 and 1861. The first group settled in southeastern Pennsylvania, while members of the second group settled in Ohio, Illinois, Iowa, and Ontario. From these early North American locations, Amish settlements have diffused throughout a much larger area of North America and also into Latin America, often moving long distances, often settling on the frontier, sometimes experiencing failures, and further subdividing into different groups. Despite this diffusion process, most settlements continue to be in the original hearth areas; thus, of 98 Old Order Amish colonies in the United States in 1976, 29 were in Pennsylvania, 11 in Indiana, 10 in Missouri, 9 in Ohio, and 9 in Wisconsin. This type of diffusion process is especially difficult to explain in general terms because many of the movement decisions reflected particular circumstances.

A Spatial Analytic Emphasis

The second approach to studying cultural diffusion is the spatial analytic approach. Although this method, as one component of the larger theoretical and quantitative movement in geography, was not popular until the 1960s, its seeds were sown much earlier in the pioneering work of the Swedish geographer Hägerstrand.

The Work of Hägerstrand

Hägerstrand studied the diffusion of new agricultural practices—innovations—by mapping spatial patterns through time and trying to understand the process of diffusion that was responsible for the changing patterns. In doing this, he introduced some original concepts and procedures of analysis. Moving beyond the primarily descriptive analyses being undertaken in the landscape school tradition, Hägerstrand developed a series of "Monte Carlo" simulation models. He chose the term "Monte Carlo" to acknowledge the fact that these models incorporate an element of chance. The basic model, which he derived from empirical work, assumed that the probability of an innovation being adopted in a particular location is related to distance from the location of first adoption. Hägerstrand thus described a probability surface—essentially a map showing the probability of the occurrence of adoption as one of decreasing probabilities with increasing distance from the site of first adoption—and then simulated the spatial diffusion process. The procedures for conducting a Hägerstrand-type simulation are straightforward:

- The diffusion of an innovation, such as a new agricultural practice, and the changing of the cultural landscape that results are explicitly interpreted as influenced both by the prevailing pattern of communication in a region and by chance.
- A probability surface is constructed on the basis of assumptions about the process of communication flow. The simplest kind of probability surface assumes that information about an innovation is passed from one person to another. It is a probability surface precisely because chance is incorporated in the process.
- **Surrogate** data, such as information on telephone calls, are used to calculate specific probabilities. The probability surface is a distance–decay surface, meaning that it shows a decreasing probability of diffusion with increasing distance.
- Once probabilities are mapped, usually in a square grid format, the simulation process is conducted and a surface of projected adoption dates is created.
- The result of the simulation is compared to the known real-world pattern of diffusion, and a close correspondence between the two is interpreted as an indication that the probability model—necessarily a simplification of reality—is a reasonable approximation of the diffusion process.

Spatial diffusion analyses typically identified empirical regularities of the diffusion process; three examples are noted in Figure 5.1, namely the neighborhood effect, the hierarchical effect, and the S-shaped curve.

There were close links between this approach and work in rural sociology, from which geographers borrowed two ideas. The first was the recognition that adopting an innovation is not immediate upon receipt of the relevant information; rather, there is an adoption process that proceeds through stages—awareness, interest, evaluation, trial, and, finally, adoption. The second idea borrowed from sociology was the recognition that individuals varied in terms of innovativeness, that is, in their willingness to adopt an innovation. The standard assumption is that individual innovativeness follows a normal statistical distribution, with a few innovators who are often opinion leaders accepting the innovation quickly, a few laggards resisting the innovation for some length of time, and the vast majority of the population placed between these two extremes. ➤➤➤ **4**

Space, Not Culture?

Studies employing the methods pioneered by Hägerstrand were quite different from culture trait studies, favoring such topics as urban settlement and

a) The Neighborhood Effect

b) The Hierarchical Effect

c) The S-shaped Curve

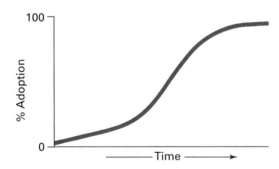

Figure 5.1 Three Empirical Regularities of the Innovation Diffusion Process. A principal concern of the spatial analytic approach to diffusion research was identifying regularities in the process of diffusion that were evident regardless of the trait studied.

In Figure 5.1a, the *neighborhood effect* refers to a situation where acceptance of an innovation is distance-biased, with adoption first achieved by a group of individuals living in close proximity to one another. Subsequent expansion occurs such that the probability of new adoptions is higher among those who live nearer the early adopters than among those who live further away. This regularity reflects a particular pattern of communications, most characteristic of rural or pre-industrial situations. It is useful to think of this regularity as similar to the result of throwing a pebble in a pond with ripples moving outward and losing intensity with increasing distance from where the pebble fell.

In Figure 5.1b, the *hierarchical effect* refers to a different interpretation of distance, with initial adoption occurring in the largest urban center and subsequent diffusion occurring both spatially and vertically down the size ladder of urban centers. Knowledge and adoption of the innovation leaps from center to center rather than spreading in a wave fashion as with the neighborhood effect. There are two key considerations affecting which of the two distance effects is more likely to occur. The first is the character of innovation itself—a new variety of corn seed will diffuse throughout an agricultural landscape, whereas a new style of fashion will diffuse through the urban system. The second is the larger cultural landscape in which the process is occurring—a face-to-face communication system favors a neighborhood effect, while a technologically more advanced system favors a hierarchical effect.

Figure 5.1c, the *S-shaped curve*, describes the expected growth through time of innovation adoption. The curve summarizes a process that begins slowly, then picks up pace, only to slow again in the final stage. There is a close relationship between the normal distribution of innovativeness in a population and this S-shaped curve—a cumulated normal distribution produces an S-shaped curve. Note that the three empirical regularities are not restricted to studies of diffusion processes; indeed, they are evident in various other areas of geographic analysis.

disease rather than specific landscape features—a classic example is Morrill's (1965) work on Swedish urban settlement. The most substantive example of a cultural geographic theme being analyzed in this manner is probably a study of settlement diffusion in the Polynesian region, which involved the use of a comprehensive simulation model to describe and explain a regional history of cultural

movement; however, this work failed to encourage cultural geographers to follow suit (Levison, Ward, and Webb 1973).

The spatial analytic approach to studying cultural diffusion incorporates processes aimed at understanding the creation of regional cultural and landscape differences. This reflects the fact that Hägerstrand was a cultural geographer in the

Swedish tradition and primarily interested in landscape change. Nevertheless, because Hägerstrand's work emphasized quantitative and modeling strategies, it was seen as lacking explicit cultural content and proved more attractive to North American quantitative geographers than to North American cultural geographers. Specifically, studies that employed the procedures introduced by Hägerstrand largely ignored the cultural content, focusing instead on the mechanics of the diffusion process. Thus, the spatial analytic approach to diffusion, although explicitly derived from traditional interests, was concerned mostly with process and model construction and less with culture and cultural landscape. This preference was in accord with prevailing interests and preferred approaches in the human geography of the 1960s.

Interest in spatial analytic diffusion research declined with the decline of interest in spatial

analysis generally after about 1970. A growing number of critics of this approach to diffusion studies identified several weaknesses, the most notable of which was the exclusion of culture, indeed of any recognition of human distinctiveness. Consequently, and as one part of the changes taking place in cultural geography in the early 1970s, a rather different approach to diffusion that reintroduced an explicit cultural emphasis came to the fore.

Diffusion, Culture, and Power

Although cultural geographers and others had long acknowledged that all members of a population are not equally receptive to an innovation—recognizing, for example, a distinction between more innovative urban dwellers and more conservative rural dwellers—the spatial analytic tradition often assumed uniform behavior for reasons of simplicity. Stimulated principally by ideas from political economy, the third approach to diffusion explicitly acknowledges that some groups, such as the poor, the less educated, the aged, and the unemployed, may be disadvantaged in having limited access to innovations. Accordingly, studies in this tradition stress the social, economic, and political conditions over which most individuals have little control. Because different diffusion processes operate in different contexts and have different causes, it is important to understand the political state and institutions. In some instances, those who are in a position to affect the innovative behavior of others may preempt valuable innovations.

The political economy approach recognizes that a process of innovation diffusion not only results in the presence of that innovation among a cultural group but also affects use of resources in space and over time. Thus, some innovations are time-saving, causing substantial shifts in the daily time budgets of household members, while other innovations, such as a village school in an agrarian society, are time-demanding. Analyses that emphasize these aspects are oriented less around the diffusion process and more around cultural change. This is not a return to the traditional approach, but it does represent a

Questions for Critical Thought

———— ⬦ ————

On the basis of this account, an important question needs to be asked. Does the spatial analytic approach to diffusion, with its emphasis on process, appear to be dehumanizing—excluding humans and their cultural identities? Certainly, most cultural geographers appear to have reached this conclusion, for they have largely ignored this work, preferring instead either to continue in the established Sauerian culture trait tradition or to seek alternative approaches. As noted, the spatial approach views diffusion as a communication process, transforming a landscape that is empty of a particular feature to one that includes that feature. Accordingly, there is limited concern with cultural origins and even less concern with cultural identities. Perhaps most critically, the simulation process effectively treats all individuals as identical to one another. Consider how you would alter this model of diffusion in order to understand contemporary processes of innovation diffusion that do not occur over material space, but rather occur through cyberspace. Is there still a need to understand processes of spatial diffusion in these contexts?

renewed concern with culture. Much work in the political economy tradition is in the context of the diffusion of agricultural innovations in the less developed world. →≫ **5**

Agricultural Diffusion in the Less Developed World

Two analyses of agricultural diffusion in Kenya illustrate the difference between the spatial analytic and the more politically based approaches. First, Garst (1974) described the diffusion of six agricultural products in one district of Kenya, indicating the way in which both spatial diffusion and the characteristic S-shaped adoption curves were affected by a wide range of physical, cultural, and infrastructural factors, including the availability of information, the location of processing plants, and various attempts by authorities to restrict some practices, such as tea growing, to particular areas (see Figure 5.2).

In the second analysis, Freeman (1985) discussed agricultural change throughout Kenya, focusing on the diffusion of coffee, pyrethrum, and processed dairy products, and concluded that preemption by early adopters was a critical part of the process, explaining both spatial spread and temporal growth (see Figure 5.3). Preemption—the process whereby those who are relatively privileged are able to adopt innovations early while effectively prohibiting others from adopting—may often be a critical factor permitting the rise of landed elites and rich peasants within a larger landscape of poverty. In many parts of the less developed world especially, those who adopt new innovations at an early stage may be entrenched elites who are able to transform initial profits related to early adoption into permanent profits. Impacts on the cultural landscape include a strengthening of the elites within a larger pattern of rural poverty and a dramatic reduction in the number of adopters of innovations. For example, Lakshman and Mayfield (1978) concluded it was not conservatism or "peasant psychology" that explained the high rate of non-adopters of high-yielding hybrid maize variety in Karnataka, India, during the Green Revolution—rather, it was the prohibitive costs of fertilizer and seed inputs that resulted in a social bias in adoption. This social bias reproduced the power of wealthier landowners, and

Figure 5.2 Agricultural Diffusion in Kisii District, Kenya, 1940-71. The Kisii district of western Kenya, occupied primarily by the Gusii cultural group, is a densely settled agricultural area that has experienced significant landscape change since about 1940 as a result of the introduction of new agricultural practices. For six of these practices, Garst (1974) discussed the diffusion process, identified the origin location, mapped spatial spread, and plotted temporal growth. The figure shows the adoption curves for coffee, pyrethrum, tea, passion fruit, grade cattle, and hybrid maize:

> Innovation diffusion among the Gusii follows a classic pattern. It is generally characterized by an initial rapid outward spread of adoption at low intensity with relatively little contrast between adopting and non-adopting areas. Later, new diffusion nodes appear at scattered locations while other nodes develop into peaks of higher adoption above the general low adoption level. As the peaks of higher adoption approach 100 per cent the adoption surface becomes highly irregular. Finally, the peaks spread out forming plateau-like surfaces of saturation acceptance that soon coalesce, producing saturated regions (Garst 1974:311).

Source: R.D. Garst, "Innovation Diffusion among the Gusii of Kenya", *Economic Geography* 50 (1974):303. This material is produced with permission of John Wiley & Sons Inc.

this miraculous technology of development, rather than trickling down to the rural masses, actually made conditions worse for some poorer farmers.

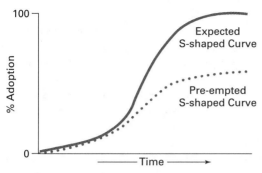

Figure 5.3 The Preempted S-shaped Curve. In a discussion of the importance of being first, Freeman (1985:17) contended that many diffusion analyses overlook "the frequent cases of preemption of valuable innovations by early adopters". With reference to the diffusion of coffee, tea, and processed dairy products in Kenya, preempted curves were explained by the presence of entrenched elites able to adopt early and then to restrict later adoption by others. Restriction was typically accomplished by means of production quotas, limiting the number of processing plants, and manipulating credit, farm extension, and social overhead capital. Source: D.B. Freeman, "The Importance of Being First: Preemption by Early Adopters of Farming Innovations in Kenya", *Annals of the Association of American Geographers* 75 (1985):18. Reprinted by permission of Taylor & Francis (http://www.tandfonline.com).

Cultural Contact and Transfer

Much cultural and related landscape change occurred as a consequence of population movement and related contact with others. Indeed, the expansion of European groups overseas between about 1450

Questions for Critical Thought

Have you noticed that this section on diffusion has not provided a substantive account of regional cultural landscape evolution? This is because most cultural geographers concerned with diffusion, regardless of the particular approach favored, have focused on specific questions rather than larger regional issues. If, as Carter (1978:56) argues, "diffusion is the master process in human culture change," what more might cultural geographers accomplish in their studies of diffusion and landscape?

and 1900 has directly or indirectly affected the lives and landscapes of peoples throughout much of the world. Accordingly, this discussion includes a general account of European overseas movement and a consideration of related Aboriginal–European contacts. A fuller consideration of links between European overseas movement and global regional concerns as developed in world systems theory is left to the following chapter. One feature of the material included in this section is the emphasis on the United States and Canada, along with some references to central and southern parts of the Americas, South Africa, Australia, and New Zealand. Before we begin to explore the era of contact initiated by European exploration, we must address the growing controversy surrounding the argument put forward by Menzies (2002) that Chinese explorers actually "discovered" North America 70 years before Columbus, circumnavigated the globe 100 years before Magellan, and reached Australia 350 years before Cook, and that China also ignited the European Renaissance by sharing its knowledge and strategies of exploration with Italy in 1434 (Menzies 2008). These explosive claims have been documented by Menzies in his popular texts, but have also been critiqued by historians as pseudo-science based on conjecture and circular reasoning (Finlay 2004). Supporters claim that Menzies's work has been corroborated by other researchers, but that historians resist his arguments because to accept them would highlight their own glaring deficiencies in uncritically reproducing knowledge deemed to be so foundational in our understanding of world history. While not venturing into this debate, it is important to highlight how it represents an important example of conflict over the production of knowledge. It also cannot go unnoticed how during a period when the balance of power is supposedly shifting away from European and North American hegemonic centers of power, the scholarship radically rewriting world history in favor of China finds such a fervent audience of supporters and detractors.

Did China Discover the New World?

Any discussion of European overseas expansion in the fifteenth century cannot help but reflect on the claims that China preempted the process of discovery. China was a great naval power in the early fifteenth century, and during his 21-year reign (1403–24) Emperor Zhu Di sponsored several voyages for the purposes of exploration, mapping, and collecting tribute from other peoples. Chinese navigators under the leadership of Admiral Zheng He reached the eastern coast of Africa and, according to Menzies (2002), may have circumnavigated the globe. But, with the emperor's death, the country entered a long period of isolation, and these expansion initiatives were not pursued further, nor were any official records of such voyages preserved.

Considering the above circumstances, a major difference between Europe and China was that the latter was a centralized empire whereas Europe comprised a number of competing political units. In China, one decision by one emperor was sufficient to terminate the exploratory process; in Europe, it would have been necessary for several countries to make such a decision at the same time—a highly unlikely circumstance—in order for European overseas expansion to end. In turn, explanations of this political difference between China and Europe often refer to basic physical geography, with China a compact landmass with a relatively short coastline in contrast to Europe's many peninsulas and islands and relatively long coastline. According to Landes (1998), the physical geography of China facilitated the spread of a single dominant cultural group and thus the creation of a centralized state.

Europe Overseas

How strange it is to find Englishmen, Germans, Frenchmen, Italians and Spaniards comfortably ensconced in places with names like Wollongong, (Australia), Rotorua (New Zealand), and Saskatoon (Canada), where obviously other peoples should dominate, as they must have at one time. (Crosby 1978:10)

Before about 1450, Europe was linked to Asia and Africa by only a few overland trade routes, while the Americas and Oceania were separate and mostly isolated regions; only the Islamic civilization had spread significantly by sea, with expansion east of the Arabian hearth to the island region of south and southeastern Asia. However, after about 1450, five European countries—Spain, Portugal, France, the Netherlands, and Britain—embarked on overseas exploration to areas previously beyond the sphere of European knowledge and influence. ➔➤➤ 6

Why Europe?

More than curiosity as to what lay beyond the horizon motivated European overseas movement. The exploration was not only scholarly; it was also political and economic—in short, it was about power and wealth. The initial desire to access trade routes and resources was the compelling motive.

Two fundamental questions, addressed primarily by historians, concern why exploration took place in the second half of the fifteenth century, and why this overseas movement started in Europe. After all, many of the advances in late medieval European culture, such as the invention of the compass and the astrolabe, were diffused from China, often via the Islamic world. Basic answers refer to a number of factors, including the advances in navigation and the increased understanding of trade winds and ocean currents already made by Portugal in the first half of the fifteenth century, the presence of government or merchant company support, the desire to expand trading activities and acquire wealth beyond the confines of Europe, and the missionary zeal of Christianity, which was especially apparent in the context of the ongoing conflict with Islam. It can further be argued that Europe was able to move overseas because of the region's distinctive physical geography and culture. The physical factors include Europe's climatic and topographic variety, its related spatial diversity of natural resources, and its extensive system of natural waterways that together prompted both co-operation through trade and territorial conflicts, which, in turn, contributed to the rise of several centers of power. The cultural factors include Europe's political fragmentation, which created a number of ambitious and competitive states; its systematized investigations in science, which led to a number of technological advances;

and its conception of individual property rights. An alternative argument might stress the proximity of Europe to North America, compared to the distance that Chinese or Islamic explorers needed to travel to reach the American continent. ➤➤➤ **7**

Ecological Imperialism

European explorers travelling overseas brought with them their conflicts, languages, religions, customs, and economic systems. At first, the Europe that was carried abroad was one characterized by feudal economic and social systems that privileged a few— many segments of the European population, including women, were seen as inferior beings—and by an attitude to nature that was essentially one of exploiting at will. Most of those who moved overseas, at least prior to the late eighteenth century, were male. Most of the colonial administrators were members of the privileged class, while many of the colonists were religious dissidents persecuted in their home areas. Their attitudes toward both the lands and the peoples they encountered were in accord with the European norms and values of the time, and the behaviors precipitated by those attitudes in many ways mirrored what was happening in Europe—a not unimportant point. Land was to be used as desired, and weaker peoples were to be used as labor—even as slaves— and moved as desired. If the indigenous population was judged unsuitable as a labor supply, slaves might be imported. In brief, Europeans, or more correctly those Europeans who moved overseas, saw themselves as superior to the peoples they met. The transition from feudalism to capitalism did little to change these general characteristics.

Of course, it was not only Europeans themselves who moved overseas but also European plants, animals, and diseases. Europeans were accompanied by what Crosby (1986:194) calls "a grunting, lowing, neighing, crowing, chirping, snarling, buzzing, self-replicating and world-altering avalanche". Indeed, many of the most aggressive plants in the overseas temperate humid regions today are of European descent, while horses, cattle, sheep, goats, rabbits, and pigs were all introduced to overseas areas by Europeans. But it was the introduction of disease, notably smallpox, that was to prove devastating to Aboriginal ways of life. The effects of European diseases were overwhelming, especially in the Americas, which had been isolated from the larger world since the land bridge between Asia and North America was reduced to the Aleutian Islands as a result of a rise in sea level following the end of the most recent ice age. Aboriginal population estimates for the Americas prior to European contact are uncertain, but Denevan (1992) estimates that the population decreased by 89 per cent between 1492 and 1650 (from 53.9 million to 5.6 million). Evidence also suggests that smallpox and other infectious diseases introduced by Europeans who first arrived on the eastern shores of North America were spread by Native traders westward along trading routes, thus moving the disease ahead of colonial expansion. As a consequence, by the time explorers and colonists reached the Midwest and western North America, Native communities had already reduced in numbers as a consequence of disease, resulting in even greater "justification" for Europeans to dispossess Native groups of their land and resources, since their population numbers seemed so small compared to the territories they inhabited (Boyd 1999; Boyd et al. 1991).

➤➤➤ **8**

The effects of disease varied spatially. Demographic collapse was not common in either Africa or Asia, whose populations, being less isolated from Europe than those of the Americas, had built up immunity to some of the diseases carried by Europeans. Nor did it occur in those more isolated areas, such as New Zealand, that were settled by Europeans after the introduction of vaccines. Even in the Americas, where some groups such as the Yahi of California and the Beothuk of Newfoundland disappeared altogether, there were others such as some Maya groups in Guatemala that survived relatively unscathed. Not surprisingly, the debates about population numbers in the Americas prior to 1492 and about the role played by disease in subsequent declines are far from resolved, with much of the debate linked to the divergent views of history that a choice of numbers implies.

Aboriginal-European Contacts: Changing Understandings

The long history of human movement has included numerous instances of different cultural groups coming into contact with one another. Contacts between foraging groups or shifting agriculturists were often an incidental consequence of the movement that was an integral part of the lifestyle of these groups. Most contacts of this kind were between groups with broadly similar cultures, including comparable levels of technology, and one result of contact was a gradual agreement, through experience, on territorial limits. Another result was the transfer of ideas and goods, a process of **transculturation**. In this way, a series of linked cultural regions came into being.

However, the context for European movement and the resulting contact with Aboriginal groups was quite different. As noted, the principal motivations for movement included the acquisition of power and wealth, while the expansion was carried out by cultures that typically viewed themselves as superior to those they met. Further, European newcomers were not simply fishermen or traders; they were representatives of Europe. The commercial system of early capitalism was both a cause and a component of contact. Aboriginal-European contacts were, then, of a different order from previous contacts between cultural groups: they involved very different cultures that had very different motivations for their behavior at the time of contact. The Europeans also possessed far superior technology, including, most significantly, weapons. The dissimilar cultural views of Europeans and Maoris are diagrammed in Figure 5.4.

Meinig (1986:205-13) supplied an overall context to clarify the specific details of cultural contact situations. He proposed a usual sequence of contact, followed by Aboriginal population loss and changes to social and ecological patterns, followed by the appearance of some form of dominance–dependence relationship, followed by some population recovery and further changes to social and ecological patterns, followed finally by some cultural stabilization.

At any given time, different areas contacted by Europeans were in different stages in this process of cultural change. The nature of interaction between different groups also varied depending upon geography, social structure, and economic imperative. For example, early fur traders depended upon Native hunters to supply them with goods.

Difference and Inferiority

There is another aspect to this discussion of European contact. The accounts written by Europeans did not present a fair and accurate picture of Aboriginal populations, who were not the unchanging peoples they were once thought to be. The obvious but typically neglected point is that the European view was necessarily flavored by ideas and assumptions that were a part of Europeans' own background and experience. It is because their cultural baggage shaped their perceptions that Europeans frequently lacked respect for Aboriginal cultures and values. Eliades (1987:33) states: "From the beginning of white–Indian contact, the Europeans, with only a few exceptions, refused to see the richness and diversity of Indian cultures. Instead of perceiving and accepting the native cultures as different, the Europeans saw them as inferior."

Sauer (1971:303) arrived at a rather different conclusion: "Throughout the eastern woodlands, the observers were impressed by the numbers, size, and good order of the settlements and by the appearance and civility of the people. The Indians were not seen as untutored savages, but as people living in a society of appreciated values." Similarly, with reference to the Canadian prairies, Friesen (1987:18) observed: "When contact was made, Europeans were quite prepared to recognize both the legal existence and the military power of the native peoples, to accept native possession of the land, and to negotiate with them for privileges of use and occupancy by the customary tools of diplomacy, including, of course, war." Nevertheless, the prevailing view in the scholarly literature continues to emphasize the negative assessments made by European newcomers. Simply expressed, Europeans equated difference with inferiority. A logical extension of this argument is that Aboriginal voices have been largely unheard because

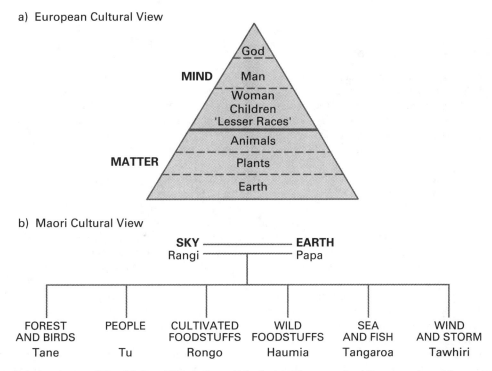

Figure 5.4 European and Maori Cultural Views. Figure 5.4a depicts the generalized European view of the world during the extended period of overseas expansion. There is a clear separation of humans and nature. As an indigenous group, Maoris are included in the "lesser race" category. Unlike the European view, the Maori view (5.4b) is premised on a close relationship between humans and nature, with nature providing humans with both material and spiritual sustenance. Source: Reproduced with permission from Anderson & Gale, *Cultural Geography 2e*, © 1999 Pearson Australia.

of the unequal power relations inherent in the process of contact and in subsequent events. ➔≫≫ **9**

The conceptual framework for much current research in this vein is based on contemporary social theories, including postcolonialism. Many Europeans overseas commented on what they observed, and these commentaries were informed by their ideological baggage. How, then, are contemporary readers to understand these commentaries? It is helpful to approach this challenge by identifying a dominant and culturally specific complex of assumptions and attitudes—usually White, male, middle class, and with a specific national identity. Perceptions and descriptions of Aboriginals are to be understood by reference to these assumptions and attitudes, to what Geertz (1973:3) describes as a "web

of significance". In other words, the accounts provided by Europeans are not to be treated as objective, factual, and truthful because such qualities cannot be achieved; rather, the accounts are mere representations, a handful among many possible understandings of reality, and they must be read with that understanding in mind. Work in the postmodern tradition often claims that knowledge is necessarily uncertain and that there is no one truth waiting to be discovered by the contemporary scholar; this idea is central to much current work on cultural contact scenarios. It is now usual to emphasize how the voices of the subaltern were usually omitted or misrepresented in official documents. Box 5.1 offers a discussion of changes in the scholarly understanding of cultural change.

Box 5.1   Cultural Change

It should not be surprising that scholarly understanding of Aboriginal–European contacts has changed a great deal since the first accounts appeared. These accounts were produced by Europeans and were necessarily from a European perspective. The characteristic view of Europeans at the time, shared by historians and anthropologists, was that civilization had encountered savagery. According to this view, Aboriginals had no history because their cultures were seen to be unchanged for long periods prior to contact, and unlike Europe, they did not possess any textual records—in other words, their history began with contact. Static cultures and their related unchanging cultural regions were often assumed to be simple responses to physical environment. This view was gradually replaced by the idea that a more complex civilization had encountered a less complex civilization, a distinction that was explicit in a 1930 study of the Canadian fur trade (Innis 1930).

Of course, the suggestion that there was Aboriginal cultural change prior to contact meant that an understanding of Aboriginals before contact could not be based on the observations of Europeans, for those Europeans were describing a dynamic, not a static, culture. Further, a traditional ethnography necessarily described cultures after, rather than before, contact, and was therefore describing a culture affected by contact. Archeological data are needed for descriptions of Aboriginal groups prior to contact, and there is today a large amount of archeological evidence to suggest that many of the apparent results of contacts were not really dramatic revisions of earlier culture, but rather logical extensions of change that was occurring before contact. For Huron groups in Ontario, for example, there was "a significant revival of intentional trade in the late prehistoric period and it was along the networks that supplied traditional prestige goods that European materials first seem to have reached the interior of eastern North America" (Trigger 1985:162). Of course, European movement and diffusion of their goods generated changes, but many of these changes need to be understood in the wider context of Aboriginal cultures that were already evolving. Further, an increasing number of scholars today recognize that it is not appropriate to assess the consequences of contact for Aboriginals in terms of such values of European culture as progress and change, as these are inherently Eurocentric. Rather, it is necessary to take into account local Aboriginal values, including spiritual beliefs and understandings of nature.

Aboriginal–European Contact: Cultural Change

The contact between incoming European populations and indigenous Aboriginal populations resulted in both groups experiencing cultural change, and in many cases it is not possible to clearly separate the roles played by the two groups in details of changes that followed contact. Certainly, the cultural change that took place involved both groups and did not simply involve the less dominant group almost unwittingly changing to adapt to the more dominant group. Thus, Europeans often adopted selected Aboriginal cultural traits judged to be advantageous in the new environmental setting.

In early French Canada, for example, the birchbark canoe was an important borrowing, as it was made from local materials, was able to carry heavy loads, and yet was light enough to be carried when necessary. More generally, much European exploration relied extensively on the knowledge provided by Aboriginal groups; in Australia, for example, inland movement by Europeans was dependent on Aboriginal knowledge of environments and of peoples encountered.

Resisting Change, Accepting Change

Although contact typically prompted some cultural change by the Aboriginal group, the specifics of such change were often the choice of the receiving group rather than an imposition by the incoming group. Raby (1973:36) offers the example of Native groups on the Canadian prairie: "The Indian bands of southern Saskatchewan . . . had by no means been converted into the competitive agrarian individualists sought by their white guardians. Much had been said to them of the virtues of agriculture. Farming, and not the supposedly demoralizing pursuits of hunting and fishing, was identified as work." Indeed, as this quote suggests, there was often resistance to Europeans' efforts to impose radical cultural change.

Aboriginal populations not only resisted many of the European attempts to impose change but in fact often accepted only those aspects of European culture they considered to be to their advantage and to involve the least disruption to their existing cultural context. For example, in the Queen Charlotte Islands of British Columbia between 1774 and the 1860s, the Haida experienced substantial population losses (because of disease), dramatic changes in their settlement pattern, and detailed changes to their seasonal cycle of activities. Changes in the Haida settlement pattern stemmed both from the abandonment and fusion of villages as a result of population loss and from the desire to enhance their participation within the fur trade. Impacts on the seasonal cycle involved changes in detail that did not affect the broader principle of seasonality. This was because the cycle served important cultural functions that proved resistant to substantive change. The introduction of a sedentary way of life was forcefully advocated by missionaries but was rejected because it was not in accord with Haida values.

➔≫≫**10** Similar conclusions can be drawn concerning the impacts of fur trading on the Canadian interior, where the continuity of traditional ways was a dominant feature of the early fur trade period. The European trade was conveniently fitted between established annual migrations that followed changes in animal location and plant resources. It is increasingly evident that Aboriginals were often very careful in their interactions with Europeans, adapting to the new challenges posed and often regarding the newcomers with disdain.

Imposed Change

In some areas there were active attempts to alter Aboriginal lifestyles. In Guatemala, for example, the military contact between 1524 and 1541 resulted in a Spanish takeover of both land and people, followed by conscious efforts to impose a Spanish way of life on the conquered groups. This forced cultural change involved both a mandatory resettlement—a process that invariably severed ties between the Aboriginal people and their ancestral lands—and heavy demands on native labor. The relationship between Spanish and Aboriginals became one of oppressor and oppressed, and institutionalized exploitation was characteristic of the colonial period; the hacienda system, for example, forced many Aboriginals into servitude on large plantations and introduced new tools, crops, and animals to the region. But such attempts to impose change were not always successful: for instance, Spanish control of Aztec populations in Mexico did not prevent those native groups from successfully retaining many of their traditional cultural values.

Dependence

One controversial issue that has emerged in recent scholarship on European contact in the Western Hemisphere concerns the view that the overexploitation of both fur-bearing animals and game by North American Aboriginal populations was caused not by a developing dependence on trade for European goods but rather by a growing belief among Aboriginal peoples that animal spirits were responsible for the ravages of disease. Some scholars argue that this new belief led Native groups not only to reject their traditional beliefs in favor of new Christian values but also to launch a deliberate extermination of animals (Martin 1978). This issue is part of two larger debates.

First, there is a debate about the characteristic Aboriginal human-and-nature relationship prior to European arrival. The conventional wisdom is that Aboriginal peoples were typically in harmony with nature, although there is evidence to suggest that some groups abused their environments and that Aboriginal peoples in general are not best described as ecological custodians. Second, there is a debate about the extent to which Aboriginal groups became dependent on the newcomers. Certainly, European culture often failed to satisfy Native aspirations. Referring to Australian Aboriginals, Reynolds (1982:129) stated that "young blacks who went willingly towards the Europeans fully expected to be able to participate in their obvious material abundance. Reciprocity and sharing were so fundamental in their own society that they probably expected to meet similar behavior when they crossed the racial frontier." This kind of expectation was rarely fulfilled.

Cultural change, **acculturation** but not necessarily **assimilation**, was an inevitable consequence of Aboriginal population losses, European attitudes of disdain and imperialism, and differences in technological sophistication, such that some version of a dominance–dependence relationship often arose. Although specific details vary from place to place, it was not unusual for Aboriginal peoples to become economically and/or politically dependent on European newcomers. Sometimes, as in many parts of Spanish America, this was effectively a forced dependence, while on other occasions, as in many North American fur trade areas, it was an unintentional effect of contact. Sometimes it occurred quickly, sometimes more slowly. Regardless, some form of dependence is a not uncommon consequence of contact that, in many instances, has carried through in some form to the present.

In the case of the North American fur trade, there is debate between those who emphasize that Aboriginal groups retained the ability to make choices about their economic lifestyles and those who see overriding forms of economic and political dependence. Certainly, in general terms, the ever-increasing exploitation of a limited resource base meant that those Aboriginal groups that became reliant on the trade often suffered once the trade declined. European cultural traits, such as guns and steel traps, facilitated the exploitation of resources that eventually resulted in the termination of the trade. It is possible that involvement in the trade may have prompted the rejection of traditional alternative ways of making a living. Further, before they began trading furs for European goods, Aboriginal

> ### Questions for Critical Thought
> — ✧ —
>
> Understanding the details of culture contact is far from easy. Facts are uncertain, and different ideologies promote different evaluations of available evidence. It is common to suggest that Aboriginals and Europeans entered into an uneven relationship characterized by misunderstandings about such matters as access to land and resources. The European desire to expand territory and to settle, in what were for them, new areas, necessarily clashed with Aboriginal understandings. In some areas forced movement occurred, while in other areas treaties were made that, from a European perspective, provided them with exclusive access to land and resources but that, from an Aboriginal perspective, merely allowed Europeans to share. Treaties signed are clear indications of the different understandings that were part of the contact process. Do these comments aid your understanding of land claim disputes between Aboriginals and government in countries such as Canada and Australia? More poignantly, do you agree that governments today need to express regret for the actions of early settlers and officials, as has happened in both Canada and Australia?

peoples exchanged among themselves only commodities that they produced, and this was a more equitable form of commerce. In the case of northern Manitoba, it is clear that the fur traders became wealthy while the Native peoples received only a marginal return for their labor.

Shaping Landscapes

Several German geographers, including Schlüter and Ratzel, advocated analyses of landscape evolution at the turn of the nineteenth century, making the concern with landscape evolution a central feature of the German cultural geography tradition. Similar ideas were evident in France with the writings of Vidal and his followers. Schlüter defined geography as the study of the visible landscape as it changed through time and "was the first to raise the landscape forming activity of man to a methodological principle" (Waibel, quoted in Dickinson 1969:132). Ratzel, meanwhile, "did not exaggerate the potency of the physical environment. . . . What saves him from such naiveté is the recognition of the time factor. . . . No one could emphasize more than Ratzel the force of past history" (Lowie 1937:120). In a similar vein, Vidal introduced the term "personality" into the geographic literature, noting that "geographic personality is something that grows through time" to create a distinct regional landscape (Dunbar 1974:28). Further, the French *Annales* school of historical research, initiated by Febvre and Bloch in 1929, has close intellectual ties with *la tradition Vidalienne*. ➔≫ 11

In the United States, Sauer built on these German and French ideas to provide an English-language conceptual basis for evolutionary analyses. For Sauer, natural landscapes when modified by humans become cultural landscapes. With this basic idea in mind, proponents of the evolutionary approach aim to understand contemporary landscapes as the outcome of long-term processes of changing relations between humans and land. The literature of cultural geography includes many studies of areas where the primary concern is the changing landscape, with one goal being to demonstrate that an understanding of past circumstances is a prerequisite to understanding the present. The following statement from Sauer, recorded by one of his students in 1936, emphasizes this focus: "We are what we are and do what we do and live as we live largely because of tradition and experience, not because of political and economic theory" (Sauer 1985:1). Interestingly, Sauer's ideas failed to attract much interest in British geography, where the early work of the leading British historical geographer, Darby, lacked both visible landscape and evolutionary content, and focused on past—rather than changing—landscapes.

Approaches to Historical Geography

How best to organise historical studies of landscape has been a question addressed by geographers for decades and there is clearly not a single, optimal solution: there is no methodological Holy Grail in historical geography, however diligently and ingeniously it is sought. (Baker 2003:133)

Historical geographers have identified and tested various approaches to the study of landscape change through time, often recognizing, however, that the differences among approaches are rarely substantive. ➔≫ 12

Box 5.2 outlines relevant approaches as noted by three reviews of work in historical geography. Three of the more popular of these are briefly discussed here, namely *narratives, cross-sections*, and *sequent occupance*.

The use of narrative in both history and historical geography is a well-established way to describe and even explain change through time. Indeed, for some historians, all history is narrative, since a narrative both describes and explains change. Although other historians disagree, seeing the narrative as failing to explain why things happen, historical and cultural geographers have chosen not to enter this debate, preferring instead to integrate a narrative approach with the cross-sectional approach.

The aim of the cross-sectional approach is to select a particular moment in time, describe the geography of that time, and then devise a narrative connecting this to another moment in time. Studies such as these do not necessarily convey details of change through time, but rather focus on amounts of change between selected times; these studies are concerned with changing geographies rather than geographical change. Some cross-sectional studies are better characterized as descriptions of past geographies in that, for reasons of data availability, they are often limited to one moment in time. The classic work in this vein was conducted by the leading British historical geographer, Darby, in several book-length studies of England in 1086 that used data from the Domesday Book to create a series of more than 800 distribution maps (see, for example, Darby 1977).

A study of Scandinavian historical geography employed a genetic and evolutionary approach to the cross-sectional method. Mead identified five stages, or cross-sections, and prepared a series of connecting narratives to link successive stages (see Figure 5.5). "The object is to pause at certain stepping stones in the stream and to look at the surrounding stones. They will be scenes of the same place, but they will

Box 5.2 Changing Geographies—Some Historical Geographic Approaches

Historical geographers have embraced a wide range of approaches to their studies, and there are many methodological writings that attempt to classify these approaches. This box summarizes three of these attempts. The approaches identified refer to the two principal historical geographic concerns, namely study of the past and study of change.

First, Smith (1965:120) notes four approaches, two of which (those indicated with an *) refer to the study of change:

- operation of the geographic factor in history
- reconstruction of past geographies
- study of geographic change through time*
- evolution of the cultural landscape*

Second, Newcomb (1969) noted twelve approaches, six of which (those indicated with an *) refer to the study of change:

- historical regional geography
- areal differentiation of remnants of the historic past
- *genre de vie*
- theoretical model
- pragmatic preservation of landscape legacies
- past perceptual lenses
- temporal cross-section*
- vertical theme*
- cross-section–vertical blend*
- retrogressive*
- dynamic culture history*
- humans as agents of landscape change*

Third, Prince (1971) noted three themes, namely the study of real worlds, perceived worlds, and theoretical worlds. With regard to real worlds, Prince identified studies of past geographies, of geographical change, and of processes of change, and listed a total of fourteen approaches, nine of which (those indicated with an *) refer to the study of change:

- past geographies
- *géohistoire* (reconstructing past landscapes in order to aid understanding of everyday life)
- *urlandschaften* (reconstructing the prehuman landscape)
- static cross-sections
- sources and reconstructions
- narratives of changes*
- geographical change
- sequent occupance*
- evolutionary succession*
- episodic change*
- frontier hypothesis*
- morphogenesis of cultural landscapes*
- agency of humans*
- rates of change*
- processes of change
- dynamics of change*
- inadequacy of inductivism* (an appreciation that facts do not speak for themselves)

Although historical geographers have found it possible to classify their preferred approaches in a variety of ways, it is clear that many of the approaches refer to change through time, including one that Prince chooses to include under the heading of studies of the past. This preference is notwithstanding the opposition that this approach faced from the then dominant regional approach to geography before about the mid-1950s. Certainly, the popularity of studies of change owes a great deal to the advocacy of Sauer and his student Clark.

present patterns and distributions particular to different times. The changes between the scenes are explained by the processes—physical and human—that have characterized the intervening years" (Mead 1981:1). ➡️»» **13**

Following Sauer, most cultural geographers who have employed cross-sections have done so in a rather specific format, namely the approach known as **sequent occupance**. Studies in this vein recognize that a landscape remains relatively stable for a period of time, but then undergoes change that is both rapid and substantial, leading into another period of stability. Thus, a succession of cultures is associated with a succession of cultural landscapes. Each period

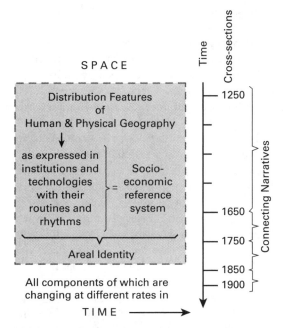

SPACE

Time

Cross-sections

Distribution Features
of
Human & Physical Geography

↓

as expressed in institutions and technologies with their routines and rhythms

=

Socio-economic reference system

Connecting Narratives

— 1250

— 1650

— 1750

— 1850
— 1900

Areal Identity

All components of which are changing at different rates in

TIME ⟶

Figure 5.5 Cross-Sectional Historical Geographies of Scandinavia. There are five cross-sectional analyses describing geographic patterns for the years 1250, 1650, 1750, 1850, and 1900. The time between cross-sections decreases because the pace of change increases through time, with, for example, more change occurring during the final 100 years studied than occurred during the preceding 500 years. Each of the cross-sections is discussed in a systematic manner, with accounts of political, population, agricultural, resource, and communication geographies. "At any given cross section, a number of people with particular sources of energy at their disposal will be viewing the options open to them and shifting the directions of Scandinavian development responsively. Simultaneously, they will bequeath a legacy of evidence for the geographical understanding of their times which will reflect their degree of literacy and numeracy" (Mead 1981:3). Connecting narratives that describe processes of change explain the transition from one cross-section to the next, and there is an additional narrative for the period since 1900. Each cross-section and each linking phase is discussed in a separate chapter. This approach is based on that developed by Darby (1973) in a study of the historical geography of England. Source: Reprinted from *An Historical Geography of Scandinavia*, W.R. Mead. (London: Academic Press, 1981): 2.

of stability reflects a particular cultural occupance, but is transformed into a subsequent period by some evolutionary or diffusionist process. One stage may be transformed into another because of the in-movement of a different cultural group or because of some substantial change in, for example, technology or economic system. In this sense, any landscape is a **palimpsest** in that it comprises the consequences of a series of different occupations. Whittlesey (1929:162) stated: "The view of geography as a succession of stages of human occupance establishes the genetics of each stage of human occupance in terms of its predecessor."

A classic sequent occupance analysis in the Sauerian tradition is Broek's (1932) study of California's Santa Clara Valley, which was designed to understand landscape change as a result of a succession of different cultural occupances within a 200-year period. The first cultural occupance was Aboriginal; the second stage was Spanish occupation, which was characterized by missions and cattle ranches; the third stage was an early American economy of wheat and cattle farming; and the fourth stage was dominated by horticultural activity. Looking back at this work in 1965, Broek (1965:29) noted that it was appropriate to add a fifth stage dominated by urbanization.

More generally, many world histories that consider human activity during the past 10,000 years or so stress the transitions from foraging to agriculture and from agriculture to industry as relatively brief periods of substantial change separated by lengthy periods of relative stability.

The differences between a historical geographic cross-sectional analysis and a cultural geographic sequent occupant analysis are worth noting. In the cross-sectional analysis, there is no suggestion that each cross-section is a stage separated from the previous and the following stages by some significant change of occupance. Rather, the cross-sections selected are intended to be representative of some larger period; relatively speaking, they are arbitrarily chosen. Thus, in a cross-sectional analysis, the cross-sections are particular moments in time and are separated by much longer periods of change. In a sequent occupance analysis, the stages described are lengthy and separated by briefer periods of rapid and significant change. Expressed another way, a cross-sectional approach emphasizes patterns at particular moments in time, albeit with process-oriented linking narratives, whereas sequent occupance places a higher priority on the processes operating to establish and maintain a particular stage of occupance.

Frontier Experiences

As part of the attempt to explain changing geographies in areas of European overseas expansion, especially the United States, cultural and historical geographers have focused on the frontier concept through analyses of the cultural and landscape changes that occurred as European groups embarked on new lives in what were, for them, new places.

The American historian Frederick Jackson Turner (1861–1932) introduced the influential **frontier thesis** in the late nineteenth century. The American **frontier** was understood as both place and process. As a place, it was defined in terms of population density, with a figure of two people per square mile suggested as marking the end of the frontier phase. In terms of process, Turner, employing terms popular in anthropology at the time, proposed that the American frontier was the meeting point between savagery and civilization, with the European newcomers initially borrowing many cultural traits from Aboriginal groups before developing their own distinctive cultural identity. Thus, the frontier was where Europeans became Americans: "American development has exhibited not merely advance along a straight line, but a return to primitive conditions on a continually advancing frontier line. . . . American social development has been continually beginning over again on the frontier" (Turner 1961:38). Within history, these ideas have been further developed in major studies of the Great Plains and other world frontiers. Despite being typically ill defined, the frontier concept has been seen as key to understanding the larger European experience overseas. →»14

Meinig (1993:258–64) proposed an alternative to Turnerian logic, namely a six-stage transition from a North American traditional system, through a modern phase, to a world system. The six stages are titled *Indian society, imperial frontier, mercantile frontier, speculative frontier, shakeout and selective growth*, and *toward consolidation*. Each of these stages is characterized by some particular cultural, economic, and political circumstances.

Transferring Cultural Baggage

"It is a striking insight", comments Harris (1977: 469), "that Europeans established overseas drastically simplified versions of European society." One possible explanation for such **simplification** is that only a fragment of larger European culture moved, and this movement was followed by an often-lengthy period of isolation in a new and different environment. This combination of a particular cultural fragment and subsequent isolation resulted in a new and simpler identity. Cultural change in Australia might be appropriately interpreted in terms of this fragment concept. An alternative explanation centers on the particular conditions, specifically the confrontation with a new land that Europeans experienced. In early French Canada, South Africa, and New Zealand, for example, Europeans with a strong sense of family and a desire for private ownership of land encountered areas of inexpensive land and limited markets for agricultural products, with the result that a homogeneous society based on the nuclear family dominated; in other words, a process of simplification occurred. The initial transplanted population might have been culturally differentiated, but the resulting settler society was culturally uniform. Rejecting the Turner thesis because of its explicit environmental determinism, Sauer (1963:49) observed that the European frontier experience overseas was affected by "the physical character of the country, by the civilization that was brought in, and by the moment of history that was involved". This observation raised three questions: What cultural baggage did an incoming population bring to an area? What happened to that baggage? To what extent did the shaping of landscape reflect cultural characteristics?

The simplification thesis suggests a loss of baggage, perhaps implying that settlers from different backgrounds in the same area did not shape markedly different landscapes. A loss of cultural traits—and, therefore, of identity—is indeed characteristic of many cultural groups after movement to a different area. One explanation for such developments relates to the fact that the motivation behind most settler activity was economic, and adopting appropriate behavior given the market economy tended to be more important than preserving some cultural traits. Thus, Swedes and Norwegians in the Upper Midwest grew wheat, while Germans in Texas adopted cotton and tobacco cultivation. Evidence from eighteenth-century southeastern Pennsylvania shows that differences in land use could not really be attributed to differences in settler source area, while evidence from Kansas points to an even more rapid loss of cultural identity than occurred in eastern North American areas settled earlier and under different circumstances. Evidence from Scandinavian converts in the Mormon region of the intermontane West suggests that the role played by the new religious identity was so important that culture traits associated with area of origin were quickly lost. ➔➤➤ **15**

But there is also a considerable body of literature demonstrating the effects of different groups in areas of European overseas settlement. For example, North American studies have often examined the role played by the ethnic background of settlers. For many cultural groups, the institution that played the largest role in minimizing trait loss was the church; indeed, in cultures primarily united by religion, the rate of trait loss was often very low.

An illuminating example of factors affecting retention or loss of cultural baggage is that of Irish settlers in nineteenth-century eastern Canada. Taking into account material folk culture and settlement morphology, the movement of Irish groups to Canada typically involved a loss of cultural traits, but the rate of loss varied substantially between three eastern Canadian locations. Retention of cultural baggage was greatest in the Avalon Peninsula of Newfoundland, and loss was greatest in the Peterborough area of Ontario; the Miramichi River area of New Brunswick represents a middle ground. The different experiences in these areas reflected two principal considerations, namely physical environment and proximity to other incoming groups. Thus, the Avalon Peninsula encouraged retention of traits because the environment was similar to that of the source area and because there were few other newcomers with whom to interact. The environment of the Peterborough area, on the other hand, was unlike that of the source area and included a number of other cultural groups. The loss of traits in the Peterborough area did not, therefore, necessarily imply a simplification of culture as there was trait borrowing from other groups to compensate for the lost traits.

In the case of German settlers in Texas, imported traits were an important factor related to agricultural landscapes, despite the fact that some of these traits were lost soon after arrival, while some new traits were adopted from other groups. Certainly, there are numerous examples of particular agricultural practices associated with certain ethnic groups and with distinctive cultural landscapes. Trait loss

---

Questions for Critical Thought

Although the simplification of some European cultures overseas is well documented, especially prior to the large-scale movements of the nineteenth century, the significance of such simplifications is debatable. Is it possible that the frontier experience played only a minor role in shaping immigrant cultures in areas of European overseas expansion? Because many frontiers were settled by few people and soon overwhelmed by the spread of a commercial economy, was the frontier itself relatively unimportant? On the one hand, think about the prolonged frontier circumstances that prevailed in a few relatively undesirable areas, such as Appalachia or much of South Africa. On the other hand, think about areas such as the Shenandoah Valley in eighteenth-century Virginia, or southern Ontario, where the frontier population, which included many opportunistic individuals, moved quickly into the new economy of commercial agriculture, shedding unnecessary cultural traits and adopting necessary new traits. Is there a history of European settlement in your local region, and if so how does it compare to these examples?

may indeed have been usual in response to economic circumstances, but it was certainly possible for some of the new traits acquired to subsequently become associated with a particular group.

In summary, it appears that trait retention by a cultural group moving to a new environment was most likely if

- the physical environments of the source and destination areas were similar;
- a dominant goal of the group was to seek a location separate from others;
- contact with other groups was limited;
- the group was socially cohesive, possessing shared values;
- the impact of external institutions was limited;
- the number of immigrants was substantial; and
- settlement was contiguous.

Evolutionary Regional Studies

Landscapes, cultural regions, or geographic personalities develop in response to cultural occupance through time, becoming, "as it were, a medal struck in the likeness of a people" (Vidal, quoted in Broek and Webb 1978:32). With this evocative phrase, Vidal captured both the concept of *genre de vie* and the Sauerian idea of humans occupying physical landscapes to create cultural landscapes through time. ➔»» 16

Writing in 1930 with reference to the American West, Sauer (1963:45) offered a clear statement of the historical geographic or evolutionary approach to landscape study: "The three major questions in historical geography are: (1) What was the physical character of the country, especially as to vegetation, before the intrusion of man? (2) Where and how were the nuclei of settlement established, and what was the character of this frontier economy? (3) What successions of settlement and land utilization have taken place?" Further: "In order to evaluate the sites that were occupied, it becomes necessary to know them as to their condition at the time of occupation. Only thus do we get the necessary datum line to measure the amount and character of transformations

induced by culture" (Sauer 1963:46). With these arguments in mind, Sauer discussed the physical landscape, the Aboriginal landscape, and the several European frontiers that moved west, resulting in a series of cultural successions on the landscape. He regarded frontiers as secondary cultural hearths.

In an earlier discussion of Mexico, Sauer (1941b) had used the term "personality" to refer to the relationship between human life and land. Because the roots of Mexican life lie in a distant past, Sauer looked to transformations that had occurred in that past in order to gain an understanding of contemporary Mexico. Thus, he identified a pre-Spanish and a Spanish past as the principal bases for the present cultural geography. He also made a critical distinction between the two cultural regions of northern and southern Mexico that were evident before, during, and after Spanish occupation (see Figure 5.6). Sauer described the north as less advanced and the south as more advanced, and concluded: "The old line between the civilized south and the Chichimeca has been blurred somewhat, but it still stands. In that antithesis, which at times means conflict and at others a complementing of qualities, lie the strength and weakness, the tension and harmony that make the personality of Mexico" (Sauer 1941b:364).

A work such as this, full of generalizations that were supported with factual detail, is characteristic of much early landscape school work—ambitious, thoughtful, and provocative. Box 5.3 notes a few of the many other examples of work in this "making landscape" tradition.

Reading the Landscape

Observing and reading a landscape is far from easy, as McIlwraith attests when he comments: "It has been said that sight is a faculty but that seeing is an art that must be cultivated and kept finely honed through persistent field observation" (McIlwraith 1997:377). There are two important traditions concerned with seeing and reading landscapes, and though they appear to be quite different, they are in fact not so dissimilar. The first is an explicitly historical approach, essentially a form of narrative, as described above, that is concerned with detailed field investigations of local landscapes. It is primarily the product of British researchers and is closely associated with the historian W.G. Hoskins. The second is a more broadly conceived endeavor, with strong historical content certainly but with some additional defining characteristics, such as a concern with landscape symbolism and with the vernacular in landscape. This is the *Landscape* magazine tradition and is closely associated with J.B. Jackson, a private scholar who described himself as a lay geographer.

The Local and Regional Narrative History Tradition

A classic study concerned with the making of landscape appeared in 1955. *The Making of the English Landscape* (Hoskins 1955) was a trailblazing book, which, although initially received without any real acclaim, led to a series of local county landscape histories and, indeed, to a television series in the 1970s. Meinig (1979b:209) quoted Hoskins as saying: "I once wrote a book with the simple title of *The Making of the English Landscape*, but I ought to have called it *The Morphogenesis of the Cultural*

Figure 5.6 A North–South Cultural Division in Mexico. This is a good example of the type of bold map produced by Sauer and other cultural geographers in the early landscape school tradition. The boundary drawn is explained in cultural, not environmental, terms, although there is also a climatic distinction in that the north is mostly arid or semiarid but includes some alluvial valleys and good upland soils, while the south is generally better watered upland and includes some rainforest. The Spanish distinguished between Aboriginals of the more cultured south, the Aztec region, and the less cultured north, the Chichimeca region. Indeed, the southern region is one of the great cultural hearths of the world. Further, the Spanish recognized the south as a region of peace and the north as a region of conflict. At the time of Spanish in-movement, the south was largely agricultural while the north was a mix of agricultural and foraging activities. Source: C.O. Sauer, "The Personality of Mexico", *Geographical Review* 31 (1941):355. This material is reproduced with permission of John Wiley & Sons, Inc.

Box 5.3 *Making the Regional Cultural Landscape*

There are several book-length works in historical geography that are closely related to the evolutionary landscape school approach; four examples are noted briefly in this box.

First, Davidson (1974) studied a small group of islands, the Bay Islands, off Honduras. Focusing on a description of the physical landscape and on a reconstruction of past landscapes, he explicitly identified culture as a cause of change. He incorporated a sequent occupance focus, noting eight cultural stages, each affecting the material landscape. He identified as the principal cause of change over the long term an ongoing conflict between English and Spanish.

Second, Williams (1974) discussed the making of the South Australian landscape, stressing the details and consequences of land clearing, such as swamp draining and woodland removal, as well as the consequences of changing technologies and incoming cultural groups.

Third, Lambert (1985), in a historical geography of the Netherlands, described the physical landscape and analyzed the impacts on that landscape, focusing on the resulting material landscape as the key concern. Notwithstanding this focus, Lambert's study displayed relatively little concern with culture as a cause of landscape change, concentrating instead on economics.

Fourth, a study of the Cistercian monastic order in medieval England (Donkin 1978) is in a rather different vein, being described as an attempt to "recreate the changing landscape of a culture rather than a cultural landscape" (Donkin 1997:248).

Environment to make the fullest impact." Certainly, the book is now considered a pioneer work in the field of local history. ➔➤➤ **17**

Hoskins wrote the book to provide a detailed analysis of the landscape as it was made over time by humans. He achieved this with a series of chronologically arranged chapters, each dealing with a particular period of landscape-forming activity. As Hoskins makes clear, landscape evolution is his primary concern and the motivation for his work:

> Despite the multitude of books about English landscape and scenery, and the flood of topographical books in general, there is not one book which deals with the historical evolution of landscape as we know it. . . . What I have done is to take the landscape of England as it appears today, and to explain as far as I am able how it came to assume its present form, how the details came to be inserted, and when. At all points I have tried to relate my explanation to the things that can be seen today by any curious and intelligent traveller going around his native land. There is

no part of England, however unpromising it may appear at first sight, that is not full of questions for those who have a sense of the past. (Hoskins 1955:13–14)

The originality of this work as a piece of historical scholarship lay in the use of the landscape itself as a source of information so much more valuable, given the aims of the study, than documents and books on local history. Hoskins wanted to read the landscape, not simply to read about it. His work demonstrates a recurring concern with the local rather than the larger regional or national scene. Further, Hoskins (1955:231) shows a concern about what he perceived to be some of the more unfortunate, even painful, recent developments in landscape: "Especially since the year 1914, every single change in the English landscape has either uglified it, or destroyed its meaning, or both."

The Landscape *Magazine Tradition*
The most substantive contribution to the literature concerned with the making of landscape, other than that associated with the landscape school itself, is

without a doubt the tradition initiated and sustained by Jackson that began with the publication of the magazine *Landscape* in 1950. Over the next 17 years, Jackson financed, published, edited, and contributed much material to the periodical. Initially subtitled *Human Geography of the Southwest*, the magazine soon outgrew its regional focus as well as the link with a single discipline to develop into an acclaimed outlet for articles adding to understanding of landscape. The distinctive, thought-provoking contents of this magazine struck a chord with many academics and others interested in landscape topics, successfully filling a void in the literature that Jackson had identified. ➡➤➤ **18**

Six principal characteristics of the tradition initiated by Jackson combined to capture the interest of readers. First, Jackson's magazine recognized that landscapes are evolving, such that there is no such thing as a static or finished landscape, a suggestion that was in accord with the ideas of cultural geographers who were steeped in the landscape school tradition. Second, Jackson stressed that to change a landscape in a desirable direction necessarily implied changing the culture of the occupants, an idea that is always a controversial matter—recall the reference to social engineering in Box 2.1, for example. For Jackson, as with Hoskins, this topic was an important one because of the sometimes undesirable tendencies of contemporary landscape-forming activity. Third, *Landscape* emphasized the vernacular, the everyday commonplace features in landscape such as houses, and it is in this context that Lewis (1975:1) used the phrase "Common houses, cultural spoor". Fourth, the tradition viewed landscapes as related to prevailing ideologies and interpreted them in relation to the values of their occupants. Landscapes were, according to this view, composed of material things but also had symbolic character. Fifth, recognizing the symbolic quality of landscape introduced the need to read the landscape in order to learn both about that landscape and about those who authored it. For Lewis (1979), a landscape was seen as an "unwitting autobiography"; more accurately, it is an autobiography that is forever being revised. Sixth, Jackson was concerned with the idea

that landscape was both a reflection of the unity of humans and nature and a means by which it was possible to understand those who helped to create the landscape.

The tradition initiated by Jackson is original, but it would be misleading to give the impression that it is quite different from landscape school writing. There are clear links with the ideas of Sauer, and not surprisingly, Sauer wrote for *Landscape* magazine on several occasions. With regard to the study of vernacular landscapes, recall that Bowman (1934:149–50), in an early review of human geography as a social science, explicitly identified the landscape school concern as follows: "Wherever man enters the scene he immediately alters the natural landscape, not in a haphazard way but according to the culture system which he brings with him, his house groupings, tools, and ways of satisfying needs." Further, the original landscape school diagram prepared by Sauer referred to landscape features including housing plan and housing structure (Figure 2.2). Sauer was by no means unconcerned with the vernacular landscape. Nevertheless, the *Landscape* magazine tradition has inspired many cultural geographers with such original ideas as its concerns with symbolism and landscape reading, ideas that clearly anticipate some of the work on symbolic landscapes discussed later in this book. Box 5.4 suggests that much of the work accomplished in this tradition might be best thought of as the geography of the everyday landscape.

Making the American Landscape

In introducing *The Making of the American Landscape*, Conzen (2010:vii) noted: "This book has its roots in the fertile bicontinental traditions of landscape study nurtured by William G. Hoskins and John Brinckerhoff Jackson, and it was written in the belief that nothing quite like it exists in the American literature, and that there is a place for it." The organization of the book reflects the complex processes that were understood to have contributed to the building of the American landscape. Thus, there is an overview of the physical environment, an account of the approximately 15,000-year Aboriginal presence, and discussions of the principal

Box 5.4 *Everyday Landscapes*

The concern in much of the writing associated with the *Landscape* magazine tradition is a concern with the everyday landscape, with the majority of people in a landscape and not with those few who are wealthy and exercise power. These everyday landscapes come into being and continually change because they reflect the changing ideas, values, and needs of ordinary people. For geographers such as Aschmann, landscapes were studied "as places where we could live and work and celebrate together" (Jackson 1997b:vii), while Clay (for example, 1980, 1987, 1994) has written a series of interpretive essays on the landscapes and character of urban areas. "Ready and waiting to be found out there in our visible landscape, there exists an observable and universal order that speaks eloquently to us, that will tell us far more of ourselves and our future than we can see at any one moment" (Clay 1987:1).

From this perspective, we only have to see what is right in front of our eyes. This interpretation sees an orderly, indeed grammatical, landscape. In a similar vein, Hough (1990) focused on the way that traditional vernacular landscapes reflected the character of an area and the identity of people, thus emphasizing differences between places, while designed landscapes tended to negate the differences between places as a result of the fact that local populations were not responsible for their creation.

There are many other books and articles that reflect this *Landscape* magazine interest in the everyday world. A few examples of titles, or phrases taken from titles, are the following:

- *House Form and Culture* (Rapoport 1969)
- *The Accessible Landscape* (Jackson 1974)
- *The Look of the Land* (Hart 1975)

colonizing cultures, namely Spanish, French, and British. Conzen argues that relatively speaking, the Spanish and French cultural occupations were spatially restricted, leading to the formation of cultural regions that are losing their distinctiveness over time. The British cultural occupation involved a series of traditions representing various population groups and the variety of physical environments they encountered. The most influential British tradition originated in New England and southeastern Pennsylvania and spread west over much of the larger American area, while the British plantation tradition expanded west throughout much of the American South. →≫ 19

These sections on principal colonizing cultures are followed in the book by accounts of the national settlement policy that was initiated after political independence, and of the effects of that policy on three principal physical environments: forest, grassland, and desert. The book identifies and discusses two landscape preferences: the first involved shedding cultural baggage, typically to create a landscape

of single-family farms and homes; the second involved retention of cultural roots. Conzen then evaluates industrial, urban, and transport processes as they have involved changes in landscape. The book concludes with sections on the role of central authority, the importance of power and wealth, and the vernacular landscape.

The multilayered organization of *The Making of the American Landscape* is thus an accurate reflection of the complexity of the subject matter, an acknowledgment of the fact that landscapes are shaped through time by a multitude of processes that are not always easily or readily distinguishable.

Making the Ontario Landscape

A second valuable addition to the body of literature on the reading of landscape is the study by McIlwraith (1997) on two centuries of landscape change in Ontario. McIlwraith's stated concern is regional, not local, and his approach is very much in the *Landscape* magazine tradition. His aim is to read landscape as a reflection of the way in which

- *Ordinary Landscapes* (Meinig 1979a; Groth and Bressi 1997)
- *Common Landscapes* (Stilgoe 1982)
- *Learning from Looking* (Lewis 1983)
- *Discovering the Vernacular Landscape* (Jackson 1984)
- *Wood, Brick, and Stone* (Noble 1984)
- *Right Before Your Eyes* (Clay 1987)
- *The Visual Elements of Landscape* (Jakle 1987)
- *Real Places* (Clay 1994)
- *The Gas Station in America* (Jakle 1994)
- *People and Buildings* (Marshall 1995)
- *Landscape in Sight* (Jackson 1997a)
- *Fast Food: Roadside Restaurants in the Automobile Age* (Jakle and Sculle 1999)
- *City Lights* (Jakle 2001)
- *Lost Diners and Roadside Restaurants* (Anderson 2001)
- *Lots of Parking* (Jakle and Sculle 2004)
- *Ordinary Affects* (Stewart 2007)
- *The Highway Experience in America* (Jakle and Sculle 2008)
- *America's Main Street Hotels* (Jakle and Sculle 2009)

Capsule phrases such as these reflect this tradition of reading landscapes created by the majority of people to serve the needs of their daily lives rather than focusing on the behavior and landscape consequences of elite populations. An interesting corollary of this focus is that unusual landscape features were necessarily of lesser concern, and *Landscape* magazine occasionally published pieces that were critical of some conservationist groups that emphasized the unusual at the expense of the everyday.

ordinary people lived their lives, and hence he shows an interest in both material and symbolic landscape. McIlwraith achieves this aim through two sets of discussions. First, he establishes a context through accounts of attitudes to land, of historical matters, of the land survey system, of the process of determining place names, and of the technologies and resources employed in the making of landscape. Second, he proceeds from these contextual issues with a consideration of particular landscape features and of the clustering of features. The book includes discussions of traditional house architecture as it reflected the consistency of vernacular behavior, as well as accounts of churches, schools, barns, fences, industrial buildings, and gravestones. Clusters of features that McIlwraith considers include farms, roadsides, transport systems, and town streetscapes.

Making the Irish Landscape

There is a tradition of historical cultural geography in Ireland that stands somewhat apart from the various approaches identified so far, but that is explicitly concerned with understanding both how the Irish landscape was shaped through time and how landscape reflects the culture of those who did the shaping. Indeed, one reason this tradition is different is that it is so eclectic in character, being based on an integration of ideas from geography, anthropology, and history. In this respect it stands apart especially from the English historical geography tradition, within which, until recently, very few attempted to integrate geography, anthropology, and history—the principal exception being Fleure (1951).

The key figure in the Irish tradition was E. Estyn Evans (1905–89), whose influential university career extended from 1928 to 1968. For Evans, an understanding of Ireland and the Irish landscape involved a wealth of interests and approaches. There were concerns shared with *la tradition Vidalienne* and the related *Annales* historical school, especially regarding *genre de vie* and *milieu*. Such interests are not surprising in light of the plurality of regional cultural identities evident throughout Ireland. Evans was also concerned with the evolution of the material cultural

---

Questions for Critical Thought

——— ✤ ———

Because landscapes are indeed shaped through time, is it possible and appropriate to observe a landscape and to seek to read the detailed history of that landscape—to regard the landscape as a text, as suggested in Box 3.5? Rather than emphasizing the cultural causes and the landscape consequences in the Sauerian tradition, why not investigate the contemporary landscape in terms of how each of the features came into being, asking questions such as: When was the feature added? Under what circumstances was it added? Why was it added? Who was responsible? Who benefits and who loses from this change in landscape?

---

landscape in response to cultural occupance, in close accord with the landscape school ideas of Sauer. As noted, Evans's willingness to incorporate these ideas was in marked distinction to the general neglect of these ideas in the more dominant English historical geography tradition. In accord with both Vidal and Sauer, Evans employed the term "personality" to refer to the idea that through a long history of human occupance, a distinctive character was impressed on the Irish landscape. ➔≫ 20

Evans maintained a focus also on fieldwork and related observation of both people and landscape. Such approaches were advocated by Sauer and practiced in the British local history tradition initiated by Hoskins. The concern with fieldwork also involved interests in common people and their way of life, and therefore in the related vernacular landscape. Further, the explicit linking of culture and landscape included recognition of the symbolic content of landscape, although this was expressly limited to the rural scene. The interests in vernacular landscapes and in landscape as symbol are, of course, central to the *Landscape* magazine tradition.

Although this eclectic approach to historical cultural geography is consistent with the related ideas of both Vidal and Sauer, the additional interests result in it being a distinctive tradition. It was practiced by Evans and others throughout much of the

twentieth century with minimal formal interaction with other scholars and with generally little recognition outside of Ireland.

Imagining Past Landscapes

In order to comprehend nature in all its vast sublimity, it would be necessary to present it under a twofold aspect: first objectively, as an actual phenomenon, and next subjectively, as it is reflected in the feelings of mankind. (von Humboldt, in Saarinen 1974:255–6)

Despite this pioneering statement, and the fact that some of the writings of Vidal and Sauer included strong visual content, the idea that both objective and subjective worlds need to be considered in studies of landscape evolution was not central either to the Sauerian school or to historical geography more generally. Even two specific proposals for a geographic study of subjective worlds, published respectively in 1947 and 1951, had minimal impact at the time, and their influence continues to be debated today.

The first of these was not clearly related to earlier or contemporary geographic literature, nor did it adequately signpost future research directions, but Wright did explicitly distinguish between the subjective and the objective:

Objectivity . . . is a mental disposition to conceive of things realistically. . . . The opposite of objectivity would, then, be a predisposition to conceive of things unrealistically; but, clearly, this is not an adequate definition of subjectivity. As generally understood, subjectivity implies, rather, a mental disposition to conceive of things with reference to oneself. . . . While such a disposition often does, in fact, lead to error, illusion, or deliberate deception, it is entirely possible to conceive of things not only with reference to oneself but also realistically. (Wright 1947:5)

Thus, a subjective view of the world might range anywhere along a continuum from complete error to complete accord with reality, and Wright proposed a new discipline—"geosophy"—to examine these true and false geographical ideas. However, these original ideas were related neither to other disciplines, such as psychology, nor to other developments in geography, and there was no substantive empirical attempt to pursue the argument.

The second contribution to the geographic study of subjective worlds, by Kirk (1951), similarly lacked influence at the time of publication, at least partly because it was published in a journal that was relatively inaccessible to North Americans, so the work was not generally known until the 1960s. Kirk's was a more explicit attempt to incorporate behavior into geographic analyses, making a distinction between the world of facts—both physical and human—and the environment in which these facts were culturally structured and in which they acquired cultural values. Kirk labeled the world of facts the *Phenomenal Environment* and the world in which they received meaning the *Behavioral Environment*. Twelve years later, Kirk (1963) developed these ideas to include the suggestion that human behavior in the Behavioral Environment was rational. These original ideas, for geography, have not become central to subsequent geographic research.

Wright and Kirk identified the presence and importance of subjective worlds, as Humboldt had a century earlier, but these contributions did not form a continuous development of ideas, nor did they stimulate research into subjective worlds. It was not until the 1960s and the rise of a behavioral geography initially associated with spatial analysis and subsequently inspired by humanistic concepts that cultural and historical geographic studies began to focus explicitly on the role played by subjective worlds in landscape evolution.

Creating Images

Much of the individual and group behavior that was critical for landscape formation occurred in situations of uncertainty, and what happened in these uncertain environments may have been more closely related to views developed from myth than to the realities of the environment itself. Indeed, as Salter (1971b:18) points out, myth may be a "prime generating factor in the decision to migrate. Though the mythical goal is seldom realized, the processes of mobility and discovery are just as real as if the myth had been realized." Although inappropriate behavior often resulted from perceptions that were significant distortions of reality, errors were normally not repeated (see Box 5.5). ➡️➤➤ **21**

Since the 1960s especially, historical and cultural geographers have shown much interest in understanding the differences between objective and subjective environments—particularly in areas of European overseas expansion during the periods of exploration and early European settlement—because of the realization that any comprehensive understanding of the evolution of settlement and the transformation of landscapes over large areas of the globe requires such understanding. The key idea supporting this type of research is that humans do not respond directly to the environment but rather to their mental image of the environment, an image that is the product of perception. Thus, human landscape-making activity is related to the human image of the environment. Although it is possible to conceive of more complex relationships and interactions, the basic logic of this approach is straightforward. Certainly, the idea that people represent environmental information in the form of an image in their heads has proven very attractive to cultural geographers. It is also well established that these images err in certain consistent ways. For example, local areas are usually known in relative detail while more distant areas are less known. Three principal issues have been addressed in the historical context:

- How were subjective worlds—images—that were different from reality created?
- What was the extent of the difference between images and reality?
- Is it possible to understand behavior and the changing landscape by reference to images?

These are not easy questions to answer either conceptually or empirically, especially given the difficulties of collecting information on imagined

Box 5.5 Learning in Landscape

Learning often occurs in environments of uncertainty, where a discrepancy between objective and subjective worlds forces people to make adjustments to their behavior. Colonization invariably imported cultural systems in confrontation with new, strange, and often inhospitable environments. Rarely were the particulars of these environments consistent with colonists' perceptions of them, and even more rarely were they totally amenable to the resolution of colonists' aims

(Figure 5.7). Perceptions and aims therefore had to be modified, and this initiated a process of active, conscious adjustment of learning that continued until perceptions were consistent and aims became realistic and attainable (Cameron 1977:1).

Further developing this idea, Cameron (1977) devised a model of adaptive learning that provides a framework for analysis of settlement in pre-1850 Western Australia. The colonization process involves the establishment of behavior patterns in

Figure 5.7 Perception, Behavior, and Landscape. This is a schematic diagram outlining the essential logic underlying many studies of perception in cultural geography (see also Haynes 1980:2). There is an external objective environment that is perceived by humans. The environment that is perceived—the subjective environment—is necessarily different to the objective environment. Further, the perceived environment is different for each individual, although individuals who belong to a group, such as a culture, are liable to have perceptions that are comparable. This is because perception is affected by both individual characteristics and group characteristics. Behavior in landscape is a response to the environment as perceived, rather than a response to the objective environment. Behavior changes landscape and thus changes both the objective and subjective environments through time.

Several attempts have been made to conceptualize these ideas more explicitly, and Kitchin (1996) reproduced nine rather more sophisticated formulations centered on the theme of cognitive mapping. There has been emphasis on trying to understand what is going on in the human mind as images are constructed, with most such work being built on the ideas of cognitive psychology and including perception as one component of cognition.

worlds in a historical context. Attempts to answer the first of these three questions have stressed the role played by those who provided descriptions. It is often appropriate to distinguish between the accounts of those who saw the area for themselves and those who produced second-hand descriptions, and it is also recognized that the impression conveyed relates to the motivations of those who are involved. In the

context of much European overseas activity, it may be useful to identify the different motivation of such individuals and groups as promoters, officials, settlers, travelers, and natural historians: each brought their particular biases to their written assessments, their various images of reality.

Promoters, such as leaders of immigration schemes and ship owners, were especially prolific a new environment, and these patterns are a consequence of a complex interplay of factors. First, movement into a new environment prompts a search for those actions that will satisfy preconceived goals. Hopefully, a colonizer can simply imitate others who are already achieving the equivalent goals, but, if this is not possible, then actions need to be determined subjectively on the basis of the image held of the area. Second, actions—for example, planting a particular crop—are taken. Third, when the consequences of actions are known, a process of evaluation is undertaken. Actions that were successful in achieving the desired goals are repeated, and those that were unsuccessful are rejected and a new search procedure begun. This sequence—search, operate, evaluate—is repeated as many times as needed until the desired goals are achieved. Through this process, the colonizer learns a set of actions, which then become repeated on a regular basis.

The model is useful, although it does not focus explicitly on such factors as prior experience of the colonizer, including any preadaptive traits, and the attitudes held by individuals or groups. Clearly, much behavior in landscape is affected by prior experience and the relative importance of individual identity and group membership, as in the example of Irish settlers in eastern Canada mentioned earlier in this chapter.

Prior experience may be a key consideration if the goals established in the new area are similar to those pursued previously, if the new environment is similar to the old, if there are limited links with other groups, if the image held of the

new environment is such that repetition of prior behavior is judged appropriate, if institutional considerations do not play a major role, and if group membership is more important than individual characteristics. The characteristics of each individual are likely to be important factors if group membership patterns are weak, if institutional considerations are lacking, if goals are unclear, and if the environment is relatively benign. Group membership will be important if individual characteristics are invariable, if institutions do not dominate, if links with other groups are limited, and if there is a shared environmental image.

As one example of the importance of perception in this larger context, consider the reaction of European settlers to North American forests. In southern Ontario, civilization and progress were deemed incompatible with the continuing presence of forest, and so colonizers attacked forests "with a savagery greater than that justified by the need to clear the land for cultivation" (Kelly 1974:64). This was despite the fact that the agricultural value of retaining some forest was understood. A lack of prior experience was also important in this situation because most incoming settlers were ill prepared for the vastness of North American forested areas. The goal of establishing commercial farming required some land clearing, but not on the scale undertaken. In this example, the behavior of settlers largely resulted from their lack of prior experience with the type of environment that they encountered, from the behavior of others, from the perception of the forest as some sort of threat to advancement, and from related attitudes.

writers and typically exaggerated the beneficial features of the environment in order to encourage settlement. The expectations of immigrant settlers were largely based on the writings of such promoters. Literature produced by officials was more variable, sometimes misleading and sometimes relatively objective. The usually optimistic—especially during the initial phase of settlement—settler literature

was taken seriously by potential settlers, but was limited in quantity. Travelers and natural historians produced the least useful accounts because they often focused on topics of very limited interest to prospective settlers.

One of the more notorious instances of a subjective environment concerned the supposed existence of a navigable Northwest Passage across the northern part of North America, linking the Atlantic and Pacific Oceans. Proposed both by explorers and by armchair geographers in Britain, the misguided image based on this fabled waterway had a profound influence on the history of the area:

> The whole enterprise was founded on a misapprehension, a geographical fiction, a fairy tale, springing out of the kind of stories sailors tell to amaze landsmen or to delude other sailors, to which were soon added the inferences, speculations and downright inventions that scholars manufactured to amaze themselves. (Thomson 1975:1)

The three examples that follow are not quite so dramatic and colorful as the example of the Northwest Passage, although they certainly illustrate the important role played by the distortion of reality, a distortion that was often intentional.

The Swan River Colony

An excellent example of how substantially incorrect images were created concerns the Swan River Colony of Western Australia during the period from 1827 to 1830. An 1827 exploratory voyage by Captain Stirling and Fraser (a botanist) was the sole substantive source of information about the Swan River area during this period. Although their visit to the area was brief, the subsequent report to the British Colonial Office supported the idea that Swan River was a suitable location for a colony and settlement because of fertile soils, a suitable climate, and adequate water.

What is remarkable about this example is the way the media presented the contents of the report to the public. As Heathcote (1972:81) remarks, "Between the 'place' and the 'people' themselves, however, lay the 'media' by which the so-called facts of the one were made available to the audience of the other." Stirling and Fraser's report was made available to the public only in the form of an article published in the journal *Quarterly Review* (1829), but this article served as the basis for later journal articles in the *Mirror* (1829), *New Monthly Magazine* (1829), and *Westminster Review* (1830). These publications were all directed at the middle class and entrepreneurs.

The first article, in the *Quarterly Review*, was written by Barrow, the acknowledged expert on the Southern hemisphere, and was aimed at an audience of ship owners, merchants, and wealthy potential settlers. Stirling and Fraser's generally favorable report was the basis for the article, but Barrow employed various strategies, such as deleting unfavorable comments and reporting optimistic conjecture as fact, to better promote the area. The three later articles continued this process of image distortion to such a degree that the *Westminster Review* article showed only superficial resemblance to the original report.

Two examples of the extent of the distortion are especially notable. First, the size of the cultivable area increased from 100 square miles (259 square kilometers) (Stirling and Fraser) to 22,000 square miles (57,000 square kilometers) (*New Monthly Magazine*). Second, the accounts of climate showed that conjecture became increasingly evident at the expense of fact as successive accounts were written. For example, in the original Stirling and Fraser report the description of climate was 90.3 per cent factual and only 9.7 per cent conjecture, while the *New Monthly Magazine* article reduced the factual description of climate to only 48.9 per cent, meaning that 51.1 per cent was conjecture. Conditions that favored such significant distortions included the short time period involved, the existence of only one basic source of information, and the pro-colony mood of the British public at the time, but the sequence by which the information was disseminated differed but slightly from the way news was communicated about other colonial areas. Undoubtedly, Swan River is a compelling instance of the misleading details that were made available to those considering investing in or moving to an overseas area. Consider the following two statements. First, the *New Monthly Magazine* reported that "the soil is a fine brown loam, alternating with broad valleys of the finest alluvial soil; the hills appeared finely timbered; the valleys produced an immense luxuriance" (in Cameron 1974:65). Second, one early settler complained: "Not a blade of grass to be seen—nothing but sand, scrub, shrubs, and stunted trees, from the verge of the river to the top of the hills. . . . I may say with certainty, that the soil is such, on which no human being can possibly exist" (in Cameron 1974:74). The subjective world presented to settlers by the media and the reality—or, more correctly, the perception of reality—that the settlers encountered were worlds apart.

Great Plains or Great American Desert?

Although the Great Plains is probably the most studied example of a subjective landscape in the historical cultural geographic literature, it is also one of the most confusing. It has often been claimed that the popular image of the plains in the period from about 1820 to about 1870 was that the area was a desert; indeed, the area was known as the Great American Desert. But this scholarly perception—held by both historians and geographers—has itself been challenged as incorrect.

The dominant scholarly view, based largely on a study of the reports of explorers and educational texts, is that the area was perceived as desert, an idea that was initiated by early nineteenth-century European explorers. Scholars supporting this view have offered three principal reasons to explain the emergence of this perception and its continuance for about fifty years. First, individuals tend to record what impresses them most and, for explorers from the eastern United States, the small areas of desert in the plains merited substantial description. Second, relative to the eastern region, the treeless plains could fairly be described as desert. Third, there was a powerful political impetus to move west to the Pacific coast as quickly as possible, and promoting a view of the extensive plains region as desert was conducive to that goal. In these three ways, the plains were being defined in eastern terms. Corrections to the myth of the Great American Desert came around 1870 with improved knowledge of the area.

But a major challenge to this dominant scholarly view claims that "the desert belief was far from universally held in America in the mid-nineteenth century" (Bowden 1976:119–20). According to this argument, the incorrect idea that the plains were perceived as desert can be understood in terms of the variable climate of the area, with some years being very dry and others much wetter: travelers passing through the region during a dry year were likely to interpret the region as a desert. Further, several influential historians, particularly, Turner, Webb, and Malin, stressed differences between the humid east and the drier plains, thus contributing to the desert image.

"A New and Naked Land"

Settlement of the Canadian prairies by Europeans between about 1895 and 1914 was not simply the next logical progression in the process of overseas movement and a response to the demand for wheat; it was also prompted and sustained by an organized campaign of intentional distortion of the geography of the region. As Rees (1988:4) comments, "The more distant and inhospitable the land, the greater the blandishments necessary to attract settlers to it. Cold, dry, treeless plains half a continent and—for Europeans—an ocean away were hardly alluring."

In the mid-nineteenth century, the region was viewed as either a northern continuation of the Great American Desert or a southern continuation of the Arctic—the area was indeed *The Great Lone Land*, as Butler (1872) called it. But this was an unsatisfactory state of affairs for both British and Canadian officials who were concerned about the lack of settlement between Ontario and the Pacific coast, especially given the possibility of American northward expansion. Accordingly, in 1857 there was a British expedition to the area under the leadership of Captain John Palliser and a Canadian expedition the following year under the leadership of Henry Youle Hind. The reports of these expeditions introduced the idea that the prairies had agricultural potential, with the exception of the grasslands area in the southwestern portion, which became known as Palliser's Triangle. This was the start of a process of image change. → ›› **22**

During the 1860s, 1870s, and 1880s, numerous favorable reports of the Canadian prairies appeared, often written with a limited factual basis. Thus, A.J. Russell, the inspector of Crown agencies in Canada, stated in 1869 (without having visited the prairies) that both of the expedition reports had probably underestimated the agricultural potential of the grassland region, while John Macoun, a government botanist and author of *Manitoba and the Great North-West*, declared in 1881 that the area was "all equally good land" (in Rees 1988:7). These and other largely unjustified claims about the physical environment encouraged massive campaigns by the two principal promoters of the area, the Canadian Pacific Railway and the government. But despite this ongoing process of image change, the results were disappointing, with little immigration from the United States, Britain, or Western Europe. Something else was needed to attract immigrants to the prairies, a fact that was recognized by Clifford Sifton, who was placed in charge of the Department of the Interior in 1896.

Sifton initiated three principal policy changes. First, he extended the range of advertising into non-traditional immigrant source areas, notably into eastern and southeastern Europe and Ukraine. Second, he substantially increased the quantity of the promotional literature. Third, he began to make major changes to the content of the promotional literature. For example, Sifton disallowed overseas publication of Manitoba temperatures. In addition, he required that the words "snow" and "cold" not be used in government publications, that snow not be included in any illustrations, that the often inadequate rainfall in the growing season be hidden through the use of annual averages, and that the great open expanses of the prairies be softened through descriptions that focused on enclosures. In this example, image distortion was intentional and substantial.

It is clear that the promoters of settlement were operating in a context of high emotion and high expectations. There was a perceived need to settle the West as quickly as possible, and it seems that this justified the use of dubious strategies to entice immigrants to the area. For many settlers, the reality they encountered on arrival was quite different from the expectations generated by the image-makers: an unliterary child from Ontario described the prairies as "a new and naked land" (in Rees 1988:36).

The conclusion reached by Christopher with reference to southern Africa applies equally well to the Canadian prairies and to many other areas of European overseas movement:

> To a large extent men believed what they wanted to believe or that which they were given to believe, rather than the truth; and there can be little doubt that large tracts of southern Africa would never have been settled by Europeans or would have been settled in different circumstances had the true state of affairs been appreciated. Immigrants' illusions were sometimes shattered soon after they reached southern Africa, but illusion had brought them, and they had to make the best of it. (Christopher 1973:20)

The discussion of subjective landscapes demonstrates that the subjective environment is often a significant distortion of the objective environment and that, as a consequence, much behavior occurs in situations of uncertainty. Certainly, humans know little of the consequence of behaviors that are not repetitive or that are not being imitated. Given an initial uncertainty, it is reasonable to suggest that a learning process usually proceeds and involves increasingly more appropriate behaviors through time.

Questions for Critical Thought

This is an opportune time to raise a very thought-provoking idea: If it is reasonable to suggest that much human behavior during the early stages of settlement in a new environment had a rather flimsy rationale, does this mean that alternative behaviors might well have occurred if better information had been available? The suggestion that different decisions might easily have been taken and then might have had different implications for landscape is explicitly addressed in the examples of subjective landscapes discussed. What is your reaction as you read these examples? Does all this seem rather frivolous, or does it read as a stimulating and seductive idea for cultural geographers to consider?

Concluding Comments

The material included in this chapter reflects the pioneering methodological statements of Sauer. Notwithstanding the various earlier European traditions that Sauer acknowledged, it is not an exaggeration to claim that he was responsible for prompting the English-language interest in both the evolutionary study of cultural landscapes and the historical geographic tradition of studying change through time.

- Accordingly, this chapter begins with a brief explanation of the need to consider changes through time in analyses of landscapes, an explanation that highlights the importance of this approach in human geography generally, not just in cultural and historical geography.
- Discussion of the landscape consequences of cultural change, as these relate to processes of diffusion, identifies three distinct approaches that mirror the history of larger human geographic interests. The three approaches are a concern with culture traits, a focus on the identification and modeling of general spatial processes, and a recognition that diffusion does not simply occur in a political vacuum but rather is tied up with issues of power and wealth.
- The account of the landscape consequences of cultural change includes discussion of the reasons for European movement overseas, but is primarily concerned with the cultural baggage that Europeans brought with them and with consequences of contact between Aboriginal and European cultures. Reflecting the available scholarship, the regional emphasis here is on North America and other mid-latitude areas of European settlement.
- The discussion of historical geographic work notes the importance of German and French traditions but again stresses the role of Sauer as a major influence. Approaches to historical geography include narratives, cross-section, and sequent occupance. Examples of empirical work discussed focus on frontier studies, transferring cultural baggage, and regional evolution.

Throughout, there is emphasis on what is often called the shaping or making of landscape. Recall that Sauer titled his major methodological study "The Morphology of Landscape".

- Evolutionary landscape studies have expanded significantly to include an interest in reading landscapes. There is an important British local-history approach exemplified by the work of Hoskins, and also the *Landscape* magazine tradition introduced and dominated by Jackson. Both of these traditions expand the Sauerian interests, with strong emphasis on vernacular features, the importance of ideology, and the symbolic character of landscape.
- The discussion of subjective or imagined landscapes emphasizes how different these might be to objectively described landscapes. Three examples demonstrate that a subjective understanding of landscape might be similar to, or very different from, reality. The examples are colonial Western Australia, the American Great Plains, and prairie Canada.
- Cultural geographers continue to be concerned with studies of changing landscapes and with the somewhat loosely defined historical geography tradition. This is evidenced by a considerable body of literature focusing on the evolution of cultural regions, especially in the United States, and on the evolution of groups of similar cultural regions—known as cultural realms—that are evident at a global scale. These contributions are discussed in the following chapter. But the general Sauerian influence is also accompanied today by a series of other concerns that are associated with developments in geography since the late 1970s. For example, the evolutionary study of cultural regions is enhanced by a consideration of the way in which humans relate to each other and thus identify themselves as members of groups. Similarly, the evolutionary study of cultural realms has been enhanced both by the arguments of world systems analysis and by the cultural studies tradition referred to in Chapter 3, especially by postcolonial theory. These contributions are discussed in subsequent chapters.

Box 5.6 What Might Have Been: Introducing and Using Counterfactual Logic

One of the most interesting yet rarely pursued questions that historical cultural geographers might consider raising is "What might have been?" Given that much landscape-making behavior occurred (and indeed continues to occur) in situations of uncertainty, it is clear that behaviors different from those that took place might have taken place, and that any such different behaviors might have resulted in different landscapes. In this sense, the basic logic behind **counterfactual** queries is simple, and their use can be justified as follows: "Every statement of causation implies the counterfactual proposition that in the absence of the causative factor the event would not have occurred" (North 1977:189). A counterfactual is any statement that is untrue in that it typically queries, "What would have happened if...?" ➔≫**23**

Because the usual object of scholarly enquiry is what is real, there is an implicit assumption that this reality is in some sense a necessary consequence of what went before; this can be described as the bias of hindsight. To think counterfactually is to acknowledge that things might have happened differently. If things had occurred differently and a different outcome had resulted, then this would suggest that the presumed causes were indeed important. If, however, things had occurred differently and the outcome was similar to the reality, then this would suggest that the presumed causes might not be critical to the outcome. One of the dangers of not thinking counterfactually is that we tend to adjust our thoughts so that the expected probability of something occurring that has in fact occurred is increased, while the expected probability of something occurring that has not occurred is decreased. Contingency may be ignored if counterfactual logic is not employed.

The principal debate about causation and counterfactuals takes place in philosophy, although historians have long taken an interest in such matters. Typically, historical analyses in this vein belong to two types. First, some counterfactual histories may be largely speculative, involving reasoned assumptions about what might have happened had some

part of the past been different; military historians have been attracted to this type of argument because they have recognized that many critical events depended on individual decision-making or on circumstances that might very well have been different, such as weather conditions. Second, other counterfactual histories are designed to specifically test ideas about the past, for example, the idea that North American westward expansion was closely related to railroads; most of these studies are in the tradition of the new economic history that developed in the 1960s and involve formal procedures of hypothesis testing by means of quantitative methods.

The following are three examples of historical cultural analyses using counterfactual logic. First, an Australian historian whose many works have inspired geographers because of their recurring concern with spatial issues made a compelling argument for counterfactual analyses:

> We easily forget that every statement we make of why events happened is in part speculation. In looking at history we have to ask: what might have happened? Every time we affirm the profound importance of a particular event—whether the finding of gold in 1851 or the failure of the North Queensland separation movement in the 1880s—we unmistakably imply that such an event was a turning point and that society—but for that event—would not have changed direction. There can be no discussion of a powerful event without realizing that it is like a traffic juncture where a society is capable suddenly of changing direction. In writing history we concentrate more on what did happen, but many of the crucial events are those which almost happened. The born and the unborn may seem completely different but essentially they have to be analyzed and discussed in the same way. (Blainey 1983:202–3)

Second, a leading North American historical geographer thoughtfully described some of the consequences for the emerging political unit of Canada that might have resulted from minor changes in the exploratory activities of Champlain during the years 1605–8 (Clark 1975). Third, as part of a larger account of the North American landscape, Meinig (1993:215) created two thought-provoking alternative political landscapes (see Figure 5.8).

Given circumstances of uncertainty and the role of chance, it is possible that cultural geographers might pursue counterfactual analyses if it seems reasonable to suggest that the details of the process that is presumed to have affected landscape creation might have been different.

A Greater United States

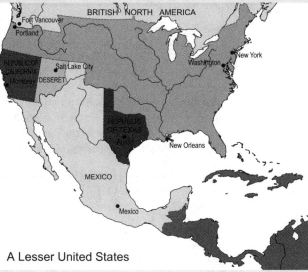

A Lesser United States

Figure 5.8 A Greater and a Lesser United States. Meinig is renowned for his evolutionary analyses of Australian and North American regional landscapes that combine incisive description and original explanations. Especially compelling are the many bold maps concerned with regional evolution. This figure suggests two alternative North American political landscapes to that which did come into being. Both were possible realities, and counterfactuals such as these are a "fresh and forceful way of reminding ourselves that there was no destiny manifest" (Meinig 1993:217). Source: D.W. Meinig, *The Shaping of America: A Geographical Perspective on 500 Years of History. Volume 2: Continental America, 1800–1867* (New Haven, CT: Yale University Press, 1993):215.

Further Reading ➡➤➤

The following are useful sources for further reading on specific issues.

1. Hugill and Dickson (1988) discuss diffusion as a multidisciplinary concern, and Entrikin (1988) provides an overview of diffusion in the context of the debate about naturalism.

2. Analyses of culture trait diffusion include Kniffen (1949, 1951a) on agricultural fairs; Kniffen (1936, 1965), Kniffen and Glassie (1966), Walker and Detro (1990), and Jordan (1983, 1985) on house types and log construction; Stanislawski (1946) on the grid-pattern town; Leighly (1978) on place names; Seig (1963), Brunn (1963), and Carter (1977, 1980, 1988) on agricultural practices; and Arkell (1991) and Carney (1994, 1996) on musical styles.

3. In addition to the 1952 book, which is a seminal work (see Conzen 1993), several innovative arguments about the origins and diffusion of agriculture are proposed by Sauer (1968, 1969, 1970). Leighly (1954), a colleague of Sauer, discusses both the traditional cultural and the spatial analytic approaches to diffusion.

4. Hägerstrand (1951, 1952, 1967) is the principal figure in the spatial analytic approach to diffusion. Gould (1969), Abler, Adams, and Gould (1971), and Haggett (2001) include comprehensive accounts of the simulation procedure developed by Hägerstrand. Rogers (1962) outlines the rural sociological approach to diffusion. Pyle (1969) analyzes the diffusion of cholera during three pandemics in nineteenth-century North America, and Haggett (2000) provides a review of the geographical structure of epidemics.

5. Blaut (1977) argues (1) that the traditional approach to diffusion provided the needed foundations for further theory development, precisely because of the breadth of the approach, and (2) that the spatial approach was too narrow because of the tendency to view innovations in isolation and to ignore the consequences of diffusion. Carlstein (1982) discusses diffusion and use of resources. Yapa (1977, 1996), Blaikie (1978, 1985), and Browett (1980) analyze agricultural diffusion in the developing world.

6. Hauptman and Knapp (1977) contrast Dutch movement and activity in the two areas of Formosa (Taiwan) and New Netherland in North America.

7. Fox (1991) discusses the rise of the modern European world; Jones (1987) outlines the European "miracle" argument, and Blaut (1993b) critiques that argument.

8. A stimulating overview of the history of the Americas is provided by Fernández-Armesto (2003). Studies focusing on Aboriginal identities and landscapes include Denevan (2001), Doolittle (2001), and Mann (2002). The October 1991 issue of *National Geographic* reviews the Americas before Columbus. Trigger (1982, 1985) and Ray (1996) are examples of studies of cultural change in North American Aboriginal groups prior to European contact. Sauer (1935), Veblen (1977), Lovell (1985, 1992), and Cook and Lovell (1992) have all contributed to the debate on the impact of disease on American population numbers.

9. Vibert (1997:3–23) and Radding (1997) discuss how the voices of the subaltern were usually omitted or misrepresented in official documents. Bauer (2003) reviews the cultural geography of colonial American literatures.

10. Ray (1974), Judd and Ray (1980), and Peterson and Anfison (1985) study Aboriginals in the fur trade.

11. Dodgshon (2008) critiques geography's consideration of time, even noting that early developments in the historical geography tradition were very unsure about the merits of crossing

the traditional disciplinary boundary between geography (the study of space) and history (the study of time). Jones (2004) notes a disciplinary trend favoring analyses of shorter time periods and argues that a lengthening of time frames enables different stories to be told. Unstead (1922) and Smith (1965:128) discuss historical geography as the study of the past.

12. Baker (2007) observes that there has been sustained debate about the practice of historical geography, but not about the practice of geographical history understood especially as changing distributions and geographical influences. Norton (1984), Butlin (1993), Conzen, Rumney, and Wynn (1993), Hornbeck, Earle, and Rodrigue (1996), and Baker (2003) detail the varied methodologies practiced in historical geography. Williams (1983, 1987) and Donkin (1997) describe the approach to historical geography initiated by Sauer. These works aid considerably in placing the landscape school emphasis in a larger historical perspective, making it very clear that Sauer initiated a "sweeping and majestic brand of culture history" (Earle 1995:457). Grey (1994) continues this tradition in a study of New Zealand.

13. Examples of the use of cross-sections include Cumberland (1949) on New Zealand, Clark (1959) on Prince Edward Island, and Darby (1973) on England. Mikesell (1976) and Conzen (1993:33–7) discuss the sequent occupance approach.

14. Mikesell (1960), Wishart, Warren, and Stoddard (1969), Eigenheer (1973–4), Hudson (1977), Christopher (1982), and Guelke (1987) consider spatial aspects of the frontier and definitional issues concerning new land settlement and cultural interaction. Webb (1931, 1964) studies frontiers in the Great Plains and in a world context. Mitchell (1977) studies the Shenandoah Valley frontier.

15. Clark (1959), Jordan (1966), Lemon (1972), Mannion (1974), Rice (1977), McQuillan (1978, 1990, 1993), Ostergren (1988), and Conzen

(2010) examine the loss or retention of cultural traits, especially agricultural practices, by immigrant ethnic groups in North America. Norton (1988) and Conzen (2010) outline general considerations relating to cultural trait loss or retention.

16. Early examples of evolutionary regional studies include those by Kniffen (1932), who focused on physical and later cultural landscapes of the Colorado delta area and recognized three stages of occupation and related landscape, and by Meigs (1935), who focused on the Dominican mission frontier of lower California, also identifying physical and cultural landscapes and the processes of change. Studies of Mexican and American landscapes are included in two valuable collections of the writings of Sauer, namely *Land and Life* (Leighly 1963), and *Selected Essays, 1963–1975* (Sauer 1981).

17. Other British work comparable to the tradition initiated by Hoskins includes the historical writing of Beresford (1957), the historical geographic writings of Darby (1940, 1956) on the drainage of the fens and the clearing of the woodland, and numerous historical geographic analyses of farms, fields, fences, and other features of the cultural landscape (see, for example, Baker and Butlin 1973; Slater and Jarvis 1982). A popular paperback book in this broad tradition is *The Penguin Guide to the Landscape of England and Wales* (Coones and Patten 1986).

18. Wilson and Groth (2003) pay homage to J.B. Jackson in an edited collection of landscape studies.

19. Lowenthal (1968) focuses on the perceptions of the American landscape that emphasized the overwhelming character, the great size, the wild and unfinished feel, and the formless and confused visual impression, especially in comparison to the landscape of England (see also Lowenthal and Prince 1964, 1965). Mills (1997) studies American physical and cultural landscapes and the way these have been

interpreted in popular culture. Other examples of landscape studies are Widdis (1993) on the cultural landscape of Saskatchewan; Ennals and Holdsworth (1981) on the vernacular architecture of the maritime province area of Canada; and Francaviglia (1991) on mining landscapes in the United States. There are many studies of regional landscape evolution, including Wyckoff (1999) on Colorado, Meyer (2000) on Illinois, Lewis (2002) on Michigan, Schroeder (2002) on Missouri, Jordan-Bychkov (2003) on the Upland South, Rehder (2004) on Appalachia, and Straw and Blethen (2004) also on Appalachia. Other studies that focus on the evolution of regional landscapes are discussed in Chapter 6.

20. The life and work of E. Estyn Evans is discussed by Buchanan, Jones, and McCourt (1971), Glasscock (1991), and Graham and Proudfoot (1993); the classic work is Evans (1973), and for a collection of relevant essays see Evans (1996).

21. Powell (1977) discusses the role of images in the transformation of global environments. Merrens (1969), Heathcote (1972), Christopher (1973), and Johnston (1979, 1981) discuss images and image-makers during European overseas expansion. Dicken and Dicken (1979) analyze historical perceptions of Oregon, and Peters (1972) studies surveyor perceptions in Michigan in the early nineteenth century.

22. Francis (1988) describes the image of the Canadian West during the settlement period.

23. There are discussions of counterfactuals in historical geography (Norton 1984:14–15; also see Norton 1995 for an example), in world politics (Tetlock and Belkin 1996), in history (Ferguson 1997), and in mostly British politics (Brack and Dale 2003). Fogel (1964) is a classic statement on counterfactuals in the new economic history.

CHAPTER 6

Regional Landscapes

The study of the earth as the home of humans. For many human geographers this simple phrase captures the spirit and intent of the discipline, acknowledging the importance of physical geography but stressing the role played by humans changing that geography. Beginning about 12,000 years ago and with increasing intensity as cultures and technologies changed and as population numbers increased, our tenure on the earth has been characterized by our practice of transforming physical landscapes into cultural landscapes. We build ever-changing homes designed to satisfy our needs and wants, and one outcome of this process of landscape evolution, an outcome that reflects both physical geography and the spatial clustering of cultures through time, is the formation and ongoing reformation of regional landscapes.

Much of the content of preceding chapters anticipated this discussion of regional landscapes in three ways. First, Sauer's methodological principles incorporate regional concepts, as did the British tradition of regional–historical geography. Second, ideas about humans and nature see region formation as one consequence of the human-and-nature relationship. Third, studies of landscape evolution recognize that cultural occupance of an area and related landscape changes often involve the creation of a cultural region. The underlying theme of this chapter is that groups of people who share some cultural characteristics and live in close proximity are likely to produce a landscape that reflects the shared characteristics—that is to say, a cultural region evolves. At this point it may be useful to indicate that cultural mapping of regions is different from political geography, which usually examines administrative boundaries, but the two are also increasingly understood in relation to each other. For example, while administrative boundaries such as states or provinces or territories are representative of certain political criteria, they are also reflective of cultural meanings attached to spatial distinctions that may have emerged as a consequence of history, economy, or social practices. Indeed, the political and cultural can co-constitute each other, for example, in cases where administrative boundaries demarcate distinctions such as different legal ages for drinking, marriage, or the rights of same sex couples. In these examples a political decision becomes part of the process of cultural differentiation between spatial units and vice versa.

In this chapter you will find

- ✤ an opening account of the regional theme, including an overview and evaluation of the approach. The several types of region—formal, functional, and vernacular—are noted, and some problems and limitations of regionalization are identified;

- ✤ an outline of a number of concepts devised to understand region creation at the subnational scale. These concepts build upon discussions of shaping landscapes in the previous chapter;

⚜ a discussion of principal examples of regional cultural geography at the subnational scale, with emphasis on the concept of regions as homelands and on the importance of a socially cohesive group in region formation. Principal examples discussed are those of the Mormon and Hispano regions in the American West. There is also an account of ethnicity and related features of material landscapes;

⚜ an account of the regional concept as applied at the global scale, including a discussion of the way in which the modern world has evolved. Central to this discussion are several major analyses of civilizations and also the concept known as world systems theory, which aims to explain the evolution of the modern political, economic, and cultural world;

⚜ a discussion of the contemporary outcome of global evolutionary processes. The pioneering 1951 global regionalization developed by Russell and Kniffen is presented, as are two other more recent attempts at defining global regionalization processes;

⚜ an assessment of globalization from a political, an economic, and, most significantly, a cultural point of view. Attempts to explain the current division into more and less developed worlds are reviewed, and the possibility that political states might play a lesser role in the future is discussed;

⚜ an evaluation of the claim that we are moving toward a world dominated by a single cultural identity.

What Are Cultural Regions?

Geographers have always been concerned with describing parts of the surface of the earth. In the mid-seventeenth century Varenius formally identified a regional approach, which was proclaimed the geographic method by Kant in the late eighteenth century and then proposed as the basis for the newly institutionalized discipline by Richthofen and others in the late nineteenth century. One consequence of this concentration of activity was that regional geography became the dominant **paradigm** in the discipline for most of the first half of the twentieth century. Indeed, it is not unusual to regard the landscape school of cultural geography as an alternative view of the discipline. Notwithstanding the possibly competing concerns of regional geography and the landscape school, cultural geographers fully appreciated the need to structure their studies of cultural landscapes in a regional framework. Delimiting regions was not the goal of cultural geography, but rather a logical outcome of the primary interest in landscape.

Culture Areas in Anthropology

In stressing the value of a focus on regional landscapes, cultural geographers turned not to the leading regional geographers, such as Richthofen, Alfred Hettner, and Hartshorne, but rather to Ratzel and to related anthropological applications of the culture area concept. Culture area analysis, although it was proposed by Ratzel, was first explicitly used in anthropology to delimit North American ethnic regions. For anthropologists, the concept was employed initially as a means to establish some semblance of order on seemingly diverse cultural phenomena. In this sense, identifying culture areas was one way to classify data. There were overtones of environmental determinism in some of this work, which identified close links between natural and cultural areas. ➔➤➤1

A culture area was defined specifically as "a part of the world where inhabitants tend to share most of the elements of culture, such as related languages, similar ecological conditions, economic systems, social systems and ideological systems. The separate groups within the system may or may not all be members of the same breeding population" (Foley 1976:104). In most of the anthropological studies the culture area concept was a tool used to recognize *cultural wholes*; a "cultural whole"—the term is virtually synonymous with *cultural region*—was an area dominated by a culture diffused from a "cultural core" that appeared in the most favorable part of the area.

Regions in Geography

It is worth noting that there is some confusion regarding the four terms used to refer to those parts of the surface of the earth regarded as internally homogeneous, and different from surrounding parts:

- *Landscape*. Sauer favored this term to characterize the extent of distribution of similar material cultural artifacts, though he also used the terms "area" and "region". Since about 1970, this term has broadened in meaning especially to refer to aspects of the non-visible landscape.
- *Area*. Anthropologists favored this term, although it is also used routinely by many geographers as a synonym for region.
- *Region*. Following Hartshorne, and as used by many geographers outside of the landscape tradition, this term was seen as the basic concept of the discipline of geography until the 1950s.
- *Place*. This fourth term, though it was also used by Hartshorne, is now associated primarily with some more recent approaches to cultural geography.

Terminological confusion aside, Sauer was explicit concerning the definition of a geographic culture area:

> The geographic culture area is taken to consist only of the expression of man's tenure of the land, the culture assemblage which records the full measure of man's utilization of the surface—or, one may agree with Schlüter, the visible, areally extensive and expressive features of man's presence. These the geographer maps as

> *Questions for Critical Thought*
> ——— ⚜ ———
>
> Terminological uncertainties are one outcome of the often deliberate practice of scholars using different terms in order to distance themselves from other scholars expressing similar ideas. As evident from the account of key concepts in Chapter 1, many instances of terminological uncertainty cannot be avoided. "Landscape" and "region" are two terms used throughout the current discussion, with landscape referring to the visual scene and region to a part of the earth's surface that has a similar landscape. From our cultural geographic perspective, the two terms are not interchangeable, but they are closely related. These comments raise an interesting question about social science academic disciplines generally. Is there a need for a more formal disciplinary definition of key terms? Terminological precision is usually essential in physical sciences, but the more discursive and often contested social science disciplines thrive on terminological debates. Is this a problem for practitioners and for students of these disciplines?

to distribution, groups as to genetic association, traces as to origin and synthesizes into a comparative system of culture areas (Sauer 1931:623).

Delimiting Regions

Having identified what is meant by the term "region", we run into the question of how regions are delimited. Alternatively phrased: *What is the regional method?* The basis for the answer that follows is the idea that delimiting formal regions (the most usual type) is essentially a process of classification, with classification being a procedure designed to impose some order on complex realities. Box 6.1 distinguishes between three types of region, namely formal, functional, and vernacular.

Classification

From a conventional scientific perspective, there are two approaches used to classify formal regions. One approach is to classify all possible locations

Box 6.1 Three Types of Region

First, a **formal region** is characterized by uniformity of a given trait or traits; it can be of any size. The culture area concept used by anthropologists and the regional concept used more broadly by geographers apply to this type. There are numerous criteria that might be used to identify formal regions. Characteristics of language, religion, and ethnicity may be used because they are reflected in the landscape. Alternatively, the concern may be more directly with the landscape, as in the recognition of a corn belt or prairie landscape. Cultural geographers typically use multiple traits in identifying formal regions. Lewis (1991:606) notes that formal regions are created through human relatedness, that is, "the notion that important commonalities unite certain groups of individuals to varying degrees, while separating them from those in other, similarly defined communities".

Second, a **functional region** is one ranging in scale from a single home to the entire world that in some way operates as a unit. Unlike formal regions, functional regions result from human connectedness rather than human relatedness—in other words, from some process of interaction between locations rather than from some shared values or characteristics. Clearly, a part of the surface of the earth that is integrated in a functional sense, such as a political unit or a religious settlement, is quite likely to exhibit some landscape similarities as well, since landscape may be affected by the functional factors unifying the region, legal system of land tenure and plot demarcation, for example. Accordingly, functional regions often overlap with formal regions in cultural analysis.

A third type of region, the **vernacular region**, is a locally perceived regional identity and name that does not necessarily align with any other formal administrative boundaries. Although a vernacular region may exhibit some distinct visible material landscape, its distinguishing characteristics relate more to a perceived or imagined sense of identity. While not widely used today by cultural geographers, the idea of a vernacular region is of interest to Geographic Information System (GIS) scholars, particularly those working on Volunteered Geographic Information (VGI), where vernacular regions as used by the public may be difficult to align with spatially demarcated regions as required for GIS mapping (De Longueville et al. 2009).

individually according to a certain set of criteria and then, where appropriate, to draw boundaries around groups of locations that are deemed to be similar, thus delimiting regions. This is described as grouping, or classification from below, and it is a procedure that has often been informally applied in attempts to regionalize a national area or, more simply, to identify a single region. An alternative approach is to treat all the possible locations as a single set and then draw boundaries between smaller groups of locations by means of some predetermined criteria. This is described as logical division, or classification from above. Classifying in this manner requires extensive knowledge of the locations so that sensible distinguishing criteria can be used. Most of the traditional world regional classifications conducted by cultural geographers and other scholars are of this second type. ➔➤➤ 2

Interestingly, some efforts at regionalization made by cultural geographers following Sauer rely on a procedure that is different from both of those associated with the exercise of classification. Rather than beginning with the finished product (that is, the region or set of regions), an evolutionary approach has been used. Specifically, the aim is to identify a cultural hearth and then trace the diffusion of the culture outwards over some larger area until a region has been identified.

Four Difficulties

There are four general issues that need to be addressed in any attempt to demarcate a **cultural region** or a set of cultural regions. First, regions are continuously evolving, and this makes the process of regionalizing difficult, since a set of regions once delimited can change. This issue follows closely from the contents of the previous chapter, especially from the idea that regions are ever-changing outcomes of cultural occupance and transformation of landscape, but it also relates to the Sauerian interest in identifying cultural hearths and tracing diffusion. This chapter includes two sections that focus on the evolution of cultural regions, one referring to the subnational scale, the other to the global scale.

Second, region identification must consider questions of spatial scale and cultural scale. Linking these two scales stresses the fact that cultural regions are, logically, regions that are usually occupied by distinguishable cultural groups, and are not just evident because of different visible landscapes. Indeed, most attempts at regionalizing the world are classifications of people rather than classifications of visible landscape. The significance of different scales, spatial and cultural, is that the tasks of regionalizing the world, regionalizing a country, and regionalizing a subnational area necessarily require use of different criteria and involve different levels of expectation concerning the internal homogeneity of the various cultures and landscapes delimited.

A third difficulty, apparent in most regionalization exercises, concerns the precise location of boundaries. Most functional regions have relatively precise boundaries, but both formal and vernacular regions are usually much more susceptible to particular interpretations made by the researcher. The reality is, of course, that most cultural regions do not have sharp boundaries separating them from other regions. Rather, the boundaries are zones of transition with a gradual merging into neighboring regions. Although cultural geographers are well aware of this situation, there is often an understandable temptation to map regions with boundaries that imply a line of separation.

Work on transborder regions is indicative of this cultural merging across political lines of demarcation; consider the American Southwest, the Canadian prairies, the Pacific Rim, or regions of Europe such as the Benelux countries (Belgium, Netherlands, and Luxembourg). In North America, transborder regions such as Cascadia in the Pacific Northwest

(typically construed as spanning from British Columbia in Canada to Oregon and Washington in the US) have been the focus of environmental and pro-business branding efforts to promote various types of integration (Sparke 2000). Part of the effort to integrate these regions comes from the shared culture of environmentalism and outdoor recreation. The North American Great Lakes region also has a long history of transborder flows, particularly in terms of migration (Widdis 2010), and integration of this region is seen as far more economically driven (MacPherson et al. 2006).

The final difficulty is the most problematic: on what basis are regions delimited? What criterion is—or what criteria are—to be used? How is this choice to be justified? Of course, there are no simple responses to questions such as these. The chosen criteria typically reflect the specific aims of the regionalization. For example, if a set of religious regions is the objective, then some religious criteria are employed. But statements such as these do not really address this issue. The key point is that multiple criteria might be legitimately employed in the act of demarcating cultural regions, and any interpretation of the resulting regionalization must consider the criteria used. There is always a danger of extending a regionalization beyond the intended purposes. What this point tells us is that there is no such thing as a correct system of regions.

These four difficulties may result in regional mapping that reflect the goals of an individual researcher, such that in many cases the boundaries of regions are located in a relatively arbitrary manner. John Fraser Hart (1982:21–2), a leading advocate of regional geography, has stated: "Regions are subjective artistic devices, and they must be shaped to fit the hand of the individual user. There can be no standard definition of a region, and there can be no universal rules for recognizing, delimiting, and describing regions." The extent to which this state of affairs is a problem is, of course, debatable, as it is dependent on philosophical persuasion. Humanist geographers are comfortable with individual researchers playing key roles, while more scientific geographers are much less comfortable because the criteria employed are seen as informally understood, rather than formally defined.

Questions for Critical Thought

Given these four difficulties, is it worthwhile to try to delimit regions? For most cultural geographers, the answer is yes, for at least three reasons. First, regions are an integral part of geographic research generally. Second, regionalization is one means of imposing some order on a complex reality. Third, the favored Sauerian procedure of identifying hearths and diffusion routes does minimize these four difficulties. Do you agree with these suggested reasons? Perhaps you are inclined to ask whether maps of regions assume a life of their own, implying order and structure where these do not really prevail. Further, does any regional exercise need to acknowledge, as Hart suggests, that regions are not real things with unique characteristics?

The Decline of Traditional Regional Geography?

Two quotations provide us with two very different opinions about the need for a regional approach to geography. First, Hart (1982:1) argued for some type of traditional regional geography, asserting that such work was "the highest form of the geographer's art". Second, Thrift (1994:200) expressed the opinion that the regional approach required "exhumation rather than resuscitation", and bemoaned the fact that it appeared to have become "an acceptable form of professional nostalgia, conjuring up memories of a golden age, now (thankfully) defunct". A reasonable assessment of the regional approach lies somewhere between these two statements.

It is commonplace in much recent human geography to make two related claims: first, that traditional regional geography has declined, and second, that a revised regional geography, one informed by recent advances in social theory, is at the forefront of much geographic research. The basic logic behind the first claim is clear: the regional approach, as advocated by Hartshorne, effectively dominated English-language geography from about the 1920s until about the mid-1950s, at which time it was challenged by the scientifically inspired spatial analysis.

Since then it has not succeeded in re-establishing itself as the geographic paradigm.

But although it is correct to say that the regional approach declined in popularity after the mid-1950s, it is also true that regional geography has remained central to much geographic research and continues to be advocated and practiced by many geographers, especially economic geographers. More important, for our purposes, the discussion of regions continues to be a popular and worthwhile approach for the study of cultural landscapes. Indeed, some of the very best cultural geography of recent years is explicitly concerned with identifying and understanding regions at a variety of scales.

Principal criticisms of traditional regional geography are that the approach is of limited value because it is concerned primarily with observing and recording facts and because most studies of regional geography implicitly assume that human geographic regions are organized according to differences in physical geography. These criticisms relate to the empiricism and naturalism evident in some regional geography, but do not really apply to the regional approach as it is used by cultural geographers who have always been concerned with a level of generalization, notably in the identification of cultural hearths and diffusion routes, and who have always rejected simplistic physical and human associations. Indeed, the more socially informed new regional geography is built partly on the ideas of cultural geography, specifically the idea that regions are related to spatial variations in culture and not to physical geography. The principal difference between traditional and new approaches, so far as cultural geography is concerned, relates to the traditional interest in material landscape and the newer interest in culture as socially constructed. Indeed, the cultural turn within human geography has seen political and economic geographers engage with the construction of regional geographies from a more critical and theoretically sophisticated position; for example, critical analysis of "area studies" and their role during the Cold War (Farish 2010), and how the relocation of wealthy business immigrants from Hong Kong to Vancouver created a transnational Asia–Pacific city (Olds 1995). ➔➤➤ **3**

Forming Cultural Regions

The processes by which distinctive areas of cultural landscape—cultural regions—come into being are complex. For the United States, circumstances that might encourage retention of cultural traits in turn might contribute to region formation. These circumstances include a substantial number of immigrants, a pattern of contiguous settlement, a set of shared values, and a lack of contact with other groups at the time of settlement.

This section details eight useful concepts developed by cultural geographers and used in the process of region identification and delimitation: cultural hearths; first effective settlement; duplication, deviation, and fusion; core, domain, and sphere; stages of regional evolution; culturally habituated predisposition; preadaptation; and cultural islands. The discussion of these eight concepts is illustrated with several examples of cultural regions that are explicitly related to the concepts described, while the succeeding section reviews a number of other major studies that focus on the idea of regions as homelands.

1. Cultural Hearths

Geographers working in the landscape school tradition and inspired by its particular interest in diffusion have paid considerable attention to the concept of the cultural hearth as an original source area that possesses some distinctive cultural attributes and from which these attributes have been diffused. Sauer stressed the importance of locating hearths from which settlement expanded as a basis for region identification, and the concept has been used in numerous cultural geographic analyses. Most of this work in region identification has been carried out on American and, to a lesser extent, Canadian landscapes. This preference is related to the fact that the European settlement of North America involved distinct cultural groups settling separate areas, thus contributing to the creation of cultural hearths and making North America particularly well suited to this kind of region identification; it also reflects the unique attachment of North American cultural geographers to the landscape school. Examples of hearth identification are discussed later in this section together with examples of some of the more elaborate conceptual formulations.

2. First Effective Settlement

"Whenever an empty territory undergoes settlement, or an earlier population is dislodged by invaders, the specific characteristics of the first group able to effect a viable, self-perpetuating society are of crucial significance for the later social and cultural geography of the area, no matter how tiny the initial band of settlers may have been" (Zelinsky 1973:13). This concept of **first effective settlement** is similar to the initial occupance proposal suggested by Kniffen (1965:551) in a diffusion analysis of folk housing. Zelinsky used the concept to map modern cultural regions, and the resulting regionalization is depicted in Figure 6.1. Zelinsky designated five regions as "first-order" settlements, namely New England, the Midland, the South, the Middle West, and the West.

The significant variations in regional size are related to differences in physical geography and in settlement experiences. For each of these five, Zelinsky (1973:119) indicated the date of first effective settlement and the date of related cultural formation, as well as the major sources of culture for region formation. Each of the first-order regions is subdivided into secondary regions, while three areas (Texas, Oklahoma, and peninsular Florida) are considered too difficult to classify given the criteria employed. All but one of the first-order regions have more than one major source of culture, the exception being New England, which has England as the sole source area. Because of this, only New England, among the first-order regions, has a date of first effective settlement noted; for the other four Zelinsky deemed it appropriate to indicate first effective settlement dates and sources of culture for secondary and, in three instances, tertiary regions. Further, in most of the cases, Zelinsky identified multiple sources. The concept of first effective

Figure 6.1 First Effective Settlement in the United States. This is the classic regionalization of the United States proposed on the basis of the concept of first effective settlement. Zelinsky preferred this concept to the core-domain–sphere model as a basis for regionalization because of the lack of isolation experienced by most of the core areas identified—this same point is made by Pillsbury. Source: Zelinsky, W. *The Cultural Geography of the United States*, 1st Edition. © 1973, p. 118. Reprinted by permission of Pearson Education Inc., Upper Saddle River, NJ.

settlement enables this regionalization, but it cannot be applied on a simple one date–one cultural source–one region basis. Certainly, the regionalization is a valuable application of the concept, but not surprisingly, the regional scene is much too complex to be represented by a single concept. ➔≫ 4

The basic logic of first effective settlement is often used in accounts of cultural landscapes. In the case of the Musconetcong Valley landscape of New Jersey, for example, many elements reflect a continuity from the eighteenth-century pioneer period to the present; thus, houses and barns along with many auxiliary structures reflect types established in the area during the eighteenth century.

The first effective settlement model has seldom been applied to areas outside of the United States. Certainly, in Europe and in many areas of European overseas expansion, where the experience was not as numerically overwhelming as it was in North America, the concept is less meaningful. This is both because the first effective settlement is much older than in the American examples, and because contemporary cultural identity is often the product of a series of contributions rather than a relatively clear reflection of a single dominant contribution. Even in the case of North America, the idea of first effective settlement obscures the complex process of colonial settlement, relying as it did on the displacement of indigenous cultural communities who were using the land before colonists arrived.

3. Duplication, Deviation, and Fusion

The Turner frontier thesis explained the formation of American cultural regions in terms of the westward-moving frontier, but there are several problems with this frontier argument. These include the assumption of an initial subsistence phase of economic activity, which was in fact not typically present, since many settlers moved quickly into commercial activity; the assumption of an initial regional isolation, which was not typically the case, since most areas were closely linked to other areas from the outset of settlement; and the assumption that American culture was derived from the frontier experience west of the Appalachians.

It is possible that the cores of the three American colonial cultural regions—those located in southern New England, southeastern Pennsylvania, and the Chesapeake tidewater—acted as dynamic cultural regions, contributing first to the creation of intermediate regions, and later to the creation of regions west of the Appalachians. Each of these regions included internal differences such that hearths and domains can be distinguished.

Three mechanisms might have been responsible for the creation of regions west of the colonial hearths: duplication of the traits evident in the hearth, deviation from the traits evident in the hearth because of different local circumstances, and fusion of traits from two or more hearths with the resulting formation of a different cultural form. Of these three, "it is the fusion process, aided by the appearance of symbols of national unity, that seems to have been most important in the cultural formation of the early trans-Appalachian west" (Mitchell 1978:67). This proposal, a development of both the core–domain–sphere model and the first effective settlement concept, allows for a variety of formative processes.

4. Core, Domain, and Sphere

A stimulating conceptual contribution to the literature on landscape evolution and related region formation is the *core–domain–sphere* model as developed and used by Meinig. The model formulation was introduced in an exemplary analysis of the emergence of the Mormon cultural region. Following accounts of the cultural group, of the movements of the group, and of the evolution of the region, Meinig outlined this generic model as a basis for both delimiting a region and identifying variations within the region. There are three components to the model (see Figure 6.2).

The first component of Meinig's model is a cultural core. This is not quite the same as the hearth area noted above. Rather, the core is the center of cultural control and the zone of most intense activity, but is not necessarily the original cultural source area; in other words, the core may or may not be the hearth. In the Mormon case, the core was founded following movement of the cultural group away

Figure 6.2 Core, Domain, and Sphere. This is the classic application of the core–domain–sphere model that Meinig developed in order to accommodate a common problem, namely mapping cultural regions that contain internal variations. The innovative solution was to employ generic concepts able to reflect such variations—core, domain, and sphere. Because cultures often spread away from a core, the transition from core to domain to sphere is likely to be suggestive of spatial and temporal patterns. The success of the model in the Mormon context seems undeniable, but it is also clear that the Mormon case is far from typical, especially as regards the isolation of the region and the distinctiveness of the group. Source: © 1965 From "The Mormon Culture Region: Strategies and Patters in the Geography of the American West, 1847-1964" by D.W. Meinig. *The Annals of the Association of American Geographers*, Volume 55, Issue 2, pp. 191-219. Reproduced by permission of Taylor & Francis Group, LLC. (http://www.tandfonline.com).

from the area of cultural origin and hence is not the hearth. Surrounding the core in Meinig's model is a domain, the area over which the culture diffused and became dominant. Although the Mormon domain lacks the intensity of occupance and complexity of development that are evident in the core, it is the area that most obviously displays the culture in the visible landscape. Beyond the domain is the sphere, an area that belongs only partly to the cultural region in question because other cultural influences from adjacent cultural regions are also present.

Certainly, the Mormon region for which this model was constructed is an unusual one because of the initial isolation of the group in the area occupied and the especially distinctive character of the Mormon cultural group. This combination of isolation and a distinctive culture is probably a prerequisite for the evolution of a core (or hearth)–domain–sphere regional identity that is as readily recognizable as it is in the Mormon case. Other applications of this model are therefore more limited, precisely because most groups are in close contact with other groups in adjacent areas, and because most groups do not actively seek to emphasize their differences from wider society, as the Mormon group has. The principal value of the model, aside from the specific Mormon application, is the explicit focus on evolution and the acknowledgment, in the reference to a sphere, that the boundaries of cultural regions are zones of transition and not lines. ➡➤➤ 5

Other applications of this model include a study concerned with urban street patterns in eighteenth-century Pennsylvania (Pillsbury 1970). The Pennsylvanian cultural hearth, domain, and sphere were mapped. A hearth rather than a core was identified, as the center of cultural activity was also the area of cultural genesis. The Pennsylvania case was probably more typical of the North American situation than was the Mormon region in that several cultural groups, rather than just one group, were present during the period of expansion. Accordingly, the cultural region developed quite differently. This argument was also extended to the larger region of the eastern United States and Quebec. Thus, the model was tentatively revised to include, for each of several cultural groups, a hearth, a domain, and a sphere.

Colonial cultural hearths were identified in the eastern United States—namely New England, Pennsylvania, and Chesapeake Bay—with a fourth hearth in French Canada. The three American hearths were previously proposed by Kniffen (1965) in a study of the diffusion of folk housing. These hearths merged as they diffused west to contribute to the formation of a national American culture. One way to interpret this suggestion that eighteenth-century North America comprised four regional cultural hearths is by reference to the simplification thesis discussed in Chapter 5. According to this thesis, the large number of local cultural regions, or *pays*, evident in Europe were simplified after being transferred to an overseas area, with one result being a smaller number of hearths. One implication of this proposal for four hearths is that some cultural groups, such as Swedes in Delaware and Dutch groups on the Hudson, became assimilated shortly after arrival.

5. Stages of Regional Evolution

In a substantial rethinking of the core, domain, and sphere logic, Meinig (1972) identified a set of six dynamic regions in the American west. For each of these regions, Meinig then identified four general stages of development in each of four categories of regional features, namely population, circulation, political area, and culture. The sequence of stages for culture was as follows:

- Stage 1: *transplant*—a selected and experimental adjustment to the new environment
- Stage 2: *regional culture*—the formation of a cohesive regional society
- Stage 3: *impact of national culture*—the consequences of improved communication and marketing
- Stage 4: *dissolution of historic regional culture*—a new awareness of regional values

The distinction between the first two and later two stages is particularly important, as the beginning of the third stage, impact of national culture, marks the end of insularity and local cultural identity and the onset of a rapid increase in the power of those forces encouraging national cultural uniformity. Much of the early regional distinctiveness has been eroded because the twentieth-century experience has typically been one of cultural integration related to the impact of a national American culture. ➔➤➤ **6**

Employing the logic of regional development resulted in the mapping of six major nuclei, or cores (the first stage), and six recognizably distinct regional cultures (the second stage), along with a number of secondary nuclei, some of which were inside and some outside the six major regions (see Figure 6.3). Because settlers in the American West came mostly from areas in the East and brought with them certain cultural traits, their behavior was more imitative than innovative. It was this behavior, combined with environmental differences, that contributed to the formation of the six regions. The six major regions identified as being present by the end of the second stage, defined as that of regional culture, are as follows.

Hispano New Mexico was the first region to form, following Spanish settlement of the upper Rio Grande Valley in the late sixteenth century. This was an isolated region of Hispanic European culture for over 200 years until in-movement by Anglos initiated change. By the late nineteenth century, it was a complex cultural area with relatively few links to the larger national system. The Mormon region formed beginning in 1847, with the initial in-movement of a distinctive religious group seeking isolation from larger American influences; a cultural region was created during about the next fifty years. Both the Hispano and Mormon regions are discussed in greater detail in the following section.

The region of the Oregon country was formed beginning in the 1840s, primarily by settlers who were part of the larger westward movement, with no one cultural group dominating the in-movement. The character of the region diverged from that of the migrant source areas in the East because of differences in location and environment. Northern California had limited settlement until the discovery of gold in 1848, an event that precipitated rapid and dramatic population growth accompanied by

related cultural and economic change. The incoming population was especially heterogeneous. The region of southern California was unaffected by the gold rush and remained as a dual Mexican and Anglo-American culture until a real estate boom in the 1880s initiated a process of rapid growth. Regional development began in eastern Colorado in the late 1850s following the discovery of gold, and the area developed as a cattle and irrigated farming area. None of these final four regions is as readily identifiable as a specifically cultural region as well as the Hispano and Mormon areas are; their identification as regions within the West is based more on economic and other criteria than on cultural criteria.

Figure 6.3 Major Nuclei and Regions in the American West. Six regions were proposed for the late nineteenth century. Hispano, New Mexico, has a major nucleus in the upper Rio Grande Valley centered on Santa Fe. The Mormon region has a major nucleus at the base of the Wasatch Mountains centered on Salt Lake City. The Oregon country has a major nucleus in the Willamette Valley centered on Portland. In all three cases, the nucleus is a prime agricultural area. Northern California is centered on San Francisco, Southern California on Los Angeles, and Colorado on Denver. In addition, several minor nuclei are located, each of which showed some degree of local autonomy. Source: D.W. Meinig, "American Wests: Preface to a Geographical Introduction", *Annals of the Association of American Geographers* 62 (1972):169.

Texas can be interpreted as a distinct cultural region that has evolved through four stages, each of which has left a mark on the present scene. The four stages are implantation, which reflects both Spanish and Mexican influences; assertion, which includes the periods of republic and early statehood; expansion, following the Civil War; and elaboration, which involves more recent developments.

6. Culturally Habituated Predisposition

A pioneering analysis of agricultural—not specifically cultural—region evolution interpreted change in three regions as a result of cultural rather than environmental or economic processes. The three regions that were examined were the American corn belt, the Philippine coconut landscape, and the Malayan rubber landscape. Although each case was different in detail, a cultural (specifically psychological) process was evident in all three cases. The American corn belt, for example, was seen as the "landscape expression of . . . the totality of the beliefs of the farmers over a region regarding the most suitable use of land in an area" (Spencer and Horvath 1963:81). Thus, formation of the corn belt followed a decline in sheep populations, which resulted from a general lack of interest in sheep rather than any crucial environmental or economic factors. The relevant cultural process was that of innovation diffusion. Once a process of change began, the mechanics of local communication networks ensured that the change became widespread; these networks were particularly effective in an area undergoing a process of first effective settlement as they were not complicated by the presence of earlier networks.

In a related fashion, the coconut landscape of the southern Philippines resulted from a particular farming mentality that was favorably predisposed to coconut planting. This psychological mindset, which was shared by most farmers in the region, was a strong force in the evolution of the coconut landscape. Expressed rather differently, it was a culturally habituated predisposition toward a particular crop capable of providing a stable return that initiated the formation of a particular agricultural region. The origin of this region was similar to, although different in detail from, that of the corn belt, particularly because all farmers in the region were familiar with the coconut and with appropriate techniques for harvesting the crop. The specific stimulus for the development of a commercial coconut landscape was a change in demand that farmers were able to respond to accordingly.

A third example of cultural causation was that of the Malayan rubber landscape. This landscape developed as a result of psychological change among the local population. Malayan farmers initially viewed rubber as an alien system of agriculture but changed their attitude toward this activity and adjusted their behavior accordingly. Of particular importance in this change was the realization that the daily work schedule for rubber, with most work done during the cooler hours, was preferable to that of rice field labour, which required a large amount of hard work during the heat of the day. ➞➤➤ 7

Of course, this understanding of landscape evolution neglects consideration of economic factors, especially the role played by market demand, and also, especially for the latter two examples, does not take into account the role of colonial relationships.

7. Preadaptation

Persistent disagreement over the nature of American frontier settlement has led to the replacement of the principal theories by a cultural theory. The American frontier of the eighteenth and early nineteenth centuries is best understood not as a time, place, or process, but as the preemption of a vast domain by one preadapted, syncretic American culture. The new culture was the Upland South of Turner and Kniffen, the processes were diffusion and migration, and the mechanism was cultural preadaptation (Newton 1974:143).

The concept of **preadaptation** was introduced specifically to explain the diffusion of culture from the American Upland South, an area stretching from southeastern Pennsylvania to eastern Georgia, covering much of the United States as far west as the plains. Preadaptation is the condition of a culture that already possesses the cultural traits required to allow it to successfully occupy a previously unencountered environment prior to moving to that environment; preadapted groups have a competitive advantage over other groups moving to the same environment. Eleven traits that enabled the culture of the Upland South to expand have been identified (see Box 6.2).

Settlers in the Upland South possessed a set of cultural preadaptations that enabled them to spread their Upland South culture beyond the Appalachians and into the plains. This essentially ecological concept was later developed to argue that seventeenth-century Finnish settlers of the lower Delaware Valley were preadapted to life on the American frontier because of their previous European experience. As such, this group can be seen as the single most important contributor to the culture of the American backwoods frontier. Although this identification of a specific, highly localized source area and of a particular ethnic group is debatable, the basic argument is clear: preadapted settlers were necessarily effective settlers, and hence, if they were the first to arrive in an area, they were the first to establish an effective settlement.

8. Cultural Islands

Expressions of group identity in landscape result from three factors in particular: the tendency of members to settle in close proximity to one another; the importation of distinctive culture traits; and the presence of particular attitudes, perceptions, and behaviors. Immigrant groups are sometimes described as ethnic to indicate that they share such important characteristics as a place of origin, a sense of tradition, a language, and a religion. For many immigrant groups, cultural inertia played a role in the creation of the landscape they first occupied, such that the presence of a regional landscape is often some measure of their group conservatism. Certainly, most groups faced two opposing forces when settling in a new environment: there was, on the one hand, a powerful sentiment in favor of maintaining shared traits and identity, while on the other hand, there was a recognition among settlers that many advantages accrued to those willing to break with their traditions. Distinctive landscapes are most likely to develop and be maintained where the group abides by certain common rules of organization. This claim is supported by numerous examples of rural landscapes of religious and other ethnic groups that display a preferred architectural style, which may be reflected in house and farm building and in church construction. In some of these landscapes, the ethnic groups themselves reflect a particular lifestyle in their choice of clothing and food.

Cultural islands are areas that are occupied by ethnic groups and that exhibit some landscape features that reflect these occupying ethnic groups. Cultural islands associated with Amish and Mennonite populations—two closely related groups—are among the most frequently studied because they tend to reflect aspects of a particular religious worldview, notably the commitment to an agricultural way of life and the preference for a closed and introverted society. The initial Amish and Mennonite groups that settled in North America favored dispersed settlement landscapes comprising farmsteads located apart from each other, while later groups that settled in southern Manitoba from Russia favored settlement in street villages with homes that had large rooms and a centrally located kitchen. ➤➤➤ 9

Some other well-researched and rather more widespread examples of cultural islands are certain Ukrainian landscapes evident in a broad belt across

---

Questions for Critical Thought

⚜

These eight ideas about the evolution of cultural regions reflect a substantial body of scholarship aimed at understanding cultural regionalization, especially in the United States. There have been analyses of landscape evolution focused on other parts of the world, but they have been less fruitful. Why might this be the case? Perhaps it is because some other areas of European overseas settlement experienced less variety in the first effective settlement, while many other parts of the world have such longstanding complex cultural geographies that these eight concepts are simply not relevant. Which of the concepts covered above do you find most compelling and why?

Box 6.2 Cultural Preadaptation

There were eleven preadaptive traits that, according to Newton (1974:152), made it possible for the culture of the Upland South to expand west to the edge of the Great Plains quickly and successfully. These traits are

- dispersed settlement, which allowed a relatively small number of people to lay claim to an extensive area;
- kin-based dispersed hamlets, which was a form that could be readily replicated and adapted to varying physical and economic circumstances;
- dispersed services, such as mills, stores, and churches, which offered numerous focuses for settlement;
- a combined stockman–farmer–hunter economy, which encouraged diverse economic activity;
- log construction, which allowed forests to be exploited;
- universal construction techniques, which allowed people from diverse ethnic backgrounds to share building tasks;
- a productive and adaptable food-and-feed complex that included cattle, hogs, corn, and vegetables and did not depend on restrictive tree or shrub cultivation;
- extreme adaptability concerning the choice of commercial crop to generate income;
- an evangelical Protestantism combined with antifederalism, which together encouraged settlements to control their own affairs;
- an open class system, which allowed White settlers to advance socially;

- a courthouse-town system that provided a focus on civil order and emphasized skills of the elite over others.

For Newton, it was these eleven preadaptive cultural traits characteristic of the Upland South culture that allowed the culture to advance westward. The culture itself was formed east of the Appalachians in the early to mid-seventeenth century, and moved west to the Great Plains, where the semiarid and treeless environments required a different set of preadaptive traits. ➡➤➤ 8

Jordan and Kaups (1989) addressed essentially the same problem, but chose to identify the preadapted culture by using the term "backwoods". Arguing that the culture was formed in the mid- to late seventeenth century in eastern Pennsylvania, they also emphasized the role that culture played in the settlement of the western United States following a brief pause at the eastern edge of the Great Plains (see Figure 6.4). Jordan and Kaups added the following traits to the list of eleven proposed by Newton:

- considerable mixing with Aboriginals
- an almost compulsive mobility, which contributed to some individuals and families moving on as many as five occasions during their lifetime
- expansion that often leaped ahead of the frontier to create islands of settlement
- no interest in conservation

the Canadian Prairies from Manitoba to Alberta; several Finnish landscapes in the upper Great Lakes, evidenced by the many Finnish place names and by the sauna that is a typical addition to farms; and certain German landscapes in the Texas hill country, where resistance to acculturation is evident in church affiliation and reflected in a landscape of distinctive ecclesiastical architecture, cemeteries, and graveyards. There are also many cultural islands

outside of North America, and cultural geographers have identified and described many of these, including Japanese areas on the Brazilian frontier and German towns in southern Brazil.

In some areas, the development of a cultural island is dependent on the ability of the ethnic group to impose their own form of land survey rather than having their landscape surveyed in common with other areas. This is because land surveys often have

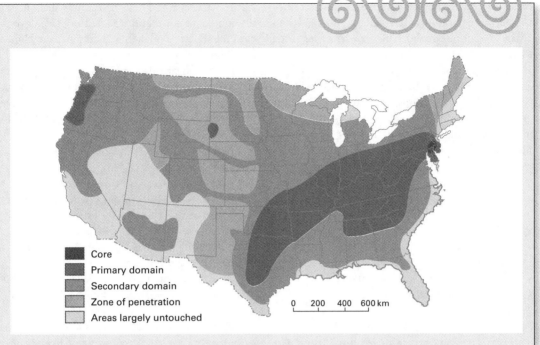

Figure 6.4 Diffusion of the Preadaptive Backwoods Pioneer Culture. This map indicates a proposed diffusion of the preadapted backwoods culture, employing the concepts of core, domain, and sphere in a modified format. There is a core area, where the culture first formed among Finnish settlers between about 1640 and 1680—the date and specific ethnic source continue to be debated. The proposed core area, the lower Delaware Valley south of Philadelphia, is much more localized than that suggested by Newton (1974). It is also smaller than and north of the cultural core proposed by Mitchell (1978), and different in detail to that described by Meinig (1993). Necessarily, then, the ethnic composition proposed by Jordan and Kaups is also different to that proposed by the several other authors, who typically identify a mix of backgrounds, usually including English, Scottish-Irish, and German. The primary domain is the area of most intense cultural transplant, beginning about 1700 and ending about 1850. The secondary domain was settled between about 1725 and 1875 and required considerable cultural adjustment. The zone of penetration is comparable to the sphere, showing even less evidence of the original culture. Source: T.G. Jordan and M. Kaups, *The American Backwoods Frontier: An Ethnic and Ecological Interpretation* (Baltimore: Johns Hopkins University Press, 1989):8–9. Figure 1.3. © 1989 The Johns Hopkins University Press. Reprinted with permission of the Johns Hopkins University Press.

lasting impressions on the boundaries of property, on road networks, and on other aspects of cultural and economic occupance. Long lot systems in the Hispano area, along the lower St Lawrence River, and in Louisiana were a marked contrast to prevailing land survey systems and are vivid indicators of a specific cultural impress. Land subdivision imposed upon an area may have the effect of minimizing potential impacts of local cultures. ➔➤➤ 10

Regions as Homelands

The Homeland Concept

The preceding section discussed several conceptual formulations relating to the formation of cultural regions in areas of European overseas expansion, and noted some applications of these concepts, primarily in an American context. This section builds on that discussion in the context of what is proving to be a particularly interesting regional concept—that of **homeland**. As detailed by Nostrand and Estaville (2001a:xv–xvii), this concept was used as early as 1971 by Carlson in a discussion of the Hispano region of the United States, but it is only in recent years that it has been widely applied to North American cultural regions. Nostrand and Estaville (1993a:1) defined homelands as "places that people identify with and have strong feelings about"; they identified five ingredients common to all homelands (see also Nostrand and Estaville 2001a:xviii–xxi):

- a distinctive, self-consciously aware group, typically ethnic in character
- a distinctive cultural regional landscape
- an emotional bonding of the group with the region
- a degree of institutional control of the region
- sufficient time for these four conditions to develop

This argument that there are distinctive, mostly ethnically based, homelands within North America also acknowledges that there are at least two other types of homeland. Most notably, for most Americans and also for residents of most other nation-states, including Canada, the national unit is itself a homeland. Further, in many parts of the world there are subnational homelands whose residents are focused primarily on achieving some degree of independence from the national unit. Nostrand and Estaville (2001a) do not suggest that homelands within North America typically have such political aspirations—the principal exceptions to this generalization are Nunavut and the French Canadian cultural region. Rather, the many examples of homelands they discuss in two edited collections (Nostrand and Estaville 1993b, 2001b) are mostly ethnically based regions with distinctive cultural landscapes that are best understood as dynamic outcomes of particular human and environment relations. In this sense, the homeland is a version of the Vidalian concept of *pays*.

In the United States and Canada, the presence of such homelands is explained especially in terms of the history of immigration and the ingrained cultural pluralism. These homelands demonstrate six human values that are rooted in place: love for one's birthplace and home, emotional attachment to the land of one's people, sense of belonging to a special area, loyalty defined by geographical parameters, strength that comes from territoriality, and the feeling of wholeness and restoration when returning to one's homeland (Nostrand and Estaville 2001a:xxiii). ➜≫≫**11**

Contrary Views

The homeland concept as detailed above is not without critics. Some cultural geographers consider it inappropriate to investigate the landscapes of North America in terms of different ethnic groups and related identities and homelands. Two substantive criticisms are worth noting.

First, Hart criticized regional cultural studies based on ethnic identity as lacking sophistication:

> Cultural geographers, who should have been able to make a special contribution to our understanding and appreciation of the importance of noneconomic values, have let us down rather badly, at least in the United States, because their approach to values has been so unimaginative. They have focused on those groups that "have worn name tags", at the expense of groups that have not been easily identifiable by such obvious distinctions as country of birth, language, and religious affiliation, and they have largely ignored the importance of region of birth, social class, and other more subtle distinctions. (Hart 1982:26)

A second criticism of the homeland concept favors use of the term "ethnic" to refer only to national identities and not to more localized identities: "There is really no serious challenge to a pervasive, if largely subconscious, code governing the proper ways in which to arrange human affairs over American space" (Zelinsky 1997:158). Such an interpretation may involve either a real or an imagined community, and is consistent with a view of ethnic groups as named human populations with shared ancestry myths, histories, and cultures, features that promote a sense of togetherness as well as an association with a place. Conceiving of "ethnic" in this fashion suggests that the cultural landscape of the United States is best understood in terms of a single dominant culture, namely the Anglo-American ethnic group. From this essentially superorganic perspective, ethnically based regional variations are acknowledged but should not be exaggerated.

The idea that homelands are best understood as applying to national groups is further developed by Conzen (2001:251), who argues that the key components of homelands involve three dimensions and nine criteria, which he outlines as follows:

- identity
  › ethnogenesis: a sense of peoplehood
  › indigenization: time to develop in place over multiple generations
  › exclusivity: promoted through geographical isolation
- territoriality
  › control of land and resources
  › dedicated political institutions
  › coherently manageable spatial unit
- loyalty
  › defense of the homeland against "alien" intrusions
  › compulsion to live within the homeland
  › production and veneration of "nationalistic" landmarks

Based on these criteria, Conzen (2001) argues that in the case of the United States, two categories of homeland predominate, namely the national homeland and a number of subnational Aboriginal homelands, although a case can be made for Mormon, Hispano, and Cajun ethnic homelands.

To clarify the meaning of the term "homeland", the following sections present the examples of French Canada and French Louisiana. Following these examples, two of the most easily recognized and well-researched cultural regions in North America, namely the Mormon intermontane region and the Hispano Southwest, are discussed in greater detail.

French Canada

No sooner permanently settled in the St. Lawrence Valley at the beginning of the 17th century than the French spread beyond the great river in quest of continental adventure. (Louder, Morissonneau, and Waddell 1983:44)

Unlike most of the American regional examples discussed in this chapter, this French Canadian hearth was quickly deserted by those who first settled there: the ambitious *coureurs de bois*, French and Métis fur traders who spread throughout much of North America from the St Lawrence to Louisiana and from Acadia to the Rocky Mountains. This expansion was doomed to failure, as the territory occupied was extensive but with few permanent locations established. Following the French defeat in North America, which was formalized by the 1763 Treaty of Paris, a gradual process of retreat to the St Lawrence occurred. Quebec was confirmed as the French homeland in North America, an insular reserve surrounded by Anglos. But the idea of expansion remained, and by about 1870 three different ideologies prevailed among French leaders in Quebec:

- A continental ideology resurfaced, prompted by substantial movement of French Canadians into New England to work in textile mills.
- A Canadian ideology was supported by some who saw the opening of the Canadian West as an opportunity to expand the French area.
- The third, northern, ideology was the one that came to fruition. This was really a retreat into Quebec, but into all of Quebec, including the extensive northern area. This commitment to Quebec ensured the survival of French culture such that Quebec today might be considered both part of and apart from the rest of the North American continent.

French Louisiana

The French Louisiana homeland has been described as a relic folk landscape "faintly scattered here and there" that, because of the Anglo domination of the larger region, is today "more myth than reality" (Estaville 2001:100). The development of this homeland can be seen as comprising seven stages. The first three stages were French and the final four English. →≫ 12

Transplantation occurred between 1700 and 1760, creating a cultural hearth along the Mississippi between Baton Rouge and New Orleans; this Lafourche hearth was characterized by the French language, Catholicism, and various French cuisine and clothing preferences; it also included a large slave population and some German settlers. Between 1760 and 1800 hearth expansion occurred and was reinforced by the in-movement of French settlers deported from Acadia by the English in 1755. (Acadia, on the shores of the Bay of Fundy, was the smaller of the two French colonies in Canada; the larger was on the St Lawrence.) These Acadian settlers formed a secondary French hearth—the Teche hearth—west of the initial location, and adjusted to their new environment in several ways. In particular, they abandoned wheat, barley, and other crops that had been grown in Acadia, switching to corn, rice, and other crops suited to the subtropical environment. Thus, by 1800, there were two rather different hearths, the original Lafourche French Creole hearth and the secondary Teche Acadian or Cajun hearth. Large numbers of Anglos moved into the area between 1800 and 1840, especially following the Louisiana Purchase of 1803, and this initiated a period of competition, especially in the Lafourche hearth, which lost its earlier isolation and purity, though the French managed to retain significant political control until 1845.

Between about 1840 and 1880, a process of accommodation and French cultural loss occurred, mostly as a result of increasing numbers of Anglos and the expansion of the rail network. These changes continued in the attraction period (1880 to 1920) such that although the French were able to cling to a sense of place, their cultural intensity was significantly diminished. The period between 1920 and 1960 was one of assimilation, involving continued dissolution of the original hearth and further Anglo intrusions into the secondary hearth. Many of these changes were linked to economic developments, such as the rise of the oil industry. Finally, since 1960 there has been evidence of a resurgence of French—especially

Cajun—identity, bringing about a process of revitalization. This final phase has involved the beautification of the Cajun identity and the Cajunization of French Louisiana.

The Mormon Homeland

In order to understand the Mormon cultural region it is important to consider both the religious background and the physical environment of the Mormons, what Francaviglia (1978:3) describes as the "bold, semi-arid backdrop which frames all cultural elements".

People and Region

As mentioned in Chapter 1, Mormons are members of the Church of Jesus Christ of Latter-day Saints, a Protestant group formed in Fayette, New York, in 1830. In 1847, following a series of movements in Ohio, Illinois, and Missouri, members of the church began a movement to an area well beyond the frontier, the Great Salt Lake area of the intermontane West. The Mormon Church is characterized by a strong sense of community related to an emphasis on co-operative endeavors, and also by a powerful administrative hierarchy; church members typically respond positively to the requests of their church leaders. The combination of these two characteristics contributed to a relative uniformity of behavior in landscape after 1847, and this facilitated the development of a formal cultural region and a homeland for the Mormon people.

On first seeing the Salt Lake Valley in 1847, Mormon leader Brigham Young declared that this was the place, a powerful symbolic assertion that set the scene for subsequent expansion of settlement, region formation, and homeland creation. The physical landscape of the region is semiarid and comprises extensive mountain areas with numerous valleys, meaning that agricultural settlement had to take place in a few selected locations. The first areas settled—valleys close to the initial Salt Lake City location—included areas of good soil well suited to irrigated agriculture. Subsequent expansion of settlement, diffusion from the core, focused on some other attractive valley locations in central

Utah and southern Idaho. By the 1880s, settlement was attempted, with varying degrees of success, in some isolated intermontane locations and in arid areas of southern Utah and northern Arizona not well suited to agriculture. Political difficulties also prompted movements into northern Mexico and western Canada. ➔⟩⟩ **13**

Despite the very real physical challenges to settlement, colonization was both rapid and successful, with almost 400 communities pioneered by 1877. This remarkable effort was largely under the personal direction of Young himself, variously called the "American Moses", the "Colonizer", and the "Lion of the Lord". The extent of the cultural region is mapped in figures 6.2 and 6.3.

Landscape

Speaking in 1874, church president George Smith observed:

> The first thing, in locating a town, was to build a dam and make a water ditch; the next thing to build a schoolhouse, and these schoolhouses generally answered the purpose of meeting houses. You may pass through all the settlements from north to south, and you will find the history of them to be just about the same. (quoted in Church News 1979:2)

Given this relatively uniform behavior, it is not surprising that a cultural region characterized by a distinctive visible landscape was formed.

One factor that contributed to the shaping of a distinctive Mormon homeland was the preference for towns—where all the local population, including farmers, resided—over more dispersed agricultural settlements. It is likely that in encouraging town settlement, Joseph Smith Jr, the first president of the church, was both reflecting his own eastern American background and anticipating the difficulty, encountered elsewhere in America, of maintaining a community in areas of dispersed farmsteads. Factors that influenced nucleated settlement include the favorable physical geography of the region, the desire to maintain the solidarity of the group, and Smith's original plan, detailed in 1833,

of the "City of Zion". Mormon towns were laid out according to a regular grid pattern and oriented to the cardinal compass points. There were square blocks, wide streets, lots as large as 2.5 acres (1 hectare) with backyard gardens, buildings of brick or stone construction, and central areas for church and educational buildings. Farm dwellings were inside the town, giving a cluttered rural appearance. Mormon towns thus both enhanced and reflected the community focus.

Within towns, the most distinctive house type was the central-hall house, a house plan that was well established in the eastern United States but one that was being modified and even replaced by the time settlement reached the plains. The plan was transferred to the West by Mormon pioneers and became a distinctive landscape feature. The two principal versions of this type are the "I" house, one room deep and either one-and-a-half or two stories high, and the "four over four" plan, which is two rooms deep. Both are symmetrical, usually with a chimney at each end.

Arable agriculture dominated the rural landscape and was supported by a network of irrigation ditches. Irrigation was necessary because the area being settled was typically semiarid, unable to support arable agriculture without irrigation, and also because the desired Mormon way of life required village settlement and agriculture together. Other distinctive features of the early Mormon landscape include the widely cultivated Lombardy poplar, which is not native to the region, the hay derrick, and the large number of unpainted fences and barns. The combination of these features clearly distinguished the Mormon landscape.

Interestingly, the Mormon landscape achieved a high level of uniformity in spite of the diverse origin areas of the Mormons, many of whom were converted to Mormonism in Europe before traveling to the Salt Lake area. The willingness of these immigrants to jettison ethnic cultural baggage is an indication of the overriding importance of their religious identity and of the role of co-operative effort in the creation of their homeland, the "Great Basin Kingdom". More so than many other immigrants, Mormons really were making a fresh start in terms

Photo Essay 6.1 The Mormon Landscape. The Mormon rural and small-town landscape is a distinctive one characterized by a number of specifically Mormon features. The exceptionally wide main street of the agricultural small town remains evident in the contemporary landscape; this photograph (*opposite top*) is of Manassa in southern Colorado. Less evident in the present landscape (as it is no longer used for its original purpose of stacking hay) is the Mormon hay derrick; this example (*opposite center*) is from Cove Fort, central Utah. In the summer of 1878, a group of Mormon pioneers, including Erastus Snow and William J. Flake, were sent on colonization missions by Brigham Young to the Lower Little Colorado River region in Arizona. They settled on the Stinson Ranch and named the townsite Snowflake. With a population of 4,230 today, the town has a Mormon temple and nine Mormon churches. The pioneer monument shown in this photo (*opposite bottom*) was dedicated in 2000. The first Mormon pioneers to cross the American border into the North-West Territories settled in Cardston, in present-day Alberta, in 1887. The town was named for Charles Ora Card, a son-in-law of Brigham Young. The Mormon temple in Cardston (*above top*), like similar temples in many of the larger Mormon settlements, provides a powerful statement of the importance of the Mormon Church in the local community. Dedicated in 1972, the Provo Temple in Salt Lake City (*above bottom*) is located on a 17-acre (7-hectare site). Modern and functional in design, the exterior finish includes white cast stone, gold anodized aluminum grills, and bronze glass panels. There is a single painted spire. Inside there are six ordinance and twelve sealing rooms. It is located close to Brigham Young University and adjacent to the church Missionary Training Center. (William Norton)

of both place and identity, and the new Mormon identity was more important than the authority of any earlier tradition.

Early experiences of the Mormon group suggested a need for political autonomy, and so church leaders, on arrival in the intermontane West, initiated the process of seeking statehood. A constitution was written, ambitious boundaries proclaimed, a symbolic name (Deseret) proposed, and a provisional government established. This tentative state was dissolved in 1851, following the creation of the smaller territory of Utah, with Brigham Young as the first governor. Further unsuccessful attempts to achieve self-government were made in 1856, 1862, 1867, 1872, and 1882. The boundaries of the hoped-for state of Deseret included all of present Utah, much of Nevada and Arizona, about one-third of California, and parts of five other states.

The Hispano Homeland

Is there a Hispano homeland region in the American Southwest? Two cultural geographers hint at an answer in the titles of their books: *The Spanish-American Homeland* (Carlson 1990) and *The Hispano Homeland* (Nostrand 1992).

People and Region

Hispanos are those people of Spanish-American ancestry who settled in the upper Rio Grande Valley of New Mexico. The physical environment of the Hispano region comprises arable floodplains and an extensive semiarid highland with a sparse vegetation cover. The core of this Spanish-American, or Hispano, region is the Río Arriba including Santa Fe, and the Río Abaja including Albuquerque. Prior to the late sixteenth century, about 45,000 sedentary and peaceful Pueblos—members of various American Indian peoples—were settled in perhaps eighty villages in this core area, while the surrounding plateaus and mountains were occupied by foraging groups, Apaches to the west and east and Navajos to the north. Spanish colonization from Mexico began in 1598, and this outpost of the Spanish empire remained largely isolated from other influences for over 200 years.

The relationship between Pueblos and the Spanish was one of Spanish domination and passive resistance by Pueblos punctuated by revolts. Following an uprising in 1680 the Spanish in 1693 regained control of the region, which subsequently grew because of increases in the Hispano population, which spread in all directions. Expansion was especially marked between 1790 and 1900. By the late 1840s, the larger region contained about 70,000 Hispanos and fewer than 10,000 Pueblos. Together these peoples effectively represented an island in a sea of nomadic Indians, linked only tenuously to European-American people to the east. In 1848 the territory became a part of the United States and was immediately converted, politically speaking, from a region in the far north of Mexico to a region in the southwestern United States. Intrusion of Anglos (a term that included French trappers and traders) began around 1821, when Mexican independence sparked changes in frontier policies, but was greatest beginning in the 1860s. These movements added another dimension—the third of three peoples—to the cultural mix of the region.

The details of regional morphology reflect both Hispano expansion and contact with other groups. The Hispano region was in its heyday in 1900, when it extended over parts of five states, with Santa Fe at its center and a population that was two-thirds Hispano (see Figure 6.5). ➝≫ **14**

Landscape

Within the larger Spanish Southwest, the Hispano region of the upper Rio Grande Valley is distinctive for two principal reasons: first, the Spanish settlers came earlier and tended to come directly from Spain, and second, the area was isolated for an extensive period following Spanish colonization. These two details are similar to factors that influenced the shape of the Mormon region, and Hispanos, like the Mormons, created a distinct region both by adjusting to their natural environment and by impressing their cultural identity on their surroundings. The principal distinction between Mormons and Hispanos is that, while Mormon behavior was largely community-focused and dictated by the wishes of church leaders, the behavior of Hispanos was more individual and family-focused, although various institutions, especially the Catholic Church, also played a role.

During the period of formative colonization, prior to substantial Anglo in-movement, a network of missions, *presidios* (military settlements), towns, farm villages, and ranches began to dominate the landscape. For the Spanish and, later, Mexican governments, this was an important frontier area where land policies were implemented to attract settlers and to ensure orderly settlement. The most distinctive landscape feature was the long lot, a system of land subdivision that involved long, narrow, ribbon-like lots, often fronting on a river. This system provided

Figure 6.5 The Hispano Homeland, 1900. This map shows the Hispano homeland during its heyday. The demographic center is Santa Fe in the upper Rio Grande Valley, the largest Hispano center and the political and religious capital. In 1900, 90% of all Hispanos lived within 150 miles (241 km) of Santa Fe. "Thus, like an island on the land, much of the Homeland in 1900 stood tall and flat, so uniformly high were its Hispano proportions. Its edges were like steep cliffs, so few Hispanos lived off the island" (Nostrand 1992:232). Since 1900, regional identity has waned somewhat, especially because of significant intrusions of Anglo populations. It remains, however, as a distinctive region today. Source: R.L. Nostrand, *The Hispano Homeland* (Norman: University of Oklahoma Press, 1992):231.

settlers equal access to irrigation water and permitted them to live close to each other in village settlements, a desirable state of affairs for both defensive and social reasons. Beyond the long lots were large pastoral areas, and transhumance (seasonal movement of animals) was practiced.

The landscape of this region is distinctive, especially because of the long lots, irrigation systems, adobe buildings and outdoor ovens, village settlement with plazas, and profusion of Spanish and religious place names. Many of these cultural traits were evident earlier in Spain and New Spain. The early Spanish colonists built buildings with sun-dried clay bricks, or adobes, reflecting the shortage of timber for building purposes in the favored floodplain areas. Houses were frequently added to over time to accommodate additional family members, resulting in l-shaped or even u-shaped forms, and they were often equipped with an outdoor adobe oven. There were also homes built of stone, while log buildings were usual only in the upland areas. Early villages comprised either houses on small lots around a plaza or rows of houses on lots along a riverfront. The latter form was especially popular after about 1850, when defense was less of a consideration. The impress of religion is also striking, with Catholic churches and religious meeting houses being characteristic features of the village landscape. The religious meeting houses (*moradas*) of Penitente chapters are of special symbolic importance. Spanish (often religious) place names are usual throughout the landscape and provide an especially compelling aspect of the regional identity.

In a discussion of regional morphology as evident in the Hispano landscape, Nostrand (1992:226–30) used the terms "stronghold", "inland", and "outland", to refer to stages of change in geographical processes. The stronghold, defined as the area that is more than 90 per cent Hispano, existed by 1790 and peaked in 1900, but had largely disappeared by 1980. The inland, 50–90 per cent Hispano, was a series of separate islands that enlarged notably between 1850 and 1900 and coalesced to replace the stronghold by 1980. The outland, 10–50 per cent Hispano, was present only between 1870 and 1930. ➔≫ **15**

Photo Essay 6.2 The Hispano Landscape. The impress of the Catholic religion is very evident in the Hispano landscape, as the buildings shown in these photographs suggest. Built in the classic New Mexico Spanish mission style in 1780 and situated on the original town plaza, this small church (*opposite top*) was originally named Santo Tomás, but the name was changed in 1881 to San José de Gracia. It is located in Las Trampas, northern New Mexico, a Spanish colonial village established in a mountain valley. It is an active parish church and was fully restored in the 1970s. The adobe church (*opposite center*) is in Abiquiú, also in northern New Mexico, and is located at one side of the old plaza. The *morada* (*opposite bottom*), a much less ornate construction than either church above, is one of several in the southern Colorado–northern New Mexico region of Hispanic settlement; the example in this photograph is on the outskirts of the small settlement of San Pablo in southern Colorado. Located south of Tucson, Arizona, the mission of San Xavier del Bac (*above top*), known as the "White Dove of the Desert", was begun in 1783 after an earlier structure was destroyed. The interior is filled with brightly painted carvings of apostles and saints and ornate décor statues that are draped in real clothing. It is an important Catholic pilgrimage site. Another important pilgrimage site is El Santuario de Chimayó (*above bottom*) in northern New Mexico. This shrine, which has been described in some tourist literature as the "Lourdes of America", was constructed in 1816 and remains an important symbolic location today. (William Norton)

Questions for Critical Thought

Homelands are understood as cultural regions that are especially closely linked to a distinctive group. But as noted, this seemingly straightforward idea is not without critics. We have already queried the possible links to claims of national identity, and we now question the rationale for the concept in a more fundamental way. Consider the specific issue of group identity. There is little doubt that questions of group identity are complex, elusive, and certainly changing—are we really able to divide people into neat parcels such as cultures or ethnic groups, or are these groupings really arbitrary creations? There is no clear-cut response to this question, and it is sufficient to raise these concerns at this time without attempting any resolution. Indeed, in addition to the elusiveness of group identity, it is now increasingly acknowledged that human landscapes, regions, and homelands are socially constructed or produced. Consider the Hispano example. Some cultural geographers have called into question both the identification of the Hispano group and the reference to a Hispano homeland, citing two "myths" involved in these ideas, namely the myth that there is a distinctive Hispano group and the myth that there is a real and discrete region occupied by the supposed group, a homeland. "It is hard to imagine anything more powerful by way of an existence-statement about a culture than to depict its map location, its boundaries, and its internal subregions" (Blaut 1984:159). Can you think of any examples that support this view of Blaut?

Shaping the Contemporary World

Cultural geographic interest in the evolution of regions has focused largely on the subnational scale, and there has been relatively little concern with the question of global divisions. This seems natural given the intellectual antecedents of the landscape school in the work of such European geographers as Schlüter and Vidal and in the culture area concept in anthropology. Indeed, there has also been little interest in world regions in the larger discipline of geography except as basic teaching aids, with world regions seen as outcomes of human-and-land relations, usually with an emphasis on physical geography as cause. A principal exception to this focus was the identification of human regions as outcomes of human energy as well as of physical geography, an approach that permitted the recognition of regions of hunger, debilitation, increment, effort, industrialization, lasting effort, and wandering (Fleure 1919).

The pioneering cultural geographic contribution to the study and identification of evolving major world regions is the textbook *Culture Worlds* (Russell and Kniffen 1951), while a more recent effort in this direction is a work by Lewis and Wigen (1997). Other than these important studies of world regions, cultural geographers have paid little attention to this topic. However, there is a substantial literature in comparative history, anthropology, and sociology on the larger topic of the rise of civilizations, as well as a concern with the evolution of the modern world as a system of related areas. Both concerns are discussed in this section.

Civilizations as Global Regions

Civilizations, or "giant cultures", have been identified by scholars, often called civilizationists, such as Toynbee, Spengler, Sorokin, Quigley, and Kroeber. Each of these writers proposed different processes, stages, and civilizations, often in significant disagreement with others, but most agreed in noting cyclical patterns in history. For most writers in this tradition, the distinction between a culture and a civilization is one of scale and level of achievement, with civilizations being larger and more complex. Indeed, civilizations usually involve an integration of many languages and religions. Civilizationists have made a major contribution to the study of world history and world geography because of their explicit concern with civilizations globally rather than with the European civilization as it formed in association with other areas. As such, they can be described as scholars of world history. ➔➤➤ 16

The best known example of this genre of world history is probably the 12-volume work of Toynbee, *A*

Study of History (1934–61). Although civilizations were not explicitly defined, they were implicitly regarded as large cultural and social groups that evolved from earlier primitive societies. The process by which a primitive society was transformed into a civilization was one of challenge and response, a form of adaptation. Toynbee used an organismic analogy, with civilizations passing through a life cycle of four stages: genesis, growth, breakdown, and disintegration.

Toynbee proposed five types of challenge or stimulus that were required for genesis to occur, namely a difficult physical environment, the availability of new territory, a military defeat, regular aggression from outside, and discrimination by others. These challenges needed to be mounted at an appropriate level—that is, sufficient to encourage a creative response without being too devastating. Following genesis, a group would grow if it continued to respond to subsequent challenges. In both the genesis and growth stages, a creative minority of the population took the lead in initiating responses to challenges. Continued growth eventually led to a situation where growth was no longer possible, and the third stage, breakdown, began. This occurred when there was a loss of unity and direction related to the demise of the creative minority. New institutions were needed but not created, such that responding to a new challenge consumed large amounts of energy, making it difficult to respond to subsequent challenges. Finally, disintegration occurred, often as a result of major social schisms.

Following these criteria, Toynbee identified 23 civilizations in world history, subsequently increasing the total to 26, of which he defined 10 as living. The 10 living civilizations were Western Christendom, Orthodox Christendom, the Russian offshoot of Orthodox Christendom, Islamic culture, Hindu culture, Chinese culture, the Japanese offshoot of Chinese culture, Polynesia, Nomads, and the Inuit, with the last three defined as living but arrested. As these regional names suggest, the basic criterion used was religion. Toynbee excluded Africa from this classification on the grounds that it was primitive and not culturally committed. Such a misunderstanding of African history and culture was not unusual among European scholars prior to the mid-twentieth

century as one component of larger racist and colonial assumptions (see Chapter 7). The typical reaction to this corpus of work has been predominantly negative, at least until fairly recently. This is not surprising given the difficulties of working at a global scale, especially given the generalizations required. Toynbee's basic naturalist idea that civilizations can be treated as organisms has been criticized by those who would prefer to see civilizations described as adaptive systems or ecosystems that can be understood as responses to ecological opportunities.

Notwithstanding these problems, it is increasingly recognized that there is some value in viewing civilizations at a grand global scale, since this may involve a useful integration of economics, politics, society, and culture, and may offer insights into both past and present that more local scale analyses are not able to achieve. Indeed, McNeill (1995) has argued for and applied an approach to world history that acknowledges circumstances of culture contact and the role of communications and that incorporates accounts of biological and ecological encounters. The idea of some dogmatic sequence of stages is not popular, but the idea that civilizations are meaningful units for analysis remains. Civilizations may be more akin to functional than to formal regions in that they are linked by human connectedness rather than by human relatedness or uniformity—recall the important distinction between types of region noted in Box 6.1.

The Evolving World System

World system ideas are a more recent alternative to approach long-term historical change. Whereas civilizationists often stress the isolation and distinctiveness of regional civilizations, the world system concept focuses on integration and interaction. (There is a more detailed discussion of the world system concept in the context of a discussion of global inequalities in Box 6.4.)

Historically, world systems covered only some part of the entire world, but the current capitalist world system, which began about 1450 with the onset of European overseas expansion, has broadened (that is, expanded spatially) to cover the entire world. While European hegemony is seen as the most important element of a truly worldwide trading system, several historians argue that well-established Asian and Arab trading systems were in place long before European expansion (Abu-Lughod 1989; Pomeranz 2000), and that Europe's rise in the 1600s did not *create* an international trading system but merely allowed Europe to *enter* the already dominant Chinese trading sphere at the time. This was particularly evident because Europe was able to extract silver from its colonies in South America, and silver was the one commodity China needed at that time (Brook 2008).

Civilizations as World Systems

Expressed rather simply, the greatest single difference between the civilizationist and world system approaches to the study of long-term historical change and the shaping of the modern world is the civilizationist interest in distinct cultures and their evolution and the world system interest in political–economy, and interconnections between regions. This distinction, which was never very clear to begin with, has become increasingly blurred, as evidenced by the discussion above as well as by recent attempts to integrate the two approaches (Sanderson 1995).

One extension of civilizationist ideas views civilizations as multiple cultures, or polycultures, which are best seen as sociopolitical units comprising multiple states linked either in alliance or conflict. Extending this general idea, world history can be interpreted in terms of a process of integrating civilizations that has resulted in a single dominant central civilization. The single global civilization is the lineal descendant of, or the current manifestation of, a civilization that emerged in the Near East around 1500 BCE, when Egyptian and Mesopotamian civilizations collided and fused. This new fusional entity has since then expanded over the entire planet and absorbed, on unequal terms, all other previously independent civilizations (Wilkinson 1987:46). According to this argument, central civilization formed about 1500 BCE, grew spatially and demographically, and is now global in scale. As such, it can be regarded as the key entity in any consideration of world systems and world economies. The central civilization is the same entity as the modern world system as defined by Wallerstein.

"The Greatest Topic in Historical Geography"

Cultural and historical geographers have not widely addressed the challenging question of the formation of global regions as one of their principal research topics (unlike economic geographers who have engaged in this work). Certainly, no equivalent to either the civilizationist or the world system approaches has been developed within the subdiscipline. This failing exists in spite of Meinig's assertion (1976:35) that this is "the greatest topic in historical geography". There have been important attempts to critically assess the type of models used to explain European expansion, for example, Blaut's criticism of the model of geographical diffusionism and European colonialism, and his argument in support of a more world systems–inspired view of the world (1993b). But, given the presumed cultural importance of central civilization or the capitalist world system to any understanding of global cultural issues, cultural geographic interest in this topic might increase significantly.

There are three aspects to the cultural geographic interest in civilizationist and world system concepts, each of which is noted briefly here and discussed more fully later in this chapter. First, civilizationist concepts link directly with cultural geographic attempts to delimit regions at the world scale. Second, according to the basic logic of both central civilization concepts and world system theory, the significance of the fact that Europe has played the critical role in the last 500 years of world history is that Europeans not only have effectively taken over the world but also have reconstructed the world in their image. This reconstruction has been political, economic, and cultural, and is variously called *westernization* or *modernization*. The significance of this situation is difficult to exaggerate, but it is also a model or phase that is rapidly being challenged as European and American economic hegemony is arguably on the decline, as countries such as China again take their place of dominance (Frank 1998). Third, the growth of a capitalist world economy can be seen as a major global transformation in cultural as well as in economic terms; indeed, it is often asserted that the world has changed from a mosaic of groups with different cultures to a more culturally unified world. This change is related to the increasing importance during the past 500 years of the nation-state and to the roles played by the three hegemons, namely the seventeenth-century Dutch, the nineteenth-century British, and the twentieth-century Americans. Thus, the transition is from a world differentiated in cultural terms to a world differentiated in political terms. ➔➤➤ 17

Global Regions

Lewis and Wigen (1997:14) argued: "We would insist that world regions—more or less boundable areas united by broad social and cultural features—do exist and that their recognition and delineation are essential for geographical understanding." This section outlines three proposed global regionalizations.

The first regionalization is the classic cultural geographic regionalization derived from the Sauerian tradition that recognized a number of *Culture Worlds* on the grounds that the "logical approach to regional geography is one that is based on cultural outlines" (Russell and Kniffen 1951:viii). This work had a considerable impact on human geography generally, as it helped replace environmental determinism with a form of cultural determinism. The second regionalization is based on the work of a political scientist who identified cultural regions as the appropriate basis for understanding the contemporary world. He justifies this approach to regionalization on the grounds that conflicts are no longer based on politics and ideologies but rather on differences in culture, and as a result the modern world is seeing *The Clash of Civilizations* (Huntington 1996). The third regionalization is an original and innovative geographic contribution that critiqued earlier work, stressed the heuristic value of global regions, and explicitly identified *The Myth of Continents* (Lewis and Wigen 1997). In addition, cultural geographers have traditionally emphasized the roles of language and religion as indicators of various cultural differences from place to place, and Box 6.3 provides a brief overview of these two variables in the global context. ➔➤➤ 18

Culture Worlds

As noted earlier, landscape and cultural region concepts were not devised, methodologically speaking, with a global scale in mind. Nevertheless, there have been attempts to divide the world into cultural regions as a means of studying general world culture. The pioneering textbook in world cultural regional geography grouped people according to culture, related culture to area in order to derive culture worlds as of the mid-twentieth century, and showed that such culture worlds had formed over a long time period. The problems of regional classification at the global scale were recognized, and various subregions were identified. The seven culture worlds, shown in Figure 6.6, were summarized as follows.

The European World was described as aggressive in the sense that Europeans typically assumed that most of their everyday cultural matters were universal, even though they may have been foreign to those outside of Europe. Although contemporary discussions in cultural geography are more theoretically and socially informed than they were in the past, it is clear that this point was not ignored in earlier discussions of cultural geography. Key European cultural traits were identified as those of field agriculture, industrialization, urbanization, and labor specialization.

Two separate cultural hearths were noted for Asia or "the Oriental World", namely northern China and the Indus Valley. Although the Indus Valley hearth showed some similarities to the European World, it was argued that it was closer culturally to northern China. Between the European and Asian Worlds, and also extending through northern Africa, is an area of semiarid and arid climate that separated and discouraged interaction between these two worlds. In this area, the Dry World, nomadic peoples developed ways of life quite different from those in the more humid neighboring areas.

The Dry World also served to separate the African world from the European. In the African world, people solved problems in a different manner than elsewhere and therefore developed different cultures. The isolated Polar World, characterized by a severe climate, extends across northern Eurasia and northern North America. Both the American and

Box 6.3 *Language and Religion— Classification and Distribution*

There are many human languages, each subject to changes in rules, content, and dialect. Interestingly, there is much debate concerning the number of languages spoken today (with the usual estimate around 6,000), although one recent major report identified about 10,000 (see Carvel 1997). To help understand the origin, diffusion, and current distribution of these languages, cultural geographers use the concept of a **language family**—a group of closely related languages that show evidence of having a common origin— although it is now widely accepted that the various language families are themselves related. Specific classifications vary in detail, but most recognize at least fourteen principal families. Within language families there are usually several subfamilies. Different languages are by definition mutually unintelligible, but each individual language may comprise several dialects, or mutually intelligible varieties of that language.

The numerically largest and spatially most widespread language family is the Indo-European, which includes most European languages and also most of those of the Indian subcontinent. The dominance of this family in the world context is related to the process of European colonial expansion. Principal subfamilies within Indo-European include Romance, which evolved from Latin and includes Portuguese, Spanish, French, and Italian; Germanic, which includes English, Norwegian, Swedish, and German; Slavic, which includes Polish, Ukrainian, and Russian; and Indic, which includes Hindi and Bengali. There are smaller

Celtic and Baltic subfamilies and also some individual languages, such as Greek and Albanian, that do not clearly belong to a subfamily. The common bond between all of these languages is their origin in Eastern Europe perhaps some 8,000 years ago in the form of proto Indo-European, itself but one branch of a much larger linguistic tree.

Classifying religions has proven more straightforward, with the two major types recognized by cultural geographers being ethnic and universalizing. An **ethnic religion** is closely identified with a particular group and does not actively seek to convert others. Numerically, the principal ethnic religion is Hinduism, the oldest of the major religions, originating in the Indus Valley about 4,000 years ago. Other ethnic religions include Judaism, Confucianism, Taoism, and Shinto.

A **universalizing religion** actively seeks converts because of the claim that the religion is proper for all people. The three principal examples are Buddhism, Christianity, and Islam, each of which can be further subdivided. Buddhism, an offshoot of Hinduism, developed about 500 BCE; Christianity developed from Judaism at the beginning of the Common Era, and Islam developed from both Judaism and Christianity about 600 CE. Historically, Christianity has been the most concerned with seeking converts, and because it was associated with Europe during the period of overseas expansion from the fifteenth century onwards, it has spread over much of the globe along with some of the Indo-European languages, notably Spanish, Portuguese, French, and English.

Pacific culture worlds developed largely in isolation from other areas until the beginnings of European overseas movement. Although both of these areas displayed significant internal cultural differences, sufficient similarities were noted to allow them to be recognized as culture worlds. →≫**19**

This regionalization, and many subsequent regionalizations included in introductory human and regional geography textbooks, were essentially attempts to classify the world into a fairly small number of discrete areas that demonstrated a degree of internal homogeneity and that were different from other areas. As such, they served as heuristic devices to facilitate larger textbook accounts of world regional geography. As attempts at delimiting meaningful cultural regions, such regionalizations

Figure 6.6 World Cultural Regions I. "Through the device of recognizing seven culture worlds, the geography of the inhabited parts of the earth is presented in an orderly manner. Each culture world is a reasonably unified subdivision of the earth's surface occupied by peoples who are strikingly alien to inhabitants of other culture worlds" (Russell and Kniffen 1951:viii). The global regions mapped are the European World, Oriental World, Dry World, African World, Polar World, American World, and Pacific World.

A transitional area is located between the European and Polar Worlds, and large areas outside of Europe are seen as affected by a New World Revolution, notably the American, African, and Pacific Worlds. Each region is discussed in considerable physical, historical, and cultural detail. This pioneering use of the cultural region concept at the world scale is a major development within the Sauerian tradition. A principal question about this contribution concerns the use of the label Dry World, as the use of a climatic term does not accord with the idea of a cultural region. There are many other schemes for world regionalization employed by textbook authors for essentially heuristic reasons, but most are similar, although different in detail, to this pioneering effort. Source: Reprinted with the permission of Scribner Publishing Group from *Culture Worlds* by Richard Russell and Fred Bowerman Kniffen. Copyright © 1951, 1969 by The Macmillan Company; 1979 by Fred Bowerman Kniffen and Josephine Burke Russell. All rights reserved.

are easily criticized, principally because the larger the area to be divided into regions, the more superficial or more numerous the regions become. As previously noted, world regional classifications typically result from a process of logical division of a large area. Such classifications require that a great deal be known about the locations beforehand so that sensible distinguishing criteria can be used.

The Clash of Civilizations

The second world regionalization discussed here grew out of concerns quite different from those of Russell and Kniffen. In an attempt to understand the principal schisms in the world following the end of the Cold War in the early 1990s, political scientist

Samuel Huntington suggested that these schisms would be located along the peripheries of the major world civilizations or cultures. The central argument he put forward was that cultures and cultural identities that can be broadly conceived as civilizations are shaping relations in the contemporary world in the sense that groups sharing cultural values are increasingly co-operating with one another. Huntington further suggested that the world since about 1990 can be seen as multicivilizational—this is an important claim that contrasts with the ideas of central civilization and of world systems, both of which see a single large area, the West, as dominant globally.

The most intriguing aspect of this argument for cultural geographers is Huntington's claim that post-Cold War conflicts, unlike most nineteenth- and twentieth-century conflicts based on economic and ideological antagonisms, are related to differences in culture. As a result, it is critically important to map cultures and thereby identify those areas (called *fault lines*) susceptible to conflict. This regionalization contains no great surprises given the earlier work of civilizationists as discussed in the preceding section,

and given the standard regionalizations proposed by cultural geographers, as exemplified by Russell and Kniffen. The key criterion used to identify civilizations is that of religion, as several of the regional names indicate (Figure 6.7).

This argument has generated substantial discussion in political science, as it represents a marked

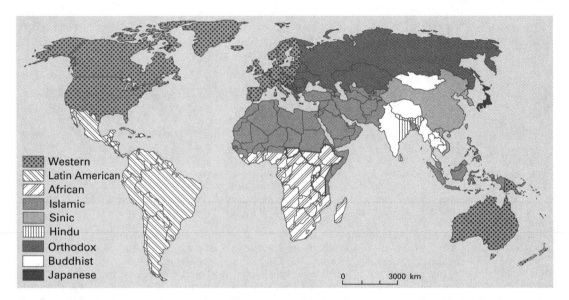

Figure 6.7 World Cultural Regions II. Although Huntington insisted on the importance of regional civilizations as the basis for understanding contemporary world conflict, the task of identifying those regions was not straightforward. Huntington (1993:25) named, but did not map, "seven or eight major civilizations. . . . Western, Confucian, Japanese, Islamic, Hindu, Slavic-Orthodox, Latin American, and possibly African civilization". Africa was listed as possible because of the divisions into Islamic and non-Islamic and into Saharan and sub-Saharan.

This figure shows the nine civilizations mapped by Huntington (1996). But even this second regionalization contains some uncertainties and apparent confusions. The map and the discussion of the map are not in complete accord. Thus, in referring to the map, Huntington (1996:21) noted that there are "seven or eight major civilizations", although nine are included on the map. What is the justification for these nine mapped civilizations?

- Five of the nine mapped are accepted on the basis of earlier civilizationist ideas; these are Sinic (also called Chinese and previously called Confucian as noted above), Japanese, Hindu, Islamic, and Western.
- It was noted that some civilizationists also add Orthodox Russian civilization.
- It was considered useful to add Latin American and possibly African.

The inclusion of the Buddhist civilization was not explained.

But, following the statement, "The major civilizations are thus as follows," Huntington (1996:45) listed and briefly discussed only seven civilizations—Sinic, Japanese, Hindu, Islamic, Western, Latin American, and African (possibly); neither Orthodox nor Buddhist were identified as major civilizations although both were included on the map—it was noted separately from the list that there is a Theravada Buddhist civilization in Sri Lanka, Burma, Thailand, Laos, and Cambodia.

These problems evident in the initial identification and mapping of civilizations are confusing from a cultural geographic point of view. However, it is noted that the extensive discussions of political and related issues that constitute the bulk of the book focus on all nine civilizations. Source: *The Clash of Civilizations and the Remaking of World Order* by Samuel P. Huntington. Copyright © 1996 by Samuel P. Huntington. (Simon & Schuster). Reprinted by permission of Georges Borchardt, Inc., for the author.

departure from the conventional concern with states as the key units in a conflict situation. There has been some broad agreement concerning the idea that future conflicts are likely to occur in boundary areas, but not necessarily the boundaries between world civilizations. The argument is particularly open to criticism for attempting to build the West around conservative values and for stressing the dangers presented to the West by others, such as Islamic fundamentals and immigrants. Perhaps the most credible alternative suggestion (again of great interest to cultural geographers) concerns the prospect of future conflicts related to discord within political units that are prompted by massive variations in quality of life, which are in turn often related to ethnic and cultural differences or, more directly, to resurgent ethnicities. ➡➤➤**20**

Although cultural geographers have not traditionally concerned themselves with conflict issues, these two suggestions about likely future conflict emphasize the close links between political, economic, and cultural themes. Good cultural geography, like good political and economic geography, needs, at times, to be broadly concerned with all other aspects of larger human geography.

The Myth of Continents

Attempts to regionalize the world using cultural criteria are problematic. It is not an easy task to delimit large areas that can be meaningfully argued to be in some sense internally homogeneous. Of course, the conventional approach, both within and outside of geography, is to use continental divisions. Most of us tend to think of Europe, Asia, Africa, North America, and Latin America as regions, at least in some informal sense. The principal alternative approach is based on literate civilizations, as discussed earlier and as used by Huntington. It is not difficult to see that the attempts at world regionalization discussed so far can be criticized for occasionally flawed logic. It may be that these essentially traditional approaches to regionalization, which are often based on presumed continental divisions, are at best misleading and at worst wrong.

One response to this problem is based on the following argument: "World regions are multi-country agglomerations, defined not by their supposed physical separation from one another (as are continents), but rather (in theory) on the basis of important historical and cultural bonds" (Lewis and Wigen 1997:13). This does not mean civilizations as discussed by civilizationists, which typically assume literacy; rather, global regions, large socio-spatial groupings, are identified on the basis of a shared history and culture. The resulting regionalization (see Figure 6.8) is based on processes and on traits, ignores political and economic boundaries, and employs cultural boundaries.

First, it is helpful to think of Africa, Europe, and Asia as comprising the supercontinent of Afro–Eurasia. Effectively, Europe can be thought of as Western Eurasia or as Northwest Afro-Eurasia. Second, note that denying that Europe is a continent does not prevent it from being treated as a cultural region because, of course, an area does not need to be a continent in order to merit cultural region status. Third, the regionalization of Asia is reconceived,

Questions for Critical Thought
———— ✧ ————

One continental division that may be particularly unjustifiable is that between Europe and Asia, given that Europe, as it is not a distinctive and separate landmass, is essentially a western peninsula of Asia. Is it reasonable to regard Europe as a continent and India as a subcontinent? The use of continents, or presumed continents, as a basis for regionalizing is certainly flawed, reflecting, as it does, a particularly Eurocentric view of the world—precisely the point made by Russell and Kniffen. Indeed, it is increasingly clear that the particular views of the world that a group holds are a reflection both of reality and of their perceived position in the world. The cultural studies tradition, especially in the form of postcolonial theory, addresses these ideas more fully. Overall, does it appear that the three maps of world regions are meaningful contributions, or are they oversimplifications of complex realities?

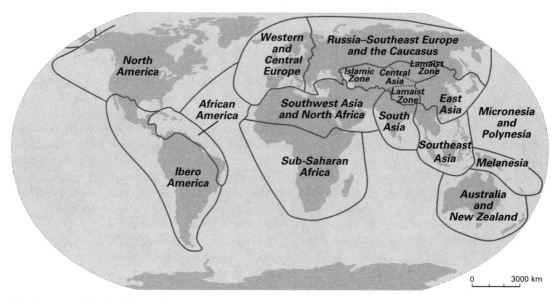

Figure 6.8 World Cultural Regions III. This interesting map of global regions followed an extended account that stressed some problems of regionalizing at this scale, but that also identified the value of regionalizing at this scale. It was acknowledged that "all human geographical entities are conventional rather than natural" and that there is accordingly a need to use regional names that are generally understood rather than names that have different meanings in different contexts (Lewis and Wigen 1997:196).

The fourteen regions recognized are those of East Asia, Southeast Asia, South Asia, Central Asia (divided into Islamic and Lamaist zones), Southwest Asia and North Africa, Sub-Saharan Africa, Ibero America, African America, North America, Western and Central Europe, Russia–Southeast Europe and the Caucasus, Australia and New Zealand, Melanesia, and Micronesia and Polynesia. Source: M.W. Lewis and K.E. Wigen, *The Myth of Continents: A Critique of Metageography* (Berkeley: University of California Press, 1997):187.

being divided into five regions, an explicit acknowledgment of both size and regional complexity. The inclusion of a Central Asian region is a distinctive addition to most regionalizations, as is the reference to Lamaism (that is, Mahayana Buddhism). Fourth, the regionalization of the Americas is similarly reconceived. Latin America is rejected as a regional name, and the Central American area is named African America.

A Cultural Geography of Our Unequal World

This section builds on the discussion of the shaping of the modern world and ways of devising a set of global regions. Here the concern is with the evolution of the current world system, especially as it includes a fundamental division between more and less developed countries. This section explores some of the reasons for this division and examines current changes related to processes of globalization.

Creating Global Inequalities

European Miracle?

One conventional answer to the question of why some nations are rich while others are poor builds on the idea that Europe is in some ways different from other parts of the world, and that the particular characteristics of the European region sparked a "European miracle" that began in the fifteenth century (Jones 1987). Forceful arguments along these lines were made by Landes (1998)—who stressed that Europe is distinctive in two respects: its physical geography and its culture—and by Diamond (1997), who outlined major global historical and economic changes over the past

13,000 years. The basic claim made by both authors is that physical geography and location do matter and are more important in determining the fortunes of a population than any purported genetic differences.

At the beginning of the Holocene period, all humans were pre-literate hunter-gatherers capable of fashioning and using tools of stone. Advancing beyond this way of life required humans to begin domesticating plants and animals, and in terms of the physical geography favoring this kind of development, different parts of the world were not equal in the opportunities that they offered. The basic logic runs as follows: in Europe, mild summers permitted physical activity, cold winters reduced the dangers of disease, and adequate rainfall permitted agricultural development. In many other parts of the world, the climate was not as conducive to physical activity. Further, not all plants and animals are suited to domestication. Indeed, only a few parts of the world had a concentration of suitable plant and animal species, and it was in these areas that agriculture developed, to be followed by increasing population densities, sedentary living, urbanization, and technological and social changes. Agricultural innovations diffused from hearth areas into environmentally suitable areas, including Europe and throughout China.

It was the technological and social changes following domestication, including the evolution of the concept of private property and the emergence of several separate political units, that allowed European guns, germs, and steel to eventually conquer the world. In semiarid areas, on the other hand, water had to be controlled with vast irrigation projects that could only be carried out by a powerful centralized state, the existence of which lessened the likelihood of private property and individual initiative.

Three other factors—culture, money, and knowledge—help to explain why Europe was the first area to industrialize, to experience **development**. The Industrial Revolution in Britain was initiated in a society with a sense of national cohesion, a competitive ideology combined with a respect for merit, and a desire to enhance knowledge. As outlined by

Weber, some of these characteristics were inherent values of Protestantism, the dominant religion in northern and Western Europe; other religious beliefs, notably Catholicism and Islam, were seen as elevating authority and limiting individual initiative. In short, culture does indeed make a difference. For Landes (1998) this cultural argument as it applies to the development of Europe can be extended to address the question of contemporary global differences in degree of industrialization. Poor areas are poor because they possess particular cultural characteristics that proved, and continue to prove, inimical to industrialization.

European Myth?

These ideas about the singularity of Europe and European technological achievement have been challenged from several directions. Most generally, critics of the "European miracle" argue that viewing history and economic change from a critical perspective does not involve seeing Europe as being more advanced than other parts of the world, at least not in the period prior to the Industrial Revolution, a phenomenon that can be explained by reference to European exploitation of other areas. According to this argument, it is the process of European overseas expansion, colonization, and exploitation of resources, the spread of European disease, the institution of slavery, and the specific character of European scholarship that can be considered causes of the poverty in the less developed world. This line of argument is a fundamental part of world systems theory as this is discussed in Box 6.4. From this perspective, ideas about the singularity of Europe are seen as reflecting Eurocentrism. ➔➤➤ **21**

Of course, such fundamental arguments cannot be resolved in the present context; the purpose here is merely to identify what is a very important area of disagreement, both conceptually and empirically. Especially relevant for cultural geographers is how ideas about cultural superiority undergird and normalize material inequality. It is worth noting, however, that in one important respect the disagreement on this matter is one of belief systems—a clash of conservatives and radicals. Thus, questioning

Box 6.4 World Systems Theory

As noted, a world system is a large social system that possesses three principal distinguishing characteristics:

- It is autonomous, meaning that it can survive independently of other systems.
- It has a complex division of labor, both economically and spatially.
- It includes multiple cultures and societies.

Although historically world systems covered only certain parts of the world, the current capitalist world system encompasses the entire world. Two types of world system can be identified: a "world empire" is an area unified in military and political terms, while a "world economy" is a more loosely integrated economic system.

The capitalist world economy was initiated around 1450 and since then has both broadened (that is, expanded spatially) and deepened (that is, evolved into a more complex system), such that it now dominates the world. The modern world system has broadened to cover most of the globe, but this process has involved an increasing separation of the world into three parts. Politically and economically, the core of the system is the dominant part, while the periphery is the weakest; the semi-periphery lies between these two extremes. The three parts are dependent on each other. Especially significant is the fact that, in order to thrive, the core needs to maintain the underdevelopment of the periphery. **Dependency theory** revolves around the idea that the more developed world, in order to become even more developed, must create a less developed world. This process illustrates the consequences of the Marxist idea of ceaseless capital accumulation.

There are five mechanisms involved in deepening of the system, or making it more complex. These are

- *commodification*, a shift from use values to exchange values in both social and economic life;
- *proletarianization*, the transformation of subsistence labor into paid labor;
- *mechanization*, the application of ever-increasing technologies to productive activities;
- *contractualization*, a formalization of human relationships; and
- *polarization*, the increasing disparity between different parts of the world system.

why some nations are rich while others are poor, and speculating about the existence of a European miracle or myth, produces answers that are colored by the preferred ideology, the intellectual baggage, of those who tackle such questions. A particular argument about the contemporary importance of culture is presented in Box 6.5.

These comments raise the much larger and more contentious issue of the legitimacy of European scholarship. It can be argued that any claim that Europe is in some way different is itself a part of the ideology of colonialism and imperialism—one of the stories that Europeans like to tell about themselves.

For Blaut (1993b:14–17) this fundamentally flawed European view of the world identifies an "Inside", namely Europe, and an "Outside", namely the rest of the world. In this picture of the world, the Inside is naturally progressive because it is seen to possess specific mental factors favoring rationality and innovation, while the Outside is viewed as naturally backward because it lacks these European factors; further, Blaut explains, this flawed European view sees the Outside as progressing only as a consequence of contact with the Inside. Blaut (1993b) criticizes this "Colonizer's Model of the World" for being a tale about Europe, told by Europeans, that

In addition to these five essentially linear evolutionary changes there are two types of cyclical change. These are the Kondratieff cycle of economic prosperity and decline, which typically lasts about fifty years, and the hegemonic cycle, in which a single political unit dominates a system politically, economically, and culturally for a period before giving way to another political unit. As noted above, a penetrating political geographic analysis argued that there have been three dominant political units (or *hegemons*) so far, namely the seventeenth-century Dutch, the nineteenth-century British, and the twentieth-century American (Taylor 1996). The future is, of course, uncertain, but it is possible that the twenty-first century will be the "Asian century", with three possible candidates for the position of dominant political unit: China, India, and Japan.

The deepening of the modern world system, like the broadening into core, periphery, and semiperiphery, can be understood in terms of the Marxist concept of ceaseless accumulation of capital. Certainly, as Hopkins and Wallerstein (1996:4) have argued, the "process of accumulating capital on a world scale required the continual development of the world's forces and means of production. This process was a very uneven one, and thereby continually reproduced and deepened what we call the core–periphery zonal organization of world production, the basis of the axial division and integration of labour processes". This is why the modern world system is the capitalist world system and also why it cannot remain unchanging. World systems theory explicitly refutes the concept of developmentalism, the idea that countries are autonomous in terms of their potential for development and that all countries proceed through a set of similar stages as development proceeds.

A conventional criticism of world systems theory is that it overestimates the importance of the world economy, thus underestimating the role of cultural and ethnic identity. Accordingly, world systems theory has been developed in a number of directions that have taken it away from this dependence on economics. Especially interesting are attempts to provide "the beginnings of a unifying framework defining a world of more than political states and economic zones", with a consideration of "people, landscapes, cultures, social institutions, economic capacities and political tenets" (Straussfogel 1997:128).

Questions for Critical Thought

There are clearly benefits and limitations linked to classificatory models that explore cultural and economic processes at the global scale. Consider the positive and negative dimensions of such global approaches to cultural and economic phenomena. Do you think such global approaches are more meaningful in some domains rather than others? Is it more meaningful to apply them to economic, rather than cultural processes?

needs to be rethought and retold from a broader cultural perspective.

Globalization Processes

"Something new, or the same old thing? A process impossible to control, or something that needs to be managed? A corporate conspiracy (a con trick pulled by the rich and powerful), or a brave new world, pregnant with emancipatory opportunities?" With these words, Short (2001) highlighted some of the uncertainties and debates about globalization. Certainly, it is common to assert that the contemporary world

Box 6.5 *Culture as Cause of Economic Growth?*

A review of economic data for Ghana and South Korea shows that in 1960 the two countries had similar levels of per capita GNP and other conventional economic measures. And yet, by 1990, the Ghanaian economy was largely unchanged, while South Korea had become an economic giant. Huntington (2000:xiii) uses this example to introduce the argument that culture matters when it comes to explaining economic growth. For the purposes of the argument, culture is understood as comprising the values, attitudes, and beliefs of a group of people.

Rejecting world systems theory and related claims about links between colonialism and dependency, Harrison (2000) focuses on five themes that are key to understanding the links between cultural and economic growth. First, there is a need to identify and emphasize links between values and progress. Second, a small number of universal values are effectively stated in the United Nations Declaration on Human Rights. Third, physical geography is an important factor—though not a decisive one—in economic growth. Fourth, culture encourages the creation of particular institutions that may promote or impede economic progress; in this context, recall the ideas of Mogey outlined in

Box 4.2. Fifth, cultures do change, but slowly. With these five themes in mind, Harrison argues that governments and development institutions must take culture into account when planning strategies for economic growth.

The experience of the Arab world might be seen as supporting these arguments (United Nations Development Programme 2002). The economic progress of this region is often regarded as being impeded by the shortage of three essential circumstances: freedom, knowledge, and gender equity. The shortage of freedom is evident in the continuing presence of absolute autocracies and a social environment that is generally intolerant by Western standards. The shortage of knowledge is evident in the generally low quality of education. The absence of gender equity is seen as a monumental waste of human resources.

The geographer Spencer (1960) proposed a comparable argument more than forty years ago, but this is not an approach favored by most cultural geographers today because it tends to see culture as cause, accepts the reality of culture in a national context, and also implies that there is a need for ideas to be imposed on countries from outside.

is undergoing a set of unprecedented changes associated with what typically are labeled as globalization processes. As Dicken (1992:1) observes, the "notion that something fundamental is happening, or indeed has happened, to the world economy is now increasingly accepted."

There are two principal components to current economic change. First, economic activity is continuing to be *internationalized*, meaning there is a spread of trade and flows of capital. Second, economic activity is becoming *globalized*, meaning there is some functional integration of dispersed activities. The increasing importance of **transnational corporations** that operate across political borders, with the result that states are no longer

containers of the production process, and ongoing industrial **restructuring** are both components of these ongoing changes. However, without denying the reality of current economic changes, it is useful to recognize that, in many respects, they can be interpreted as evolutionary rather than revolutionary. Bayly (2003) uses a wealth of world historical data from different societies to demonstrate just how interconnected the world was already prior to the current globalization processes.

It is also commonly argued that the contemporary political world is one of considerable uncertainty, a circumstance that is related to the nature and pace of change since about 1945. The period from 1945 to the end of the twentieth century witnessed a substantial expansion of the global economic system, an expansion that placed much stress on traditional political units; this occurred along with a process of decolonization that resulted in the creation of a number of independent nation-states. There have also been environmental impacts linked to increasing levels of technology and population growth. Further, the fundamental structure of global political power changed dramatically in the early 1990s with the demise of the former Soviet Union and the more general transition from communist rule to capitalist democracy throughout eastern Europe. Another dramatic transition was sparked by the terrorist attacks on the United States in 2001, on Bali in 2002, and on Madrid in 2004. All of these changes have had far-reaching implications for the contemporary world, and there is a broad consensus that the world is entering a new epoch. Box 6.6 identifies some of the key terms and ideas related to this apparent transition.

Despite the fact that both economic and political structures are changing dramatically, it is the issue of cultural change that seems to be most threatening. Many critics of globalization see the greatest threat as being one of a loss of local and national cultural identities leading to a more commodified, less colorful, and increasingly homogeneous world.

With the idea that economic, political, and cultural transformations are necessarily interwoven, three topics are now considered. First, is the gap between rich and poor increasing or decreasing? Second, what is the future of the political state, our current dominant basis for organizing political life? Third, will our

---

Box 6.6 *Naming the Transition*

There is general agreement that the world is a changing place. Webber and Rigby (1996:1–5) have identified the following new circumstances as representing relatively sharp breaks from earlier circumstances:

- Industrial economies are in disarray, with many having huge debts.
- A transition from manufacturing to service employment has forced increasing numbers of people into unemployment and many others into unstable, part-time, or contractual work.
- Major corporations are expanding globally, resulting in increasing amounts of money being transferred across the globe.
- There are several new industrial countries, especially in South and East Asia, with southern China emerging as a dominant manufacturing region.
- Finally, the very value of growth is being questioned in light of damage to the environment.

Recall that in Box 3.5 Jameson (1991) described postmodernity, using Marxist terms, as a mode of production, a period of late (taken-for-granted) capitalism characterized by multinational capitalism and the commodification of culture. Numerous terms have been proposed to identify the current global shift in economics, culture, and politics. The most general terms introduced so far in this book are "modernity" and "postmodernity". But it might be useful to describe the transition from modernity to postmodernity as one from

- **Fordism** to **post-Fordism**,
- **organized capitalism** to **disorganized capitalism**,
- mass production to **flexible accumulation**,
- national economies to a single global economy,
- growth to stagnation,
- nation-states to region states or ethnic territories,
- local cultures to a global village, or
- the golden age to the less-than-golden age.

Of course, labeling in this way increases the danger of creating a reality, such as the golden age, which never really existed. It is with these linguistic concerns in mind that, as noted in Box 3.4, poststructuralist philosophers, such as Derrida, Deleuze, and Foucault, emphasize the heterogeneous and plural character of reality.

cultural future be characterized by a single global culture? Three important questions, each of which prompts several and varied responses. ➙➤➤ **22**

The Gap between Rich and Poor

Debates about globalization often focus on whether inequalities in wealth are increasing or decreasing. Specifically, there is a substantial debate concerning whether or not the gap between rich and poor parts of the world is widening. Again, this is a debate with powerful ideological overtones, with catastrophists viewing the contemporary world in essentially negative terms and cornucopians arguing that global problems are exaggerated and that human technology will solve any existing problems. Certainly, there is evidence to suggest that the gap between rich and poor—with respect to both countries and individuals—is increasing; at the same time, however, there is evidence that the gap is decreasing. Much depends on what data are used and how they are interpreted. Cultural geographers have not focused much attention on this important topic, but there seems every likelihood that this situation will change given the growing recognition of the need to consider economic, political, and cultural issues together rather than separately.

It is not enough, of course, to talk of such matters as income and wealth strictly in terms of economic criteria. The gap between more and less developed worlds can also be measured in terms of such fundamental matters as population fertility and mortality. Here again, much depends on how data are collected and interpreted. On the positive side, it is clear that differences in fertility are decreasing as the less developed world proceeds through a fertility transition. On the negative side, deaths from AIDS and related reductions in overall quality of life are especially concentrated in parts of Africa.

Taylor (1992:20) stated: "We need to construct a new global geography that focuses on the world map of global inequalities. We should very consciously proclaim this to be the most important map in the world and focus our Geography accordingly." Klare (1996:354) argued for a new cartography that reflects the "increased discord within states, societies, and civilizations along ethnic, racial, religious, linguistic, caste, or class lines", while noting that these stresses, along with ever-increasing environmental damage, represent the greatest threat to global security. Kaplan (1996) offers what is perhaps the most compelling argument along these lines, although it has been criticized for making unwarranted assumptions about the maintenance of historical patterns of conflict.

The Demise of the State?

Is the institution of the state and the related conception of nations at risk? Certainly, there are three general reasons why it can be argued that the nation-state is increasingly becoming a dysfunctional unit for organizing human activity and economic endeavor.

The first is the rise of multinational organizations that, in this era of increasing globalization, serve to replace the state in some specific matters. Political groupings such as the United Nations, defense organizations such as the North Atlantic Treaty Organization, and economic groupings such as the World Trade Organization and European Union all function to reduce the role played by individual states. Second, while many nations are joining together in political and economic associations that diminish the power of individual states, many groups in many parts of the world, basing their nationalist arguments on claims of ethnic distinctiveness, are laying claim to subnational territory. Third, the state is challenged by current attitudes to environment, both because environmental impacts do not honor national boundaries and because resources are limited. Taylor makes this argument:

> The threat to the state comes not from the cause of globalization, an economic one world, but the consequence, the destruction of the environmental one world. It is not only the fact that pollution is no respecter of boundaries: the whole structure of the world-system is predicated on economic expansion which is ultimately unsustainable.

And the states are directly implicated as "growth machines"—it is unimaginable that a politician could win control of a state on a no-growth policy. (Taylor 1994:184)

Similarly, Wallerstein notes that ecology is now a principal political issue:

Over 500 years, the accumulation of capital has been predicated on the vast externalization of costs by enterprises. This necessarily meant socially undesirable waste and pollution. As long as there were large reserves of raw materials to be wasted, and areas to be polluted, the problem could be ignored, or more exactly considered not to be an urgent one. (Wallerstein 1996:225)

There are no obvious answers to questions about the future of the nation-state or about cultural identities as these relate to the presence of nation-states. It may be that nations will become less than, but also more than, nation-states, meaning that nations are of increasing importance in the world, but that they do not need to be states in order to be meaningful units. From this perspective, nation and state can be delinked without one being subordinate to the other. With reference to Canada, Penrose (1997) detailed three stages in the process of state evolution culminating in a delinking of nation and state. First, the state is constructed through the combined processes of reinforcing state structures, building a nation, and the actions of hegemonic groups. Second, the state is deconstructed through the impacts of both globalization and fragmentation. Third, it is possible that the state might be reconstructed in a form that is more appropriate to contemporary circumstances. Croucher (2004:112) reaches a similar conclusion, noting that "it is the very malleability of nationhood, both in content and form, that explains its persistence."

Interestingly, the value of counterfactual thinking—that is, of proposing alternative scenarios—is widely acknowledged as a way of understanding the emergence and transformation of global systems. "The increasing complexity emerging with the transformation of world-systems through time means that there was a great deal of chance involved in what types of structures eventually emerged. That is to say, just because a particular sequence of historical events did happen, does not mean it had to happen" (Straussfogel 1997:128). This argument is similar to the ideas presented in Boxes 5.6 and 7.4.

Cultural Globalization

Globalization is broadly understood as a set of processes that increase links between people and places. Thus, "information, ideas, and people are moving and interfacing worldwide more freely, rapidly, frequently, and at greater distances than ever before" (Croucher 2004:15). Certainly, for cultural geographers, the single most interesting aspect of current global changes concerns the oft-stated claim that the world is becoming culturally homogenized, along with the counterclaim that it is not. Needless to say, then, there are varying assessments concerning what is happening to cultures globally. If we think of culture in terms of who we think we are—our identity—then it is helpful to note that such perceptions of identity have changed through time in accord with larger cultural changes.

Scales of Identity

It has commonly been suggested that prior to the rise of the nation-state, cultures and identities were essentially local, community-based, but that the rise of the nation-state gave individuals the option of viewing themselves as part of either a local or national identity. It can be argued that the nation-state acted as a type of cultural integrator, with a common political system and, in some cases, a common civil religion creating a cultural harmony. In some other cases, however, the creation of a nation-state resulted in much cultural uncertainty, and the fact that this circumstance has not played itself out satisfactorily in much of the world is evident from the earlier discussion of ethnicity and also, more generally, from any consideration of contemporary political events. Finally, there is the possibility that a global culture is now emerging, which necessarily adds another layer of potential identity confusion or, expressed in a more postmodern constructionist tone, contributes to the creation of multiple identities. Any discussion of this possible third stage is complicated by the presence of several quite different claims and interpretations. Three possibilities are noted.

Toward a Global Culture?

A first position on the existence of a global culture focuses on the claim that an emerging global culture is related to the erosion of local cultures. This argument is premised on the belief that the diffusion of industrialization outwards from the West involves a diffusion also of Western culture and Western values, and is typically expressed as an argument against Westernization or Americanization.

To date, the principal areas affected by industrialization are in Asia, and so the question as to whether parts of Asia are becoming more Western in cultural terms is often raised. Zelinsky (1992:156) discussed this general issue using the heading "Transnationalization of Culture". Of course, such questions are not entirely innocent—as Blaut (1993b) stressed, there is little doubt that people in the European world are used to thinking of the West and the rest, and that some of the traditional distinctions that those in the West identified are becoming less and less clear. Economically and politically, Western hegemony is being challenged, especially by Asian economic growth. But what about culture? And what about values? Are Western characteristics being accepted elsewhere such that it might be proper to speak of an emerging global culture? It is not difficult to recognize that any suggestion of Western values being accepted elsewhere is a controversial suggestion.

Interestingly, the spread of American values in particular is challenged even in some Western countries, such as France and Canada; these challenges are premised on the idea that there is and should be something that can be defined as a national culture. What is possibly distressing about policies that try to prevent supposedly national cultures from being diluted is that they can be seen as just one step away from the policies of totalitarian states that, "fearful that any opening will destroy them, close themselves off and issue all types of prohibitions and censures against modernity" (Llosa 2001:68).

"Jihad versus McWorld"?

A rather different interpretation of the cultural consequences of the diffusion of technology suggests that as technology brings humans closer together, the cultural barriers between groups become more evident (James 1964). This second position sees a confrontation between the two processes of parochial ethnicity and global commerce. It can be argued that some cultures choose to be excluded from the globalization process because of the association with science and technology, and that such exclusion results in an assertion of cultural identity that is expressed through various localisms. According to Barber (1995:6), the tendencies of local ethnic affiliation (which he labels "Jihad") and the tendencies of global integration (which he labels "McWorld") are both at work, both visible sometimes in the same country at the same time and working with "equal strength in opposite directions, the one driven by parochial hatreds, the other by universalizing markets, the one re-creating ancient subnational and ethnic borders from within, the other making national borders porous from without". Both of these forces may negatively affect the nation-state.

For Croucher (2004) and many other commentators, the 2001 terrorist attacks confirm this cultural division, with the conflict being not between states but between different sets of values.

Overlapping Identities?

A third position is based on the belief that global processes are more likely to produce ambivalence or multiple identities than any coherent identity, a situation that will result in overlapping scales of culture and identity. According to this argument, "spatialized communities are the real 'containers' of culture, of meaning and identity, not the 'virtual' communities created by forms of electronic communication, and the networks built around flows of goods and services" (Axford 1995:164). This argument attempts to distinguish between, on the one hand, the possible creation of global or world spaces in technological and economic terms and, on the other hand, the possible globalization of meaning structures, of identity.

Clearly, this third interpretation can be regarded as one aspect of the modern-to-postmodern transition. It is proving particularly influential in cultural geography and is considered further in several parts of Chapter 8, with discussions of sexuality and emerging cultures. However, it may be useful to note that most of these discussions are themselves products of a particular time and place, namely the English-speaking academic setting at the end of the twentieth century and beginning of the twenty-first.

Concluding Comments

- Cultural regions develop at a variety of scales in response to the many different criteria that serve as bases for human identification with others.
    - › The relatively small, local, ethnically defined region and the much larger, global cultural region are parts of a single dialectic.
    - › The presence of a cultural region demonstrates the critical role in landscape creation played by individuals as members of groups, as well as the critical role played by those members of a group who are able to affect the behavior of others to ensure some consistency in behavior among group members.
- A regional approach to cultural geography derives both from the established tradition of geographic regional study and from the specific incorporation of regions in the landscape school.
    - › Notwithstanding some confusion regarding the words "region", "area", "landscape", and "place", cultural geographers have embraced the regional approach.
    - › "Landscape" and "region" are the terms preferred in this chapter, with landscape as

a general term referring essentially to the visible scene, and region as the more specific term referring to a part of the surface of the earth that has a similar landscape.
    - › Applying the concept at the subnational scale is the favored approach, with many cultural geography studies centering on American regions.
    - › Other approaches have explored transborder regions, where similar cultural history, practices, and outlooks have contributed to the construction of bi-national regions.
    - › Applying the regional concept at a global scale in order to aid in identifying and mapping cultural regions has a venerable and controversial tradition and is currently experiencing something of a resurgence.
    - › There are two different, perhaps contradictory, approaches to global regionalization, namely the civilizational and the world system approaches. The resurgence of interest in global issues is related to the dramatic changes occurring in the contemporary world, especially concerning globalization.
- A key issue raised on more than one occasion in this chapter has not been addressed directly. Recall that some regions can be described as homelands, a term that is used to reflect an especially close link between people and both the physical and built landscape of their region. This idea has sparked some controversy and is a useful entry point into a larger debate about the legitimacy of distinguishing distinct groups of people and related distinct regions.
- Notwithstanding some methodological concerns, and as the discussion of regional studies in this chapter makes clear, many cultural geographers both favor and continue to work in a regional tradition that does not place priority on matters of social production. The success of this continuing and dynamic regional perspective, as demonstrated in this chapter, is undeniable.

Further Reading ➞ ➤ ➤

The following are useful sources for further reading on specific issues.

1. Mason (1895) and Wissler (1917, 1923) are early anthropological applications of the regional concept, a tradition that culminated with the publication of a major work on the cultural and natural areas of North America (Kroeber 1939). According to Carter (1948), the anthropological contributions attracted much geographic attention. A survey of the culture area concept as used in anthropology and a comparison with geographic applications is found in Mikesell (1967:621–4, 626–8).

2. Hartshorne (1939, 1959) authored the classic works on geography as the study of regions. Butzer (1989a) discusses the methodological conflict between Sauer and Hartshorne. Harvey (1969) details regions as an example of classification procedures.

3. Gilbert (1988), Pudup (1988), Johnston (1990), and Lee (1990) are discussions of the new regional geography. Wishart (2004) discusses the concepts of region and period as similar types of generalization.

4. Uses of the first effective settlement concept include the following: Gastil (1975), in an account of areas of cultural homogeneity; Elazar (1984, 1994), on the political scene in the United States with reference to the westward migration of cultural traits but without any explicit use of the work of cultural geographers; Wacker (1968), on the Musconetcong Valley of New Jersey; and Wacker (1975) and Wacker and Clemens (1995), on the larger area of New Jersey.

5. There are few examples of cultural region formation and transformation outside of North America that have employed core, domain, sphere concepts, notwithstanding the generic flavor of the model. For Wales, two principal cultural regions can be delimited according to the relative strength of the English and Welsh languages, and a bilingual zone between the two can be equated with the Meinig domain (see Carter and Thomas 1969; Pryce 1975). Also see Withers (1988) for an account of the Scottish Highlands cultural region.

6. Harris (1978) addresses the historical geography of North American cultural regions.

7. Cross (2007) highlights links between Amish groups and dairying in North America. Roinila (2006) examines Finnish landscape imprints in North America.

8. Aschmann (1965) uses the preadaptation concept, without using the particular term, in a discussion of five Apache cultural traits that made their occupation of the American Southwest possible.

9. Kollmorgen (1941, 1943), Augelli (1958), Lehr (1973), Raitz (1973), Gerlach (1976), and Gade (1997) discuss aspects of ethnicity and cultural islands.

10. Jordan and Kaups (1989) offer a discussion on the linking of specific ethnic groups with specific environments, and Garrison (1990) provides a critical commentary on this idea. Francaviglia (1994) discusses the understanding of the American Southwest as it has been continuously redefined through a series of cultural occupances. Luebke (1984) and Shortridge (1995) explain many developments in the Great Plains in terms of distinct incoming cultures. Meinig (1969) examines Texas as a cultural region; Hilliard (1972) and Nostrand and Hilliard (1988) treat the American South; Cobb (1992) looks at the Mississippi Delta; Wyckoff (1999) focuses on Colorado; Hausladen (2003) includes accounts of the American West as place and myth. Jordan, Kilpinen, and Gritzner (1997) discuss the folk landscape of the North American mountain West. West and Augelli

(1966:11–16) identify cultural regions and related landscapes in Central America, emphasizing the interplay of physical geography and history as reinforced by isolation. Jordan-Bychkov and Bychkova Jordan (2002) map regions in Europe based on particular variables, such as language and religion, while attempting a broadly defined cultural regionalization of the continent in terms of North and South, East and West, and core and periphery. Ostergren and Rice (2004) provide an overview of the cultural geography of Europe. Rose (1972) and Jeans (1981) address the possibility of delimiting cultural regions in Australia. Christopher (1984:193) notes that, throughout much of Africa, "New European societies were formed which were able to impress their ideas upon the landscape and create an image of France or England overseas"; nevertheless, there has been little interest in identifying any resultant cultural regions.

11. A Special Issue of the *Journal of Cultural Geography* edited by Nostrand and Estaville (1993b) outlines the homeland concept and includes several regional studies. The concept is further detailed in *Homelands: A Geography of Culture and Place Across America* (Nostrand and Estaville 2001b), and additional regional examples discussed. See also Hurt (2003) for a further refinement of the homeland concept and for the example of the Creek nation.

12. Trépanier (1991) and Estaville (1993, 2001) discuss French Louisiana.

13. Meinig (1965) is the pioneering cultural geographic analysis of the Mormon region, although the presence of the region was generally acknowledged prior to that analysis (Zelinsky 1961); Francaviglia (1978) provided the seminal landscape analysis (see also Jackson and Layton 1976, Jackson 1978, Bennion 2001, and Starrs 2009); the symbolic importance of the region to Mormons has been described by Jackson and Henrie (1983). Smith and White

(2004) discuss the case of Mormons located in Chihuahua, Mexico, in terms of their detachment from the homeland. Toney, Keller, and Hunter (2003) and Yorgason (2003) analyze aspects of twentieth-century change in the Mormon homeland.

14. Meinig (1971), Carlson (1990), and Nostrand (1992, 2001) examine aspects of the formation of the Hispano cultural region and the characteristic landscape; Smith (1999) focuses on cultural change occurring at the northern edge of the Hispano homeland. Nostrand (2003) provides an unusually detailed and insightful account of a single village, while Wright and Campbell (2008) discuss recent landscape change in some New Mexico villages.

15. Barcus (2007) identifies and analyzes Hispanic settlement beyond the traditional region in parts of Appalachia.

16. Melko (1969), Sanderson (1995), and Wilkinson (1987) discuss the writings of civilizationists.

17. Hugill (1997:348) notes that the "the most vital debate in social science" is the attempt to develop a theory of the history of the capitalist world, and possibly even a theory for all human history. Political and economic geographers have made major contributions to this debate (Dicken 1992; Taylor 1996).

18. Mosely and Asher (1994) is a comprehensive atlas of language; Renfrew (1988) stresses the relationships between language families; the role of language as a basis for group delimitation and communication between group members has been noted especially by Wagner (1958b, 1974, 1975); cultural geographic overviews of religion include those by Sopher (1967, 1981), Gay (1971), and Park (1994).

19. Sauer (1940) outlines a global regionalization that stresses the historical origins of regions in early civilizations.

20. Ó Tuathail (1996:240–9) notes possible failings of the Huntington world regionalization, and

Kaplan (1994, 1996, 2000) suggests that conflicts might be based on differences in qualities of life and access to resources.

21. Wolf (1982), Gills (1995), and Sanderson and Hall (1995) address aspects of world systems. The principal contributions to world systems theory are by Wallerstein (1974a, 1974b, 1980, 1989). Hopkins and Wallerstein (1996:1–10) describe six "institutional domains, 'vectors' of the world-system" that are "distinguishable but not separable". These are the interstate system, structure of world production, structure of the world labor force, patterns of world human welfare, social cohesion of states, and structures of knowledge. Bayly (2003) presents a world history that explicitly avoids a Eurocentric perspective that sees Western values as diffusing around the world. He achieves this through a variety of strategies, including showing how many of the supposedly original Western ideas had antecedents or equivalents elsewhere. Nevertheless, he does point out that the process of European overseas expansion served to make much of the world a vast agricultural hinterland for western Europe. McLynn (2004) identifies 1759 as a climactic year with the defeat of the French in North America facilitating the rise of the British Empire.

22. Nisbett (2003) is an intriguing and controversial book that reports a substantial body of research highlighting the different ways in which Westerners and Asians think. Rankin (2003) provides a detailed overview of globalization processes through an account of anthropological and geographic perspectives.

Kofman and Young (1996) consider aspects of globalization including the development of theory, the role of the state, and gender implications. Some accounts of globalization tend to be highly critical. Thus, Renner (1996) argues for an increasing gap between rich and poor countries. Other accounts interpret globalization in a positive light. Thus, Norberg (2001) and Legrain (2002) argue that poverty, the environment, and democracy all benefit from globalization, while Dollar and Kraay (2002) focus on details of the poverty reduction resulting from globalization. For arguments that interpret cultural globalization as a form of creative destruction, see Cowen (2000), and for a critical review see Geertz (2003). The edited volume by Berger and Huntington (2002) focuses on issues of cultural diversity in the contemporary world, stressing, for example, that global culture may be diffusing but that it always does so with localized modifications. Turner (1993), Axford (1995), Albrow (1997), and Lee and Wills (1997) discuss the need to consider economic, political, and cultural issues as one. Smil (1993:5) identifies the all-too-common ill-informed reporting of environmental issues. Dalby (1996) outlines some criticisms of Kaplan (1996) who offers a powerful comdemnation of the state of the world. An original survey of the geopolitics of the world system is provided by Cohen (2003). Ohmae (1993) identifies economic challenges to the nation-state. Various scholars argue that the key characteristic of globalization is not economic but rather the political process of deterritorialization (Ruggie 1998; Scholte 2000).

CHAPTER 7

Power, Identity, Global Landscapes

The principal concern of this chapter is how identities relate to the myth of races within the human species and to related ideas about ethnic identity. This conceptual content informs empirical accounts of inequalities at global and national scales, especially as they relate to the consequences of European overseas movement and globalization processes. In the following chapter the discussion of identity is further developed to include such topics as gender, sexuality, and disadvantaged groups generally, and these are discussed mostly with reference to inequalities at community and local scales.

The discussion of global inequalities included in this chapter highlights some critical cultural differences linked to ideas of race and ethnicity, but there are, of course, many other directions that such a global discussion might take. Population geographers highlight global patterns of, for example, fertility, mortality, and life expectancy, and it is clear that these matters are closely tied to cultural circumstances; declining fertility, for example, is tied to the weakening of male authority in patriarchal societies. Economic geographers address global inequalities in analyses of, for example, employment opportunities and income levels, while political geographers identify differences in, for example, global patterns of human rights and democratic institutions; all such matters are also linked to larger cultural contexts. Certainly, any comprehensive review of our unequal world needs to take all of these, and other, circumstances into account, and it might be argued that cultural geographers, keen as they are to incorporate more than just cultural content in their studies, might usefully extend their analyses in such directions.

In this chapter you will find

❖ a discussion of the idea that we live in one world, but one world that is significantly divided by cultural variations;

❖ a historical account of the illusion of races within the human species, which concludes that the biological idea of race is not a valid means of classifying individuals into groups. Despite the fact that the idea of biological race has no scientific validity, the reality is that "race" is a powerful social construction and is a politically salient term that is widely employed;

❖ a consideration of the apartheid era in South Africa (1948–1994), and several instances of genocide, which illustrate that although biological race is a myth, racist thought and practice are an everyday reality;

❖ an account of ethnicity and ethnic identity that considers the roles played by language, religion, and national identities.

Linking Identity and Power

There is a recurring interest in much current cultural geography with questions of human identity, especially with how identities relate to the exercise of power and the many social and spatial inequalities that prevail in the contemporary world. The link between identity and inequality is often explained by the fact that a principal basis for distinguishing one group identity from another is in terms of relative **power**, an idea that is central to Marxist, feminist, and postmodernist ideas, as discussed in Chapter 3. The interest in identities is not new, as cultural geographers have longstanding interests in such characteristics as way of life, language, religion, and ethnicity (Box 7.1). Nor is the interest in inequalities new, as many of the concepts devised by cultural geographers to explain the rise of regional landscapes and homelands—concepts such as first effective settlement—explicitly acknowledge that some groups are able to impose their identity on landscape while others are not. But the current concerns with identity and inequality are distinctive for at least two reasons.

First, cultural geographers are now making use of more sophisticated theoretical ideas than those belonging to the landscape school tradition, ideas that explicitly consider identity formation and inequalities. Second, there has been a broadening of the culture concept, from the superorganic entity used in the landscape school tradition to one that encompasses a more meaningful view of human identity and agency. Ideas about culture emerging from the new cultural geography acknowledge that many aspects of identity cannot be subsumed under a single general concept (such as class or nationality), but are rather socially constructed, intersectional, and continuously changing. Most notably, there is recognition that dominant groups are typically able to assert their chosen identity while imposing identities on subordinate groups, thus creating a cultural world that is structured on their terms. This is, of course, precisely the argument that Marxist geographers use in their critiques of capitalist society and economy, that feminist geographers use in their critiques of a male-dominated world, and that postcolonial geographers use in their studies of oppressed and excluded groups. More generally, another component of this broadening of the culture concept is contemporary cultural geographers' willingness to incorporate traditional social geographic subject matter, such as accounts of class, into their analyses, a circumstance that has at least partly removed the rather artificial distinction between the two subdisciplines of social and cultural geography. This has also opened up interesting channels of exchange between cultural and economic geography.

The Mistaken Idea of Race

Race is a biological term that, when applied to humans, is a myth. This section uncovers ways in which the term is incorrectly used to describe human groups; anticipating our conclusion, not only is the term meaningless but, further, it is so laden with tragic baggage that it is extraordinarily unhelpful to use it today. Nevertheless, although race is a myth, so many people continue to harbor negative views of others based on presumed racial identity that it is important to ask why, and understand how to categorize it using such terms as "racism", "**racialization**", "prejudice", and "stereotyping". →»»1

The Unity of the Human Species

A single and simple fact to begin: there is only one species of humans in the world, namely *Homo sapiens sapiens*. It is correct to say that this human species displays variation among its members, but it is biologically incorrect to say that these differences are sufficient to merit using race—that is, subspecies—to make distinctions.

The biological differences that do exist among humans result principally from chance and from the fact that human groups migrated globally and were isolated from each other for an extended period of time, specifically between about 100,000 years ago and about 10,000 years ago. It was this separation

Box 7.1 One World: Multiple Divisions

A pioneering world regional textbook of human geography that was first published in 1964 highlighted a concern that is at the forefront of much of the current work in cultural geography, namely that human groups and their landscapes exhibit the characteristics of *One World Divided* (James 1964). The text was written with the premise that the great problems requiring attention by geographers were those related to inequalities, as these are evident in patterns of poverty, hunger, overpopulation, relations between peoples, environmental destruction, and violence. The logic behind his argument was as follows.

The world is divided both by differences in physical geography and the availability of related resources and by differences in human geography. Differences in physical geography, such as differences in climate, landforms, and soil, are significant but essentially fixed. Differences in human geography, such as differences in language, religion, and way of life, are, on the other hand, constantly changing. In reviewing global changes, especially during the past 200 years, James asserted that there was one particularly tragic circumstance:

Man himself has created divisions and barriers far more difficult to cross than those caused by nature. Man has brought into being mountains of hate, rivers of inflexible tradition, oceans of ignorance. And as technology brings men closer together the man-made barriers loom larger and more forbidding. (James 1964:3)

Since James invoked these powerful metaphors in 1964, most evidence substantiates his claim that cultural divisions are prominent. The long list of regions involved in recent or ongoing conflicts that at least partly reflect cultural divisions includes Northern Ireland, Quebec, the Middle East, Cyprus, the Kurdish region, Uganda and other countries in central Africa, Nigeria and other countries in West Africa, and the former Yugoslavia. All of these conflicts occur in spite of improved communication technologies, which James could not, of course, have anticipated in 1964. At this time there seems little reason to believe that ongoing globalization processes will reduce the significance of human-made divisions. Indeed, globalization-linked economic exploitation and resource extraction have extended and intensified social divisions and inequality, for example, in the cases of oil extraction in the Niger Delta (Watts 2007) and gold mining in Guatemala (Nolin and Stephens 2010).

Geographers' concern with human-induced inequality has continued and intensified in the twenty-first century. In *Violent Geographies*, a book edited by prominent human geographers Derek Gregory and Alan Pred, the materiality of this violence is exposed:

Now landscape is more than a metaphor . . . the sheer physicality of landscapes can become saturated with political violence. We know that "landscape" is a freighted term and that its cultural formations do significant ideological work. Contemporary deconstructions of the "lie of the land" have shown that aestheticized landscapes typically produce their effects by artfully concealing the very work—the embodied labor—that has been invested in their production. But landscapes can embody other traces of blood, sweat and tears, and sometimes they are not hidden at all. They are deliberately written on the face of the earth to inspire fear on the faces of those who inhabit them. (Gregory and Pred 2007:4)

Much of the content of this book highlights this divided world. Although James was concerned with divisions at the global scale, it is clear that divisions are apparent at all spatial and social scales. Gregory and Pred's book reveals the *intimacy* of these violent inequities, how they are carefully and calculatedly meted out through personal confinement, exclusion, terror, dislocation, and dispossession. While James was concerned mostly with highlighting such fundamental cultural variables as language and religion, today all kinds of differences are manipulated to enact socio-spatial exclusion, including gender, sexuality, and disadvantage generally.

that allowed some selective breeding and some adaptation to physical environments to occur. Together these two biological processes facilitated the evolution of some biological differences. What is important, however, is that the period during which groups were separated was not long enough to allow races, in the biological sense of genetically distinct subspecies, to form.

The most obvious adaptation to physical environment—obvious in the sense that its results are clearly visible—was the evolution of several different skin colors, with lighter skins associated with temperate environments and darker skins associated with tropical environments. This adaptation has to do with the pigment melanin, which develops in the hair, skin, and iris of people (and animals) as a defense against the ultraviolet radiation that can cause cancer; humans and animals in tropical environments produce more melanin than those in temperate regions. But neither this nor any other difference means that humans can be neatly classified into racial groups. Indeed, there is no one biological trait that can be used to classify people into such groups. If we take three of the ways in which humans vary biologically—blood group, hair texture, and skin color—we discover that each is influenced by different genes and is thus inherited independently of other characteristics. What this means is that the identification of such typically defined groups as Negroid, Mongoloid, Australoid, and Caucasoid is fundamentally flawed and therefore meaningless. Indeed—and this is an important observation—there is typically more genetic variation within one supposed race than there is between supposed races.

In summary, humans are a biologically variable species, subject to the same processes that have produced variation in other living things, namely chance and natural selection in different physical environments, but humans cannot be divided into biological subspecies or races. The statement that all humans are members of the same species is biologically confirmed by the fact that any human male is able to procreate with any human female. In the context of humans, race is not an objective fact.

The Illusion

There are no races within the human species, and yet until quite recently it was widely accepted that races exist. How and why did this myth arise? What purpose did it serve? Why is it necessary to discuss this myth in the present context? Answering these questions and understanding the context for the emergence of the race myth is simple in principle, but difficult in practice.

Simply expressed, the notion of race, misguided though it was, became an important component of nineteenth-century scientific and political discourse in part because it seemed logical to many observers trying to understand the apparent diversity of humans. If we step back from this statement, it is not too difficult to appreciate that emphasizing difference in this way was but one of two directions these observers could have taken. The other direction, the one they did not take, was to realize that humans have many more characteristics that unify rather than separate them.

But the misconception of race arose also because it suited the aims of many nineteenth-century Europeans. The principal purpose was a scientific one: race enabled physical anthropologists to devise formal classifications using what appeared at first sight to be sound scientific grounds. But there was another important and much more dangerous purpose, namely that of using presumed physical differences to explain cultural differences. It is therefore necessary to discuss this myth of race because it has been routinely used (perhaps a better word is "abused") to justify rankings of groups on the basis of perceived abilities and to justify discriminatory thoughts and behaviors. In other words, notwithstanding the fact that there are no distinct human races (only one human race), the supposed existence of different races is an idea that became influential in the nineteenth century and remains important in the contemporary world, and it is this idea, this socially constructed myth, that is the subject of the present discussion. ➔≫ 2

It is often claimed that the idea of race has a long history, and that many groups in the past perceived themselves and others as having racial status. This claim lacks credibility, since most recent evidence favors the idea that the myth of race was essentially a nineteenth-century creation. There is no word corresponding to the English word "race" in Jewish, Greek, or Roman writings, although there is evidence of divisions between groups, especially the political and cultural divisions between citizens and barbarians. Archeological evidence from Roman sites in Britain, though, does indicate that Nubians from Africa held high office within the Roman military (Leach et al. 2009). For Hannaford (1996:12), the Greek "political idea involved a disposition to see people not in terms of where they came from and what they looked like but in terms of membership of a public arena," and it therefore "inhibited the holding of racial or ethnic categories as we have come to understand them in the modern world". Historically, both Islam and Christianity, in their capacity as universal religions, welcomed converts. They saw themselves as communities of believers, not as racial groups.

The word "race" dates only from the fourteenth century, and it was rarely used before the sixteenth century. It seems likely that there was no significant racial thought in the European world prior to the late seventeenth century, and that the claim that there was such thought is itself largely a product of the nineteenth century. The myth of race is not some timeless concept; rather, in historical terms, it is a fairly recent invention.

Creating the Illusion

Seven principal developments contributed to the rise of the European idea of race. First, following the Reformation, polygenetic theories of human origin appeared; these theories contradicted the previously dominant Augustinian view of a single human origin and became a key element of the myth of race. Second, during the extended process of European overseas expansion, beginning about 1450, the idea that races existed seemed logical to Europeans because some of the people they encountered really did look different. It requires little imagination to appreciate how perplexed both sides of a cultural encounter might be on experiencing a first sight of

the other. Third, with the rise of a global system of commercial capitalism related to colonial expansion, Europeans in some overseas areas, notably the southern United States, began to rely on slave labor; their growing dependence on slave labor fueled an extremely lucrative industry for slave traders and effectively institutionalized slavery as an accepted social relation, thereby fixing Africans in a position of inferiority.

A fourth development that contributed to the rise of the race myth involved the creation of republics following revolutions in late eighteenth-century France and the United States, which served to integrate the concepts of nation and people. This raised new ideas about nations existing as racial communities. A fifth development had to do with major advances in the scholarly classification of humans—such as those by Carolus Linnaeus (1707–78), Johann Friedrich Blumenbach (1752–1840), and Buffon—which were one part of larger scientific advances. It should be pointed out that these understandable attempts to understand human variability and what it meant to be human did not attach any values to the classes identified. Sixth, Enlightenment philosophers such as Rousseau and Montesquieu paved the way for environmental racial theories with their ideas about civilization and savagery. Overall, the period from the late seventeenth to the early nineteenth century witnessed a rise in popularity of the idea that civilization advanced not through the "public debate of speech-gifted men and the reconciliation of differing claims and interests in law but through the genius and character of the *Völker* naturally and biologically working as an energetic and formative force in the blood of races and expressing themselves as Kultur" (Hannaford 1996:233). The seventh development contributing to the European idea of race was the emerging field of ethnology, which developed as both a natural and human science concerned with identifying the biological and historical origins of particular groups. It was in this context that the myth of an Aryan race emerged.

Thus, it was primarily nineteenth-century Europeans who formulated the myth of race and created a history of racist thought through a reinterpretation of the past using the new language—discourse—of national identity, progress, and social Darwinism. Barthold Georg Niebuhr (1776–1831) rewrote the history of Greece and Rome using racial rather than political classes, while Augustin Thierry (1795–1856) performed a similar task for French history. Together, these works prompted the idea of an original European Aryan culture superior to those of both Greece and Rome. More generally, the tendency for most cultures at most times to dislike and even fear strangers and those who are in some way different was conflated with racism. Critically, the new doctrine of **nationalism** was supported by the idea of race, with national group membership contingent upon such characteristics as language and skin color, and with nation—whether real or imagined—replacing religion as a primary basis for group identity. It was in this context of the rise of the myth of race that racial classifications, now including hierarchies, were proposed by many scholars.

Classifying People

Western academics and intellectuals of the nineteenth and early twentieth centuries generally accepted that it was appropriate to classify people into racial categories and, further, that the classes identified reflected not only physical criteria but also intellectual ability and cultural characteristics generally. For example, Count Arthur de Gobineau (1816–82), in his *Essay on the Inequality of the Human Races* (1853–5), classified Europe into races with Germanic people as the superior group. Gobineau doubted the ability of Africans and Native Americans to become civilized, and generally saw race as the sole explanation for progress. For Gobineau, the existence of discrete races, with definable physical, intellectual, and moral characteristics, was not just a fact but the most important fact for any understanding of humans. And, given the existence of fixed races, he expressed considerable concern about mixing races and the loss of racial purity that would ensue.

Probably the most influential and widely accepted work on races was the 1899 study of civilization *The Foundations of the Nineteenth Century*, by Houston Stewart Chamberlain (1855–1927). Chamberlain argued that race explains everything, and he claimed civilization as the sole prerogative of the German race with its Aryan roots. Jesus, he argued, was a member of the Aryan race. Overall, Chamberlain was much less scientific than Gobineau, referring frequently to race souls, to the power of pure race, and to the importance of remaining racially pure.

A typical minutely calibrated sliding scale of human groups equating ability and culture with physical characteristics placed Nordic or Teutonic peoples at the top, followed by Alpine, Mediterranean, Slavic, Asiatic, and African groups. One influential racial classification was developed as recently as the second half of the twentieth century (Coon 1962). This classification distinguished five principal presumed groups: Australoid, Mongoloid, Caucasoid, Capoids, and Congoid (or Negroid). In principle, each group was distinct from each other group, and Coon asserted his belief in the reality of racial purity.

It is important to acknowledge that many scholars accepted the idea that humans could be divided into distinct and pure racial categories, and even the idea that races could be ranked was widely viewed as correct. Certainly, by the early twentieth century, race was a dominant organizing idea, and it was generally believed that humans could be properly classified and ranked according to appearance. In other words, the idea of race as both a scientific category and a value judgment about people was part of the discourse of the time.

In some instances, racial classifications were taken to especially unpalatable extremes. For example, one regular feature of racial classifications and rankings was the explicit distinction made between German Christians and Jews. Jews in the West were a well-defined group traditionally excluded from the state, but political anti-Semitism developed only in the late nineteenth century, in relation to race studies. In Germany, ideas about the racial inferiority of Jews contributed to the policies of the National Socialist Party, which declared that only members of the supposed German race could be citizens of the German state; these policies were used to justify the mass murder of Jewish and some other groups during the Second World War.

Of course, racist ideas, with roots in Enlightenment and nineteenth-century thought, were not influential only in Germany. Rather, they were widely accepted and implemented in various ways, both practical and scholarly, throughout much of the world, especially in Europe and areas of European overseas expansion. Indeed, nationalism, colonialism, and imperialism were all supported by the idea of race, as **neocolonialism** continues to be at the present time. In the late nineteenth and early twentieth centuries, racist immigration policies designed primarily to limit or totally exclude immigrants from Asia were introduced in such countries as the United States, Canada, Australia, and New Zealand.

It is significant that the various social science disciplines were institutionalized at a time when race was an important organizing idea. Certainly geography, anthropology, sociology, and psychology were all caught up in the idea of race. In geography, Ratzel was a founder of a German journal devoted to the study of race. He was also responsible for popularizing the idea of *lebensraum*, a related concept referring to the living space required by a political state. Several works by Taylor, beginning with *Environment and Race* (first published in 1927), applied what was understood as the fundamental biological and physical anthropological fact of race to the study of larger human geographies. But Taylor did not proceed to rank races, noting, "The science of ethnology is so largely a product of Western European peoples that it is natural for all classifications of the races of man to accept as a fundamental principle that this group is the highest type in human evolution. In this study we are concerned with ethnical not ethical status" (Taylor 1937:43). Other early twentieth-century geographers concerned with grand global-scale issues acknowledged race but did not attach cultural importance to what they saw as a biological fact. In outlining the principles of human geography from two quite different perspectives, both Huntington (1920) and Vidal (1926) saw race as largely irrelevant. But many commentators in the early 1900s conflated geographical expansion with civilizational advancement, as this 1926 quotation from Thurston demonstrates (one in which he clearly overlooks the history of China's immense trading empire of the fourteenth through sixteenth centuries):

> Although yellow men have been leading settled civilised lives in the valleys of China quite as long as the white men elsewhere, they have not made the great strides that the white man has made in modern times, so that although they are almost as numerous as the White race they have remained in their original homelands, leaving the New World to be developed by the white people. (quoted in Lambert 2002:297)

Race was also an important area of analysis in the discipline of statistics. In particular, the study of eugenics, a term referring to the use of selective breeding designed to promote certain characteristics while breeding out others, was popular in the early twentieth century. The term was coined by the British statistician Francis Galton in 1881, but it was promoted especially by Karl Pearson, who argued that eugenics was essential if human degeneration was to be avoided and human progress maintained:

> How many centuries, how many thousands of years, have the Kaffir or the negro held large districts in Africa undisturbed by the white man? Yet their intertribal struggles have not yet produced a civilization in the least comparable with the Aryan. Educate and nurture them as you will, I do not believe that you will succeed in modifying the stock. History shows me one way, and one way only, in which a high state of civilization has been produced, namely the struggle of race with race, and the survival of the physically mentally fitter race. (Pearson, quoted in Weitz 2003:40)

However breathtakingly arrogant such a statement may seem today, it is critical to understand that such logic was part of the discourse of the time. The illusion of race was so powerful that it was a part of the taken-for-granted world both in the scholarly context and in popular understanding.

Demolishing the Illusion

The first explicit suggestion that this myth of race was unscientific came from the anthropologist Boas (1928). His claim was more fully detailed by Huxley and Haddon (1936), who argued that the term was fundamentally flawed and should not be used. But the most important statement was Montagu's treatise *Man's Most Dangerous Myth: The Fallacy of Race*, first published in 1942. In this landmark study, written at the height of racist policy and practice in Nazi Germany and also at a time when racial segregation was the norm in the American South, Montagu (1942) explained that, notwithstanding almost universal acceptance of the reality of race, the term was without scientific credibility and thus invalid. He also noted that any myth is most effective when it is not recognized as such. Montagu understood not just that race was a myth but that the word itself was racist. In the introduction to the sixth edition of *Man's Most Dangerous Myth* (1997) he wrote as follows:

> I have never before put what is wrong with the idea of "race" in the form of a formula. Let me do so here. What the formula shows, in simplified form, is what racists, and others who are not necessarily "conscious" racists, believe to be the three genetically inseparable links that constitute "race": The first is the phenotype or physical appearance of the individual, the second is the intelligence of the individual, and the third is the ability of the group to which the individual belongs to achieve a high civilization. Together these three ideas constitute the concept of "race". This is the structure of the current concept of race to which most people subscribe. *Nothing could be more unsound, for there is no genetic linkage whatsoever between these three variables.* And that is what this book is designed to discuss and make clear. (Montagu 1997:31)

It is an astounding commentary on late twentieth-century society that Montagu found it necessary to state this some fifty-five years after first publication of the book, at a time when the myth was exposed by irrefutable biological facts and following so much

evidence of political and social misuse. Such is the seductive quality of the myth.

As stated by Stanfield (1997:27), "Race as a myth is a distorting variable that convolutes and in other ways distracts attention from the variables that really matter in understanding how and why human beings think, act, and develop as they do. . . . If we can learn race, we can unlearn race." This is precisely the approach adopted in this text. ➝➤➤ **3**

Perpetuating the Illusion?

The great irony, indeed tragedy, surrounding the race myth is that studies exposing the falsity of the idea of race did not prevent the rise of scholarly race relation studies that typically were critical of racism but that at the same time accepted the idea of race as a legitimate basis for analysis. Thus, sociological analyses of race initiated in the 1920s by Park were continued by the Swedish economist Gunnar Myrdal (1898–1987), who produced a major study of race relations in the United States (Myrdal 1944). In this way, the study of race relations became established as a legitimate component of the social studies curriculum, which included human geography, and it has remained so through to the present. Social science studies also influenced the development and implementation of the race relations legislation in many Western countries and the racial labels routinely used for various official purposes. ➝➤➤ **4**

Thus, although it is consistently acknowledged today that race is a fallacy, the term continues to be used in mainstream social science, albeit sometimes framed by quotation marks. The term is also commonly used in the mass media, as any casual review of newspapers or similar sources indicates. And in sociology especially it continues to be quite usual to conflate the terms "race" and "ethnicity". A recently published text states: "In Canada, ethnic and racial diversity is a fact of life," with race referring to a "group that is defined on the basis of perceived physical differences such as skin colour" (Rosenberg 2004:403, 405). Thus, the term is used even though, at the same time, the author stresses that "the existence of distinct and definite biological races is largely a myth" (Rosenberg 2004:405). Similarly, the term is regularly used in geography, although not, of course, in the context of meaningful biological and cultural differences. Rather, geographers tend to treat the term the way Jackson (1989:177) does when, in a discussion of racism, he defines race as a "social relation across deeply entrenched lines of inequality".

Concern with "race and racism" remains a principal research tradition of both cultural and social geography, dating from the 1960s and linked to sociological analyses of race relations. The approach includes descriptions of immigration, residential segregation, **ghetto** formation, and assimilation, especially with reference to North American and British cities. Certainly, notwithstanding the inherent flaws, the myth of race linked with the concept of ethnicity has proven to be a difficult idea for cultural geographers to ignore. Probably the most substantial area of work linking the two relates to the concept of a **plural society**. Initially developed for the study of colonial societies in southeast Asia, this concept distinguishes between societies that are culturally homogeneous with normative integration and societies that are deeply divided on an ethnic basis and held together only through strong state control.

Why is it so usual to uphold a term that is both unsound as a fact and that has been used to justify so many recent atrocities? As Gates (1986) observes, scores of people are killed on a daily basis in the name of differences ascribed to race. The answer appears to be that some social scientists consider it helpful to use the word "race" in discussions of racism. While it has been discredited from a scientific perspective, by invoking race one could argue that it takes seriously the damage race thinking has done, and continues to do. Continuing to use the term might highlight the socio-political and cultural salience and legacy of racial thinking; it might be considered a commitment to expose, understand, and challenge the *ongoing* material and symbolic damage racism has done to generations. While this makes some sense, it has been persuasively argued that the dangers of perpetuating the myth are simply too great. As explained as early as 1942 by Montagu, the term is so powerful and dangerous that it needs to be relegated to

Questions for Critical Thought

Commentary on the election of Barack Obama as President of the United States in 2008 often employed the word "race". Do you think that a principal difficulty with many of our current well-meaning attempts to combat racism is the simple fact that, at the same time, we persist in using the term race as though it had some meaning? Use of the term is especially prevalent among, but by no means restricted to, more conservative scholars (see, for example, Sowell 1994). Even if we understand that it is a myth, perhaps the simple act of using the term can be interpreted as implying that it is an objective reality. Do we need to abandon this dangerous word altogether? Are you inclined to think that it is not enough to note that it is "a social construction rather than a natural division of humankind" (Jackson 1987:6), and that we ought instead to accept the fact that race applied to humans is a myth, a fallacy, an illusion, and thus abandon all use of the word? Or, do you think that in order to acknowledge the ongoing power of racial thinking, the legacy of racist institutions, and structures of exclusion, using the term race (while acknowledging it is a scientific fallacy and social construction) is a meaningful way to debate current issues?

the garbage can of history. But Montagu was writing in the 1940s; in the twenty-first century we still see the power that race thinking exercises, and thus we need to do more than just deny the relevance of the word—we must recognize its continued salience as a powerful socio-cultural construct. This section has described the history of the myth, while also acknowledging that race continues to be a taken-for-granted fact by far too many people.

The Reality of Racism

Today racism remains a fact of life for many people. ➤➤➤ **5** It assumes a variety of forms, from mild silent disapproval of another person to mass murder. Race may not be a fact, but racism most certainly is. Racism is a particular form of prejudice that typically incorporates prejudgment of others. **Prejudice** refers to the holding of negative opinions about another group, negative opinions that are based on inaccurate information and that are extremely resistant to change. Racism similarly incorporates a schema or blueprint about others, and a **stereotype** is a set of beliefs and expectations about another group that is difficult to change even where there is clear evidence that these beliefs and expectations are incorrect. Box 7.2 outlines the divergent political attitudes toward both past and present racisms that were highlighted at a 2001 conference sponsored by the United Nations.

Our concern with racism in this section is with some tragic twentieth-century examples that were applied at national scales, that affected large numbers of people, and that have had overwhelming

cultural geographic relevance. The first example is **apartheid**, a set of racist policies introduced and practiced in South Africa by Europeans to ensure that they would be able to retain power in a country in which, unlike most other mid-latitude settler colonies, they were a minority of the total population. Although there has been considerable debate about the reasons for the emergence of apartheid, especially between liberal and Marxist scholars, the present account situates the issue in the larger context of the discussion of the myth of race as presented above.

Apartheid in South Africa, 1948–1994

During the extended period of European overseas expansion between the fifteenth and nineteenth

Box 7.2 *Debating Racism Today*

The United Nations-sponsored conference against racism, racial discrimination, xenophobia, and related intolerance, held in Durban, South Africa, in 2001, provided some insights into current attitudes about racism, both past and present. Conceived as an opportunity for creating a new world vision at the beginning of the twenty-first century, the conference focused especially on the causes and forms of racism and on the provision of redress in the form of reparations for persecuted groups. Unfortunately, different countries interpreted these two matters differently.

Before the conference began, the American and Israeli delegations withdrew on the grounds that the intent of the conference was itself racist. The principal concerns of the two delegations related to two issues. First, the agenda included language, inserted at the insistence of some Arab countries, that described Israeli treatment of Palestinians as genocide. Second, the agenda called for a full apology for transatlantic slavery to be issued. Many European countries as well as the United States objected to this demand, which was inserted at the insistence of some African countries. Their objections were

based on the claim that slavery was legal at the time it was practiced, but also reflected their concerns that a flood of lawsuits might ensue if an apology were issued. From the perspective of the US and Britain especially, the emphasis on the ongoing conflict between Israel and the Palestinians and on past circumstances of slavery threatened to overshadow such current instances of racial discrimination as China's treatment of Tibetans, Zimbabwe's actions in Matabeleland, Russia's war in Chechnya, and the caste system in India.

Following extended and often bitter negotiations, the two controversial positions were rejected. The final declaration acknowledged that the transatlantic slave trade was an appalling tragedy, especially in light of the way it dehumanized victims, but did not include an apology. There were no direct criticisms of Israel, although the Palestinians' right to self-determination was included. Perhaps more than anything else, this conference succeeded in exposing some bitter differences of opinion within the world community. On balance, these differences prevented the much anticipated coming together from being achieved.

centuries, there were numerous encounters between incoming European groups and indigenous populations. The indigenous population of South Africa included several Khoisan groups located primarily in the western part of the country as well as Bantu groups in the east and north. The Dutch East India Company established a settlement at the southern tip of South Africa, now Cape Town, in 1652, but there was little subsequent immigration. The area came under British administration in 1806, and although there was some British immigration to the colony, the area never proved to be as attractive a destination as the various other British mid-latitude possessions. In the late seventeenth century, Dutch settlers had begun to move inland across the western area occupied by the Khoisan group, and this expansion and the related introduction of diseases prompted a decline in the indigenous population that was similar to declines in other mid-latitude areas such as the United States. But the Bantu groups in the north and east were less susceptible to imported diseases, and a series of conflicts occurred between the Bantu and both the migrating Dutch (known as Afrikaners) and the incoming British. There were also ongoing disputes between Afrikaners and British colonial authorities, especially related to the Afrikaner republics in the north, which had formed during the nineteenth century. Further, following a practice carried out in several other British colonial areas, indentured laborers were brought from India. Most of these laborers worked in the sugar cane plantations in Natal colony in the northeastern part of South Africa.

Thus, South Africa comprised several groups involved in a series of hostilities, and the creation of the Union of South Africa from four British colonies in 1910 was really a compromise between two powerful but quite different European groups. In 1910 the population was about 5 million, of whom about 1 million were European, although the proportion of Europeans steadily declined to about 12 per cent by the end of the apartheid era in 1994. The apartheid era began in 1948 with the election of the Afrikaner-dominated Nationalist political party, and at that time about 55 per cent of the Europeans were Afrikaners. →≫≫ **6**

Group Identity in South Africa

A key consideration in any account of racist geographies is the self-conception of the dominant group and the identities assigned by that group to subordinate groups. In this instance, it is important to ask: Who were the Afrikaners, the self-declared "White Tribe" who assumed control of the country in 1948?

As noted, Dutch settlers, beginning in the late seventeenth century, slowly moved inland over considerable distances, establishing an extensive farming economy, such that a distinctive Afrikaner identity is generally considered to have been initiated during an almost 250-year period of movement. Factors that helped to shape this identity include the process of land acquisition and settlement, the opposition to both Dutch and British colonial authorities, isolation from others, a Calvinist religious tradition, the Dutch-based Afrikaans language, conflict with Bantu groups, the formation of republics, and war with the British (1899–1902). Lester equated the rise of Afrikaner identity with nineteenth-century nationalist movements, noting that there was an

> implicit externally derived, but internally manifested, threat to the culture, especially religion and/or language of a body of people; a constructed but widely perceived history, the interpretation of which rests on the shared experience of "the group" as opposed to alien "others"; a threat to the economic fortune of most members of this historically defined group; and the creation of a set of symbols by which the group is demarcated and exhorted to act. (Lester 1996:60)

It is important to stress that for Afrikaners the self-conception included both a distinctive culture and a pure race, although the introduction of apartheid after 1948 necessarily involved grouping all White people—Afrikaners, British, and other Europeans—together.

Quotes from two ministers in the new Nationalist government serve to present the Afrikaner rationale for implementing apartheid. First, in 1953 the Minister of Justice stated:

If a European has to sit next to a non-European at school, if on the railway station they are to use the same waiting rooms, if they are continually to travel together on the trains and sleep in the same hotels, it is evident that eventually we would have racial admixture, with the result that on the one hand one would no longer find a purely European population and on the other hand a non-European population. (quoted in Christopher 1994:4)

Second, in 1954 the Minister of Labour stated:

I want to say that if we reach the stage where the Native can climb to the highest rung in our economic ladder and be appointed in a supervisory capacity over Europeans, then the other equality; namely political equality, must inevitably follow and that will mean the end of the European race. (quoted in Christopher 1994:2)

Both of these powerful political statements are premised on the completely incorrect notion that there are such things as races within the human species. As discussed in the previous section, there are not, but the power of racial thinking does create material inequalities that endure through the generations. The two principal intents of apartheid were, first, to separate groups seen to be different and, second, to allow Whites to continue dominating the country.

Dividing People, Dividing Space

It is not easy to convey the enormity of the apartheid endeavor, this "uniquely selfish and degrading concept", as Christopher (1994:8) calls it. Box 7.3 outlines the key legislation designed to ensure complete social and spatial separation so that the politically and economically dominant White group could retain power over the country in spite of increasing demands for political rights by the Black majority. The basic mechanisms devised to guarantee the survival of the White group were the removal of anyone not considered to be White from those areas designated as being for the occupation or enjoyment of the White population, and the elimination, so far

A beach sign during the apartheid era made it plain that non-Whites would not be welcomed. Such signs were commonplace throughout the landscape of apartheid South Africa, being found on washroom doors, park benches, many entrances to buildings, and so forth. It is difficult to exaggerate the humiliation suffered by those who were restricted to inferior beaches, facilities, and services. Often described as "petty apartheid", this segregation of landscape was in fact far from petty, as it was both insulting and dehumanizing. (William Norton)

as was feasible, of all contact between Whites and others. →≫≫ 7

The key first step in the implementation of apartheid was to formally classify all individuals according to their supposed racial type, hence the 1950 Population Registration Act. Inevitably, the task of assigning a racial identity to every person living in South Africa was controversial in many specific cases. Classification was based on physical appearance and social acceptability, with the general intent of ensuring that those who were definitely not White not be classified as White. Appeals were permitted, and there were about 7,000 successful appeals between 1983 and 1990.

As the legislation listed in Box 7.3 indicates, there were three levels or spatial scales. First, there was an ambitious attempt to redraw the political

Box 7.3 Segregation in South Africa— Legal Milestones

In South Africa, as in most other colonial areas, forms of spatial and social segregation were evident in both rural and urban areas from the early days of European involvement. As a result, it is fair to say that apartheid was rooted in the colonial era. This colonial segregation was intensified and formally institutionalized by the dominant European group beginning in 1910, after the colonial period, when the four former colonies united as the Union of South Africa. The principal acts were the following:

- **Natives Land Act, 1913**: This act, which was regarded as the first stage in drawing a permanent line between Europeans and the indigenous population, identified those rural areas assigned to the indigenous population.
- **Natives (Urban Areas) Act, 1923**: This act required local authorities to establish separate locations for the indigenous population.
- **Native Administration Act, 1927**: This act effectively made those areas designated for the indigenous population subject to a political regime that was separate from that governing the rest of the country.
- **Slums Act, 1934**: This allowed inner-city suburbs to be demolished and any residents of indigenous ancestry to be removed.
- **Native Trust and Land Act, 1936**: This act extended the rural area assigned to the indigenous population.

In addition to these acts and numerous amendments, there were other acts for specific regions designed to limit the locational choices open to Asian and Colored populations.

Beginning in 1948 with the election of two allied parties that united to form the National Party in 1951, the policy of apartheid—total segregation—was introduced, building on the segregation of the colonial period and the Union period. The principal acts were the following:

- **Prohibition of Mixed Marriages Act, 1949**: This act prohibited marriages between Whites and others.
- **Immorality Amendment Act, 1950**: This banned extramarital relations between Whites and others.
- **Population Registration Act, 1950**: The cornerstone of apartheid, this act provided for compulsory classification of all individuals into what the government considered to be distinct racial categories. Initially three groups were identified: White, Black, and Colored, with the Colored group subdivided into Cape Malay, Griqua, Indian, Chinese, and Cape Colored.
- **Group Areas Act, 1950**: This was designed to effect the complete spatial segregation of the designated population groups within cities, with respect to both residences and businesses.
- **Reservation of Separate Amenities Act, 1953**: This act was designed to create separate social environments for the various population groups—the so-called petty apartheid.
- **Promotion of Bantu Self-Government Act, 1959**: This created a hierarchy of local governments for rural reserve areas.

The end of apartheid came as a complete surprise to most observers. During the 1980s, there was a gradual dismantling of petty apartheid. A series of political developments led in 1992 to a referendum, in which over two-thirds of voters (all designated as Whites) accepted the principle of an undivided South Africa and rejected apartheid. The first democratic government was elected in 1994.

map by removing all-Black areas from the state and creating independent Black homelands. The population movements involved in this process, which were still incomplete when apartheid collapsed in the early 1990s, involved about 1.7 million people being resettled in designated homeland

During the apartheid era, residential segregation was enforced with areas of African housing typically located outside of the main town area. This housing for Africans, which lacked many basic services, is just outside the small town of Grahamstown in eastern Cape Province. (William Norton)

areas between 1960 and 1983. Most of those moved were Blacks who were regarded as surplus to the labor requirements of White farming areas. When it was clear that the apartheid system was crumbling, right-wing White political groups made several proposals to partition South Africa to include two specifically Afrikaner states. These were Orandee, an isolated desert state in the northwest of the country, and Afrikanerland, a state that included much of the agricultural and industrial heartland. Neither of these came to pass.

The second spatial expression of apartheid affected urban areas. The aim was to restructure these to ensure complete segregation and to restrict Blacks' access to urban areas. Certainly, the colonial period and the Union period before 1948 witnessed the evolution of a partially segregated city, but the aims of apartheid were much more comprehensive. This urban restructuring was an enormous project that was incomplete when apartheid ended.

Third, personal/social apartheid included a myriad of regulations intended to minimize contact between members of the White group and others. Two examples of many suffice to make the aim of this form of apartheid clear. The first, the floor plan of a typical White home included separate quarters (usually one small room and a washroom) for a Black servant, thus ensuring residential segregation was enforced within the confines of each individual White residential lot. Second, most of the beach areas, including the better beaches, were reserved for Whites. Together, these two examples are suggestive both of the absurdity of the apartheid project and of the humiliations it involved.

Overall, Whites manipulated space in order to manipulate South African society. It is clear that apartheid policies were introduced by and for Whites, who derived many benefits from measures such as the Group Areas Act. While Whites benefited from such measures, the several other designated groups experienced real disadvantages. The apartheid era was one of ever-increasing hostilities between Whites and others. By the early 1990s it was clear to the Nationalist government that the

Questions for Critical Thought

———— ⚜ ————

A distinguished South African historian published a magisterial volume, *The Afrikaners: Biography of a People* (Giliomee 2003), which attempted to answer the seemingly unanswerable question: why apartheid? Afrikaners were a deeply religious people, proud of their values and cherishing their independence, but why did they implement such a perverse ideology that had such devastating effects on others? Through a detailed historical analysis, Giliomee argues that the Afrikaners fell victim to a delusion—that they had to retain their cultural identity, specifically their presumed racial purity, at all costs. Through time, he argues, they developed a persecution complex, accumulating grievances against others and becoming increasingly embittered. Their motive for implementing apartheid was survival of their group, or what they understood as their race. If this argument is correct, do you think that similar concerns might be articulated for other groups in other parts of the world today? You might consider the rise of right-wing extremism in Europe, and concerns about neo-Nazi groups in Russia. Might these beliefs have been inspired by the same kind of "persecution complex" or as a result of something else?

experience was not sustainable, and apartheid ended peacefully in 1994. ➝➝➝ **8**

As a concluding comment, it is a mistake to think that the extreme conditions of apartheid could only have occurred in South Africa. Box 7.4 details some provocative ideas concerning the possibility that the United States might have adopted segregationist and apartheid policies similar to those employed in South Africa had the demographic circumstances been comparable to those in South Africa.

Genocide

Overall, remarkably little work has been accomplished by geographers, cultural or otherwise, on the geographic dimensions of **genocide** or **ethnic cleansing**. This is despite the very substantial transformations of national, regional, and local geographies that result from the selective population deportations and exterminations involved in these horrendous activities. Most of the relevant literature is historical, political, or sociological. ➝➝➝ **9**

Definition

The term "genocide" was coined during the Second World War and formally defined in the 1948 United Nations Convention on Genocide as follows:

> In the present Convention, genocide means any of the following acts committed with intent to destroy, in whole or in part, a national, ethnical, racial, or religious group such as
> (a) killing members of the group;
> (b) causing serious bodily or mental harm to members;
> (c) deliberately inflicting on the group conditions of life calculated to bring about its physical destruction in whole or in part;
> (d) imposing measures intended to prevent births within the group;
> (e) forcibly transferring children of the group to another group.

For many scholars, this definition is inadequate both because it is too broad, including acts other than mass murder, and because it is too narrow, excluding such possible victim groups as the mentally ill, homosexuals, a particular class, or a particular political group. Chalk and Jonassohn (1990:23) proposed the following definition, which is more useful for our purposes: "*Genocide* is a form of one-sided mass killing in which a state or other authority intends to destroy a group, as that group and membership are defined by the perpetrator." The mass killing is "one-sided" in that there is no reciprocity: the victim group has no wish to destroy the perpetrator group. The reference to "group" is intentionally nonspecific, with no indication of how group membership is determined.

Comprehending the Incomprehensible?

Why are genocides committed? One important reason was discussed in the earlier section on the illusion of race: one of the preconditions for genocide

Box 7.4 What Might Have Been: Segregation in South Africa and the American South

Separation of population groups was a feature of many areas of European overseas expansion during the colonial era and also subsequent to that era if a European group achieved political dominance. In all cases a key rationale for separation was the maintenance of European privilege and the furthering of European interests at the expense of other groups. A brief consideration of the southern United States as it compares to South Africa is interesting in this context.

In both South Africa and the American South the term "segregation" came into popular usage in the early twentieth century, but the details of the way the idea was implemented were quite different. In the case of the southern US, there were two principal elements to segregation before the 1960s: legalized social separation, comparable to the unfortunately named petty apartheid of South Africa, and exclusion of one group from the electoral system. There were few American equivalents to the acts identified in Box 7.3, but the US and Canada did control issues of belonging and citizenship along racial lines. Indeed, after the Civil War the southern United States, despite much discrimination and both de facto and de jure separation, was characterized by considerable integration. The main reason for the different experience in the southern United States was that the subordinate group was a minority and not an overwhelming majority, was culturally similar to the dominant group, and was a part of a democratic state and not a conquered people.

There were, however, some proposals in the southern US to adopt a segregation system similar to the one developed in South Africa. One proposal following the Civil War called for an explicit spatial separation of groups, while a second proposal favored implementing policies along the lines of the South African 1913 Natives Land Act. While official segregationist policies in the US South were overturned with the civil rights movement, the ending of official segregation policies did not mean the end of segregation practices. Most notably, the Federal Housing Administration's redlining approach to

is a distancing or separation of one group—the perpetrator "in-group"—and another, the victim "out-group". The victim group is often given a derogatory label, further emphasizing that they do not belong. Victim groups may be perceived by perpetrator groups as something less than human; they are dehumanized, vilified, stereotyped, scapegoated, and stigmatized. All victim groups are seen in negative terms, certainly as outsiders, but possibly also as worthless, immoral, a threat to the perpetrator group, and perhaps even as subhuman or animal-like. In the eyes of the perpetrator group, the victim group's identity is predetermined and fixed, with individuals slotted into distinct categories. Necessarily, all such attempts to fix identities are flawed. Although specific genocides have specific causes, it is clear that the illusion of race combined with the rise of national identity foster the insider–outsider perceptions that have contributed to many of the genocides of the twentieth and twenty-first centuries. Identities are often linked to place, and in most cases genocide is justified by the perpetrators in terms of their group right to occupy a particular place without others present.

For genocide to occur, the perpetrator group must be able to exercise power over the victim group. Totalitarian states have been especially prone to commit genocide because they function without real constraints on their ability to exercise power. The worst mass killings of the twentieth century were committed by dictatorial states in which most aspects of cultural, political, and economic life were controlled by the government. It is much more difficult for a democracy to commit genocide because this would require the perpetrators to act in secret. Genocide requires the participation of many people mortgage financing has led to spatial outcomes similar to those developed in South Africa.

In terms of North America's indigenous community, processes of spatial segregation occurred through the reservations system. The circumstances of the two cases point to the value of counterfactual thinking as this line of reasoning was outlined in Box 5.6. Frederickson's comments with reference to American Indians are interesting here:

> Assume for a moment that the American Indian population had not been decimated and that the number of European colonists and immigrants had been much less than was actually the case—creating a situation where the Indians, although conquered remained a substantial majority of the total population of the United States. After the whites had seized the regions with the most fertile and exploitable resources, the indigenes were consigned to a fraction of their original domain. All one has to envision here are greatly enlarged versions of the current Indian reservations. Then suppose further that Indians were denied

citizenship rights in the rest of the country but nevertheless constituted the main labor force for industry and commercial agriculture. It is hardly necessary to continue; for one immediately thinks of the kind of devices a white minority might then adopt to insure its hegemony under conditions where a majority of the Indians in fact work off the reservation and even outnumber the whites outside these designated areas. The twin objectives of white supremacy would then be, as in South Africa, to maintain direct minority rule over most of the country and some kind of indirect rule over the reservations, while at the same time providing for a controlled flow of Indian workers for industry and agriculture in the white regions. (Fredrickson 1981:246–7)

In brief, what happened in South Africa might easily have happened elsewhere had the demographic circumstances been comparable. Instead, what did happen was seemingly less overt, but did enact somewhat similar processes of spatial segregation and eventual material disadvantage.

in the perpetrator group, and this level of participation is more likely to occur when the actions taken are formally authorized by the state, such that participating in mass killing is legitimized. Further, much evidence suggests that most of those who participate in genocide are "relatively ordinary people engaged in extraordinary behaviors that they are somehow able to define as acceptable, necessary, and even praiseworthy" (Alvarez 2001:20).

Victims of genocide are killed because of who they are, not in an individual sense, but in terms of the group to which they are seen to belong. Furthermore, their group identity is typically determined by the perpetrators. In many cases, how victims see themselves is irrelevant; it is how they are viewed by the dominant group that matters. They are not killed because of what they have done, although they may often be blamed for all kinds of ills.

Most cases of genocide occur at times of severe national tension. In the cases discussed below, genocides are associated with wars, post-revolutionary circumstances, tensions stemming from the defeat or withdrawal of a colonial power, or collapse of a larger political unit.

A Century of Genocide

Although the fact is rarely highlighted in conventional accounts, much of human history has been characterized by violence and brutality, and the world has witnessed many horrendous events that could be characterized as genocides (Chalk and Jonassohn 1990:5–8, 32–9). This account focuses on those genocides that involved assumptions about race. Together, the myth of race and colonial activity enabled nineteenth-century Europeans to see themselves as naturally dominant over others, Africans especially. The assumption of natural superiority is evident in Winston Churchill's 1880 report of a British defeat of the Sudanese as one of those "spectacular conflicts whose vivid and majestic splendour has done so much to invest war with glamour", and his description of such battles generally as "only a sporting element in a splendid game" (quoted in Weitz 2003:47). Brutal colonial regimes were established especially by Belgium in the Congo and by Germany in South West Africa, with the Germans killing about 65,000 of the 80,000 Herero people between 1907 and 1911 (a first formal apology was offered by Germany in 2004). But the most tragic genocides resulted from the combination of the myth of race, the concept of national identity and related desire to create uniform societies, and the rise of powerful and corrupt totalitarian states capable of imposing horrific policies.

Armenians in Turkey

With the decline of the Ottoman Empire prior to the First World War, the new state of Turkey, motivated by nationalism and related ideas about creating an exclusive and homogeneous Turkish population, initiated genocide against the Armenians, the largest non-Turkish group in the country. Armenians and Turks had shared the area for several thousand years, but Armenians were different from Turks both linguistically and religiously, having been one of the first groups to adopt Christianity. In 1916 an American diplomat in the region described forced movements and mass murders, recognizing that the issue was "nothing less than the extermination of the Armenian race" (quoted in Weitz 2003:1). Estimates suggest that about 1 million Armenians out of a total population of 2.1 million were killed.

The Soviet Union

Following the 1917 revolution, the Soviet Union attempted to modernize rapidly in order to establish a utopian nation with great-power status. One aspect of this objective was an obsession with categorizing people as a means to distinguish between friends and presumed enemies of the state; the category of "enemy of the people" was basic to this process. There was never a well-developed racial ideology; instead, people were categorized according to other identifying characteristics that could not be objectively determined, such as class, national background, and political view. As a result of the complex set of officially recognized social classes that was created, people came to be seen not as individuals but as members of a class—the favored class was the proletariat. Nationality, the second basis for classifying people, reinforced then widely accepted assumptions about the reality of nations, and Russian national identity was valued above others. The third principal basis for classifying people was presumed political ideology. Needless to say, only those who accepted the principles and policies of the governing Soviet regime were tolerated. In addition to these three identifying characteristics, particular behaviors such as alcoholism and vagrancy were seen negatively and typically resulted in exile or execution.

With these bases for classifying people in place, the Soviet Union engaged in population purges, forced starvations, forced labor, and deportations from the 1920s to the 1950s, substantially changing the cultural geography of the state in terms of population distribution and local cultural identities. Cossacks were the first victims in 1919–20, with perhaps 500,000 of a population of 3 million either killed or deported. But most acts of genocide occurred between 1929 and 1953, when Stalin was leader. Between the late 1920s and the mid-1930s, a campaign against small peasant proprietors resulted in about 6 million deaths from famine. The Great Terror of 1936–8 involved concerted assaults on a wide range of undesirables, including people of aristocratic descent, minority national groups, and those with contacts outside the Soviet Union. Also in the 1930s most national identities—the list is far too long to itemize—other than Russian were targeted for purges, with millions of people deported or killed either intentionally or by neglect. The number killed through execution or neglect and in prison camps is not known; traditional estimates range between 20 and 40 million, although some recent historical research suggests a lower total. These activities were part of what today is called "ethnic cleansing" and resulted in a dramatic transformation of the cultural geography of the state. What is important to remember is that in all cases, individuals were targeted because they were seen as members of groups, not because of any individual characteristics or because of what they had done.

The Holocaust

While Stalin's Soviet Union was based primarily on the slippery concept of class, Hitler's Germany was premised on the even more slippery myth of race and especially on the belief that Jews were a racial enemy that had to be eliminated in order to permit the German state to progress. Although anti-Semitism prevailed throughout much of Europe, both the intent and scale of the Holocaust are beyond rational understanding. As an ideologically motivated genocide, it is without precedent. Overall, between 6 and 8 million Jews, Gypsies, and other undesirables were murdered between the late 1930s and the end of the Second World War (Box 7.5).

Race thinking in general and anti-Semitism in particular were typical of the times, but Hitler was especially vitriolic, using the basest of terms to describe Jews in his political treatise *Mein Kampf.* The great tragedy is that someone holding these beliefs achieved a position of power enabling him to surround himself with like-minded people to carry out such ruthless policies.

The principal goal of the Holocaust, articulated in numerous pseudoscientific writings, was to save Germany and the purported Aryan race from racial pollution. Aryans, or Germans—the terms tended to be used interchangeably at the time—were represented as the great creators of culture, while Jews were the great destroyers. Jews were also portrayed as carriers of various diseases, and German society was to be purified through the elimination of all Jews. Some other groups were also judged expendable, including Gypsies, Blacks, the mentally and physically handicapped, homosexuals, communists, and others variously seen as outsiders. Indeed, a program of euthanasia, in place in the 1930s, was responsible for the deaths of several thousand mentally and physically handicapped Germans. Further, during the war, German military successes in eastern Europe involved distinguishing different types

of people and eliminating those who were judged physically or culturally inferior. Strategies used included enforced birth control, sterilization, and abortions, together with forced movements, neglect, and murder.

Following his appointment as chancellor in 1933, Hitler introduced a series of policies designed to limit the daily lives of Jews. Implementing these policies required clarification as to who was and was not Jewish, a vexed question given that Jews had been a part of larger German society for many years. Definitions of degrees of "Jewishness" were first formulated in the Nuremberg Laws of 1935; like any attempt to place individuals into identity categories, this exercise was flawed from the outset. But neither practical nor scientific absurdity prevented the Nazi regime from implementing the presumed racial classification. ➡➤➤ **10**

By early 1941, "Jewish life in Germany had been reduced to a confined, impoverished, frightened experience" (Weitz 2003:124). But the prevailing discrimination and random acts of terror were soon to become genocide. By mid-1941, Jews were explicitly targeted for extermination, with mass killings becoming commonplace in areas of German occupation, including Poland, Lithuania, Ukraine, and Serbia, where these killings often met with the approval of local non-Jewish populations. But with deportations proving difficult in wartime and with the rate of killing seen as far too low, the Third Reich looked for other mechanisms to exterminate Jews. At the Wannsee Conference, held in January 1942, procedures for the organized mass murder of Jews in concentration camps were developed, and the policies were implemented over the next three years (see Figure 7.1).

It is not easy to understand either the chaotic population displacements or the mass murders that were routine in Germany and German-occupied areas during the late 1930s and the Second World War. Armed with the illusion of race, certain as to the need to preserve the purity of an Aryan racial identity, embittered by perceived grievances against non-Aryans and especially Jews, empowered by widespread acceptance and even active involvement by many ordinary Germans, the National Socialist

Box 7.5 The Holocaust

Concentration and death camps were located in Germany and in areas of German occupation during the Second World War, including six major camps in Poland. The comments here give details about some of the camps shown on the map in Figure 7.1.

Dachau was the first Nazi concentration camp, established in 1933. It was used mainly to imprison German political prisoners until late 1938, after which members of many other groups that were seen as enemies of the state were imprisoned here, including Jews, Gypsies, Jehovah's Witnesses, and homosexuals. Dachau was liberated in April 1945.

Buchenwald was a concentration camp opened in 1937 and liberated in April 1945.

Auschwitz-Birkenau was a complex consisting of concentration, extermination, and labor camps. It was established in 1940 as a concentration camp and included a killing center in 1942. Between 1 million and 2.5 million people were killed here, mostly Jews but also Poles, Gypsies, Soviet POWS, and others.

Chelmno was set up in 1941 as an extermination camp. Between 170,000 and 360,000 Jews were murdered here between late 1941 and 1944. Chelmno was technologically primitive, employing carbon monoxide gas vans as the main method of killing. The Nazis dismantled the camp in early 1945.

Majdanek was a camp and killing center opened in late 1941. Originally a labor camp for Poles and a POW camp for Russians, it was reclassified as a concentration camp in April 1943. At least 200,000 people were killed here. The Soviet Army liberated it in July 1944, and a memorial was opened there in November of that year.

Belzec was initially a forced labor camp but became an extermination camp in 1942. About 600,000 Jews were murdered here in 1942 and 1943. The Nazis dismantled the camp in the fall of 1943.

Sobibor was an extermination camp opened in May 1942 and closed the day after a rebellion by its Jewish prisoners on 14 October 1943. At least 250,000 Jews were killed here.

Treblinka was an extermination camp opened in July 1942. Between 700,000 and 900,000 persons were killed here. A revolt by the inmates on 2 August 1943 destroyed most of the camp, and it was closed in November 1943.

Bergen-Belsen was set up as a concentration camp in 1943. Thousands of Jews, political prisoners, and POWS were killed here. It was liberated by British troops in April 1945, although many of the remaining prisoners died of typhus after liberation.

party took advantage of their position of power to perpetrate the unthinkable.

Cambodia

The communist Khmer Rouge, led by Pol Pot, ruled in Cambodia (renamed Kampuchea in 1976) from 1975 to 1979 and engaged in mass killings designed to purify Cambodian society and create a pure race, a unified nation, and a classless society. At least 1 million people were killed, with some estimates as high as 2 million. Groups were identified as undesirable based on racial criteria and communist logic about

classes. The first group targeted was the intelligentsia, including artists and ballet dancers. Subsequent groups targeted included people of Vietnamese and Chinese origin, Muslims, and selected social classes. As with the other genocides described, people were defined not by their individual characteristics or their behavior but by their presumed group membership. Strategies for eliminating undesirable groups included deportation, intentional starvation, forced labor, and murder.

The first major purge designed to help homogenize the population was the forced evacuation of

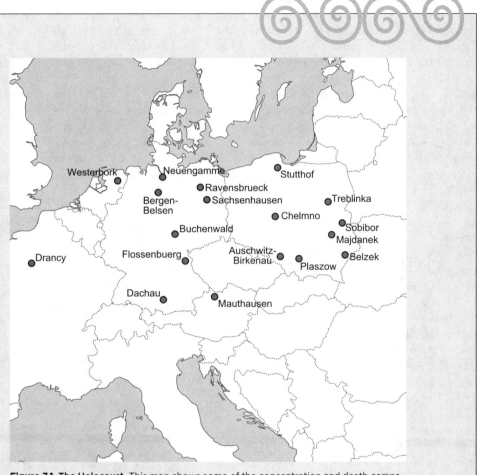

Figure 7.1 The Holocaust. This map shows some of the concentration and death camps located in areas of German occupation during the Second World War. Source: Various.

the capital city, Phnom Penh. This was followed by a series of ethnic purges against the close to twenty ethnic minorities in Cambodia; these were intended to produce a single nation with a single language. The attempt to purify the population, to eradicate all difference, was doomed to failure and ended with members of the Communist party turning on each other.

Rwanda

Rwanda comprises two principal tribal groups, Hutus and Tutsis. Belgian colonial rule, which lasted from 1916 until 1962, favored and ruled through the minority Tutsi group. This caused considerable resentment among the majority Hutus. Crucially, all Rwandans were issued with an identity card stating their tribal affiliation.

Since independence, Rwanda and several adjacent countries have been beset by ethnic tensions. Following the 1994 murder of the Hutu president, Juvénal Habyarimana, extremist Hutu militia initiated a systematic massacre of Tutsis. In less than four months, about 800,000 Tutsis and moderate Hutus were killed, sparking massive refugee movements. The violence was so severe and the country so divided that news reports at the time described situations in which members of a Catholic Church congregation murdered others who attended the same church. The situation was further complicated by instabilities in neighboring Democratic Republic of Congo (known until 1997 as Zaire). Although there are many differences between this genocide and the others discussed, the basic circumstance is the same, with people being killed because of their presumed group membership, in this case Tutsi or Hutu. Tragically, this pattern of tribal conflict leading to genocide has occurred in recent years in a number of African countries, including Burundi, the DR of Congo, Ethiopia, Liberia, Nigeria (Biafra), Sierra Leone, Somalia, and Sudan.

Serbia

Yugoslavia was established in 1945 as a socialist federation of six culturally diverse republics with Tito as leader. Following Tito's death in 1980, various nationalist sentiments began to surface, anticipating the political collapse that began in 1991 with the secession of Slovenia, which was quickly followed by that of Croatia. During this tumultuous period there were numerous conflicts between members of different cultural groups, and atrocities were common. Serbs, the most powerful group, were embroiled especially in two bitter conflicts: first in Bosnia, beginning in 1991, where Bosnian Serbs clashed with Croatian Muslims, and later in Kosovo, in 1999, where Serbs fought with Albanian Muslims. The extremely complicated pattern of ethnic distribution in the former Yugoslavia is suggested in Figures 7.2 and 7.3. In cases of state-sponsored ethnic cleansing or genocide, the presentation of data such as this in maps can be used as propaganda in the campaign against certain groups, since such seemingly definitive data presentation erases the complicated reality of social heterogeneity that often exists within village and regions. ➜➤➤ 11

As Yugoslavia disintegrated, each republic sought to clarify the nationality of its members, but it was Serbia, the largest of the six republics, that developed a particularly virulent form of nationalism. Serbian leader Slobodan Milosevic resolved to preserve socialism and turned to a nationalist and racist ideology to justify creating a homogeneous and racially

pure Serbian state. In retrospect, many commentators suggest that the pivotal event in the rise of a virulent Serbian nationalism was a 1987 speech in the southern autonomous Serbian province of Kosovo, whose population was about 80 per cent Albanian. In this speech, given at a time when tensions between local Serbs and Albanians were rising—and near the site of the powerfully symbolic 1389 battle of Kosovo Polje, where Ottoman Turks defeated Serbs—Milosevic stressed the presumed purity of Serbian identity and the right of Serbs to occupy what he claimed was their homeland. Subsequently, Serbian leaders began the

Questions for Critical Thought

——— ✥ ———

Although it is not difficult to summarize the character of genocides, any summary is obviously going to fall short of answering all of our questions. One question that still seems necessary to ask is *why*? It isn't enough to point to a few sadistic leaders, for, after all, there is much evidence to suggest that many ordinary people become directly or indirectly involved in genocides. Wilmer (1997:4) asks: "How is it that individual people are persuaded to abandon civility in favor of brutality? Why were elites able to so readily mobilize them to undertake acts of inhumanity on the basis of appeals to identity?" Hobsbawm (1993:62–3) responds by noting that "history is the raw material for nationalist or ethnic or fundamentalist ideologies, as poppies are the raw material for heroin-addiction." One reason cultural geographers focus so intently on identity construction and "othering" is because this process of dehumanization plays an important role in socially distancing those deemed "different". Creating social hierarchies also allows for inequality to be naturalized; therefore, exposing how certain groups maintain and protect their power goes some way to questioning how inequality is produced. We must also recognize that normalizing such inequalities (the "natural order of things", or blaming the victim) permits the erosion of empathy, setting in place the precursor for inhuman treatment of others. Consider these points and identify a situation you are familiar with where elements of such social practice are in place.

complex, indeed impossible, process of categorizing people into fixed identities in order to facilitate creation of a pure state.

The several conflicts going on as republics strived to assert their distinctiveness and gain independence, the racially motivated belief in a pure Serbian identity, and the fact that Serbia was the most powerful republic together provided the necessary preconditions for several genocidal incidents that were motivated by the desire to make specific locations racially pure. The town of Vukovar, located on the border with Croatia, was the scene of a massacre of about 300 Croats in 1991. In Srebrenica, a supposed United Nations safe haven, Bosnian Serbs massacred about 7,000 Bosnian Muslims in 1995. Further atrocities were committed by Serbs against Muslims in several areas of Kosovo in 1998 despite United Nations intervention, including air strikes. Ethnic violence resurfaced in 2004 as Kosovo moved painfully toward becoming a viable multi-ethnic region.

Figure 7.2 Ethnic Groups in the Former Yugoslavia. Although this map may appear very complex, it is only an approximation of a much more complex distribution. Indeed, it is not possible to accurately map cultural distributions at this scale because the patterns are so remarkably complex. Source: A.J. Christopher, *The Atlas of States: Global Change, 1900–2000* (New York: Wiley, 1999):191.

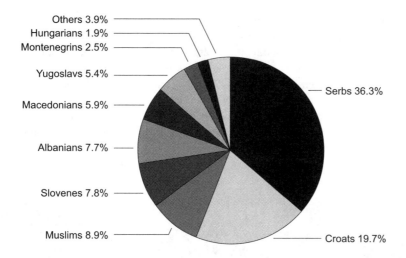

Others 3.9%
Hungarians 1.9%
Montenegrins 2.5%
Yugoslavs 5.4%
Macedonians 5.9%
Albanians 7.7%
Slovenes 7.8%
Muslims 8.9%
Serbs 36.3%
Croats 19.7%

Figure 7.3 Ethnic Composition in the Former Yugoslavia. This diagram showing the relative sizes of cultural groups is a simplification because the specific cultural identities are only some of those that might be included. Source: Various.

The examples of genocide presented above all follow a common pattern, which can be summarized as follows. Genocide is an extreme behavior, usually taken as one part of larger efforts to dramatically reshape the geography of peoples and places. Racist and nationalist ideas form the basis for categorizing populations into groups, with perpetrators of genocide labeling victim groups in the most negative of ways. Victim groups are typically blamed for any and all social and economic problems and are generally seen as impediments to progress and prosperity. The perpetrators, using powerful metaphors of "purifying" and "cleansing", justify genocide on the grounds that victim groups are something less than human. A powerful state is usually a prerequisite to the introduction of the policies designed to categorize people and to the subsequent practices of deportation and mass murder.

Ethnicity and Nationality

A central consideration in contemporary social science is the issue of group membership as it relates to aspects of identity. This issue has been raised earlier in this chapter as well as on several prior occasions in this textbook. The tendency in earlier chapters was to treat group membership in essentialist terms—that is, as a given (an exception to this is the discussion in Chapter 6 of the Hispanic group initiated by Blaut [1984]). The current interest in identity

raises a very challenging question that reintroduces the debate about essentialism and constructionism. Recall that essentialism involves the claim that a specific identity is basically unchanging, whereas constructionism involves the claim that identity is fluid, contested, and negotiated. Much work in contemporary cultural geography accepts that identities are socially constructed, an idea that was at the forefront of the preceding accounts of the myth of race and the related discussions of apartheid and genocide. Although the perpetrators of apartheid and genocide employ essentialist ideas of identity, it is clear in the discussions of these topics that such essentialist assumptions are necessarily flawed.

In the process of inventing and reinventing group identities it has been quite usual for political leaders and others to use the words "race", "ethnic", and "nation" interchangeably. Perpetrators of apartheid and genocide, prone to sloppy thinking designed to serve their own purposes, have abused these terms to suit their objectives. Given our explicit intent to historicize, contextualize and problematize the term "race", it makes sense to consider "ethnicity" and "nation" as useful alternatives. Both, however, need to be carefully defined.

Understanding Ethnicity

Several of the studies that identified the unscientific status of race use "ethnicity" as an alternative.

But this is also a term that many scholars and others use interchangeably with "race". So what does ethnicity mean?

Unfortunately, but not unsurprisingly, the concept of ethnicity has a rather difficult scholarly origin, and there remains considerable debate about the meaning of the term. According to Rupesinghe (1996:13): "Ask anyone to define ethnicity and the problem begins. We are left with a host of interpretations. The difficulty in defining ethnicity is that it is a dynamic concept encompassing both subjective and objective elements. It is the mixture of perception and external contextual reality which provides it with meaning." Given these difficulties, and the longstanding links to the myth of race, it is hardly surprising that an analysis of 65 studies of ethnicity noted that 52 of them offered no explicit definition (Isajiw 1974). Broadly speaking, minority groups are often defined as "ethnic", especially in the North American immigrant context, and Raitz (1979:79) notes that in this capacity, "ethnics are custodians of distinct cultural traditions." This is the interpretation traditionally used by some governments; in Canada, for example, the Census includes a question on ethnic ancestry, not ethnic identity. →≫ **12**

Six Components of Ethnicity

For cultural geographers, the six components of ethnicity identified by Smith (1986:21–31) are most helpful. First, an ethnic group has an identifying name. This is a means by which people are able to distinguish themselves from others and to confirm their existence as a group. The symbolic importance of naming is clear, suggesting the reality of an identity and its difference from others. (The related question concerning the importance of naming places as one means of indicating ownership and authority is noted in Chapter 6.)

Second, an ethnic group has a common myth of descent. This may take the form of myths of temporal origins and spatial origins. The various perpetrator groups in the preceding accounts of genocides, especially Germans in the 1930s and Serbs in the 1990s, each invented and reinvented a myth of descent. Germans under the Nazi regime promulgated the idea of an original Aryan race, while Serbs portrayed the 1389 battle against Muslims as a decisive historical moment that justified their territorial claims in the 1990s.

Smith's third component of ethnicity, closely related to the second, is a shared history. An ethnic group is a historical community that typically relies on shared memories to unite successive generations. Each new experience is added to the previous history and interpreted and understood in terms of that history. Necessarily, such a history is but one reading of the past, a reading designed to reinforce the identity of the group.

Fourth, an ethnic group has a distinctive and shared culture that is reflected in lifestyles and values of the group. This may take the form of a common language and religion along with, for example, particular customs, institutions, folklore, architecture, dress, food, and music.

Fifth, an ethnic group is linked to one or more places. This link is most powerful when it is to a particular territory, a place that is seen as a homeland. The group may occupy this territory, may aspire to occupy this territory, or may only have recollections of this territory. Especially meaningful places may be contained within, or may even lie outside, this homeland.

Sixth, an ethnic group is a community that possesses a real and meaningful sense of self-identity and self-worth, such that other ways of dividing people—for example, according to social class—are not able to create meaningful divisions within the ethnic group. Group identity also involves a belief that all members share a common fate, and that individual identity is inseparable from larger group identity—circumstances that facilitate the process of mobilizing group members and strengthening group solidarity as needed. Such ethnic solidarity is most evident during times of stress, when the identity, perhaps even the very existence, of the group is challenged.

Smith (1986:32) synthesized these six criteria to define ethnic groups as "named human populations with shared ancestry myths, histories and cultures, having an association with a specific territory and a sense of solidarity". Incorporating the conventional cultural variables of way of life, language, and religion, along with a spatial component and a sense of identity, ethnicity emerges as a useful, almost catch-all, concept. Needless to say, however, the term does have some unfortunate overtones. The most obvious of these, already stressed, is the fact that it tends to be used as a surrogate for the myth of race. Related to this is the often explicit use of the term to delimit groups in opposition to others. Most ethnic groups are ethnocentric, seeing the in-group beliefs and behaviors as natural and desirable and the out-group beliefs and behaviors as unnatural and perhaps as abhorrent. Box 7.6 discusses the complex example of ethnic identities in Macedonia.

The genocidal circumstances discussed earlier are products of states constructing national identities for themselves and to exclude undesirables. But there are also examples of regional groups constructing identities of nationhood for themselves. In many cases, these constructed identities are manipulated by people in positions of power in order to seek national status and thus achieve territorial status, as Knight (1982:523) explains:

> Ethnic or cultural integrity and uniqueness, minority status, the desire for national freedom, the natural right to independence through self-determination—these are the terms used by regional ethnic group elites as they formulate separatist political ideologies, which are, in turn, often rooted in a variety of emotional, historical (generally from a revisionist perspective), and politico-economic realities, as perceived and understood by the regional (ethnic) group itself.

Language and Religion

Language and religion are two of the bases that groups employ to assert their ethnic distinctiveness and, in some instances, their national identity. Certainly in cultural geography it has been usual to regard language and religion as the most important variables that unite groups (for a brief overview of language and religion in a global context, see Box 6.3). The presumed importance of language as an indicator of cultural, ethnic, or national identity

is noted in Box 7.7 in the larger context of the rise of English as a global language. Language is one of the principal means by which certain cultures identify themselves through time, while religion is often the key repository for important beliefs, attitudes, and traditions that serve to unite a group.

There is often a close relationship between a religious creation myth and the origin myth of an ethnic group, and as noted in Box 6.3, there are some religions that are classed as "ethnic", meaning that they are associated with a specific group of people; principal examples include Hinduism, Judaism, Taoism, Confucianism, and Shinto. Further, the structure and organization of a religion may provide mechanisms to communicate ethnic myths and symbols. There has been little explicit discussion of these matters in the cultural geographic literature, although Stump (1986:2) has noted that religion was "important in identifying and understanding the diffusion, distribution and character of groups defined by their social, ethnic or regional identity". ➙➛➛**13**

Links to Ethnicity?

It is interesting that although language and religion have long been seen as distinguishing markers of ethnic groups, it can be argued that they are becoming less relevant. This suggestion is in accord with the emphasis that contemporary cultural geographers place on the idea of multiple and flexible identities. It is also a suggestion that implies that policies favoring a **minority language** at the expense of some other language, as in the case of Quebec, may be misguided if the ultimate goal is to retain group identity. In brief, although traditionally they have played an important role in establishing the identity of ethnic groups, both language and religion may not be critical components of ethnic identity in the contemporary world.

Sharing a common language is not always essential to a sense of ethnic community. In the case of Wales, for example, English-speaking people in southern Wales may sense a Welsh identity similar to that of the Welsh-speaking people in parts of northern Wales. Similarly, Scotland has a distinctive English dialect in the Lowlands, known as Lallans

Box 7.6 *Ethnic Identities in Macedonia*

Today, Macedonia is a small landlocked country in southeastern Europe, with a population slightly over 2 million. It borders on Albania, Bulgaria, Greece, and Serbia and Montenegro. Part of the border shared with Serbia and Montenegro separates Macedonia from the Serbian province of Kosovo.

Historically, the region has always been of strategic importance, both because it is located on major thoroughfares connecting different parts of Europe and connecting Europe and Asia, and because it has often been seen by major powers as a key to the larger Near Eastern region. These factors have contributed both to a very complex pattern of ethnic distribution and, critically, to numerous different and contested interpretations of the specific pattern and identities of people.

With these facts in mind, Wilkinson (1951) conducted a remarkable survey of about 200 ethnic maps of Macedonia, most of which were prepared after 1840. In total about 150 different groups were

identified on these maps, a fact that makes any attempt at ethnic mapping perilous to say the least. In describing ethnic maps prepared by representatives of a number of interested parties, including Greeks, Bulgarians, Serbs, Albanians, Romanians, Germans, Austrians, British, French, and Americans, Wilkinson (1951) showed that the maps reflected the interests of their creators, with each map "becoming the vehicle of the ideas of its particular compiler— ideas which were largely the product of their age and environment and which differed accordingly. It is patent that even if the population of Macedonia had remained unchanged and static, the variety of ethnographic maps would still have remained comparatively rich" (Wilkinson 1951:321). Figure 7.4 is an example of one such map. Wilkinson's work is very detailed and full of insights that anticipate many of our current interests in the social construction of identity and, especially, the ways in which people in positions of power manipulate identities.

Serbo-Croats, including Greek Orthodox, Muslim and Roman Catholic (these subgroups were separately distinguished in the original)

Albanians, including Greek Orthodox, Muslim and Roman Catholic (also separately distinguished in the original)

Romanians, Pindus Vlachs and Moldo-Vlachs

Bulgarian Pomaks

Greeks, including Orthodox and Muslim (separately distinguished in the original)

Exarch Bulgarians

Circassians, Tatars, Turks and Turcomans (each separately distinguished in the original)

Graeco-Vlachs (separately distinguished in the original and classified as Romanians)

Serbo-Bulgarians (separately distinguished in the original but classified as Bulgarians)

Graeco-Bulgarians (separately distinguished in the original but classified as Bulgarians)

Graeco-Albanians (separately distinguished in the original but classifed as Greeks)

Figure 7.4 One Interpretation of Ethnic Identities in Macedonia. This is one of the about 200 maps reproduced by Wilkinson. It was drawn by an Austrian official, Sax, in 1877, a time when Austria wished to consolidate political power in the region. In compiling this map, Sax emphasized religious rather than linguistic facts and incorporated what he described as an element of group consciousness, by which he did not mean the myth of race or folklore but rather a deep-seated sense of community. This map served Austria's interests, stressing heterogeneity at the expense of simplification and thus combating claims by other powers, especially Bulgaria at that time, to have some rights to the region. Source: Adapted from H.R. Wilkinson, *Maps and Politics: A Review of the Ethnographic Cartography of Macedonia* (Liverpool: Liverpool University Press, 1951):80–1.

and sometimes considered to be a separate language, and a Gaelic-speaking Highlands, such that a continuing sense of Scottish ethnic identity may be explained in terms of the Presbyterian religion and the particular Scottish legal and educational systems rather than in terms of language.

Further, recent census evidence for Scotland suggests that there is currently a decline of the Gaelic language. But this decline in the number of Gaelic speakers—the apparently objective indicator of ethnic identity—is accompanied by a Gaelic cultural revival that can be seen in television programs, educational changes, and attendance at Gaelic festivals. This seeming contradiction might be explained by reference to measures of cultural identity other than language: in other words, it is unclear whether language really is an appropriate indicator both of the health of Gaelic culture and of identity with the cultural group.

Similarly, religion is not necessarily a good indicator of the strength of a group's ethnic identity. Although it is important as a symbolic code of communication and as a focus for social organization, particularly prior to the onset of an industrial way of life, religion often transcends conventional ethnic boundaries. This is especially so in the case of Christianity, a universalizing religion that has spread spatially and grown numerically, notably in association with the larger political and economic contexts of imperialism and colonialism. With over 2 billion Christians worldwide, the religious identity is characterized by ethnic diversity. Indeed, few Christian denominations, large or small, are limited to a particular ethnic group identity.

Understanding Others

In some instances, a particularly sensitive issue concerning religion and ethnicity is the perception that one religious group has of another. Recall the suggestion in Chapter 6 that the contemporary world is characterized by a clash of civilizations, especially as these are associated with religious beliefs and attitudes. In particular, many commentators contend there is currently a conflict between Islam and Christianity, a conflict that assumes cultural, political, and economic dimensions.

According to Mazrui (1997:118), the Islamic world is perceived by the Christian world, especially by the United States, "as backward-looking, oppressed by religion, and inhumanely governed", although it is in fact "animated by a common spirit far more humane than most Westerners realize". In this context, recall the reference in Box 3.4 to the ideas of Said concerning the way in which dominant groups perceive other groups. Of course, the myth of race is also worth considering here.

It can be argued that some differences between the Christian and Islamic worlds concerning, for example, population fertility policies reflect the specific timing of cultural change rather than fundamental disagreements about quality of life. Mazrui identified numerous ways in which the Islamic world might be considered more humane than the democratic Western world: for example, Islam has always been more protective of minority religions; it has not been hospitable to either fascist or communist governments; it has not witnessed policies of genocide at the same scale as has the Christian world; and, it has been resistant to racist policies such as apartheid. Certainly, it can be argued that there is a difference between democratic principles and humane principles.

National Identities

Since becoming, along with race, a dominant concept in the nineteenth-century European world, nationalism and its corollary, the nation-state, have become influential globally. The idea of the nation-state is rooted in the belief that the world can be conveniently divided into national groupings, each associated with a particular territory. The concept of nationalism is the belief that each of these groups has some inalienable right to occupy and govern its territory. One possible consequence of these two related ideas is that the world cannot be without conflict until all national groups have their own nation-states. However, the contemporary world is a long way from fulfilling this nationalist goal. A mid-1990s account identified 210 nations without states but noted that these were only a fraction of the total, which could be as high as 9,000. This account asserted: "In varying

Box 7.7 *English as a Global Language*

Why a language becomes a global language has little to do with the number of people who speak it. It is much more to do with who those speakers are. (Crystal 1997:5)

For a language to achieve global status, it needs to be spoken by large numbers of people as a mother tongue in some countries, to be an official language in other countries, and to be a priority in foreign-language teaching in other countries. English qualifies on all three counts: it is spoken by many people as a mother tongue in Australia, Britain, Canada, Ireland, New Zealand, South Africa, the United States, and some Caribbean countries; it is an official language in the sense that it has some special status in over seventy countries (more countries than for any other language), including India, Ghana, and Nigeria; and it is taught as a foreign language in over 100 countries, again more countries than for any other language. Together, these three circumstances are producing increasing numbers of English speakers, about 1.5 billion today.

Language, of course, does not have some existence independent of those who speak it, such that any decline or increase in numbers of speakers has much to do with the cultural, economic, and political success of the speakers of the language. In the case of English, Britain was a leading world power during the critical period of global migration, when capitalism and industrialization were spreading. The global spread of English is seen by some critical linguistics as the result of Linguistic Imperialism, in that Britain's colonial rule imposed the language throughout colonized regions, especially through educational systems. English is also the mother tongue of the US, the more recent global hegemonic nation. Through the spread of cultural products (Hollywood) and the power of the US-dominated global financial systems (IMF and World Bank), English continues to dominate. However, there is the argument that rather than just replicating the language, many countries actually appropriate the language for their own needs and purposes.

Is it possible that the rise of English as a global language is contributing to the death of some other languages? The evidence here is contradictory. Through the course of human history, as many as several thousand languages have appeared and disappeared, but most of these disappearances were related to the demise of a particular group of people rather than to the spread of a global language. Certainly, it has often been argued that retaining a language is central to the survival of a distinctive cultural or ethnic group, and there has been a great deal of interest in the circumstances of a **minority language** in this context. Aitchison and Carter (1990) have stressed the threat to the Welsh language posed by English. Similarly, with specific reference to Irish, Kearns (1974:86) has claimed: "That a separate language contributes to the imprint of a distinctive nationality is widely accepted." However, a rather different interpretation of the role played by language in the preservation of a distinct identity is possible, as noted in the text discussion of Scottish Gaelic.

---

degrees these people resent their subservience, and they struggle, sometimes covertly, often overtly, to achieve political independence and hence, in their view, to control their own lives and destinies" (Doob 1996:xiii). ➜≫ **14**

But of course the nationalist endeavor is fundamentally flawed. Put simply, the world is not conveniently organized into groups of people, nations, waiting to be packaged into nation-states. The very idea is nonsensical given the long history of human migrations, European overseas expansion, the diffusion of languages, religions, and other cultural criteria, and, of course, the fact that there are numerous other ways that people use to identify themselves and to associate with others. Reflecting a constructionist rather than an essentialist view, nationalism is now usually seen as a concern with an imagined community (Anderson 1983). Hudson and Bolton (1997:1) described the Australian national identity as a "fabulous beast", noting that any one supposed identity was necessarily exclusionary because it ignored the reality of multiple identities, including those of groups defined ethnically and sexually. To give credence to the concept of national identity alone is wrong. But, of course, this does not prevent nationalism continuing to be a potent force in the contemporary political world, as evidenced by the plethora of separatist and independence movements.

The idea of nations as imagined communities, while attractive, is usually obvious and often empirically unsatisfactory—obvious in the sense that all feelings of belonging are socially constructed, and empirically unsatisfactory as it appears to ignore the fact that many nationalisms, such as those of the United States and Japan, are products of shared history and shared projects. But there is another problem with thinking about nations, whether viewed as real or imagined, which is that the nation-state may not be in a privileged position as a unit of analysis, as Wilmer (1997:3) suggests:

> The state appears to be more movable, malleable, and contestable than ever. The state consists of a set of institutions that are authoritative only because other sets of institutions recognize them; it has been socially constructed within historical and interpretive social processes and practices.

Given this debate about the identity of the nation, two further ideas might be helpfully added to the six components of ethnicity identified earlier in this discussion. First, the idea that a group has an identifying name implies something more than that the group has a sense of self, of identity; it also implies that the group is in some sense secure enough that any threats to the sense of identity may elicit a defensive response in order to avoid psychological or physical annihilation. Some psychoanalytic ideas suggest that the self is bounded and identified with reference to positive images of who we are along with negative images of who we are not. Because this is an idea that can be applied at both individual and group scales of analysis, it is important in the context of ethnicity. Thus, national and other more local identities may be framed in terms of opposition to others, an idea that is discussed further in Chapter 8. In addition to this group perception of self, there is also tension between national ethnic identity and other more local ethnic identities, a circumstance that is reflected in some accounts of English and Scottish identity.

Second, nationalism can also be seen as a form of cultural capital—that is, the knowledge, attitudes, behaviors, and styles that characterize a group. This cultural capital occupies a space, a homeland. It might be argued that homelands can be seen as *motherlands*, in patriarchal terms, because they offer group members protection and emotional security—interestingly, the concept of a *fatherland* is more likely to be used with reference to matters of law and order, not homeliness. ➔➤➤ **15**

Concluding Comments

- This chapter includes some critical comments about the myth of race that merit repeating. All humans are members of a single species, *Homo sapiens sapiens*, and there are no biological subspecies or races.
  - › The idea that there are races, which has been highly influential in Western thought especially beginning in the nineteenth century, is factually incorrect.
  - › Logically, then, the idea that presumed races can in some way be ranked in terms of cultural and other abilities is also factually incorrect.
  - › Notwithstanding that race is an illusion, the assumption that races exist and can be placed in a hierarchy has contributed to the social construction of identities, leading in turn to such horrors as apartheid and genocide.
  - › One way to interpret the illusion of race is to note that it was the belief that human difference is rooted in the body.
- The concept of ethnicity, although like so many other social science concepts difficult to pin down, is a useful term.
  - › As discussed by Smith (1986), it includes characteristics—such as language, religion, and nationality—and circumstances—such

as a shared history and vision for the future—that together often prove to be meaningful bases for identity formation.

> Nevertheless, it is an unfortunate term in some respects, as it is closely tied to the myth of race.

- Discussions of the myth of race and of the concept of ethnicity highlight the important role that group identities play in our understanding of the world.
- It is worth asking, however, whether or not the recognition of group identities is problematic in the sense that such identities are necessarily reductionist and dehumanizing, as was painfully evident in the accounts of apartheid, genocides, and racism generally.
- But to point to such abuses of identity is not to deny the very real importance that most people

place on a sense of belonging to a group or, in many cases, to several different groups.

- Ideally, we might conceive of a world where people associate with others for positive reasons, without such associations also involving negative perceptions of those who favor other group identities.

The basic logic of this chapter, namely that identities are socially constructed, typically reflects prevailing power relations, and evidence of these in landscape is further discussed in the next chapter. The spatial scale changes, with increased focus on local and community scales, while the basis for identity formation is broadened to include gender, sexuality, and ability/disability. There continues to be emphasis on landscape, not only as it reflects identity but also as it contributes to identity.

Further Reading ➡➤➤

The following are useful sources for further reading on specific issues.

1. Montagu (1997) is the seminal statement on the fallacy of race, while Hannaford (1996), Fredrickson (2003), and Weitz (2003) provide excellent overviews.

2. Anderson (2007, 2008) presents a critical discussion of the social construction of race.

3. Nash (2003) considers the need for anti-racist geographies. There is a substantial literature on the related topic of anti-racist education (for example, Dei 1996). Bonnett and Nayak (2003) comment on the cultural geographies of racialization. Bonnett (2000) stresses that the colors white and black are labeled with positive and negative connotations respectively. See also Sibley (1995).

4. Jackson and Penrose (1994) provide discussions of race and nation that challenge the social construction of these identities while working uncritically within the socially constructed discipline of geography. In their social geography text, Pain et al. (2001) include a chapter

on race (used without quotation marks) and ethnicity (see also Panelli 2004). An edited collection by Dwyer and Bressey (2008) includes chapters on the meaning of the myth of race, some specific examples of racist geographies in the British context, and analyses of relevant everyday experiences.

5. Forrest and Dunn (2006) discuss racism as a fact of life in eastern Australia.

6. Lester (1996:1–14) outlines the debate between liberal and Marxist scholars concerning the rise of apartheid. Lester (2000) provides a succinct historical geography of South Africa.

7. Western (1981) analyzes the impact of apartheid on Colored populations in the Cape Town area.

8. Dodson (2000) provides a clear account of the dismantling of the apartheid landscape (which is described as a dystopia, a bad or evil place) and outlines the need for the cultural geography of South Africa to be understood in terms of power relations and social constructions, not maps and material artifacts.

9. Geographers have not conducted major regional studies of genocides, using the spatial perspective to record and explain population movements, locations of prison camps, and regional and national consequences of mass murders. One insightful article, by Wood (2001), compares geographies of the Bosnian and Rwandan genocides. Most regional textbooks—for example, Jordan-Bychkov and Bychkova Jordan (2002) on Europe—refer to relevant genocides, but necessarily do not include substantial primary research. A similar observation applies to political geography textbooks, for example Taylor and Flint (2000). There are, however, various articles and book chapters that consider specific aspects of genocides, most of which reflect interests in such issues as contested and memorial landscapes. Charlesworth (for example, 1992, 1994, 2003), Cole and Smith (1995), Ó Tuathail (1996, Chapter 6), and Doel and Clarke (1998) are examples. Smith (2000) discusses the Holocaust from a moral geographic perspective.

10. Arendt (1951) and Steiman (1998) describe the nineteenth-century rise of anti-Semitism. Goldhagen (1996) wrote about *Hitler's Willing Executioners*, stressing the complicity of ordinary Germans in the Holocaust.

11. Wilmer (1997) provides a thoughtful discussion of identity and culture in the Balkan region. Bell-Fialkoff (1993) outlines the history of ethnic cleansing. White (2000) outlines nationalist tendencies in southeastern Europe. Sowell (1996) looks at the historical migrations of selected culture groups.

12. Gil-White (2001) tackles the difficult question of why it is that, despite the myth of race and social construction of ethnicity, ordinary people routinely accept that these categories are real. Marger (2003) provides a discussion of ethnicity based on the idea that ethnic groups are groups within larger societies that are characterized by some unique set of traits, a sense of community, ethnocentrism, membership that is basically fixed, and a link to territory. This book includes detailed discussions of ethnic circumstances in several regional contexts, namely the United States, South Africa, Brazil, Canada, and Northern Ireland. All of this is accomplished in the context of an account of "race and ethnic relations". Hutchinson and Smith (1996) include discussions of ethnicity in conceptual, theoretical, historical, and contemporary terms, along with accounts of the relationships between ethnicity and religion, language, the myth of race, and nationalism. The book concludes with several chapters on the possibility of moving beyond ethnicity as a marker of identity. Heffernan (1998) provides a good account of European identity through a discussion of the historical experience. Eller (1999) reviews ethnic conflict from an anthropological perspective, with discussions of various regional examples including Sri Lanka, the Kurdish region, Rwanda and Burundi, and Quebec.

13. Valentine, Sporton, and Nielsen (2008) discuss links between language and identity in the context of globalization processes that involve increased movements of people. Zelinsky (1992:149–53) discusses the rise of a common civil religion.

14. Jones (2008) discusses the role played by national elites in the reproduction of nationalist discourses and practices. Anderson (1983) is the seminal statement on nations as imagined communities; McLeay (1997) describes the construction of the Australian national identity; Castells (2004) offers a rather different interpretation. With reference to national identity, Dijkink (1996) refers poignantly to "maps of pride and pain". Juergensmeyer (1995) describes some dangers of religious nationalism.

15. Dittmer and Larsen (2007) discuss aspects of Canadian nationalism. Taylor (1991),

Withers (1995), B.J. Graham (1997), Brace (1999), Pittock (1999), Darby (2000), Morley and Robins (2001), and Hardill, Graham, and Kofman (2001) discuss aspects of geography and identity in a British context. Rogerson and Gloyer (1995) discuss Gaelic culture. Subnational movements in South Asia are discussed by Mitra and Lewis (1996). Herb (2004) analyzes the construction of national identities in post–Second World War divided Germany. Hage (1996) discusses homelands as motherlands. One of the ways in which one group establishes authority over another group during times of conflict is by assaulting women in order to demonstrate the inability of the fatherland to protect its daughters (also see Ó Tuathail 1995:260). Agnew (1997a) provides a clear account of the geographies of nationalism and ethnic conflict. In a discussion of the breaking of nations, Cooper (2003) argues that the twenty-first century might be the worst in European history.

CHAPTER 8

Power, Identity, Representation

This chapter explores issues of power, identity, and representation. Much of this builds on the newer theoretical formulations informing cultural geography since the cultural turn. In this chapter you will find a discussion of four interrelated topics that have emerged within feminist and poststructuralist writings, and that are particularly relevant to the broader cultural turn, namely

⚜ discourse and power;

⚜ identity and the construction of identity;

⚜ intersectionality; and

⚜ the practice of research, knowledge production, and representation.

Discourse and power have to do with the idea that power is exercised through the **production of knowledge**, in that our knowledge about "reality" is not in fact independent of language, but constructed through its use, interpretation, and representational effects. As a result, dominant discourses are produced within systems of power relations, and the two work recursively to maintain each other. This is fundamental to the relationship between power and the construction of truth regimes.

Identity and identity construction is the notion that identities are fluid and negotiable, but certain aspects of identity are emphasized within the context of dominant discourses of the time. At any time certain elements of identity, whether based on class, **gender**, religion, or sexuality, and the behaviors they become linked with, will be constructed as "normal", marginal, or deviant depending upon dominant social norms of the time and place. As such, identity must be read through the context within which it is being given meaning, and, in

the process, how it relates to ideas of difference can be determined.

Identity and intersectionality: We all embody various cultural traits simultaneously. We all have some kind of gendered identity, some kind of ethnic background, some kind of class position, and so on. At any time one or other of these multiple identities we possess may be the dominant means by which we express ourselves, or through which others interpret us. Intersectionality refers to our attempts to understand how a person's social location is expressed through the intersections of these different aspects of identity, and how this changes across space and time. Many cultural geographers argue that intersectionality *multiplies* the nature of advantage/disadvantage some people face; for example, being of low income, a racial or religious minority, and an immigrant woman might create multiple overlapping forms of discrimination (just as White, male, middle class, and of the majority religion might create multiple advantages in North America). Intersectionality means that such factors do not just add up, but they multiply each other.

The critical interest in **research, knowledge production**, and representation is perhaps one of the most radical departures from earlier cultural geography, because here the very idea of how "truth" is produced and advanced is questioned. This approach has many intellectual roots, but the contribution of feminist scholars is noteworthy. During the 1980s feminist intellectuals critiqued and rejected the masculinity of knowledge production (the dominant universal assumption of Western scientific rationalism), and instead argued that no one has the ability to see everything from nowhere (the "god trick" as the social theorist Donna Haraway called it). Instead Haraway argued that we are all somehow *positioned* in relation to the knowledge we create (our social location or positionality), and, as such, the knowledge we produce can therefore only ever be partial. In light of this, academics' interpretations of the world can only reflect and prioritize their own positionality, even if this is not explicitly revealed or problematized in their analysis and argument.

In this chapter we will further examine these four key issues using various examples. These issues have all been touched on in early parts of the book, but here we explicitly engage with them in turn, and highlight their relevance to cultural geography today. The division between these four topics assists us for purposes of order, but these factors overlap and inform each other in multiple ways. We will also find numerous points of connection between the information presented here and material in earlier chapters. This chapter aims to assist you by providing an overview of these conceptual frameworks in terms of how they inform contemporary cultural geography.

Discourse and Power

Once we problematize the idea of power, it means we need to examine where it comes from, where it resides, who has power, and also what allows certain forms of power to be maintained over time and across space. ➡️➤➤**1** This later point can be effectively demonstrated through a critical examination of colonialism. Colonial power is based on domination

and dispossession of one group by another. In the process of colonial exploitation, resources of one region, often including people, are extracted in order to advance another. Colonialism is a form of imperialism, which creates cores and peripheries in terms of resource access and use. This fundamental sociospatial system of inequality has been conceptualized in the world systems approach (Chapter 6). However, the actual operation of such colonial power is complex, and its outcomes depend upon multiple factors, including the power of discourses that create and maintain a sense of inherent superiority between the colonizers and the colonized. This process does rest on violence in many cases, but it can also be constructed and maintained through effective discursive structures of difference and hierarchy that work to justify systems of colonial difference and inequality. ➡️➤➤**2**

The cultural turn has witnessed a great deal of research that examines colonialism and colonial relations. Such research must not, however, be interpreted as merely historical reflection. Rather it is important as an assessment of modern forms of power relations, since the legacies of colonialism are understood to have profound contemporary significance. Assessing these legacies demands an understanding of how various forms of colonial power came about, the nature of such power, and the continuation of its influence in terms of current comparative development (see Box 8.1).

Edward Said and Postcolonialism

The postcolonial work of Edward Said (1978, 1993) focuses on the way in which colonial powers construct the identities of the colonized to reflect a series of binary relationships that reinforce relative positions of power and powerlessness. Known as Orientalism, Said argued that Europeans constructed the concept of the Orient as a counterpoint to the West. As Shurmer-Smith and Hannam (1994:19) explain, "It is the function of the Orient to play 'Other' to the West's 'Same'." Based on this discourse, Westerners developed an image of themselves as having the right to intervene elsewhere and to use others and their places as they chose. Through this imperial

Box 8.1 Legacies of Colonialism

Economists Acemoglu, Johnson, and Robinson (2001) have argued that current global comparative development differences have their roots in the nature of colonial settlement and the resultant form of institutional governance that emerged, rather than purely natural climatic or geographic reasons. They argue the main factor that determined whether purely resource extractive or colonial settlement occurred had to do with the rate of mortality faced by initial European colonizers. For example, Belgian colonization of the Congo came to rely on the extraction of rubber and copper resources using violent methods of indigenous control. Relatively few colonists settled in the Congo due to the very high disease-linked mortality rates, and some economists argue this is why few institutions were formed beyond those that supported the violent and extreme form of colonial extraction meted out by Belgian government officials. When contrasted to colonial outposts such as Canada, New Zealand, and Australia, the authors argue that their comparatively better developmental indicators today reflect the legacy of colonial institutions that led to stronger property rights and financial markets. Such institutional foundations were arguably more apparent in British than French, Spanish, or Portuguese colonial settlements. Differences in mortality rates explain the higher number of colonials who settled in place, and thus the subsequent investments made in building effective institutions. It is the legacy of this process of institution-building, borne of initial mortality rates faced by colonizers, that Acemoglu, Johnson, and Robinson (2001) say partly explains the longer-term emergence of comparative development difference between colonial outposts. Of course such arguments have been challenged, with Bhattacharyya (2009) claiming that the level of malaria is a more powerful explanatory factor in the level of development in the case of Africa, yet he concurs that "one should not forget that the process of long-run economic development is complex and cannot be adequately captured using reduced form models" (774). While this refers to the work of economists, the point is to recognize that colonialism took on different forms in different places, and the form of power relations embedded in these colonial systems differed due to economic, cultural, and environmental factors. Cultural theorists and geographers have played an important role in theorizing the nature of such power, its spatiality, and its longer-term effects. ➡️➤➤ 3

imagining, places outside of Europe were colonized both conceptually—for example, through mapping and renaming processes—and practically, in such ways as slavery, resource extraction, colonial administration, and plantation agriculture. The processes of European imaginings of the Orient developed in three ways: a) through scholarly research and study of the Orient, b) through a focus on cultural and aesthetic elements of Oriental culture within the larger grid of power/knowledge at work at the time that was controlled by European interests, and c) through the creation of a discourse—Orientalism—that allows for both the construction and domination of the object of its study, that of the Orient (Gregory 2009). ➡️➤➤ 4 Mapping was an important tool in this process of colonial expansion (see Box 8.2).

Said traced the origins of Orientalism to Napoleon's occupation of Egypt in the late 1700s, and the process by which Egypt became known and made available to European interests with little or no input from the subjects of the region itself. This process of *knowing* was how European colonial power was able to construct the Orient as an essentialized place outside of, and fundamentally distinct from, Europe. Based on this process of objectification, the region and its population was seen as needing the kind of order only colonial imposition could deliver. Colonial powers denied their own role in constructing this image of the Orient, and in the process paved the way for colonial order to be imposed. This process allowed for the space to be seen as exotic and bizarre, and as a consequence in need of order and

Box 8.2 *Cartography as Discourse and Power*

Mapping, or cartography, is a key example of how power can be exercised and transcribed from systems of thought to material reality. The map contributes to the formation and reproduction of power in numerous ways. Brian Harley (1989), one of the most famous critical scholars of cartography, argued that "maps are too important to be left to cartographers alone." Why? Because he believed cartography belonged to the terrain of the social world in which it was produced, and he argued that scholars needed to subvert the naturalness of the images of the world displayed in maps. In its place he demanded that the *cultural* rules of cartography should be exposed. Much as other critical scholars of the time were arguing, Harley felt we needed to break the assumed link between reality and representation dominant in cartography at the time. Harley was particularly interested in colonial maps, and he identified different cultural rules of colonial cartography that reflected and reproduced dominant social, political, and economic understandings of the world at the time:

- *Ethnocentricity* is evident for all cultures, which tend to place themselves at the center of the universe as they construct it. Even in the Renaissance a new myth of European ideological centrality was evident in the maps that were produced. By placing one's own culture at the center the mapmaker gave geopolitical force and meaning to representation.
- Rules of *Social Order* are inserted into the smaller codes and spaces of cartographic transcription; for example, the location of religious or imperial buildings, the placements of royal crests and coats of arms and flags of possession. Cartographic techniques such as the width of lines and color display hierarchies of **authority** in line with dominant orders of the day. In many colonial maps the indigenous population is portrayed in subservient roles, or are absent altogether. Such cartographic techniques are a mechanism for strengthening the social rules and values of the colonizer.

- *Intertextuality* refers to other images present on the borders and edges of maps. The seemingly decorative art on European colonial maps actually acts to reinforce political and cultural power relations. Even more "scientifically" accurate maps add to the inevitability and permanency of colonial conquest and possession.
- *Absences are as important as presences.* The issue of power in the cartographic process is also evident in the relationship between the mapmaker and the person commissioning the map, even if this is not visually evident in the physical map itself. Power is both internal and external in so far as who is mapping what (internal), and who is the patron or sponsor (external). Mapmakers construct the maps while negotiating between the needs of the clients and their own cartographic skills. They may try to impress the client by demonstrating new methods of representation not requested or imagined possible by the client, or the client may demand that certain features be contained or prioritized, or even ignored depending on their own requirements (Wood 2002).
- *Naming.* As we have examined in other parts of this book, the very act of naming places creates a new colonial space, and allows for material possession of territory. Rivers and mountains are particularly important sites to name since exploration and naming create a form of material and symbolic possession once names are recorded. In effect, explorers *invented* rather than *discovered* places. Naming is a failure to represent what is empirically there, but also an actual erasure. Space is reduced to a stage for imperial history to be acted out. Current and past land claims protests in colonial settler nations such as Canada, New Zealand, and Australia demonstrate the continued importance this process of naming has in terms of colonial dispossession. As Harley argued (1992: 532), "Maps were the first step in the appropriation of territory. Such

Continued

visualizations from a distance became critical in choreographing the Colonial expansion of early modern Europe."

Harley's call for critical evaluation of maps was in many ways ahead of the cultural turn for geography, especially cartography. His arguments can be seen as prescient in terms of the transformations that occurred in the discipline through the late 1980s and 1990s. His critical research illustrates how maps formed part of the larger discourse of colonial power. Maps were a vital piece of knowledge during the age of exploration in a practical sense, but they also became powerful tools of expropriation during and before the colonial period. These powerful discourses can also become manifest in how social relations are formulated, and how social identities are constructed.

normalization. Orientalism was therefore both a precursor to, and mode of, colonial domination and control. Arguably today we live in an age of "neo-orientalism" driven by the "war on terror," and also "practical orientalism", the everyday experiences that shape relations between different religious and ethnic communities, such as can be seen in the rise of Islamophobia and the wave of anti-Islam rhetoric that has swept Europe and parts of North America (Gregory 2009).

A competing idea, **Occidentalism** (Buruma and Margalit 2004), has also been suggested. Occidentalism is the view of the Western world—especially the United States—that has come to the fore in much of the Middle East in recent years. Those viewing the West from an Occidentalist perspective see it as corrupt, arrogant, militaristic, materialistic, and secular (for example, consider episodes of the talk show *Oprah*, post-9/11, that asked, "Why do they hate us?"). Such a view comprises a set of generalizations that, although not without some factual basis, combine to paint an oversimplified and distorted picture of what is actually a complex and varied Western world. ➡➤➤ **5** One key difference, however, between this idea of Occidentalism and Orientalism is, of course, who has the power to actually impose their imaginary ideas onto the space of the other. In the case of Orientalism, colonial and postcolonial systems of power and control permit the insertion and application of European and American logic into the social, legal, political, and economic systems of "the Orient", for example, in terms of American foreign policy that involves support for Israel as well as political and military

involvement in Afghanistan and Iraq (Gregory 2004). How Occidentalism might be transformed into a system that actually affects the operation of Euro-American space in terms of daily existence and social, political, and economic systems is certainly not evident (the politics of "terror threat levels" notwithstanding, since this is complicated by the intersection of domestic security–military corporations and political systems in the West). Therefore the *practical* impact of Occidentalism's mode of imagining the West appears strikingly limited when compared to the material impact (neo)-Orientalism has demonstrably had in parts of the Middle East.

Said's research was an inspiration for postcolonial studies—a critical approach examining the cultural legacies of colonialism and their continuing impact on identity formation and political relations centered on European dominance or Eurocentrism. Postcolonialism offers a political as well as an academic attempt to resist European narratives that were constructed to justify and rationalize various forms of colonial exploitation and superiority. In the place of scholarship that normalized colonial power structures, postcolonial scholars presented research

Questions for Critical Thought

What possible modes of power (terrorism being one) do you think exist that might illustrate the power of Occidentalism to affect (materially and/or symbolically) the space and people of Europe and North America?

that highlighted how indigenous groups resisted or altered European power, creating instances of hybridity, where the colonizers were transformed by the colonized to some degree. This attempt to shift the focus away from obsession with European superiority and control can be seen as central to postcolonial approaches. One important genre of work that exemplifies this corrective to the colonizers' dominance of historical knowledge production, and thus came to be associated with postcolonial studies, is **Subaltern Studies** in India (see Box 8.3).

Identity and the Construction of Identity

Edward Said's Orientalism, Brian Harley's critical cartography, and Subaltern Studies all focus on how those in power attempt to control space and the populations inhabiting those spaces. As colonial authorities developed their methods of dispossession over the colonized, forms of social difference were often used as a means to implement and justify modes of power and control that materially disadvantaged certain groups over others. Those in power created laws and rules that subjugated groups had to learn to operate within. These rules governed what

people could do and how they should behave. If some practices were deemed outside of the "normal" rules that had been constructed by those in power, then such groups came to be seen as deviants, outsiders, and "others" to the powerful "self". This process of demarcating what is inside and outside of "normal" structures of society can be seen in numerous colonial and contemporary cases. They are important not just because of their relevance to how social inequality is formed when such rules are imposed, but also because such structures of exclusion leave legacies that contribute to generational disempowerment. These are just some reasons why identity and identity construction has become increasingly important to cultural geographers during the post–cultural turn period. As Said stated, "The construction of identity is bound up with the dispossession of power and powerlessness in each society, and is therefore anything but mere academic woolgathering" (1993:332).

Constructing Identities of Exclusion

Why is it important to understand the construction of social identities? What are the material consequences of such symbolic categorization processes? David Sibley (1995), in an influential conceptual and empirical study, attempted to explain why and show

---

Box 8.3 Subaltern Studies

Subaltern Studies emerged in the late 1970s, and focused on constructing strands of Indian history that escaped the "colonialist and bourgeois nationalist elitism" of Indian history writing at the time. Emerging during a period of great political turmoil in India (Indira Gandhi's 1975–77 period of emergency rule), Subaltern Studies appeared to break with the Indian nationalist elite who offered histories of India that reproduced readings representative of the socially and politically privileged segments of Indian society. The Subaltern Studies Collective attempted to move the focus away from elites in India toward subaltern low-caste/low-class communities, and in the process advanced the idea of peasant autonomy and agency. Such a focus on the peasant community as one possessing agency

and elements of autonomy countered colonial and elitist reading of historical development of India under colonialism, as well as during the independence phase. This approach was based on a form of Marxist scholarship that deliberately focused on peasant consciousness, but one that, at the time of its emergence, was at odds with more structuralist readings of capital and class exploitation. The scholarly focus of Subaltern Studies moved from material aspects of subaltern experiences to deeper theoretical engagement with the present significance of subaltern forms of power/knowledge. By the late 1980s, the school of Subaltern Studies had gained significant influence outside of India and was seen as part of the wider postcolonial school of thought with which Edward Said was associated (Chari 2009).

how the process of identity construction occurs by focusing on "geographies of exclusion". This concern builds on selected work in psychoanalytic theory that sees human behavior and experience as deriving at least partly from the unconscious mind. Psychoanalytic theory posits that humans have an unconscious desire to establish order in their world and purify their environment in order to minimize personal unease and discomfort. This argument is seen as helping to explain the importance of a series of dualisms between, for example, clean and unclean, tidy and untidy, private and public, inside and outside, and, most crucial for our purposes, self and other. In this process of creating positive and negative dualisms, the transference of identity to a non-human form can be used to consolidate the association of non-humanity with devaluation. Calling a certain group of people "rats" or comparing them to some other animal removes their humanity; it objectifies them and paves the way for other forms of violence to be enacted upon them (as noted in the discussion of genocide in Chapter 7). ➔➤➤**6**

Following this logic, one can examine the tendency for people to devalue, reject, marginalize, and even exclude those who are deemed "different" in any one of a number of ways, including skin color, supposed race, ethnicity, language, religion, social class, gender, sexuality, mental ability, and able-bodiedness. This desire to "purify" in line with a normalized and regulated sense of identity requires some kind of political authority and power to convert this desire into a material reality—be it through force, hegemonic acquiescence, or a mix of the two. When a desire to purify is activated, it permits and advocates conditions for intolerance and hate to flourish. To gain agreement from the masses to advance such a policy of exclusion requires tools of discursive construction to do the work of isolation and separation, and most importantly naturalize or normalize the idea that there is some kind of fundamental and irreconcilable difference that cannot be managed without resorting to a mode of purification. Take, for example, this description of the Nazi policy of Jewish genocide from the novel *Fugitive Pieces* by Anne Michaels (1996:165), which demonstrates how this process of dehumanization works:

Nazi policy was beyond racism, it was anti-matter, for Jews were not considered human. An old trick of language, used often in the course of history. Non-Aryans were never to be referred to as human, but as "figuren," "stücke"—"dolls," "wood," "merchandise," "rags." Humans were not being gassed, only "figuren," so ethics weren't being violated. No one could be faulted for burning debris, for burning rags and clutter in the dirty basement of society.

While we may think such sentiments belong to the past, the reality is that anti-Semitism is still prevalent today (see photo of May 2004 memorial vandalism).

A similar process of objectification allowed for slavery in eighteenth-century America to exist alongside the 1776 United States Declaration of Independence, which stated clearly "that all men are created equal." Objectifying slaves as not human, or as a different species or "race" from the "White man", meant that White American society could reconcile the ethics of the nation and the reality of its economic and social system (recall Chapter 7). Such examples tell us that the process of objectification is fundamental to the construction of conditions that permit, support, and justify extreme marginalization of others. ➔➤➤**7** This is a process that is constantly operating throughout society in terms of derogatory jokes and social understanding and constructions of "others" in everyday exchanges. We may think this is an extreme example, but the process of othering is constantly underway in mainstream media. Consider public and media discourse today regarding refugees and asylum seekers. Seen as "others" through the application of state regulations, refugees are objectified as a "problem" that is seen as threatening to society and the nation at large. Or consider how welfare reform always suggests that people who work hard can be successful, implying that if people are poor, it is their own fault (based on some essentialized aspect of their identity which is conveniently hived off to explain why they are not able to support themselves and their families). ➔➤➤**8** These are not historical but contemporary examples of how powerful discourses, regulatory practices, and media representations can objectify humans and elements of their identity to justify subjecting such people to a wholly different level of treatment than we would expect and demand for ourselves. This process of social distancing allows for the dehumanization of the subject.

There is a long history of the social construction of difference; we read about this in Chapter 7 in terms of race and racialization. The reasons behind it are variable; sometimes the economic and political benefits the dominant group gains from marginalizing and subjugating entire populations is clear, but in other cases the process of differentiation becomes so normalized that deep hatreds emerge as a result, and these are rationalized across society through assumption and categorization. Racial

Questions for Critical Thought

❖

Consider your own experiences of engagement with processes of social exclusion. These may appear to be only "small" events, such as bullying at school, a neighborhood dispute about residential issues, or local government debates about changes in land use for religious groups. Using these cases, can you highlight examples of identity construction aimed at subjugating, categorizing, differentiating, and marginalizing certain social groups? Did such debates raise community objections and widespread support, or were they just ignored by those not directly involved?

In May 2004 this memorial to French Jewish soldiers who died during the 1916 Battle of Verdun was defaced with swastikas and Nazi slogans. The 80-foot high monument is located in Fleury-devant-Douaumont in eastern France and is adjacent to France's principal site of homage to those killed during the First World War. Such acts of vandalism targeted at Jews and other groups are all too common today. Monuments and cemeteries are often favored by those responsible for these actions as they are important symbolic locations linking the past with the present—desecration of a memorial site is also an assault on the memory and identity of the group being attacked. (AP Photo)

discrimination, xenophobia, genocide, and Islamophobia are all examples where assumptions and stigmatization drive processes of distrust, social distancing, and possibly hatred and violence. In this section we explore three examples of excluding others, Islamophobia and veiling, the practice of redlining in the US, and the experience of Roma in Europe.

Islamophobia and Veiling

In 2010 the French government banned women wearing the burqa or niqab in public (types of veils that cover the full face with an opening for the eyes), arguably in support of France's strong state secularism. In 2011 Switzerland banned the construction of minarets by a vote of 57 to 42 per cent, and polls in London found 75 per cent would support the same policy. In Canada, Quebec's Bill 94 invokes a ban on accessing public services for women wearing the niqab. Canada is also debating a ban on face coverings at citizenship ceremonies and while accessing public services. FIFA, the international governing body of football, also banned women from playing soccer while wearing head scarfs (though this ruling has since been reversed). These policies single out women for different treatment, even though one could argue the issue is about religious expression, and therefore should be protected in most developed economies.

This "othering" of Muslim women is highly debated by scholars, who take different views regarding their concern about veils—either as a material and symbolic representation of the repression of women's rights (the veil is a patriarchal imposition that symbolizes control over women) or as a representation of women's rights to express themselves (modesty, religious devotion and practice). Critical scholars have questioned the whole focus on the Muslim "other". Whichever perspective is taken, it is clear that the veil is a powerful symbol of otherness. In these examples one can argue that WMDs (women in Muslim dress) are seen as a threat just as "weapons of mass destruction" are (Khan 2009). The WMD reference places the process of othering Muslim women within a wider geopolitical debate over issues of security, such as those explored in Huntington's clash of civilizations (explained in Chapter 6). The complex geopolitics of the Middle East and Islamic states are transferred onto these "obviously" Muslim bodies in the West—women in Muslim dress. What this highlights is how the niqab and other items of Muslim dress for women are being used as the visual repositories of multiple complex political and cultural processes. Exactly why this is the case and how it plays out will be different depending on the context. Islamophobia is differently constructed in France, Canada, and Switzerland. Similar processes of objectification and exclusion may occur, but the micropolitics and public engagement with these processes of identity construction will be informed by cultural and political norms and attitudes.

The complexity of this practice is explored by Gökariksel and Secor (2012) in their focus on "veiling fashion" in Turkey, where women exercise both religious and fashion statements by embracing veiling. To wear veils in a way that expresses fashion and style, the women interviewed by Gökariksel and Secor highlighted the importance of *Kaldirmak*, "to handle" this style of dress both in terms of style and bodily deportment (Gökariksel and Secor 2012:848). In finding a balance between the conflicting meanings of fashion and religion, Gökariksel and Secor argue that "these women form themselves as subjects of ethico-politics" (849). "Ethico-politics indicate the relationship between various elements of the self . . . and a moral code" (849). Turkey is a Muslim majority country, but veiling was initially discouraged in the early and mid-1900s, and even banned in public spaces in an effort to promote a modernizing state. By the 1980s Islamic movements gained ground in the nation, and veiling emerged as a politicized issue. Many women began to embrace the veil; in some cases this was seen as advancing particular political messages. Gökariksel and Secor focus on the veil as an ethico-political issue in the context of new forms of Islamic consumption, which mark the rising wealth of Islamic professionals who then want to express their faith through these consumption practices.

A veil can express fashion sense in how the scarf is tied, whether the brand name of the scarf is showing, how it fits with the outfit worn, if it is worn symmetrically or asymmetrically, if it is secured with or without a pin, or if a thin cotton scarf is worn under

This image represents the immense variation possible in forms of Muslim dress. The women are of different ages and portray different veiling fashions. Based on how a woman wears her veil, it might be possible to determine her class background. In this regard the veil must be seen as an "open-ended" and ambiguous signifier, in that it can be read in many different ways. Under conditions of Islamophobia, though, the bluntest of readings may likely occur. This is a clear example of how identity is framed by geography; wearing a veil will be read differently in Istanbul than in Dallas. (© Valery Voennyy, 123RF.com Photo #9205618)

the main scarf (a bonnet). As Gökariksel and Secor demonstrate, the irony of such fashion sense is that it draws the attention of others, confounding the religious point of wearing the scarf in the first place (to enhance modesty). Gökariksel and Secor argue that women have to regulate their bodies or *nefis*, their souls, or carnal desire, through their use of the veil, but they also engage in the practice of fashion and consumerism. The ambiguity of managing their *nefis* and of regulating their bodies to meet the needs of religious observance as well as making a fashion statement can be achieved by some women through the adoption of different forms of veiling in different places. Rather than being simply placed in the debate

about the veil as a political tool of Islamist movements in Turkey, Gökariksel and Secor highlight the immense complexity and ambivalence that surrounds veiling: "veiling fashion, in all its moral ambivalence, is part of the administration of an ethical project of the self that takes the body as its substance, desire as its rupture, harmony as its ethic, and Islam as its moral telos" (Gökariksel and Secor 2012:860).

Such debates about difference, in this case the veil, speak to the intersectionality of religion, gender, generation and ethnicity. To deal with the immense complexity and politicization of the veil, we can think about it dialectically. This approach relies on uncovering the contradictions that emerge between different processes, and how they are constantly implicated in the opposite view. If we place an issue into some kind of relative or contradictory situation, some social issues become partially reconciled in people's minds. For example, we may appreciate and possibly agree with the argument that women who are compelled to completely cover themselves represent some form of patriarchal imposition on female freedom, but if we take a critical look at what appears to be the opposite case—a woman's freedom to wear anything or nothing—evidence of *Western* patriarchy begins to emerge. Take, for example, the case of female sexualization. Based on the findings of the American Psychological Association (APA) Task Force on the Sexualization of Girls, sexualization occurs when

- a person's value comes only from his or her sexual appeal or behavior, to the exclusion of other characteristics;
- a person is held to a standard that equates physical attractiveness (narrowly defined) with being sexy;
- a person is sexually objectified—that is, made into a thing for others' sexual use, rather than seen as a person with the capacity for independent action and decision making; or
- sexuality is inappropriately imposed upon a person (especially relevant to children).

All four conditions need not be present; any one is an indication of sexualization. Self-motivated sexual exploration, on the other hand, is not sexualization by APA's definition, nor is age-appropriate exposure to information about sexuality.

If we accept that sexualization, especially of young children, is a form of exploitation and repression (in that it objectifies a person only through their sexual attractiveness), we begin to see how the construction of WMDs as a threat to Western society and the extreme sexualization of young girls and women within Western society can both be interpreted as a form of male exploitation aimed at controlling women. At first glance the issue of fully covered and uncovered women and girls seems contradictory, but using the idea of exploitation and control, we can begin to reconcile these differences and collapse

Questions for Critical Thought

What is your perspective on the issues of veiling and sexualization of women and girls? Does viewing these issues in relationship to each other make you think about each of them differently?

them through a process of dialectical transformation (see Figure 8.1).

Even if we do not necessarily reconcile the practices of sexualization and covering, the fact that we have complicated our perceptions of both means that we may be able to appreciate some of the arguments about the importance of female modesty and the limits of publicly displayed sexuality for Muslim women and society. This process of moving toward greater understanding, and perhaps even empathy, for others means that our ability to categorically differentiate and make judgments of others is complicated and restrained.

Redlining in the US

Another example of excluding others based on identities can be seen in the Jim Crow laws in the United States, which stipulated segregation between Blacks and Whites in terms of where people could live, work, and be educated. Once those laws were repealed, explicit, state-mandated segregation gave way to different forms of spatial segregation. Under the New Deal program of domestic economic development initiatives in 1930s USA, the Federal Housing Administration (FHA) was set up to offer federal home loans (mortgages), and part of this process required the need to determine mortgage risk for different neighborhoods. This mortgage appraisal system effectively tied risk to "race", and determined that mixed-race neighborhoods were financially more risky to lend to than White neighborhoods. The process of identifying mixed neighborhoods as higher risk in terms of mortgages became known as **redlining**, since the lenders would draw red lines around the neighborhoods they deemed to be of a higher financial risk. ➤➤➤ **9**

Figure 8.1 Thinking Dialectically about Patriarchy, the Veil, and Female Sexualization. This cartoon offers a very sophisticated problematization of how we perceive the issue of Muslim dress and women's rights. In many cultures a woman's right to dress as she pleases is usually supported, but the exploitation of the female bodily form is still pervasive in such societies. In many western societies the naked or semi-naked female form is used to sell beer, sporting events, vehicles, burgers—just about anything. Such a process of objectification carries a high price when it comes to women's rights more generally, and it has the effect of sexualizing all kinds of banal spaces, such as public washrooms and bus stops. (Malcolm Evans)

After the Second World War new suburbs emerged and families benefitted from federal funds that permitted millions to own their own homes. In light of the consequences of redlining, many developers prevented Black families from buying in these new settlements. In neighborhoods where Black Americans lived, the movement of White families to other neighborhoods created the phenomenon of "White flight". As White families and, in some cases, middle-class Black families had alternative housing options, they left inner-city neighborhoods. The potential market for such inner-city real estate was already limited in terms of race and class; add to this the difficulty such buyers would face getting an FHA mortgage, and we can understand how the value of these homes declined. Poorer families moved in and more homes became rentals. The tax base declined and so the amount of taxes each homeowner had to

pay increased and services declined, prompting more families to move out if they had the ability to do so. Their neighborhoods became increasingly isolated and homogenous in terms of socio-economic class as suburbs became the focus of increased wealth in the post–Second World War period. To generate a local tax base, polluting and undesirable industries located in the inner city, and neighborhoods were cleared to make way for freeways and other urban developments. The practice of the Federal Housing Administration created the structure for this process of inequitable wealth generation to occur. The real estate market was not necessarily racist in an explicit way, but the structure of the market, the actors, and the legislation that formed it allowed wealth generation to benefit Whites at the expense of Black families. Families who bought in the suburbs saw the value of their homes increase, while Black families who lived in the city saw their neighborhoods decline in terms of property value and service delivery. The long-term consequence of such power structures meant that the racialized nature of spatial containment and segregation of the urban landscape represented a transfer of wealth from Blacks to Whites, a transfer that has significant intergenerational consequences. A generation of White Americans experienced the benefits of inheriting their parents' wealth in the form of real estate. Even today home ownership rates in the US vary to the degree that in 2010, 71 per cent of White Americans owned their own home compared to only 45 per cent of Black Americans (US Census Bureau).

This example illustrates how the interaction between government laws (mortgage insurance) and processes of identity construction (perception of race and risk) produce the very conditions that frame the basic assumption underlying the construction of identity in the first place. Race was seen as a "normal" way to assess mortgage risk, and once the process of limiting access to credit and devaluing homes in mixed-race neighborhoods was set into play, it became a way to justify the practice in the first place. It reinforced the very conditions that permitted its creation in the first place.

The legacies of redlining and the ongoing racialization of space are still central to any focus on identity and place in urban North America. As Darden and Wyly (2010) clearly show, the subprime mortgage financial crisis of 2008 hit racialized and ethnic communities much more significantly than White Americans. Their analysis of the degree of subprime lending practices by race and location show that in some parts of the US, non-Hispanic African-Americans were much more likely to have been sold subprime loans. This has resulted in significant rates of foreclosure: "Subprime mortgage lending is the most recent market innovation that promised to transcend blatant racial inequality (race based redlining and exclusion) but it only creates a more pernicious regime of racial stratification (racialized divisions between fair lending and risky, predatory exploitation)" (Darden and Wyly 2010:432). The authors conclude that "our maps of subprime racial disparity at the height of the boom are being redrawn into a cartography of crisis—default, foreclosure, and

eviction, but also protest, legislation, and litigation" (432). Darden and Wyly draw a connection between the "Troubled Asset Relief Program" that committed almost $3 trillion of public funds to support the Wall Street banks that created the crisis in the first place, and the financial and personal misery millions of working and middle-class Americans now have to face. Such analysis is a pointed reminder of the importance of understanding the long-term consequences of power and identity construction. ➜➤➤ **10**

Roma

Certain discourses can be more blatant in how they construct some groups as "outsiders". In Europe today the group that is most profoundly discriminated against in an often open and blatant fashion is the Roma. Romani people are said to have originated in the Indian subcontinent and now reside across Central and Eastern Europe. There are about 12 million across Europe, with 6 million living in the European Union (European Commission 2012). The Roma have long experienced marginalization and violence, often existing on the fringes of society making a living through itinerant work such as collecting scrap metal. Currently in Europe the Roma continue to face discrimination in access to education in the Czech Republic and Slovakia. (Roma children are often placed in schools for the mentally challenged.) They face discrimination in Italy and France and forced evictions in Serbia, and have been forcibly returned to Kosovo (Amnesty International 2009). The long history of Roma identity as constructed as an unwanted group is explored in David Sibley's (1995:68) book *Geographies of Exclusion*. He examines the popular association of Roma (Gypsies) and dirt: "Here, the problem is Gypsies' dependence on the residues of the dominant society, scrap metal in particular, and their need to occupy marginal spaces, like derelict land in cities, in order to avoid the control agencies and retain some degree of autonomy." Places occupied by Gypsies are avoided by members of the dominant population because they are seen as threatening and sites of criminality. As Sibley (1999a) notes, a fear of the other becomes a fear of place. Further, the apparently disorganized character of Gypsy space compared, for example, to a homogeneous suburb, results in the devaluation by the dominant culture of that space, and of the people occupying that space. In the city of Kingston upon Hull, England, where Gypsies have lived for about 100 years, numerous conflicts eventually prompted the building of two permanent locations for what was widely regarded as a deviant group. One was in a heavily polluted industrial area and the other was in an old quarry that was used for dumping garbage. We see here how social identity and spatial location become conflated as efforts to resettle Roma communities reinforce the idea of marginality and refuge.

In Europe the experiences of Roma communities have become increasingly politicized as governments seek solutions to the problems of Roma marginality. In 2010 the French government announced it would round up and expel illegal Roma and destroy their encampments. Similar efforts occurred in other EU nations in their attempt to deport Roma to Romania and Bulgaria, despite reports of extreme human rights abuses to Roma in these countries. In many cases EU states imposed draconian measures even when individuals had EU citizenship, and therefore the right to freedom of mobility and to seek work across the EU. In addition, when Romania and Bulgaria joined the EU, their nationals (including Roma), unlike other EU member states, had to apply for work visas, and even then they were only allowed to work in certain sectors of the economy. The crisis that emerges in light of these measures reveals that states such as France are asserting their sovereignty to prevent certain groups from having the same rights other Europeans have. French authorities argued that their actions were against all residents who did not meet EU residency rules, but leaked interior ministry memos revealed a specific targeting of this ethnic group: "Three hundred camps or illegal settlements must be evacuated within three months; Roma camps are a priority. . . . It is down to the *préfect* [state representative] in each department to begin a systematic dismantling of the illegal camps, particularly those of the Roma" (Willsher 2010). Critics argue that the EU should do more to pressure states in central and eastern Europe to reduce the blatant discrimination Roma face there, and work harder to protect

> ## Questions for Critical Thought
>
> The Roma and other marginal "traveller" groups are portrayed in popular media in multiple ways, but often their social marginality is one of the key features emphasized. Consider films that portray such groups, and contrast how their marginality is explored differently. Examples include *Chocolat* (2000), directed by Lasse Hallström, *Snatch* (2000), directed by Guy Ritchie, and *The Way* (2010), directed by Emilio Estevez.

their human rights across the region. In 2012 when the French presidency passed from Nicolas Sarkozy to socialist François Hollande, he promised to work harder to integrate Roma into French society, but he too has resorted to bulldozing camps and referring to Roma camps as "sources of filth and crime" (Sayare 2012). The plight of Roma communities is one of the most severe tests of human rights in Europe today.

Identity and Intersectionality

Identity construction is tied up with issues of power and rights, as revealed in the previous section, but what comprises our identity? We need to think about our identity not as just one essentialized dimension, but as axes of identities that are contextualized in different spatial and social settings. We may be treated in a certain way because one or another of our identities is prioritized due to the spatial or social context we inhabit at a certain time. We may also choose to emphasize a particular dimension of our identity in different ways in different places; for example, we might dress in more "feminine" attire in certain situations, or more professionally in others. We may not even be aware of our class position until we meet someone from a very different class background, when suddenly our differences become clear in the way we speak and dress, our educational background, our parents' occupations, the income we earn, and the properties we own. If we can, we may even decide to emphasize only one element of our identity in certain situations, but suppress it in others. This is known as strategic essentialism. Before we examine the issue of the intersectionality of identity in more detail, this section explores the various axes of identity. We have already explored the issue of supposed race in some detail in the previous chapter, but below we consider class, gender, sexuality, disability/able-bodiedness, and age. Then we consider intersectionality in more detail.

Class

Arguing from a Marxist perspective, Blaut (1980) notes that classes are an appropriate variable for delimiting cultures, and that certain individuals or classes typically exert power over others, a situation drastically affecting the behavior of groups. Most of the work concerned with class focuses on conflict between classes. Harvey (1993) details a powerful example of class power relations that transcended differences of supposed race and gender by demonstrating how capital was prepared to exploit rural poverty and minimize expenses at any cost. His analysis was of a fire in a chicken-processing plant, the exit doors of which were locked, resulting in the death of twenty-five workers and seriously injuring fifty-six. Harvey used this example to argue that class relations determined the position of those people in the factory more than any other aspect of their identity.

It has been common in social science to link resistance and conflict with social class, although, not surprisingly, cultural geographers working in the Sauerian tradition have shown little interest in class as a basis for delimiting groups and describing landscapes. Rather more surprisingly perhaps, class was not a key concern in the postmodern fascination with identity and place. It might be argued that the rise of **feminism** effectively displaced the traditional Marxist concern with class in many sociological accounts of conflict. Other theorists leading the cultural turn questioned the explanatory power class possessed in past Marxian analysis. For example, Stuart Hall, through the Birmingham Centre for Contemporary Cultural Studies, examined how the formation of identity in the UK was shaped by more than just class location, and pioneered a focus on articulation of identity, media reception,

> ### Questions for Critical Thought
> ——— ✥ ———
>
> Class is clearly on the agenda as economic turmoil and change result in winners and losers. How does this development influence the power of cultural geography to explain what is happening? Is class always the most important aspect of these debates? Consider how class intersects with other axes of identity in your own region in terms of who is doing well and who is losing out in the current economic phase. How do geography, gender, immigrant status, age, and ability relate to processes of economic change?

and consumption as a means of identity formation. According to Hall (1996:5), the process of identity creation is a strategic and positional activity such that identities are "the products of the marking of differences and exclusion". Together with other theoretical developments such as postcolonial and postmodern approaches in the 1980s and beyond, class became reduced in analytical importance as gender, supposed race, sexuality, and other dimensions of subjectivity were more fully theorized. However, class as a locus of analysis has received greater attention in light of the 2008 financial crisis, together with the rising consciousness of the growing divide between the rich and the poor (see Box 8.4).

Class has always been one of the key categories of analysis for cultural geographers, even through the height of the cultural turn, but its relative importance waxed and waned, and in the 1990s and 2000s it was obscured by other elements of difference. The 2010s suggests class is now back on the agenda, but it is rarely examined in isolation; rather it is placed in relation to other axes of identity.

Gender

Inspired by feminist concepts, cultural geographers have recognized gender as an important marker of identity. Before we explore the issue of gendered identities, let us recap what we mean by "feminism". Feminism offers a theoretical approach that suggests

Box 8.4 *The Rising Importance of Class Analysis*

The financial crisis of 2008 and the response of many governments to bail out the banking sector (seen by many as the reason for the crisis in the first place) have reignited a protest movement that draws on traditional Marxist class analysis. David Harvey, a prominent Marxist geographer, presents his view of the situation in a powerful 10-minute animated YouTube video, explaining that the inherent crisis of capitalism results in a continual geographical expansion of production in order to maintain high levels of profit at the expense of the majority wage labor population (http://www.youtube.com/watch?v=qop2v_np2co).

The Occupy movement is one example of a social movement that, while it called on many protest movements, emphasized the idea of being "part of the 99 per cent" of Americans who are not the owners of major capital and do not enjoy adequate social protection in an advanced country such as the United States. Politically, the concern over such inequalities has in the past been rebuffed as raising the specter of "class warfare". To such arguments the billionaire mutual fund manager Warren Buffett responded by stating: "There's class warfare, all

right, but it's my class, the rich class, that's making war, and we're winning" (Stein 2006). Other such sentiments against the excesses of capitalism have also emerged in light of the financial crisis and the level of protection banks gained from governments, as opposed to working and middle-class individuals. One of the main aspects of a Marxist class interpretation of the current crisis is the intersection between the owners of capital and political authority. Those with access to greater resources (property, capital), are seen to have greater political influence, and thus able to use political power to their advantage. Recent Occupy protests have highlighted the role of state protection agencies (the police, the courts) in protecting the interests of the capitalist class. This protest movement is driven through socio-cultural as well as political-economy change. In New York City the Occupy movement has used the courts to charge the police with false arrest, and civil liberty groups have long demanded that New York Police Commissioner Ray Kelly resign over increased police brutality and surveillance of poor and racialized communities (Devereux 2012). Protest has taken many forms, including court action, but also street art (see photo).

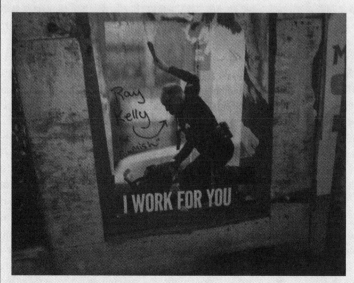

Poster in Greenwich Village, New York, February 2012. This image is publicly displayed to raise awareness of what many protestors see as increased securitization of urban spaces in North America. The image refers to New York Police Commissioner Ray Kelly, who has been criticized for his management of the police and the rising incidence of police brutality. The reference to "he works for you" is a sardonic comment on the disconnect between publicly funded services and the degree to which they provide a public good for all elements of the local population. Instead there is an inference that the police do not work for everyone, only certain (class) interests in the city. (M. Walton-Roberts)

sexual differentiation operates as an organizing principle and axis of social power. Feminism represents an academic and political movement since it challenges and seeks to dismantle all inequities, but feminism does not imply a united field of theory, political position, or perspective. Gender is not conceptually isolated from, or prioritized over, other axes of difference. There is a clear geography to gender relations. At the national scale we can see how nationalism, militarism, and development represent qualities that are indicative of gendered action, control, and authority. At the regional scale the issues of economic relations, migration control, political representation, and access to resources become sites of contestation that are again gendered in terms of who accesses these resources and whose decisions and needs are prioritized. At the local scale, home, social reproduction, and intimate partner violence evidence the shift to feminist political concerns traditionally linked to private space. Hierarchical gender relations are affected by, and reflected in, the spatial structure of society, which entails both material and symbolic relations. Spaces are given meaning through social practices, which are produced through power relations. When we begin to assess this process we see there is nothing "natural" about the gendered nature of social organization. ➔➤➤ **11** Gender relations are linked to both material and symbolic conditions. The documentary *The Gods of Our Fathers* (1994), directed by Anne Henderson and released through the National Film Board, traces the different kinds of gender relations through the evolution of social organization, from foraging to agricultural to urban to industrial systems of social organization based on different political arrangements. This exploration of social organization over time and space demonstrates that gender relations are not the result of some kind of "natural order of things", but rather are the product of historical change, political and economic control, social regulation and other processes of social ordering.

Gendered Identities

Feminist geographers have usually accepted that gender is a social construction derived largely from the natural category of biological sex, but this category need not be understood as a binary (male/female), rather it can be seen as a spectrum. Gender, as a binary identity, is formed initially through the differential treatment of girls and boys and continues to be reinforced through the life cycle. This differential treatment is accompanied by different societal expectations of the values, attitudes, and behavior of boys and girls. Feminist theorists emphasize that the differential treatment of boys and girls is a nearly universal, cross-cultural circumstance that applies regardless of place and time. Thus, gender is constructed through various social norms based on biological sex; it is not arguably caused by biology.

But it is not sufficient to note that girls and boys are raised differently and expected to be different throughout their lives. It is also clear that they are raised unequally, with boys socialized to be aggressive and to assume leadership roles, while girls are socialized to be passive and to be compliant followers. It is possible that such characteristics have some initial biological cause linked to environmental factors, but even if this is so, it is clear that the socialization process emphasizes and increases any natural differences. Overall, this process tends to work to the advantage of men.

Working with gender, as with any other basis for identity formation, is problematic because it can be taken to imply that one identity characteristic is always more important than any other characteristic. Specifically, focusing on gender alone may wrongly imply that all members of a gender are more alike than they are different. Recognizing this danger, feminist geographers have increasingly concerned themselves with bases for identity formation in addition to gender, especially ethnicity, class, and sex. In this context, Radcliffe (1994) stressed the need to consider how different femininities, and masculinities, are created in different times and places. Feminist theorists continue to debate these issues, as there is clearly a contradiction between arguing the importance of gender while at the same time acknowledging that gender is interwoven with other bases for identity. In fact it was feminist work in the area of overlapping forms of oppression in the 1960s and 1970s that led to the idea of intersectionality, in part because the feminist movement was struggling with the idea that gender would offer a common factor for examining the experiences of women living in vastly different geographical contexts. Using gender as a common lens for studying oppression experienced by White professional women and poor Black women was seen as too simplistic; such vastly different social locations had to be understood in terms of multiple contexts that framed the lives of these women, the challenges they faced, and the social hierarchies they had to contend with. In the late 1980s these challenges led feminist thinkers to explore such articulations in terms of intersections of structural inequality.

Researchers who choose to stress the similarities of women's experiences regardless of ethnic and other differences are said to be practicing strategic essentialism. ➜➤≫ 12 Although it is the constructionist view and not essentialism that frames much contemporary social thought, it is clear that such constructionism weakens the political commitment and agenda of some groups. Strategic essentialism offers a conceptual resolution to this difficulty by setting aside—perhaps only temporarily—the constructionist argument in order to benefit politically from the essentialist argument. Whether or not it is appropriate, intellectually speaking, to reject the constructionist position for pragmatic political reasons is clearly open to debate.

Feminist ideas and practice have contributed to widespread social change, especially in the Western world, where these ideas have brought about increased involvement of women throughout the workforce and greater understanding of the need for women- and family-friendly workplaces, to cite just two examples. Nevertheless, at the global scale most observers agree that the basic structures of patriarchy continue to function. As genders are defined in terms of each other, changing ideas about women necessarily imply a rethinking of men, and so there is an ongoing re-evaluation of masculinity as well (see Box 8.5).

Sexuality

Since about 1990, cultural geographers have shown much interest in studying sexuality—particularly sexualities other than heterosexuality—as an expression of identity. One obvious contemporary example is the degree to which some societies have evolved in terms of recognizing the legitimacy of **homosexuality** or the rights of same sex couples. In the early twentieth century, the dominant view of homosexuality in the Western world was negative. This negative view was evident in society generally and reinforced by both the legal and the medical establishment—indeed, the American Psychiatric Association once identified homosexuality as a sexual deviation. That formal designation was eliminated in 1973 on the grounds that same-sex relationships are just one of the ways in which sexual preference is expressed.

While heterosexuality (male/female sexual relations) is still considered by many as the only "normal" form of conjugal arrangement—known as heteronormativity—in most developed nations, restrictions against same sex unions are increasingly recognized as discriminatory. Nevertheless, in parts of Africa, the Middle East, and Asia homosexual acts are still deemed illegal.

Geographers' research on sexuality has been conducted mostly within the larger framework of feminist geography, but researchers have increasingly turned to a set of ideas labeled **queer theory**, a poststructuralist critical theory emerging from the fields of queer and women's studies. Teresa de Lauretis (1991) has been identified as the first to use the term "queer theory". Queer infers that which is not considered the norm or benchmark, but in the process it challenges absolutes or categorizations that might define a person or an action as one thing, but not another. In this regard queer theory attempts to destabilize normalized assumptions about sex, gender, sexual identity, and the **body**. Queer theorists such as de Lauretis, Judith Butler, Diana Fuss, and Adrienne Rich were all inspired by Michel Foucault's three-volume work, *The History of Sexuality*. Foucault's argument traces the historical evolution of sex and sexuality from the seventeenth century, through the Victorian age of sexual repression, and into the modern period when "the legitimate and procreative couple laid down the law" (Foucault 1978:3). Foucault also argued that the spatiality of sexuality was also increasingly limited: "A single locus of sexuality was acknowledged in social space as well as at the heart of every household, but it was a utilitarian and fertile one: the parents'

Box 8.5 Changing Understandings of Masculinity

In a discussion of the cultural politics of masculinity, Jackson (1990) identifies the rise of new versions of masculinity as a logical consequence of political and social changes associated with feminism and the rethinking of femininity. New versions of masculinity arise, Jackson explains, because masculinity, like femininity, is an identity that is constructed through gender relations. But according to Jackson, these identity changes have not disturbed the fundamental imbalance of power between men and women. Some other authors disagree, however.

According to Nathanson and Young (2001), hatred of men—"misandry"—prevails in contemporary Western society, with men increasingly patronized, even demonized. The authors argue that the increased involvement of women in traditionally male pursuits has caused men to be emasculated through loss of their traditional functions of protecting and providing for women. The financial crisis of 2008 has exposed severe changes in the landscape of work for men and women, but some argue it has hit men harder. To cope with job losses, home foreclosures, and increased bankruptcies, many young families had to resort to moving in with their parents, and men changed roles with their partners, who were able to find work. The changing role of men in the home creates new social challenges in terms of gender relations and parenting. The increasing crisis of precarious employment has in many ways undermined the social identity of men and masculinity in developed and developing

nations. What has been termed "jobless growth" in emerging economies such as India reveals a continuing trend toward flexible and casual labour, and the traditional method of forming one's identity around occupation, which has been so important to the traditional male breadwinner of the past, is becoming more and more fraught. This all highlights the fragility of masculinity as good blue-collar and middle-class jobs disappear in developed nations, and are not even forming in emerging nations. For example, Jeffery et al.'s (2008) *Degrees without Freedom* reports on extensive ethnographic work with young rural middle-class men in northern India enrolled in various education programs. These young men were seemingly embracing the notion of modernity and social mobility through their investments in education, yet it was not being transformed into good employment prospects, and economic growth in India generally was not translating into security for such young middle-class men. Without good employment prospects their ability to secure good marriages was curtailed. The work of Jeffrey et al. offers a profound critique of current economic models of development, where precarious self-employment is one of the few channels left open to men who cannot access the limited formal employment sector. The consequence this has on male identity and gender relations need to be further mapped out, but these changes speak to profound re-evaluations of masculinity and male identity.

---

bedroom." Anything that deviated from this norm was silenced by the increasingly restrictive social imposition and control of sexuality that deemed what sexual practices were appropriate and "normal".

It is now generally accepted that sexual behavior is not simply the product of some instinctual drive designed to ensure the continuity of the species; rather, sex is a multifunctional behavior that can be fully understood only in a larger cultural context. This is why a particular sexual behavior may

be unacceptable in one cultural context, and yet acceptable in another; and why cultural attitudes toward sexualities other than heterosexuality change through time. Distinguishing between what is in some sense "right" or "wrong" clearly depends upon what criteria are being used to make such judgments. Most critically, it is clear that consensus judgments about particular sexual behaviors usually reflect power relations, with most feminists arguing that the critical power relation in this context is patriarchy, which is, in turn, based on heterosexuality and the normative control function the nuclear family structure imposes in terms of relations and responsibilities from the household to the global scale.

Today, societal attitudes to same-sex couples are changing, as are relevant legal issues. Increasingly, relationships involving same-sex partners are being legally recognized as legitimate and socially acceptable. Rather than see lesbian, gay, bisexual, and transgender (LGBT) people as "abnormal", many societies have recognized such identities and relationships as deserving of equal rights to such things as marriage, adoption, immigrant sponsorship, and legislated rights to access spousal social protection—in effect to have all the same rights heterosexuals have. Some societies have used the formal legal framework to promote sexual equality, even if in some sectors of society heteronormativity actually remains the dominant social (though not legislated) norm. In such cases the official discourse has transformed (or, as President Obama's recent admission during the 2012 presidential race suggests, "evolved") over time, and this is reflective of power shifts within society more broadly. Of course the acceptance of LGBT rights varies spatially even within jurisdictional units that legislate equality (for example, between rural and metropolitan regions).

Such transformations are an evolving process, and globally LGBT communities still face immense discrimination and threats of violence. One only has to think of the violence committed against those practicing non-heterosexuality, such as that portrayed in the film *Boys Don't Cry* (1999), directed by Kimberly Peirce, in order to realize how heteronormativity can be a violent regulatory structure imposed and enforced through various informal and institutional practices. There still remains considerable opposition to the recognition of same-sex rights in some quarters, especially from right-wing conservatives, certain Protestant denominations, Catholicism, and Islam. In many parts of the world lesbians and gay men continue to experience homophobia, homosexual sex remains a criminal activity, prejudice and discrimination are routine in many workplaces, and people in same-sex relationships have reduced rights to pensions and other social protections.

One cultural geographer, Valentine (1998), chronicled a personal history of harassment and persecution that may be all too typical of the experiences of gay men and lesbians in many walks of life. More generally, with reference to geography, Binnie and Valentine (1999:175) contend that "sexuality as an object of study has become assimilated into the discipline while homophobia remains deep-seated." ➔➤➤ **13**

Rethinking Gender and Sex

As noted earlier, the conventional understanding of gender is that it is rooted in the biological distinction between female and male, with femininity and masculinity being the social constructions imposed respectively on female and male. ➔➤➤ **14** But another interpretation of gender and sex is possible, one that asserts that the two sexes, female and male, are actually part of an identity spectrum. The argument follows primarily from consideration of three circumstances:

- intersexuality, when a child is born of uncertain sex;
- transsexuality, when body and gender do not appear to match; and
- instances where a person has XXY chromosomes (men have XY and women have XX chromosomes).

The typical social and, perhaps, medical response to these circumstances is to follow appropriate procedures to make that person either biologically male or female, a form of institutional normalization rolled out using medical procedures, forms of parental or peer pressure, or other therapy used to "correct" what is read as an abnormality. With these circumstances in mind, it can be argued that

> the notion that we all fit into either a male or female body is just that, notional. It cannot be sustained, either over time or space. Simply put, there is nothing "natural" about "male" and "female" bodies. There is nothing natural about everyone being forced into one sex or the other. Rather, our belief in the existence of two, and only two, sexes

is structured by our ideas about gender. What this means is that our understanding of gender (man and woman) is not determined by sex (male and female) but that our understandings of sex itself are dictated by an understanding that man and woman should inhabit distinct and separate bodies. So, sex does not make gender; gender makes sex. (Gregson, Rose, Cream, and Laurie 1997:195)

Despite changes in the legal rights of LGBT communities, those who consider themselves transsexual are immensely disadvantaged in a range of ways. Even the most basic information forms demand categorization as male or female, but in some places this exclusionary process is being rectified. In India since 2005 passport applications allow the applicant to choose "other" as their sex as well as male/female, and in 2009 electoral law allowed transgender people to register as a third sex. Tamil Nadu, a state in southern India, recognizes a third sex category on all state application forms (APCOM:2012).

This constructionist view of sex and the body involves what may be described as a destabilizing of sexual identities. Much work in this vein proceeds to develop these ideas further by building on Butler's (1990) work regarding the idea of sexual identity as performance, or as performativity, especially when focused on the body and embodied experiences and representations; that is, marking the gender of a body through things such as how the body moves, how the body is dressed, and other bodily representations, all of which become part of the process of identity expression and interpretation (see Box 8.6).

Questions for Critical Thought

Geography influences our gender performativity. In your own case, how is your gender performed differently depending on geography—that is, is your behavior different at your parents' home, in your own dorm, in the classroom, in the gym, at the hockey rink/soccer pitch? How and why is your gender performed differently based on geography?

Able-bodiedness and Disability

Earlier accounts of gendered and sexual landscapes cite as one of the basic premises behind much cultural geographic research the claim that landscapes in general are made by and for heterosexual men; indeed, much feminist work asserts that this is intentional in order to perpetuate patriarchy. But it is clear that landscapes also "presume able-bodiedness, and by so doing, construct persons with disabilities as marginalized, oppressed, and largely invisible 'others'" (Chouinard 1997:380). In this account the term "disabled" is used in a conventional way to refer to people who are not able-bodied, and as such it includes, for example, those whose mobility, vision, or hearing is impaired. The assumption that people with disabilities are in some sense inferior to others and might need to be treated differently is known as **ableism**. Table 8.1 identifies some stereotypical perceptions about able-bodied and disabled people.

There are two ways of thinking about disability. The traditional approach focuses on medical issues, arguing that disabilities are specific and individual departures from normal bodily and/or mental conditions. It is this perspective that is used, implicitly or explicitly, when people with disabilities are defined as different and inadequate others. A more recent approach focuses on social issues and sees disability not in terms of individual medical circumstances but rather in terms of society's response to those who are "differently abled". From this perspective disability is understood as those disadvantages and restrictions on activity that are imposed by dominant society on those who have physical and/or medical impairments. One problem with this social perspective is that there is a danger of trivializing the very real problems that some people with disabilities face in their everyday lives.

Built environments have not typically taken into consideration the problems faced by people with disabilities and might be appropriately described as hostile, unwelcoming, and exclusionary landscapes. At least implicitly, the assumption made by those responsible for the built environment is that it is the responsibility of those with disabilities to find ways

Box 8.6 *The Body*

Cultural geographic studies of identities and of human relationships with places traditionally incorporated accounts of bodily characteristics, such as skin color, but these studies were not framed in terms of bodies themselves. The current cultural geographic interest in the body was stimulated especially by Bourdieu, Foucault, and Butler.

For Bourdieu (1994), social class and the body are inextricably related. Thus, compared to poor people, wealthy people are able to display their bodies in more varied and exotic ways, especially through choice of clothing and ornamentation, and they are able to engage in a greater variety of bodily activities. Class distinctions are also evident in bodily mannerisms (how you eat, greet people, and "carry yourself") and your communication skills.

Foucault (1978) focuses on ways in which the body is produced through prevailing discourses, noting, for example, that attitudes to different sexual preferences change through time and are different in different places. Foucault (1977) also stresses that the body is subjected to discipline, a set of rules of conduct that carry different expectations for the body in different times, places, and social contexts. Consider, for example, what are seen as acceptable body displays in different workplaces, or in a typical workplace as compared to a vacation setting.

Butler (for example, 1993) discusses the body in relation to the concept of performativity. This is a particularly influential idea in much feminist geography focusing on the production of gendered and sexual identities. From this perspective, these identities are not only contested and unstable but they are also performative, meaning that rather than determining our behaviors, these identities (or other people's readings of them) are formed by our behaviors, by our bodily performances that repeatedly communicate who we are. Put another way, gendered and sexual identities are what we do, and not what we are. In this sense, the body is an act, a style, and the repetitive performances of everyday behavior construct and consolidate certain identities. ➡➡ **15**

How does this idea play out in practice? There are at least two possibilities. First, performances are affected by the dominant discourse, specifically by the understandings of how different gendered and sexual identities ought to behave. For example, McDowell (1995) describes the workplaces of female merchant bankers in London as being defined by a dominant form of male masculinity that affects the performances of those who do not conform to that dominant identity; specifically, female merchant bankers use their bodies in gendered performances, playing out a role that is expected of them, but sometimes in strategic ways that can advance their own interests. The second possibility is that performances are freely chosen by the individual and reflect a form of agency. Thus, a person might wake up each morning, make a decision on what their gendered or sexual identity will be for that day, and then behave—perform— accordingly. Regardless of the specific interpretation, gender and sexuality are not only constructed identities—they are fluid and negotiable, but they can also be policed informally and formally. The discussion about veiling fashion indicates how these processes of agency, performance, and external constraint intersect.

As noted earlier in this chapter, although much cultural geography has centered on relationships between humans and environment, and between humans and other humans, it is only recently that some of the work has been situated in the context of the body. Much contemporary cultural geography understands bodies in at least two fundamental ways. First, the body is each individual's personal space and contributes to identity construction in that it incorporates characteristics such as sex, physical appearance (including skin color), physical ability, and mental ability. Second, the body functions as a boundary between self and other, contributing to circumstances of social inclusion or exclusion and to related struggles and contestations (Valentine 2001:15).

Table 8.1 Constructing Disabled Identities

Able-bodied	Disabled
Normal	Abnormal
Good	Bad
Clean	Unclean
Fit	Unfit
Able	Unable
Independent	Dependent

Source: R. Imrie, *Disability and the City: International Perspectives* (London: Paul Chapman, 1996):37. (Reprinted by permission of Sage Publications Ltd.)

to overcome their disabilities rather than the responsibility of dominant societies to accommodate people with disabilities. This attitude no longer prevails in many parts of the world and, increasingly, both public and private spaces are being modified. The most obvious example is the construction of ramps to facilitate wheelchair access to buildings. But, of course, people with disabilities continue both to be treated differently and to have different experiences in many places. Table 8.2 highlights one example of intersectionality, in the overlapping of two forms of structural oppression, the fears of a sexual attack experienced by those women who have a physical disability that limits mobility. →≫ 16

Age

Cultural geographers concerned with the identities and behavior of people in particular age categories do not have any meaningful body of conceptual work to turn to, nor have they developed relevant concepts. The contrast with cultural geographies of ethnicity, gender, and sexuality is clear. Most of the relevant research has uncritically accepted and applied the traditional psychological and sociological distinctions that identify such categories as children and elderly people. This focus on groups positioned at the beginning and end of the life cycle reflects the interest in studying those who are understood to be on the margins and perhaps excluded from larger society as a consequence. Although there are obvious and fundamental physical differences between age categories, most research on groups identified by age focuses on the social construction of identities, arguing that prevailing stereotypes and assumptions about children and elderly people reflect **ageism**. →≫ 17

The Elderly

The idea of an elderly identity is gaining attention from cultural geographers, especially as a consequence of the demographic changes of falling birth rates and increased life expectancy that are resulting in rapidly increasing numbers of elderly people throughout most parts of the world. Increasing numbers mean that elderly people are collectively playing

Table 8.2 Effects of Fear of a Sexual Attack on Women with and without a Disability

Always or sometimes do the following because of fear of sexual attack	% of respondents with a physical disability	% of respondents with no physical disability
do not answer the door	45.5	35.0
put off routine calls	23.8	10.6
feel unsafe with strangers	59.1	61.8
feel unsafe with people I know	18.2	4.8
do not go out	27.3	8.3
do not go out alone	59.1	33.7

This table, presenting results of questionnaire surveys that focused on fear of violent crime in parts of Edinburgh, Scotland, indicates some of the effects of fear of a sexual attack as these relate to whether or not women have a walking or other physical disability that limits their mobility. Source: Adapted from R.H. Pain, "Social Geographies of Women's Fear of Crime", *Transactions of the Institute of British Geographers* NS 22 (1997):241.

more active roles in society and in politics—in many countries elderly people constitute a significant special interest group.

The perception that the elderly represent a problem group is usual in Western societies, where they are often viewed in terms of their financial and physical dependence on the larger population, or their increasing social and political influence "grey power". In many Asian cultures, however, the elderly enjoy positions of greater hierarchy within the extended family, and care for elder family members is a traditional responsibility of children, often sons. Indeed in South Asian cultures the son's own status is read through his treatment of his elderly parents. Instances of ageism, however, are prevalent in Western societies, especially in the media, which tend to portray the elderly as immobile, incapable of coping, helpless, and vulnerable. Certainly all humans experience some loss of physical well-being as they age, and some elderly people are supported by the state, but the typically negative perception of the elderly—for example, the presumed association of aging with a dramatic loss of physical ability—is often inappropriate and merely serves to reinforce misunderstandings. The fact is that, as with other identity categories, there is not one elderly identity, but rather many and fluid identities. Indeed, age is a relative term, one that can be understood in chronological, physiological, and social terms. Aging means different things for different individuals and is interpreted differently in different social contexts. →≫ 18

Geographers have focused attention on the spatial behavior of elderly people as it is enabled and constrained by personal and social circumstances, including available funds, degree of mobility, and contacts with family members and peers. These accounts have paid particular attention to the geography of fear that elderly people may experience both inside and outside the home. Most of these studies also take into account gender, ethnicity, and other relevant characteristics. Again the intersection of age and gender can be seen to play an important role in the perception of fear, as explored in Table 8.3, below. →≫ 19

Traditionally, elderly people lived with other family members in extended family units, but increasingly in the Western world the elderly live as couples, alone or in sheltered accommodation. One spatial stereotype about the elderly is that they "gradually become prisoners of space as physiological deterioration and environmental constraints necessitate physical, social, psychological and, by implication, spatial withdrawal" (Rowles 1978:xv). A presumed consequence of imprisonment is the tendency to remain in an old and possibly inappropriate house and neighborhood. This stereotype is rooted in fact, as many elderly people are constrained in their residential mobility for essentially negative reasons, such as financial limitations. But, the reason that many elderly people have for not moving may be related to a more positive consideration, namely love of place, or topophilia, as discussed by Tuan (1974).

Table 8.3 Women's Fear in Public and Private Spaces as Related to Age

	Age in Years			
% very worried or fairly worried about each incident	18–30	31–45	46–60	60+
Sexual assault outside by a stranger	83.2	71.9	53.5	39.2
Physical assault outside by a stranger	65.1	63.8	48.0	44.4
Sexual assault in your home by someone you know	28.0	28.0	18.4	21.7
Physical assault in your home by someone you know	23.7	29.5	18.2	20.1

This table, presenting results of questionnaire surveys that focused on fear of violent crime in parts of Edinburgh, Scotland, shows relationships with age, gender, and type of space.
Source: Adapted from R.H. Pain, "Social Geographies of Women's Fear of Crime", *Transactions of the Institute of British Geographers* NS 22 (1997):240.

Intersectionality: Putting It All Together

Having considered some of the different axes of identity, it is clear that understanding how people experience their surroundings is dependent upon the degree of structural inequality they face as a result of the social hierarchies in place, and their position within them. The idea of intersectionality is not that one aspect of these identities is added to another, but that the types of inequality (or advantage) faced by a person are multiplied. Each of the above sections has used one aspect of identity as a "way in" to discuss issues of social location and related forms of inequality (or privilege) associated with that form of social identity. It was clear, however, even from that discussion that other factors would intersect or articulate to certain degrees, and the outcome of such interactions are important to understand. This is an expression of intersectionality. Two examples of intersectionality described below expand the components of identity to location and the state.

Intersectionality and Geographical Context: Migrant Labor in the Gulf

A job vacancies poster (see photo at right) was posted in a shop window in the city of Dubai, in the United Arab Emirates. It asks for a "Filipina lady" for a cleaning position. There is a gender requirement: a "lady" is needed. There is a national requirement: someone from the Philippines. Since this job ad was posted in Dubai, we know that person will be a migrant worker. The UAE is one of the Gulf Cooperation nations, and this region as a whole depends upon millions of migrant workers to fill service and construction jobs for the wealthy domestic population, who tend to benefit from state oil revenues. The region's labor market is highly segmented in terms of the jobs migrant workers can access, the rights they can claim in terms of status and state support, and the income and respect they earn. In this case the location of a "Filipina lady" requested for a low-skilled cleaning job in Dubai represents a form of intersectionality that speaks to supposed race, gender, nationality, and citizenship status. The labor of cleaning is also reflective of a form of segmented labor market that is so finely determined in

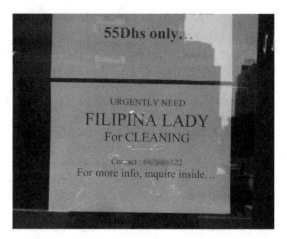

A sign posted in a store window in Dubai, UAE, November 2008. This job vacancy advertisement seems banal in the context of Dubai's labor market, but it represents a form of everyday labor market segmentation based on gendered and racialized expectations. These include what the pay will be, what the employer can expect the employee to do as part of her job, and other conditions of employment. In some cases these expectations do not even have to be contractually agreed upon; they are just assumed. Again, this is an example of social and spatial intersectionality. (M. Walton-Roberts)

this particular labor market that one does not merely advertise for a cleaner but indicates the intersectionalities as a matter of normal expectation. ➜➤➤ **20**

The poster is completely explicit about the processes of identity at work in this labor market, yet such practices of racial and gender segmentation are completely normalized. The cultural identity of the employee is pre-supposed; indeed that is the desired outcome of the job ad placement. The hiring of the "Filipina lady" will satisfy what this employer needs since the assumption is that a certain pay structure, a certain expectation of what work needs to be done, number of hours, and level of work are all built into the "Filipina lady for cleaning" job wanted sign.

Intersectionality and the State: Super-Diversity

There are other examples of intersectionality that include forms of identity that are explicitly imposed by the state, including another dimension of identity construction that we need to recognize. For example, the proliferation of different categories of visas immigrant-receiving nations such as Canada and the UK administer. This has led to the term "super-diversity" (Vertovec 2007).

Super-diversity is the term Vertovec (2007) uses to describe the interplay of multiple variables that go beyond ethnicity. Super-diversity presents another dimension of intersectionality, which demands that academics be more in tune with the specifics of diversity and the intersections between different forms of identity—ethnic, class, or gender identity, for example—but in combination with forms of identity imposed by state-defined categorization of legal status. These multiple variables inform the nature of demographic change occurring in migrant-receiving nations such as the UK and Canada. They include the following:

- Differential immigration statuses and their related entitlements and restrictions of rights depending upon the class of migrant visa the individual enters under. For example, different classes of immigrants have different rights to access routes to permanency.
- Divergent labor market experiences, depending upon their category of entry. Some workers can move between jobs, whereas some workers may have to remain with one employer or in one sector.
- Discrete gender and age profiles. Each of the different classes of visas attracts different types of migrants. For example, international students tend to be younger and seasonal agricultural workers tend to be male, whereas live-in care-givers tend to be female.
- Different patterns of spatial distribution in terms of settlement (that is, smaller rural communities host temporary agricultural workers, larger cities require higher-skilled, high-tech, and creative class workers).
- Mixed local-area responses by service providers and residents to these immigrant groups depending upon their social and economic location, as well as the nature of rights they have been awarded based on their "class" of visa. For example, seasonal migrant workers do not have access to the same kinds of services permanent skilled workers do (Vertovec 2007:1025).

While historical evidence exists of such cosmopolitan processes of mixing occurring throughout history, Vertovec maintains that "it is the emergence of their scale, historical and policy produced multiple configuration and mutual conditioning that now calls for conceptual distinction. 'Super-diversity' is proposed as a summary term" (1026). How relevant is this view in the Canadian case?

Over the last ten years in Canada there has been a trend toward the kind of super-diversity Vertovec highlights. In particular we have seen an increase in the number of international students and temporary foreign workers who enter Canada (Figure 8.2).

Rapidly changing landscapes of immigration and ethno-cultural diversity are evident in many Western nations today, and Canada's long-held position as a leader in terms of building the immigrant nation through permanent migration has been eclipsed lately in light of the growing role of temporary migrant labor. The last several years have seen substantial selection flexibility emerge in Canada's immigration policy. These recent developments include various restrictions to the permanent skilled worker program, the expansion of various programs that grant some selection control to the provinces and territories, significant expansion of Temporary Foreign Worker (TFW) programs, and new policy routes to allow the conversion of some temporary workers to permanent residents (Alboim and the Maytree Foundation 2009). Rather than a long-term demographic nation-building exercise based on incorporating all migrants as permanent citizens, Canadian immigration policy increasingly works as a tool to capture selected human capital under temporary and permanent migration streams (Fudge and Macphail 2009). The most substantial change has been a massive increase in temporary foreign worker streams. These are migrants who are selected for specific employers and jobs, and who are granted a work visa for a limited time period, and in many cases only for a specific employer in a specific region of the country.

These two examples of intersectionality both illustrate its value to analytical assessment of current understandings of changing cultural identity, and illustrate how different dimensions (in this case geographical location and state policy) add additional

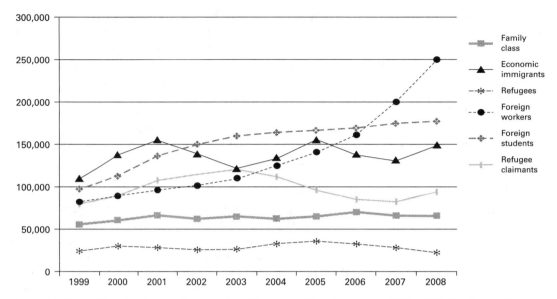

Figure 8.2 Super-Diversity: Canada Permanent and Temporary Class Immigrants, 1999–2008. This figure offers a graphic representation of the changing nature of migrant flows to Canada. The emergence of this pattern of temporary migration is driven by both employer demand and government policy response, which makes the process of hiring temporary workers easier. The long-term outcome of these migration streams is variable, since some higher-skilled workers and students are given an opportunity to convert to permanent status, but many will have to return home after their contract period is over. In this regard the image reflects policy directions in line with Vertovec's observations of "super-diversity". Source: CIC Facts and Figures 2008.

dimensions of identity to current hierarchies of social power and rights.

It is no longer possible to assess cultural processes based on essentialized singular categories; we cannot make assumptions about someone's experiences using only one dimension of their identity (such as, male or female). How we conduct research on intersectionality will differ depending upon our research questions and goals, and our underlying philosophy. When we work with intersectionality the range of permutations we might have to consider becomes extensive, especially once we consider the multiple dimensions of identity that might be relevant in specific contexts. And bear in mind that the more in-depth a study becomes, the more significant seemingly minor differences become to the analysis of social relations and hierarchies. There are, however, some options in terms of how we research intersectionality.

McCall (2005) suggests there are three approaches to intersectionality that can be under-

stood as part of a spectrum. They range from what she terms "anticategorical", where all types of categorizations are deconstructed, to "intracategorical", where a category or particular group is explored in relation to a relatively underexamined dimension of identity, to "intercategorical", where certain categories are strategically adopted in order to document and illustrate specific inequalities. Her framework is demonstrated in Table 8.4, illustrating the directionality of the methods in terms of the degree of critical deconstructivist (all categories are deconstructed), to a critical realist position, where some categories are retained and given analytical power in light of their strategic value for the goals of the research.

This discussion of approaches to the underlying philosophy and methodology of research on intersectionality leads us into the final section of this chapter on research, knowledge production, and representation.

Table 8.4 Approaches to Intersectionality Research

Name	Methodology	Limitations	Spectrum of analysis
anticategorical complexity	Critically deconstruct all analytical categories.	Difficult to contrast or compare experiences.	
intracategorical complexity	Focus on particular social groups at neglected points of intersection.	Becomes a form of case study: more extensive comparison to any other group may be limited, or unbalanced in terms of analysis.	From deconstructivist ↕
intercategorical complexity (or categorical approach)	Adopt existing analytical categories strategically to document relationships of inequality between groups.	Can this approach capture all the complexity of intersectionality, while also relying on discrete categories?	to critical realist

Source: McCall 2005.

The Practice of Research, Knowledge Production, and Representation

Cultural geographers employ a range of qualitative research methods (research tools) and methodological justifications (philosophical or theoretical underpinnings) in their research. Cultural geographers tend to produce qualitative research, which does not offer a descriptive measurement and prediction of phenomena, as statistical description/analysis does, but is more attuned to examining subjective understandings and the interpretation of meaning (hermeneutics).

Since the 1980s in particular, feminists, many inspired by the work of Donna Haraway, sought to not only expose and critique how masculinity shapes research activities and truth regimes but also embrace the idea of partial knowledge and the situatedness of their own knowledge claims. Many feminist geographers also explicitly identify the need to represent the interests of marginal groups in society and advance a social justice agenda that takes seriously the need to move economies and societies toward greater equity.

The idea of situated knowledge has been particularly influential in the work of feminist scholars, but also cultural geographers more broadly. Situated knowledge is linked to Haraway's critical feminist objectivity, an epistemology attuned to social relations and contextual underpinnings of all knowledge production. This approach to research understands that everybody experiences the world from their own particular social location (which is framed by the intersectionality we have discussed above). Becoming aware of our own social location engenders sensitivity to forms of epistemic violence, or the "god trick". The god trick refers to the idea that we can imagine ourselves as outside of our social location, as if we are observing the world from some privileged position outside of the social hierarchies that define our world—the idea that some researchers (traditionally the most privileged, White, male, upper class) have the pleasure of seeing everything from nowhere. Instead of this artificially detached location, Haraway maintains that researchers should recognize and critically analyze their own positionality, and this might inform their perspective on the world they are researching. Such critical self-engagement is termed "reflexivity". It refers to critical self-reflection: "how do I see the world differently from other people and why is that?" Such an effort to locate ourselves in relation to others, especially those who face greater structural inequalities than ourselves, offers us a way to situate our knowledge and attempt to overcome power imbalances that exist in the (research) encounter, but that may not be obvious to us. ➔»» 21

Such efforts to undermine the epistemic violence wrought by academic research in the past can also be defined as a way to see oneself through the "gaze" of others. By the "gaze" feminist scholars are referring to a unidirectional process of observation, where power resides with the owner of the gaze, not the object it is being directed at. If we become conscious of the gaze, we can try to reverse it, and make ourselves, or other researchers, the object of the gaze; this reverses the origin of power. The gaze can be seen as the masterful gaze of detached authority (an example of the god trick). It is a process of converting things into objects of consumption through the vision of a privileged subject. The process is unidirectional (power relations), but can also be ambivalent. The gaze is a masculinist form of knowledge/power (Haraway 1988).

Within this attempt to understand how knowledge is produced, the issue of representation has been profoundly examined. This is a complex form of critical reflection as well as an acknowledgment of the problems of colonial research of the past. The **crisis of representation** emerged in terms of critically questioning the right of mostly Western academics to appropriate the voices of often-subjugated communities. While this crisis of representation emerged initially in the field of anthropology—where researchers in the past did abuse their power and exploit the "objects" of their research—it spread to other disciplines in the social sciences. Instead of research built on "colonial" forms of power domination and non-reciprocal data extraction, research is now subjected to greater regulations in terms of codes of research ethics. Methods of engagement such as community-based research have become more widespread, and "giving back" to the "subjects", not "objects", of research is an approach that is being advanced in many disciplines. For cultural geographers this debate has changed the nature of the research we engage in. For example, there is more consideration of the appropriateness of working with different ethnic groups or marginalized communities. Questions about how research might help the communities that are studied have become legitimate issues of concern. In this regard, "research" has changed in that it is not only what is produced,

as in an end *product* (such as truth or knowledge) that matters, but also the *process* of research (such as questions of social justice, community collaboration, inclusion, and empowerment) (Pratt 1992). These are profound changes that are still contested and questioned across human geography, and are not necessarily the dominant practice across all social science disciplines. Even within cultural geography, these practices occur more along a spectrum than as binary opposites.

The reasons for the critical attention to research practices emerges, in part, from a rather long and shameful history of how social scientists incorporated marginalized communities into their research programs. Some examples of this in anthropology have been captured in the film *Secrets of the Tribe* (see Box 8.7).

In Canada the politics of representation and misrepresentation of First Nations communities by academics has a particular resonance. Past cases of communities not being properly consulted about research projects, and benefits not being shared with the communities, has resulted in a revised approach to research ethics.

Researching Elites

The identities selected for study by cultural geographers and discussed in this book have three general features in common: first, they are identities that are often discriminated against by the majority; second, they are usually associated with a political agenda; third, they are often identified with by the researcher (it is not unusual for researchers to explicitly state their insider status). As a result, it is possible that cultural geographers, in their desire and willingness to focus on particular disadvantaged others, neglect those others who tend to discriminate rather than be discriminated against and who either do not have a political agenda or whose political agenda is less palatable to researchers.

Cultural geographers have engaged in "researching up", that is, looking at groups that are considered privileged. For example, Linda McDowell (1995) looked at issues of gender in the case of merchant bankers in London. Tickell and Peck (1996) focused on government planners and urban developers in Manchester, England. Also, many researchers have researched urban gentrifiers, those fairly privileged residents who move into depressed inner-city neighborhoods and begin the process of urban transformation. These subjects are relatively privileged in relation to their neighbors, and much work has focused on this section of the population. ➔➤➤ **22**

While there is some research engagement with economically elite groups, there are some areas

---

Box 8.7 The Politics and Power of Representation

The film *Secrets of the Tribe* (2010), by José Padilha, focuses on the Yanomamö Indians, an untouched tribal group in the Amazon basin who were the object of anthropological study in the 1960s and 1970s. Anthropological research on this group resulted in contradictory conclusions about the social organization and practices of the group. Napoleon Chagnon, a University of California anthropologist, contended that the Yanomamö were an extremely violent warlike society, but others argue that it was Chagnon's own role in bringing in outside goods and distributing them among the Yanomamö that created the conflict within the community he commented on. These disputes led to academic conflicts and debates, but later interrogation of what anthropologists were actually doing to the Yanomamö Indians highlights the colonial and exploitative nature of academic research during this time. Medical experiments with measles caused harm to the Yanomamö, and in some cases researchers were accused of sexual abuse and engaging in sex with children. The film presents these examples of the academic engagement with primitive groups in a way that clearly highlights the abuse that processes of research can enact upon others. It is episodes such as these that contributed to the crisis of representation that shook the anthropological world, but the ramifications were also felt within geography, especially cultural geography.

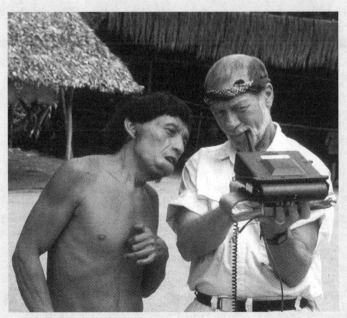

Napoleon A. Chagnon, once called the "most controversial anthropologist" in the United States, is here shown interacting with members of the Yanomamö, a society of indigenous tribal Amazonians living in the border region between Brazil and Venezuela. His research extended over several years from the 1960s to the 1990s, and his book *Yanomamö* is now in its sixth edition. The controversy over Chagnon and other anthropologists who have worked among the Yanomamö is still very active, and gained prominence most recently in 2005, when the American Anthropological Association rescinded their highly critical 2001 task force report on Chagnon's work. The subject is sure to retain importance and speaks to ongoing concerns regarding research, representation, and knowledge production. (Photo by N. Chagnon)

where relatively less research has been conducted. For example, conservative groups, including some religious cults and certain North American militia groups, are less likely to be studied because their different identities are not in sympathy with the identities of cultural geographers attracted to this type of work, which typically involves insider research. Part of the issue with these groups might rest with the difficulties of access to the community. There is also an increasingly bureaucratized research ethics process for social scientists that is modeled on the biomedical field, thus imposing increasingly restrictive requirements on qualitative researchers (Martin 2007). One aspect of this increased oversight is a negative view of deception, something that would be necessary to get those with racist, sexist, or xenophobic views, for example, to share them with researchers. There are some examples where deception has been used in the past; for example, with police offices in the UK (Fyfe), or with employers in order to assess levels of discrimination in hiring practices (Arai and Thoursie, 2009). But conducting research on the practices of those who are in relative positions of power, and who generate or enforce the frameworks of inequality, is not easy. Certainly, it can be argued that, largely for ideological or social justice reasons, cultural geographers have chosen to focus on disadvantaged groups. Overall, the interest shown in difference might be considered selective rather than all-encompassing.

Concluding Comments

- Much of the content of this chapter reflects the cultural turn by focusing on four themes:
  › discourse and power,
  › identity and the construction of identity,
  › intersectionality and the practice of research, and
  › knowledge production and representation.
- It is clear from reading this that the cultural turn represents a significant break from the traditional Sauerian view. Key components of the cultural turn include a rejection of all theories that purport to offer some single correct answer and an appreciation of human differences. In its

place researchers are more conscious of issues of positionality, reflexivity, and the politics of research.

- The emphasis on identity that is at the heart of this chapter is a key component of contemporary cultural geography, but it is not without criticism. In addition to those cultural geographers who prefer to conduct research along more established Sauerian lines, perhaps accepting particular identity formations and focusing on questions of landscape change and regionalization, contemporary social geographers have raised concerns about the focus on identity. Thus, Gregson (1995:139), raising concerns about the cultural turn, worries specifically about the preoccupation with matters of meaning and identity and the related lessening of interest in the facts of social inequality. Certainly, there is a possibility that some accounts of identity that are intent on identifying differences at the micro level lose sight of more macro structural issues. This has been rectified to some extent in light of the financial crisis of 2008, which has precipitated a renewed focus on class inequality, the contradictions inherent in capitalism, and neoliberal forms of social policy and organization. These are arguably becoming of more interest to cultural and human geographers more broadly.

- Discussions of human identity necessarily raise questions about how people associate with each other and about how they construct and reconstruct their identities in opposition to others. As we have seen, answers are based on the idea that identities are not given to people at birth. Rather, they are generated by people for themselves, or they are constructed by others and imposed using various systems of power and control. It is increasingly popular for cultural geographers to study human identities as the people involved understand them rather than as they are understood and assigned by the researcher. But in the application of what we term "intersectionality", researchers can and do engage with categories of identity along a spectrum from complete anticategorical deconstruction, to strategic analytical adoption of certain categories in order to expose how social hierarchies create inequality.

- Related to the issue of identity are the links between identity and landscape. The humanist interpretation of place stresses that place is not so much a location, a thing, but rather a setting for human behavior, a relationship. From a humanist perspective, the meaning of a place cannot be understood without awareness of the identity or identities of those who occupy the place, but there are also more recent interpretations of place that are conceived in terms of the way places are controlled by those in authority and challenged by others. This connection between place and identity can be considered as recursive relationship, and it is this issue we turn to in the next chapter.

Further Reading ➡➤»

The following are useful sources for further reading on specific issues.

1. Barnes and Duncan's (1992) *Writing Worlds: Discourse, Text and Metaphor in the Representation of Landscape* provides an edited collection that explores these different forms of representation across various contexts. Drawing on the work of Foucault, the editors argue that discourses derive their power from the material institutions they are embedded within, legitimizing and normalizing discourses, thus bestowing them with power. Michel Foucault is the leading theorist whose writings have been profoundly influential in this area.

2. Anne McClintock (1995) provides an important analysis of colonial power relations through the lens of gender, race, and class using discourse analysis and other social theories that interrogate structures of colonial power and resistance. Michael Taussig (1987) examines the extreme violence of colonialism using the case of Colombian rubber plantations.

3. There are good accounts of postcolonialism by Blunt and McEwan (2002), Shurmer-Smith (2002, Chapter 7), Clayton (2003), and Nash (2004).

4. Said's (1978, 1993) original works have inspired much cultural geographic work on colonialism and orientalism. Kay Anderson's (1991) work on Vancouver's Chinatown is one important example.

5. Harris (2004) discusses reasons for the rise of anti-Americanism in the early twenty-first century. Dalby (1991) discusses the concept of otherness as it applies in a political context.

6. In addition to the seminal work on geographies of exclusion by Sibley (1995), see also Hodge (1996), Wilton (1998), and Sibley (1999a).

7. Keith and Pile (1993), Pile and Thrift (1995:13–51), Pile (1996), Longhurst (1997), Sibley (1999b), and Kingsbury (2004) discuss selected recent advances in psychoanalytic theory that offer insights into the relations between the social and the individual; there is particular focus on object relations theory as it facilitates accounts of geographies of difference.

8. Hodge (1996) identifies the construction of "otherness" in the western part of Sydney, Australia, an area of neglect, disadvantage, unemployment, lack of services, and high levels of criminal activity, while Dear and Wolch (1987:8–27) discuss the "social construction of the service-dependent ghetto".

9. The PBS program *Race: The Power of an Illusion* provides an excellent material on redlining and White flight. http://www.pbs.org/race/000_General/000_00-Home.htm.

10. Dunn and McDonald (2001) describe geographies of racism in New South Wales, Australia.

Bonnett (2000) discusses the implications of taking whiteness for granted.

11. There are several social geography textbooks that include substantive content on gender and sexuality, including Blunt and Wills (2000), Holloway and Hubbard (2001), Pain et al. (2001), Valentine (2001), and Panelli (2004). Bondi and Davidson (2003) discuss gender in cultural geography.

12. Bell and Valentine (1995:23) and Gregson et al. (1997:74) discuss the merits of strategic essentialism.

13. Bell and Valentine (1999) and Phillips (2004) review geographies of sexuality. Boone (1996) and Binnie (1997) discuss geography and queer theory; see also Pain et al. (2001, Chapter 6). Brown and Knopp (2003) discuss queer cultural geographies.

14. For the argument that sex is socially constructed, see Cream (1995). Hubbard (2000) describes the way that heterosexuality has created landscapes of oppression in Western cities. Sumartojo (2004) analyzes contested landscapes with reference to gay and lesbian hate crimes.

15. Accounts of the body and performativity by cultural geographers include Valentine (1993), Crang (1994), McDowell (1995; 1999, Chapter 2), Longhurst (1997), Callard (1998), Teather (1999), Gregson and Rose (2000), Nash (2000), Hubbard, Kitchin, Bartley, and Fuller (2002), Moss and Dyck (2003), and Landzelius (2004).

16. Kitchin (1998) and Gleeson (1999) discuss geographies of disability, and Golledge (1993) describes the experiences of people with disabilities in landscape. The edited work by Butler and Parr (1999) includes studies of disability in varied geographic contexts.

17. Matthews and Limb (1999) observe that cultural geographers have typically ignored children as one of the identities to be recognized and analyzed. Holloway and Valentine (2002) and Gagen (2004) provide reviews. Two comprehensive reviews of the geographies of youth cultures are the book edited by Skelton and Valentine (1998) and that by Aitken (2001). Leary (1999) provides an Australian example of spaces created by youths. Pain et al. (2001) provide an overview of children and the elderly in the context of social geography.

18. McPherson (1998) provides a detailed sociological perspective on the social construction of the elderly. Rowles (1978) provides a classic humanist understanding of elderly people with the key idea that they are prisoners of space.

19. Valentine (1989) and Pain (1999) provide analyses linking gender and landscapes of fear, while Day (1999) maps fear in public spaces in Orange County, California, as it relates to gendered, ethnic, and classed identities. Wekerle and Whitzman (1995) include many examples of women working to make neighborhoods more women-friendly.

20. Hanson and Pratt (1995) focus on the way local labor markets discriminate against women, and Ehrenreich and Hochschild (2003) focus on nannies, maids, and sex workers in the emerging global economy.

21. There are several texts on research methodology that explore these issues in more detail, for example, DeLyser et al. (2009) and Hay (2010). Donna Haraway's (1985, 1988, 1991) work has also been important in framing questions of positionality and knowledge production.

22. McDowell (1997b) examines gender and class with regard to bankers in the city of London. Tickell and Peck (1996) explore local business councils and the remaking of Manchester. Ward and Jones (1999) provide an interesting article that examines the methodology of elite interviewing.

CHAPTER 9

Living in Place: The Socio-Spatial Relationship

This chapter focuses on everyday places, some ordinary and some extraordinary. Ideas about identity and power resurface throughout this chapter, and the reader is reminded that relationships between people and place are complex and recursive. We can call this a socio-spatial relationship, between the spatial (place) and the social (people). In many ways this approach is similar to the idea of landscape and identity relations, but, increasingly, cultural geographers who examine socio-spatial relations are attuned to power, discourse, and identity construction, not just material cultures. Rather than a static identity read off material culture onto the landscape and vice versa, there is now greater attention focused on the constantly shifting politics and practice of claiming space, asserting oppositional identities, and resisting dominant discourses, power structures, and relationships. To illustrate the socio-spatial approach, this chapter provides some key case studies that encourage you to reflect on how place is constructed through social relations and how social relations are formed by place. ➝≫ 1

In this chapter you will find

❖ an introductory section that highlights the meanings of "place" and "people";

❖ a discussion of the recursive relationship between the social and the spatial;

❖ a section that examines landscape, place and identity, and "reading the landscape";

❖ a socio-spatial portfolio that explores the link between place and identity using a selection of case studies including
  • place and literature,
  • gender and place—suburbs,
  • consumption, place, and identity—the mall,
  • monuments and public art—The Angel of the North,
  • urban places and power—Sydney, Australia, and green bans,
  • power, place, and exclusion—migrant workers in UAE,
  • identity, place, and protest—Bhopal, India, and Union Carbide;

❖ concluding comments.

Revisiting Place and People

Place

Cultural geography has long been concerned with the related concepts of landscape and *pays*, or local regions. This concern stems most notably from the Sauerian interest in landscapes in a regional context and from the Vidalian interest in *pays*. But for many cultural and also social geographers today, the preferred concept is that of *place*. Since being introduced in the context of humanist approaches in the early 1970s, the concept of place has generated a tremendous amount of interest.

Certainly the phenomenological concept of "place" added another dimension to established concerns of humanist geography. Rather than using the term to refer to where something is located—as a container of things—or emphasizing visible features of a landscape, humanist geographers used place to refer to a territory of meaning and to where people live or inhabit. In this sense, "place" is qualitatively different from terms such as "landscape", "space", and "region" in that it involves the idea of being known by and knowing others, of having an investment and connection to place that is lived every day. In this tradition, places are interpreted as experiential and social phenomena: place is *intersubjective*, that is, shared in the sense that the meaning of place can be communicated to others, and is also generated by the presence and actions of those others. Relph (1985:26) explains the meaning of place in this way: landscape and space "are part of any immediate encounter with the world, and so long as I can see I cannot help but see them no matter what my purpose. This is not so with places, for they are constructed in our memories and affections through repeated encounters and complex associations." ➙➤➤ 2 Understood in this way, places are created through the human occupation of space and the use of symbols to transform that space into place, but then the place itself becomes a part of this process and the recursive relationship between place and people is registered in how communities evolve. Box 9.1 explains this meaning of place in more detail.

Place is an important element of the geographer's toolbox, and for those approaching the concept from feminist and poststructuralist perspectives, place is explicitly read through the power of difference. This process is caught up in identity politics, which Agnew (1997b:249) describes as being about "struggling to establish the recognition of collective differences in identity within a society in which those differences are either not acknowledged or involve negative evaluations and sanctions". In seeking to validate identity, place becomes an important site of assertion and conflict. "The meaning ascribed to the role of place in identity politics is highly contestable" (Agnew 1997b:254). In this chapter the socio-spatial portfolio will explore various cases that highlight how the politics of identity and place intersect. Before we illustrate this intersection between people and place, however, we explore the category of people in more detail below.

People

Broadly speaking, scientific perspectives view humans as passive objects, while humanist perspectives view humans as active subjects. This simple distinction reflected the contested character of geography in the early 1970s, when the positivist-inspired spatial analysis that had dominated the discipline began losing ground to other approaches, including those based on humanist philosophies. The dispute was often framed in the context of the debate about structure (society) and agency (individual), and occasionally in the comparable humanist terms of context (society) and intentionality (individual).

The conceptual underpinnings for a discussion of the model of humans have, however, shifted considerably with the emergence of poststructuralist thinking. Indeed, for many cultural geographers the debate about the relative merits of more scientific and more humanist emphases has been replaced by concepts that try to integrate the two. In addition, a revised and as yet unresolved debate has developed around the question of who or what is the subject. Thus, Philo (1991) comments that cultural geographers might profitably consider "the temporal and spatial variations in what

Box 9.1 *Focusing on Place*

Of course, "no one lives in the world in general" (Geertz 1996:262). Although most people are able to live in local landscapes in which they are comfortable, many minority identities are obliged to lay claims to a place where they have to compete with other more dominant claims over that place. This simple observation has had a powerful impact on social science generally in recent years, causing several disciplines to "discover" geography and undergo what is sometimes called a "spatial turn". Reflecting on this development, Agnew (1989:9) could not help but wonder "why we should have taken so long to arrive at a theoretically coherent concept of place as a defining element of geography and key idea for social science as a whole". The answer, he concluded, had to do with place being overlooked in favor of social science's traditional focus on class and community, which was in turn related to the longstanding preference given to evolutionary and naturalist emphases.

Places, as Johnston (1991:67–8) explains, structure "how people tackle problems, both the small and usually trivial problems of everyday life and the large, infrequently met, problems which call for major decisions". Given this importance of place, Johnston proposed a reorientation toward a concept of place that comprises the following six points:

- The creation of places is a social act; places differ because people have made them different.
- Places are self-reproducing entities because they are the contexts in which people learn, and they provide role models for socialization and for nurturing particular sets of beliefs and attitudes.
- No regional culture exists separately from the people who produce it.
- Within a capitalist world economy, places are not autonomous units whose residents have independent control over their destinies.
- Places are seldom the unintended outcomes of economic, social, and political processes but are often the deliberate products of actions by those with power in society, who use space to create places in the pursuit of their own interests.
- Places are potential sources of conflict.

Johnston's interest, as reflected in these points, is primarily with places as sites of conflict rather than with the humanist understanding of place. In this regard the interest with power and identity is clearly evident in this understanding of place.

selves are and in how selves conceive of themselves", and this view is echoed by Pile (1993:122), who argues that a "search needs to be instigated into alternative models of the self, as a means of understanding the position of the person within the social".

In addressing this question, cultural geographers have turned to a variety of sources, but especially to some recent advances in psychoanalytic theory that offer insights into the relations between the social and the individual. Pile (1996), for example, argues that psychoanalysis provides a basis for understanding the spatiality of everyday life. Essentially, the psychoanalytic argument is that identity in place is explained principally by reference to the subconscious. In particular, object relations theory is used to unravel connections between individuals, social worlds, and material environments. A related source of inspiration from psychoanalytic theory is the work of Jacques Lacan (1901–81), which recognizes three different types of space—real, imaginary, and symbolic—and which sees the self formed in relation to otherness. This psychoanalytically inspired approach is also seen in contemporary work on **affective geographies**, in approaches that rely on performance geographies and **nonrepresentational theory**. These psychoanalytically inspired approaches to the self tend toward what has been termed the "affective turn", which is taking place across the social sciences. "Affect" in psychological terms is the instinctual or bodily reaction that precedes conscious thought; it is practice that comes before thought (for example, reaction to extreme heat does not require the stimuli—heat—to be represented in your mind before you react; your practice or response is to withdraw from it unconsciously). The affective turn draws the connection between bodies and unconscious reactions/practices that shape our ability to engage in and with the world around us. In geography the affective turn can be seen in "emotional geographies" and "**non-representational theory**" (Thrift 2007), which attempt to expand the focus of cultural geography beyond representation and human subjectivity, toward practice, what bodies do, or the "taking place of life as a composite of embodied practices" (Anderson 2009:505). ➔➔➔ **3**

This focus on the body can also be traced to the work of Michel Foucault, and his interest in **governmentality**, or how the state governs not just over territory (territorial sovereignty), but also over people by managing their bodies and their practices and behaviours (**biopower**). ➔➔➔ **4** In this regard the people, or the subject, are formed in relation to the state, so today we might say that public health programs promote healthy bodies that are nonsmokers and practice safe sex. This is an example of biopower or biopolitics, since the state is shaping subjects to be these kinds of bodies through policies of medicalization (Rose 2006). We might agree these practices can save governments money in the long run, because the state is expected by many people to take on the responsibility for public health, but this process of governance depends upon the subject (us) abiding by certain behavioral norms, and this has important implications for the formation of subjectivity, society, and the self.

Another form of state governance over the individual can be seen in tendencies that conceptualize "the self" in neoliberal political terms. Neoliberal refers to the restructuring of collective state responsibility for social welfare, and in its place ideas of responsibility, entrepreneurialism, self-help, and personal reliance have become the discursive and governance mechanisms the state can employ to construct "the self" in certain terms. "Responsibilization" (Shamir 2008) is one term used to reference this

process of self-governance, and this practice of envisioning the individual as the key "self-reliant" agent in systems of social organization is itself seen as a new form of governmentality, since governments use various means to get the population to internalize the assumption that one must care for oneself, rather than rely on collective forms of social welfare in order to maintain social protection. This may sound like common sense, but we have to remember that some people and communities face problems that are not of their making; they face structural problems they have no control over. Making these people internalize the idea that they alone must solve the problems that are not of their making, and where they have little to no control over the context of their existence, is problematic. Such changes mark fundamental shifts in state–citizenship relations that are generally recognized as part of what has been termed "neoliberal political change". ➔➔➔ **5**

This process of self-reliance plays a role in what people might expect of themselves and others in terms of their community, their homes, and the places where they live. It also references how difference (for example, sexual, class, supposed race, gender, ability, age) might be made tangible in place by the nature of the relationship with the state (be it local, regional, or national), as well as alternate forms of collectivized or privatized methods of action and support. For example, it is no surprise that the quality of schools differs according to socio-economic features of the neighborhood, even in cases of public school systems (such as Canada), where the educational system is ostensibly supposed to be the same across space. North American and European education systems are increasingly differentiated by the neighborhood. (In the US the role of charter schools, where parents take financial control of the public funds the school receives and determine hiring and educational practices, is one case in point.) This case of educational inequalities that are perceived as the result of what the community does or does not do to help itself deflects attention to *systematic* inequalities (poverty, racism, class bias) that are left to reproduce themselves. In light of this case we can see how people might expect the state to exercise its

Questions for Critical Thought

Consider your views on university education. Do you believe governments should provide free or subsidized university education? Who should pay for your education, and why? Is your subject formation controlled through systems of biopolitical control, that is, does the state communicate to you what kind of subject you should become, and what you should do to achieve that expectation? What happens should you resist the pressure to become the subject the state wants you to be? Do all communities face the same kinds of expectations to engage in education? If not, why not?

responsibility to provide equal access to public goods such as education (liberalism); or they might see it as their own responsibility to provide for their children's education, even if it is at the expense of others (neoliberalism). Whatever form of subject formation, both these examples suggest the positive value and necessity of education for children is already internalized (biopower). There are those, however, who do not or are not able to engage in the education system; for those people opportunities will be constrained. For a good analysis of this in the US context, watch *Teached*, a series of short films that assess the so called race-based "achievement gap" in the US (http://www.teached.org/).

The Recursive Relationship between the Social and the Spatial

Cultural geographers frequently argue that places are constituted by people and that, in turn, people are constituted through place. Whereas Duncan (1980) critiqued Sauerian approaches as superorganic (where culture is read as an all-encompassing homogeneous status), current cultural geographers specifically highlight forms of difference, diversity, and contradiction through culture and its explicit relationship to dominant power structures. Cultural geographers today tend to have a more expansive awareness of different vectors of control or regulation, so that in addition to governments and other institutions of legal and regulatory control, other influencing processes such as the media, peer groups, and global and transnational culture are recognized for their role in shaping place and identity through spatial change and vice versa.

Following this logic, people interpret themselves and are also interpreted by others according to the places they inhabit. For example, when someone is identified as having come "from the wrong side of the track", immediate assumptions about their character and behavior are generated that are "read off" the place they come from. Following the comments made earlier, think about the preconceived ideas someone might develop about a person based on which high school they attended. In this process of "placing someone", a whole series of complex socio-cultural, economic, and political factors are invoked to produce or construct a certain cultural identity. These identity markers are then deployed in different ways depending upon who is constructing them and for what ends. Consider an employer reading someone's CV or résumé and the assumptions they might make based on the university the applicant attended and the program they completed.

People are also "placed" in terms of their regional identity. Such geographical markers become relevant when they are relational; when you live and have always lived in a region, as have your neighbors, friends, and co-workers, few people might think about what being from that region might mean. In fact you may consider a different scale in your interpretation of others; rather than the region it becomes the street, the school, or the town that person comes from that shapes how you view them. However, once you move out of the region, the terms of comparison change, and the scale at which your identity is constructed changes. One interesting example of this differentiation is India, a country composed of diverse groups based on a number of socio-cultural and linguistic factors. When members of these different communities become part of an international diaspora, their identities (both individually and relationally) may become more diverse and complex as the scale of interpretation becomes multi-scaled (local, regional, and national difference almost coalesce). Or their different identities may be collapsed by the receiving society into a macro regional identity such as "South Asian". In some cases the receiving community has little recognition of the internal diversity represented within the immigrant diasporic group. Receiving communities may just rely on interpretative frameworks such as "foreigners", "immigrants", or "people from elsewhere".

Box 9.2 Regional Identities: The Indian Diaspora and the Politics and Practice of Constructing Identities

This box is based on the work of Huzan Dordi, graduate student at Wilfrid Laurier University.

India is often understood as a land of regions and religions. No other country in the world embraces such an astonishing mixture of colors, creeds, and castes, and diversity of culture and customs. The same can be said of the Indian diaspora, those people who have left India to reside overseas (see Figure 9.1).

The Indian community that resides abroad is as diverse and dissimilar as India itself. The Indian diaspora is a culmination of different phases of immigration to a myriad of regions for over a hundred years. Whether through colonialism, mercantilism, or globalization, the Indian diaspora has left a mark in the countries in which they have resided. Estimated at 25 million people, India possesses the second-largest diaspora in the world (Ministry of Overseas Indian Affairs 2009).

Atypically, the Indian diaspora cannot be considered as a unified population. The regional differences in India and its populace are quite apparent. The overseas Indians emigrate from a land of 28 states and 7 union territories, each of them having their different and distinct customs, traditions, and languages. Having a common language, religion, or set of traditions does not bind Indians or the Indian diaspora in any way. As Shashi Tharoor explains, a Hindu from the southern state of Tamil Nadu is closer to a Muslim from the same state than a Hindu from the northern state of Uttar Pradesh (Tharoor 2007). The two have the ability to speak in Tamil, while the Hindu from the north speaks in Hindi. Through syncretic histories in India, it is very common for a Muslim wedding in the northeastern state of Bihar to imitate Hindu nuptials. The wedded may adhere to separate religions, but the language and ceremonies are very much the same.

Such interplay of history and geography allows many Indians and Indian immigrants to associate not with India but with the state/region within India to which they originally belong. For example, Jazzy B., a Punjabi DJ only relatively famous in India, draws packed crowds in Punjabi immigrant–dominated communities of Surrey, BC, in Canada, and Southall, in the UK. A South Indian construction worker in Saudi Arabia considers himself more amicable toward the policies of the southern state of Kerala than India (Tharoor 2007).

The relationship between the diaspora and their respective states is also a symbiotic one, and often more symbolically important to the individual migrant than their relationship to the nation. Many states like Gujarat, Kerala, and Punjab, whose economies depend heavily on financial remittances sent home by migrant workers, adopt laws that make it simpler for migrants and NRIs (non-resident Indians) to send their remittances. The Indian diaspora is also encouraged by state governments to invest and establish businesses in the state/region of their origin. Often the government of India and other state governments will use attractive rates of return and other standard financial incentives to attract the investment of the diaspora, but they also rely on emotive and persuasive cultural tactics to encourage the "sons of the soil" to invest in "Mother India". Such rhetorical strategies demonstrate the importance of cultural and geographical attachment to place in the economic development agenda of the sending nation (Walton-Roberts 2004).

Some Indian states hold large functions and gatherings outside India that celebrate their respective states and their people. *Chaalo Gujarat* (on to Gujarat) is one such gathering. Held in a city with a significant Gujarati immigrant population,

The nature of socio-cultural difference represented within a single national or world regional grouping may be completely ignored or simplified in these contexts. Box 9.2 examines this in more detail in relation to the Indian diaspora.

Landscape, Place, and Identity

We now have a sense of what we mean by people and place, but we also need to understand how place relates to landscape, as it has been imagined since the

this three-day extravaganza celebrates "all things Gujarati" (Chaalo Gujarat 2012). Such gatherings present the opportunity for entrepreneurs, investors, politicians, actors, and artists (belonging to the state and the diaspora) to gather under one tent and forge links and networks. As a result, Gujarati immigrants residing in the US or Canada may be drawn to invest in India. However, that investment will be derived from affiliation to Gujarat, and not due to allegiance to India. For some communities

migration also creates a distinct form of identification through spatial connections that transcend the national space. For example, Kerala, in southern India, is a traditional migration route to the Gulf nations, which created such a strong orientation and connection with that region that for those in Kerala, places such as Dubai in the United Arab Emirates comprise a much stronger part of their everyday geographical imagination than more remote parts of India (Osella and Osella 2008).

Figure 9.1 Location of the Indian Diaspora. This map illustrates the location of people of Indian origin globally. The data come from the Ministry of Indian Overseas Affairs and show only those countries where over 50,000 people of Indian origin are located. In many of these locations the first Indian populations arrived through processes of colonialism, where Indians were employed by the British colonial government to work in plantations, such as Fiji, or as managers and bureaucrats, such as parts of East Africa. In many cases these communities remained after colonialism and form part of the population of these countries today. Source: Ministry of Indian Overseas Affairs data, 2012; cartographer, Pam Schaus, Wilfrid Laurier University.

cultural turn. As we have seen, the superorganic claim that cultures authored landscapes was challenged by Duncan (1980). Since that time landscape studies have covered a range of ideas, three of which come up in the following discussions. First, Cosgrove (1985, 2003) proposed an influential humanist perspective of landscape as a particular "way of seeing"; this implies that there are also other ways of seeing, leading to the suggestion that we might talk about the "duplicity of landscape" (Daniels 1989). Second, Duncan and Duncan (1988) identified the important role of power relations in shaping landscapes, which could be understood as texts and as outcomes of discourses. Third, Cosgrove (1997a) argued that landscapes could also be viewed as theater, spectacle, or carnival. Commenting on these three developments, Demeritt (1994:167) jokes that cultural geographers today are setting "aside the hiking boots preferred by Sauer for the patent leather shoes more appropriate to fieldwork in the cafés and art museums now of empirical interest". ➔⟫ **6**

Six Tenets of Cultural Landscape Studies

Observing that—despite the centrality of Sauerian and related *Landscape* magazine traditions—there is no one monolithic approach to studies of the cultural landscape, Groth (1997) identified six widely held tenets of such studies.

- First, the **ordinary landscape** is important and worthy of study because our everyday experiences are essential in the formation of that cultural meaning. Logically, then, the landscapes in which we are interested are not restricted to those that are in some way different, but include those in which we live and work. While it is common to trace this idea back to the introduction of *Landscape* magazine in 1951, the traditional Sauerian approach by no means excluded the ordinary, although it certainly emphasized the rural and did not encourage studies of the ordinary in urban areas. As Groth (1997:6) comments, "For every forty studies of barns and fields, there has been only one about urban factories, workshops, offices, or corner stores as workplaces." ➔⟫ **7**

- Second, contemporary research into cultural landscape is as likely to be urban as it is to be rural, and as likely to be concerned with consumption as with production. These are substantive changes to traditional interests, and they represent more than just a changing scene: the interest in urban landscapes particularly is linked to questions of power and inequality.

- Third, cultural landscape studies sometimes choose to stress the uniformity of a given landscape as a reflection of some overarching national considerations, and sometimes stress local identities and differences from place to place as a reflection of ethnic variations. In this regard the landscape alternates between being viewed as "one book" or as "multiple, coexisting texts" (Groth 1997:7).

- Fourth, writing about landscape emanates from both scholarly and more popular sources, reflecting both academic and literary styles. Indeed, many of the articles that appeared in *Landscape* magazine exemplified the literary style, even when written by academics, and the collection of landscape readings edited by Thompson (1995) continues this tradition of incorporating writing by both academics and non-academics.

- Fifth, the fact that there are numerous and quite different approaches to the study of cultural landscape is a consequence of the interdisciplinary character of the enterprise. Some studies of landscape stress description while others stress interpretation.

- Sixth, most landscape analyses are based on spatial and visual data, meaning that landscape is often studied directly. Landscapes are, of course, accessible to us and can be experienced differently from place to place. One longstanding complaint about the landscape studies carried out by cultural geographers relates to their apparent overemphasis on what can be observed as opposed to what is recorded in written texts. Notwithstanding such concerns, the emphasis on seeing, and recording and interpreting what is seen, remains fundamental.

Landscape as Text

Regardless of conceptual inspiration, there is general agreement that cultures express themselves in the landscapes they create. This idea invites an interpretation of landscape as something more than material and visible; namely, as something symbolic.

Cultural groups assert their identity on landscape, intentionally and otherwise, and the following discussion of **iconography**—the description and interpretation of landscape based on uncovering its symbolic meanings—is premised on the idea that the identity of a landscape is expressed through symbols. Cultural groups adapt to physical environments, change those environments through the impress of their culture, and create a sense of place—their landscape. Clearly, such landscapes can also be read as texts.

Reading the landscape is an important tradition in cultural geography, and the metaphor of landscape as text—as a manuscript that has been written upon by people over time—is not a new one. Before about 1970, it was quite common for historical and cultural geographers to conceive of landscape as a document, written on by successive groups of people with each entry overlapping/erasing some of what was written previously while adding something new. In this Sauerian tradition, the writing was accomplished by cultural groups and was typically seen as resulting in the formation of cultural regions. But this traditional view of landscape as text has been further informed by humanist ideas, with landscapes being studied as "our unwitting autobiography" (Lewis 1979:12), but that biography is not necessarily written by all members of the community, and the communication system may be appropriated by a dominant group.

Given the renewed emphasis on the metaphor of landscape as text, studies by cultural geographers are increasingly being informed by work in literary theory that considers how texts are read. In particular, there is interest in **semiotics**, in the study of the way landscape meanings are built, and in uncovering the various writings through deconstruction. Recall that deconstruction theory is used to uncover what is not said in a text, such as the role of power relations, as well as what cannot be said because of the constraints imposed by language, especially given the limitations imposed by a particular discourse. For some cultural geographers, reading a landscape is often an exercise in reading social relations and in understanding individual and collective action.

The metaphor of landscape as text, though well established, continues to be a source of critical comment and debate. A particular source of contention is the relative emphases placed on the metaphoric and symbolic content of landscape on the one hand and on the "real" content of landscape on the other hand. On this, Mitchell (1996:95) comments, "Because landscape is partially a text or a representation we have often tried to understand it entirely as a text or representation." Certainly, as Peet (1996b) observes, much contemporary cultural geography emphasizes the reading of landscapes "as though they were texts" (Peet 1996b). But for many contemporary cultural geographers, reading a landscape involves both reading the text and also reading interpretations of the text—not necessarily a straightforward task. Further, understanding landscapes in these terms requires us to appreciate that they are related to a number of usually competing discourses, defined by Duncan (1990:16) as the "social framework of intelligibility within which all practices are communicated, negotiated, or challenged". ➜➤➤ **8**

For the remainder of this chapter we will examine a number of examples that illustrate how place, identity, and landscape intersect. These examples will illustrate both imagined and material examples, but the most effective way to understand socio-spatial or place and identity interactions is to appreciate how the imagined and the material intersect in the production of place, how people experience being in and out of place, and how place and identity constitute, or make up, each other.

A Socio-Spatial Portfolio: Illustrations of the Link between Place and Identity

Place and Literature

Since first being introduced as a source of geographic understanding by regional geographers in the early twentieth century, creative writing has generated an uneven amount of interest. By the 1970s, humanist geographers were attracted to the idea that literature "calls up within the reader essential images of the world" (Salter and Lloyd 1977:1), while more recently, cultural geographers inspired by the cultural turn have taken an interest in literature as a "material artifact which fulfills a role designated by its position in various social and economic processes" (Sharp 2000:327). From both a regional and a humanist perspective it can be asserted that literature contains universal truths that speak to the human condition, such that no other sources of knowledge are necessary. As Salter contends,

> There is no need to write additional textbooks in cultural geography. All the messages of the profession are already committed to ink. The motivations, processes, patterns and the consequences of human interaction with the landscape have all been discovered and chronicled with grace and clarity. . . . We fail, however, as scholars to make adequate use of these data for the simple reason that this material is labelled "fiction". (Salter 1981:142)

Salter's claim may be over-ambitious, but it does capture the motivation and spirit of much of the work in geography that uses literature to help capture the essence of place. Thomas Hardy's Wessex, the Brontes' Yorkshire, D.H. Lawrence's Nottinghamshire, James Joyce's Dublin, Charles Dickens's London, Salman Rushdie's India, Mark Twain's Mississippi River, Stieg Larson's Sweden, Anne Rice's New Orleans, John Irving's New Hampshire, Douglas Copeland's Vancouver, Alice Munro's Ontario—for all of these authors the setting is integral to the story being told, not merely an incidental backdrop. Readers learn not only about the pleasures and perils of London from Dickens and postcolonial India from Rushdie, but also about social relations and working conditions in these places during times of significant change.

Novelists also share how place is understood and remade by immigrants. Monica Ali's *Brick Lane* demonstrates the fear an immigrant woman faces living in and leaving her Tower Hamlet apartment block in the East End of London, England. Novels can also powerfully communicate the nature of urban centers and the micro geography of place as it is demarcated by ethnic, supposed racial, class, and other social differences. For example, *The Apprenticeship of Duddy Kravitz* by Mordecai Richler speaks to Montreal of the 1940s, where religious and ethnic differences demarcated neighborhoods.

Other books have themselves become powerfully emblematic of specific socio-spatial intersections. Richard Wright's novel *Native Son* vividly detailed life in 1930s Chicago through the story of Bigger Thomas, an African-American living in utter poverty in a South Side Chicago ghetto. In constructing the character of Bigger, Wright set out to expose the structures of exploitation and suppression that would eventually dictate Bigger's life because he was Black and living in a racist society. Wright's "protest" novel has been vigorously discussed and debated, but there is no doubt that it contributed to a profound change in understanding what life was like for Black men living in the US at that time. The novel and the debate surrounding it also add to the powerful symbolism the city of Chicago holds for urban scholars and artists. After all, the Chicago School was the model of urban–ethnic transformation and evolution based on the ecology model (see Chapter 4), and this line of study emerged from Chicago's distinctive cultural context. The great migrations of the 1930s and the movement of African-Americans from the American South to the new urban centers in the American Midwest radically altered people's sense of self and belonging. While some critical assessments of Wright's novel would later suggest it merely reinforced the idea of Black male criminality (Bigger Thomas commits and is charged with murder in the novel), at the core of Bigger Thomas is the violence and alienation of White racist American society of the 1930s. →»» **9**

Literature in Context

A possible difficulty with geographic studies based on literature is that they tend to treat literary source material as unproblematic, as being in some way able to reflect the essence of place truthfully. Such an assumption is dangerous, as our earlier discussions of situated knowledge (see Glossary) and writer positionality made clear. Literary texts are no different from academic texts in that both are part of larger discourses that are imbued with power relations. Literature needs to be interpreted with an awareness of both the author's biases and the circumstances of the time, as Richard Wright's *Native Son* makes clear. It is not possible to assess novels, or other creative writing, for use in geographic studies without taking context into account.

Inevitably, as part of a larger discourse, literature both reflects and creates people and place. For example, novels written for boys during the late Victorian and Edwardian periods "leave us in no doubt as to what manly boys were supposed to think and how they should act" (Hugill 1999:335). Imperialism, racism, social advancement, new technologies, individualism combined with team spirit, Christian values, and chaste sexual behavior were all character traits valued in young men and promoted in the literature of the time.

The way that literature is necessarily bound up with prevailing attitudes and behaviors is exemplified by the writings of authors such as the late nineteenth-century British writer H. Rider Haggard. Haggard's fiction features soldier heroes that offered readers the "assurance of a clearly recognizable gender identity and, through this, the security of belonging to a gendered national collectivity that imagines itself to be superior in strength and virtues to others" (Dawson 1994:282). Through the construction of an imagined Africa Haggard's novels also reinforced the key Victorian British perception of Africa as the other, reflecting the prevailing view of the "dark continent" as the "White Man's Burden" in an age of imperial expansion. MacDonald (1994:113), outlining the invention of Rhodesia by the British in the late nineteenth century, notes that "the historical facts of conquest were rearranged, dramatised, and mythologised almost as soon as they happened, epitomising, for imperial apologists, the triumph of civilisation over barbarism." It was precisely this image of Africa as a savage land that was perpetuated by Haggard. But the Victorian British also maintained a romantic image of Africa, as seen in both fiction and exploratory literature of the time, which contains descriptions of Africa's natural beauty, limitless natural resources, and peaceful population. ➡➤➤**10**

Gender and Place

Landscapes can reflect the interests of those in power. Feminist geographers have argued that the cultural landscape is made by men and for men, both as a reflection of patriarchy and as a means of maintaining the privileged position occupied by men. There is little debate about the claim that landscapes reflect patriarchy, but the assertion that these landscapes are created with the deliberate aim of maintaining male societal privilege is clearly more controversial. Certainly, however, although relationships are typically complex, all landscapes reflect gendered identities.

Homes

Consider the design of homes. The home is an essentially private domestic space, constructed at least partly as a reflection of the assumption that heterosexuality prevails, that men interact with the public domain through their work while women focus their attentions on maintaining the home and raising children. As such, the home can be understood as a warm and welcoming place, filled with love. Traditional humanist understandings of home emphasize positive interpretations that focus on family, security, shelter, and companionship. But some feminist geographers have stressed that a home can also be understood as a site of unpaid domestic work, oppression, and sometimes sexual and domestic violence. Indeed, for many women the simple distinction between home and work is fundamentally flawed because their work is in the home. It can further be argued that space inside the home is gendered. Certainly, it has been quite usual in the Western world to locate the kitchen, designed as a separate area for the unpaid work of women, at the rear of the home, and this might be interpreted as devaluing the work of women. In many other cultural traditions, homes are explicitly structured with different spaces for different genders.

On the other hand, feminist-inspired research on such gendered social behaviors as child care and eldercare identifies the home in more positive terms, as does some work on domestic open spaces. Of course, homes are neither all good nor all bad; rather, the meanings attached to home tend to result from social interactions and are generally characterized by fluidity. →≫≫ 11

Suburbs

Consider the design of suburbs. The suburban expansion so characteristic of urban areas since the turn of the twentieth century and especially in the post–Second World War period, involved a separation of workplace and home that can be understood as reflecting gender norms. Suburban expansion involved several factors that limited the ability of women to work away from home, including an absence of child care facilities and limited public transportation. Rose, Kinnaird, Morris, and Nash (1997:149) observe, "As well as the design of individual living units, then, the design of housing estates also articulates particular assumptions about who will be living in this environment and what they will be doing there."

Suburbs as a built form reflected the social relations of the day. In the post–Second World War period, men returned to a society where family relations became normalized as nuclear heterosexual couples with children. The community of Levittown, for example, was a classic homogeneous, White, middle-class American suburban community built during this period.

Social and spatial organization reflected the Fordist economic structure of the day; domestic consumption was the "heart and lungs of modern industrial society" (Silverstone and Hirsch 1992:3). Society enjoyed an unprecedented period of sustained economic growth from the 1950s to the 1970s known as "the long boom". This was an era of culturally mandated masculinity, where the male breadwinner was the head of household, and the "family wage" was the norm. Mothering and being a housewife was seen as a vocation, and "experts" gave advice on maintaining a successful home. As opposed to home as a site of paid work, as homes were during the cottage industry phase before the Industrial Revolution (and as they would become again in the age of telecommuting

and home-based businesses), the postwar suburban home was primarily a site of domestic reproduction and consumption that expressed modernity and the modernization of domesticity.

This modernization of domesticity, however, held at its centre a paradox that would be part of its unraveling. At the core of this new form of femininity and domesticity, the modern home aimed to relieve women of the burden of heavy domestic labour. Postwar suburbs emerged alongside the industries that built the washing machines, the refrigerators, the stoves, the food mixers, and the air conditioners. This production system provided employment for the men of the house, and the devices made life easier for women in the home. These labor-saving devices meant women could focus on other aspects of keeping home, beyond the sheer physical toil of domestic work that women of an earlier era, and contemporary working-class women of the day, had to endure. This meant that female selfhood was firmly rooted within the domestic sphere, making housewifery an occupation. But, during the 1950s, "a collision between femininity and female agency was bound to emerge from these curious and contradictory enfoldings of the public and private" (Lloyd and Johnson 2004:271). This arrangement arguably began its cultural decline with the publication in 1963 of *The Feminine Mystique*, by Betty Friedan. She saw "the problem" that women were increasingly facing, despite all the labor-saving devices and consumption they enjoyed, was the isolation, dissatisfaction, and lack of personal achievement they faced in their suburban life.

By the mid-1970s society was changing; divorce surpassed death as reason for marital dissolution in the US (Stacey 1998:9). Contraception and the revolutionary potential of the birth control pill entered Western society, changing the moralistic inevitability of the modern family arrangement that had dominated the postwar period. The swinging sixties saw young people flood university campuses seeking a different life from their parents. The landscapes of the age were no longer the Levittowns and television episodes of *Leave it to Beaver*, but city neighborhoods like Haight-Ashbury and the rise of the hippie counterculture during the "summer of love".

Questions for Critical Thought

---- ✤ ----

What kind of neighborhood did you grow up in? How does it relate to the type of community you live in now? Think about how your behavior is shaped by the environment you live in.

Times have changed since the 1960s, of course, but the focus on suburbs and the home during this period indicates the connections between subject formations—how people's identities are formed along axes such as gender, supposed race, sexuality, and the places that come to dominate social imagination and economic production. Today population censuses are increasingly demonstrating what many scholars have argued for some time—there are all kinds of households, and how they manage their domestic arrangements is always constrained by dominant systems of spatial arrangement.

Consuming Places

Some places are created as cultural resources and as sites of consumption, while other places—and sometimes people—are recreated or manipulated in order to increase their appeal to consumers. Cultural geographers are taking an increasing interest in the consumption of places and other cultures as a part of tourism, the consumption of the activities of others by attending sporting events, festivals, and carnivals, and the consumption of food and drink. Many sites of consumption—places such as hotels, malls, and theme parks—falsify place and time to create both other places (what might be called *elsewhereness*) and other times. Consider, for example, the many "Polynesian" hotel lounges outside of Polynesia, or Rainforest Cafes outside of the rainforest, or "Beefeater" hotel lounges far removed from London. Consider a place like Disney World, where children are transported into a world they have visited only in books and films. Consider the Abraham Lincoln Presidential Museum in Springfield, which has been criticized for being more like "Abeworld" or "Six Flags over Lincoln" than a museum, and is a

place "where the unreal blurs with the real and ultimately upstages it" (Kamin 2005).

All consumption is cultural (Slater 2003). The seemingly mundane act of eating, for example, involves numerous cultural decisions such as what constitutes food and how that food needs to be prepared, while shopping similarly involves decisions about identity formation and needs. As a cultural practice, consuming has two theoretical implications. First, consumer culture needs to be understood by reference to both the institutions of consumption, such as the shopping malls and tourist sites where consumption occurs, and the advertising that mediates between our needs and wants. These worlds of consumption are continually being socially created, and the pleasures of consumption are social pleasures. Second, consumption is a symbolic practice that may be interpreted. Much contemporary cultural geography is building on these ideas to generate analyses of consumption and the sites of consumption.

Many landscapes in urban areas, including sites of consumption such as malls and sports stadiums and sites of spectacle, reflect popular culture, and are increasingly similar from place to place with only minimal evidence of any local characteristics. In this regard, they have little reference to distinctively local culture and place: they are effectively placeless. This situation is seen as resulting primarily from the fact that such landscapes reflect non-local processes of investment and consumption. From a critical humanist viewpoint, Relph makes the claim that

> the postmodern logic of places is that they can look like anywhere developers and designers want them to, and in practice this is usually a function of market research about what will attract consumers and what will sell. In postmodernity it is as though the best aspects of distinctive places have been genetically enhanced, then uprooted, and topologically rearranged. (Relph 1997:220)

It is helpful to think about this claim in a larger context. For example, as Sutter (1973) points out, traveling through North America often means that the "next place you come to" is essentially the same as the place you came from. Recognizing the sameness of small-town North America also highlights the powerful bond between those towns and the people who live there or who visit on a regular basis. Apparent sameness does not necessarily translate into placelessness. ➔➤➤ **12**

In some cases, however, the distinctiveness of place is commodified to promote heritage tourism. Some selected past is exemplified to present characteristics of specific people and places, creating an experience as well as heritage products that can be consumed by tourists. In many cities, areas no longer used for their intended purposes, perhaps wharf areas or industrial sites, are packaged as tourist attractions that often incorporate opportunities for education and participation. Most such heritage sites are necessarily contested, both in the sense that there may be debates about the merits of preservation and/or reconstruction and, more importantly, in the sense that most heritage sites are partial and involve an intentional directing of the **tourist gaze**.

A particular category of tourist experience and event is termed "**spectacle**". These are the most prominent sites for contemporary tourism and recreation; they are landscapes of consumption associated with postmodernity and related rise of a global culture. There is intense competition to stage these major world events, such as world fairs and the Olympics, as they contribute to local economies, provide opportunities for urban renewal, and enhance the marketing or "branding" of place. Sporting events especially attract both visitors and television-viewing audiences. Major sporting occasions, such as the Summer Olympics, the World Cup of soccer, and the Super Bowl, are spectacles or theater, which, at least temporarily, unite diverse cultures in a shared activity. ➔➤➤ **13**

Landscapes of Consumerism

Retailing sites in the contemporary developed world offer much more than the opportunity to purchase goods and services—increasingly they offer opportunities for leisure, recreation, and other experiential activities. The development of the shopping mall is one consequence of the rise of consumer culture and the success of advertising that promotes the shopping experience as something more than merely acquiring needed goods and services. Currently Western society is criticized for being too consumption-driven; our economies are dependent upon our consuming goods, often far more than we need. In order to promote this activity we see the postindustrial landscapes of the shopping mall proliferate all over North America. Various forms of marketing now seep into all spaces (product placement within film and television shows; advertising on school and college campuses, on buses, in airplanes, and even in public washrooms). One dominant form this consumption drive takes is in places of "leisure" built around the "experience" of shopping. This experiential promotion creates in us the subjectivity of the consumer. Arguably we have become habituated to shopping not only as necessity and as a pastime but also as a means of demonstrating our identity; "I shop at Costco" or "I shop at Wal-Mart" says something about who we are. Our social identity converges around these places of consumption. The normative pressure of this arrangement is such that it becomes difficult to resist; we would be seen as odd if we wore the same clothes, refused to engage in any shopping beyond bare necessities, or insisted on repairing clothing and shoes, and only buying secondhand clothing, for example. Children are bullied at school for what they wear or don't wear, and society increasingly normalizes and advocates that we shop on the weekend and on important holidays. Landscapes of the mall and shopping complexes proliferate, and this is one example where the social and the spatial constitute each other. This process is upheld by an immense discourse of consumption articulated by various forces of advertising, social pressure, and tacit or overt state support. For example, until the credit crisis, people were constantly encouraged to take on debt to feed their desire to consume; after 9/11 President Bush encouraged Americans to "keep shopping", despite the financial debts that subsequently piled up (Bacevich 2008).

Malls highlight one of the dominant features of the contemporary world, namely the tendency to commodify everything. This trend complicates an already complex politics of identity and raises questions about what is authentic and what is inauthentic. The Mall of America, for example, is one of the largest in the world. First proposed in 1985, it is located in the Minneapolis–St Paul urban region on a 78-acre (32-hectare) site close to the airport, two freeways, and numerous motels. The mall opened in 1992 with a 70 per cent retail occupancy rate, but by 1997 the rate was 99 per cent and it was attracting more visitors annually than Disneyland, Disney World, and Sea World combined. The mall houses extensive and varied retailing and recreational facilities, like the much-discussed West Edmonton Mall in Edmonton, Alberta, which was the first mall to explicitly incorporate substantial recreational activities; this example, the prototypical megamall, is detailed in Box 9.3.

Monuments and Memorials

Landscapes reflect a complex blend of dominant and subordinate cultural identities, with processes of inclusion and exclusion constantly at play and largely determined by those in positions of power. A popular strategy to assert identity and authority in landscape is the erection of memorials in the form of monuments, statues, or informational plaques. A part of the iconography of landscape, such structures represent a particular version of the past: they tell the story that those erecting the structure wish to tell. But there are always other versions of a story, and indeed, there may well be other quite different stories. Monuments are key points of hegemonic meaning in landscape, and

for this reason the interpretation of monuments is often contested. Many monuments commemorate leading political and military (usually male) figures, thus memorializing a sense of the past that is presumed to be meaningful to the present. This is why changing political circumstances, such as the defeat of an oppressive regime, can prompt monument destruction. ➙➤➤ **14**

With reference to the national scale, Meinig (1979c:164) notes that a "mature nation has its symbolic landscapes. They are part of the iconography of nationhood, part of the shared set of ideas and memories and feelings which binds a people together". These places may be monuments, government or religious buildings, sites where key events occurred, or larger regions that in some way capture the presumed essence of the national being. They may be intentionally constructed to reflect and reinforce national identity, or they may be places that become symbolic over time.

New nations also strive to promote places that reflect some desired national unity. Israel achieves a sense of identity at least partly through the Zionist rediscovery of places that were important in ancient Jewish history and that are interpreted as foundational to the state (Azaryahu and Kellerman 1999). In the notorious case of Nazi Germany, the medieval town of Rothenburg ob der Tauber exemplified new ideas about landscape and national identity through efforts to reorient tourism and solidify the town's international fame as "the most German of all German Towns" (Hagen 2004).

Other nations that experience a transformation in their political circumstances, for example, through decolonization or the removal of an oppressive regime, frequently make explicit attempts to highlight the change. They may do this by renaming places and destroying monuments. The celebratory toppling of a statue of Saddam Hussein in Baghdad captured the sense of relief and newfound freedom experienced by many Iraqis following the American-led intervention in 2003. More generally, after about 1990 most Eastern European countries that were previously under Soviet influence made conscious efforts both to eradicate the evidence of an oppressive past and to build signposts for the future.

Box 9.3 *The Mall as a Postmodern Consumption Site*

The West Edmonton Mall's "Map and Directory: The Wonder of It All" (West Edmonton Mall n.d.) leaves little doubt that this place is intended to be more than a collection of retail establishments:

> How to shop West Edmonton Mall. It's a lot easier than you might expect . . . and even more fun. Whether you're shopping in a hurry or in the mood to explore, it's all here under one roof. With over 800 stores, there's always one nearby that has what you're looking for. And there's always something exciting to see or do.

Certainly, the inclusion of a hotel, an ice rink, a deep-sea adventure setting that includes submarine rides and dolphin presentations, a water park that has artificial waves, a bungee jump, a large amusement park, theaters, clubs, a casino, a bingo hall, numerous eating places, and financial and other services—all in addition to the over 800 stores—leaves little doubt that this is indeed much more than a shopping mall. It is also a tourist attraction, an entertainment center, and a business center, and it is therefore concerned with more than one aspect of consumption.

How are cultural geographers whose research is inspired by the cultural turn studying the mall? One answer is that they are reading it as a communicative text. According to Goss (1993:19), developers are "designing into the retail built environment the means for a fantasized dissociation from the act of shopping" as one means of manufacturing "the illusion that something else other than mere shopping is going on". The mall is being interpreted as a postmodern consumption site, "a landscape of myths and elsewhereness" (Hopkins 1990:2). As with theme parks such as Disneyland, the mundane is seen as unacceptable in the megamall.

Gregson (1995:137) has expressed concerns about the direction of such research, questioning specifically "the omission of the fundamental and the mundane (notably shopping) from these readings". Noting that most shopping is accomplished by women and that women comprise the majority of retail workers, Gregson (1995:137) contends: "When we look at geographers' readings of the megamall what we find then are masculine and masculinist representations masquerading as universal and homogeneous tendencies in the world of consumption."

Similarly, aspiring nations identify places that are interpreted in national terms. Raento and Watson (2000) discuss the contested significance underlying the town of Gernika/Guernica as a center of meaning for Basques in northern Spain; the town played a vital role in the Spanish Civil War, was represented in a famous Picasso painting, and is commemorated in the Americas by Basque immigrants.

Many monuments are constructed as a means of creating or reinforcing some ethnic or national identity. War memorials are often both a readily visible and an easily understood symbolic landscape feature and, in the case of Australia, the building of numerous memorials to those who died during the First World War has made a significant contribution to national identity and unity. However, while war memorials

are dominantly interpreted as celebrations of bravery and national identity, they can also be seen in terms of a national failure to protect citizens, so, as with so many other landscape features, their meaning is contested through competing discourses.

In some cases monuments not only reinforce some story about the past but also contribute to the making of the future. For example, the Voortrekker Monument in Pretoria, South Africa, tells a story of heroic Afrikaner migrations north from the Cape into an often difficult interior. At the same time, it helped create a particular version of Afrikaner identity, which in turn facilitated the rise of the apartheid state. Neglected in this selective Afrikaner interpretation of the past are other groups, especially Africans, and their stories.

Controversy may also erupt when changes to a memorial landscape are proposed. As home to a series of Confederate statues, as well as many expensive houses, Monument Avenue in Richmond, Virginia, is a powerful symbolic location. In 1994, a proposal to erect a statue on the Avenue of the Black tennis player Arthur Ashe provoked uproar among some members of the White community on three grounds. First, there were aesthetic objections, based on the claim that the symbolism of the Avenue would be destroyed, especially as a casually dressed tennis player would simply not fit in with Confederate soldiers in full military dress. Second, there were objections based on doubts that Ashe's status was sufficient for him to merit a statue alongside Confederate war heroes. Third, there were explicitly racist objections. But there were also objections from members of the Black community: some argued that Ashe's accomplishments as a Black man were far superior to those of Confederate soldiers who had fought to preserve the institution of slavery in the South; others believed that the impact of

the statue as an inspiration for young Black people would be diminished if it were placed on Monument Avenue. After a long public meeting with about a hundred people speaking, City Council voted in favor of the Monument Avenue site, and despite subsequent attempts to have that decision reversed, the statue was unveiled in July 1996. This example highlights the complex intersection of the power of place, the social construction of identities, and landscape iconography. Figure 9.2 shows the 1996 location of statues along Monument Avenue.

Public Art and Place: The Angel of the North

Increasingly urban development and city boosterism depends upon the commissioning of what has been termed "public art"—art that is conceived and commissioned to be displayed in the public domain. Antony Gormley is a public artist whose work is well known in the UK. In 1994 he began work on a 20-metre- (66-foot) high steel sculpture called the

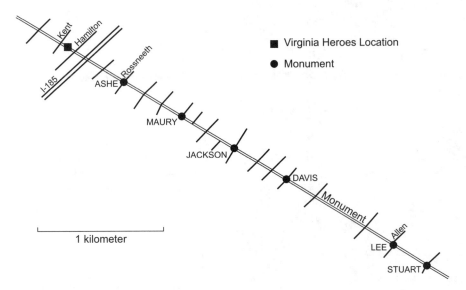

Figure 9.2 Location of Monuments along Monument Avenue. Monument Avenue is one of Richmond's most prominent streets. The Ashe statue is located along with statues to five Confederate heroes (Maury, Jackson, Davis, Lee, and Stuart) that were erected between 1890 and 1929. Source: Adapted from J.I. Leib, "Separate Times, Shared Spaces: Arthur Ashe, Monument Avenue and the Politics of Richmond, Virginia's Symbolic Landscape", *Cultural Geographies* 9 (2002):297.

Angel of the North, funded by the National Lottery Association and costing approximately £1 million. This structure sits high on a hill that is an old mining site at the intersection of two busy highways, the A1 and A167, in Gateshead, northern England, and is seen by approximately 33 million people every year. Gateshead sits across the River Tyne from Newcastle-upon-Tyne, and together they form Tyneside. As early as the 1400s the region was the site of coal mining, and later extensive steel works. Underwater cable production, steam locomotives, and the electric light bulb were some of the technological developments of the day that emerged from this region. While industrial development brought prosperity to the region,

"The icon of England"? The Angel of the North statue designed by Antony Gormley, constructed of steel, located at Gateshead, Tyne and Wear, in the UK, 2010. (M. Walton-Roberts)

by the 1970s the decline of the mining and steel industries and the rise of the financial service sector in the south of England saw the region slip into long-term unemployment and decline.

The Angel of the North represents a piece of public art that was conceived, designed, engineered, fabricated, and, finally, located in a manner that speaks to the region's industrial heritage as well as the hope for its technological future. Built of steel, the massive structure was designed to fully utilize the resources and industrial skills present in the local region. The engineering company Arup, located in London and Newcastle, was responsible for overseeing the construction and erection of the Angel. Turning Antony Gormley's artistic concept into material reality demanded immense engineering skills. The process had to produce a massive sculpture that could endure extreme weather in an exposed location while remaining in a fixed position. The process relied upon state-of-the-art computer design, engineering, and craftsmanship. The fabrication was completed at Hartlepool Steel Fabrications Ltd (a local firm), using high-quality steel products, as well as teams of welders and other industrial workers of the type that have long been central to local economic development in the region.

The Angel of the North was located on top of decommissioned mine shafts, referencing the region's rich history and tradition of mining. When asked about why he named the piece "the Angel", Gormley stated:

> People are always asking, why an angel? The only response I can give is that no-one has ever seen one and we need to keep imagining them. The angel has three functions—firstly a historic one to remind us that below this site coal miners worked in the dark for two hundred years, secondly to grasp hold of the future, expressing our transition from the industrial to the information age, and lastly to be a focus for our hopes and fears—a sculpture is an evolving thing. . . . We are evolving a collective work from the firms of the North East and the best engineers in the world. (Gateshead Council 2008)

This statue expresses the connection between place and people. The history of the northeast of England is one shaped by industrial working-class development. The mining industry, steel works, and engineering talents have played a pivotal role in numerous industrial advances—shipbuilding, early mining technology, and now advances in high-tech design and materials engineering. The piece of work Antony Gormley conceived of and embedded into place is an emblem of this industrial history, which he aimed to achieve in a manner that arguably advances the future prospects of the region by embracing the complex, difficult, and cutting-edge technology of industrial design.

When this project was first proposed, its construction met with opposition from some local government officials. The public was also suspicious of the value it would bring to the region. But, once constructed, the Angel was named by one organization an "icon of England". The region also saw tourism spending increase, contributing over £1.6 billion to the local economy in 2008 (The Angel Symposium 2008). Despite these apparent successes, there is still significant criticism of this type of landmark art, both from the perspective of what happens to the artistic process when engineers become so central to design and also when huge sums of money are invested in such landmark art projects when the industrial regions they are meant to represent face unemployment and stagnant wages. As one art commentator highlights:

> Don't stare in despair at your landmark doodle; just hand Arup the envelope and they'll work out the "complex geometries" and come up with a professional "assessment of overall stability and buildability". Arup is the Jim'll Fix It of the landmark sculpture business, a one-stop-shop for local enterprise partnerships wishing to lay the ghosts of the industrial past by commissioning monstrous beacons of false hope for the future. . . . The question is, do monuments need engineers? The answer is, only when they are unjustifiably big. There are good reasons for making enormous dams, but there

are almost never good reasons for making enormous sculptures. In fact, the reasons are almost invariably bad: jingoistic, militaristic, imperialistic, narcissistic or just plain hubristic. (The Jackdaw 2011)

Ten years after the construction of the Angel of the North, the BBC ran a retrospective to determine if the level of objection that met the initial launch of the structure was still present in the area. Brian Sewell, an art critic who called the piece a monstrosity in 1998, thought it was still a monstrosity in 2008, but one with "fascist undertones". But a local woman living near the site, who had argued against its construction in 1998, ten years later declared, "I love it." A government minister and a local councilor from the area, who both objected to it in 1998, still objected to it in 2008, citing that the money spent on its building could have been used for investment in local job creation (BBC News 2008). In 2008 a YouGov public opinion poll listed the Angel of the North in the top ten landmarks of Britain. In a survey by *Car News* on the top ten most distracting landmarks, the Angel of the North was second after Stonehenge. In a national survey in the *Yellow Pages* the angel was voted one of the "wonders of Britain".

It is clear that this kind of cultural branding of a region through landmark art projects is contested. It is not obvious how power informs the process, since the Angel provided some work for local tradesmen and developed important industrial design skills. Local people who opposed the idea initially in many cases came round to it later, and some officials argue it has boosted tourism revenue, but others still object to funds being spent on "cultural consumptive" rather than "economic productive" functions.

Questions for Critical Thought

❖

What do you think is the value of public art, especially landmark pieces such as the Angel of the North? Should their value be determined only in relation to economic outcomes?

Power and Urban Places: Sydney Harbour and Green Bans

Many landscapes and their meanings need to be understood as parts of larger discourses. Schein (1997) conceptualizes the landscape of one neighborhood in Lexington, Kentucky, as the intersection of several discourses, namely those of landscape architecture, insurance mapping, zoning, historic preservation, neighborhood associations, and consumption. Each of these discourses includes ideals, guiding principles, and often even explicit and institutionally mandated rules that function to discipline the residents of the neighborhood.

In an account of urban redevelopment in central Christchurch, New Zealand, McBride (1999) outlines a series of different readings of the city. A traditionalist reading sees the city as English in character, with Cathedral Square as the symbolic heart, and argues for landscape preservation in the face of changes prompted by capitalism. An alienated reading sees central Christchurch as having been taken over by tourists, with others excluded. In explicit opposition to the traditionalist reading, a market reading interprets urban change in solely economic terms, without any consideration given to questions of preservation and heritage. Those who are responsible for selling the city to tourists provide yet another reading, as they typically strive to find a balance between preservation and development.

In many instances, powerful individuals and institutions, including the state, act as agents to shape the landscape, and landscapes are being read accordingly. In a discussion of a planning dispute surrounding proposals for urban expansion in Lexington, Kentucky, McCann (1997) identified two interested groups. Opposing expansion were environmentalists, preservationists, and some horse-farm owners; favoring expansion were developers, members of the construction trade, and the chamber of commerce. These two groups—textual communities—read the symbolic landscape surrounding Lexington differently, the former seeing "an interconnected, living entity, the existence of which benefited the population of the entire region", and the latter seeing "a series of individual, privately owned land parcels, the use of which was primarily a concern of individual landowners" (McCann 1997:647). The debate between these two groups—involving two very different readings of the landscape—played out in the context of the planning process, and McCann (1997:660) concluded: "The morphology of landscape in capitalist societies is, therefore, mediated by the contexts and discourses of the planning process which make use of the legitimacy and power associated with institutional sites in order to maintain a liberal ideology of equality and fairness."

While many researchers are concerned about the rise of neoliberal values shaping what happens to urban environments, there are still some examples that act as a testament to how people can resist changes to the landscape through protest and political action. Sydney, Australia, provides a very rich example of how one community led a protest against redevelopment and gentrification of The Rocks, a poor working-class neighborhood, and in the process created a culture of public involvement that crossed class and ethnic differences. This culture of civic protest was reactivated at various times to protest the Australian government's plans to sell a number of decommissioned military sites across Sydney Harbour to the private sector. The story of public involvement in urban change in Sydney provides an informative example of how the social and the spatial constitute each other.

Today The Rocks is a well-known historic tourist district in downtown Sydney next to the harbor. It is part of Sydney Cove, the site of the first European settlement in Australia in 1788. For much of its history the area was known as a slum, since its proximity to the harbor made it a popular site for sailors seeking prostitution and alcohol on shore. In 1900 an outbreak of bubonic plague in the area led the government to take it over with a plan to demolish the buildings, many of which had fallen into disrepair. The arrival of the First World War put an end to this redevelopment plan. After the war a number of buildings in the area were demolished to make way for the Sydney Harbour Bridge. The Second World War delayed any further redevelopment plans, and it was not until the late 1960s that the Cove Redevelopment Authority took control of the process and planned to demolish the historic rundown buildings of the Rocks and replace them with high-rises.

In 1971 The Rocks Resident Action Group (RRAC) formed to oppose the redevelopment of The Rocks because there were no plans to rehouse the mostly poor aboriginal tenants of the area. The RRAC aligned with the New South Wales (NSW) arm of the Builders Labourers Federation (BLF), who at the time were a national trade union representing the interests of the construction site laborers. The BLF had previously used "black bans", strikes or industrial action against their employers, but they also had a history of using "green bans" to object to projects for environmental, social justice, or heritage conservation reasons. The BLF had first used a green ban in the 1960s when the NSW BLF joined a protest over the proposed development of a prized green space called Kelly's Bush. Though this area was part of a wealthy neighborhood, the leader of the NSW BLF, Jack Mundey, persuaded the union to stand up for the right of all citizens, regardless of their class, to have access to green space (Mundey 1981).

When the demolition of The Rocks was proposed in the 1970s, the NSW BLF supported the RRAC through the use of a green ban. They refused to allow their union members to demolish any buildings in the area, in part because of their historical value but also due to the fact that there was no plan to rehouse any of the current tenants. Unlike the earlier Kelly's Bush protest, which was an alignment of labor and environmentalists, this protest saw a new alignment between labor and historic conservationists. These two groups were distinct in terms of socio-economic class, yet once they saw their common interest in protecting The Rocks, they worked together for several years, until eventually, in 1975, the state government abandoned its intention to redevelop The Rocks.

This strategic intersection between these groups shows how the contentious "politics of identity" can be made effective as a protest movement when groups with different interests and identities come together around a common place. In addition to the RRAC and the BLF, The Rocks protest also saw local heritage conservationists get involved, and they tended to be mostly White, wealthy, middle-class elites. The power of this protest movement became

less of a single identity protest, but a confluence of shared interests in promoting a sustainable, socially just community that protected the historic value and meaning of a specific place (Iveson 2010).

Such protests against urban redevelopment during the 1960s and 1970s were widespread in a number of communities around the world. Like The Rocks, some were successful, such as protests over the redevelopment of the Downtown Eastside in Vancouver (Ley 2008). But others were not so successful, such as protests over redevelopment of the tenement blocks in Glasgow (Bryant 1982) or immigrant neighborhood clearances in the West End of Boston (Wilson 1963). Even the story of The Rocks is not a full success, since today the working-class neighborhood has given way to a tourist heritage site devoid of a resident community (Karskens 2008).

This story of public activism and redevelopment in Sydney does not end with The Rocks protest. The ability of various groups to effectively protest urban redevelopment was reactivated a decade later when the Australian federal government proposed to decommission a number of military sites across Sydney Harbour, including Cockatoo Island, the biggest island in the harbor, which had been the site of military naval shipbuilding yards. At first the Australian government's intention was to sell the land to private interests, including Australian millionaire Kerry Packer, who was rumored to want to build a casino on Cockatoo Island. In response, a number of groups lobbied to protect specific sites in their community, including the Friends of Cockatoo Island (Shelley Clark 2004) and Mosman's Headland Preservation Group. Both of these groups turned to and were supported by Jack Mundey (previously the leader of the Buildings Laborers Federation who objected to the destruction of the Rocks) in their efforts to lobby to keep the lands under public control. These protest groups argued that the lands were effectively owned by the Australian citizens, and since this was merely a change in use, they should remain in the public's ownership. In response to this public input, in 2001 the Sydney Harbour Federation Trust was formed by the federal government as a government agency with the responsibility to manage the former defense and other Commonwealth sites in and around Sydney

Harbour. This covers eight main sites, including various headlands and bays, bush lands, islands, and historic buildings. Today most of these sites are publicly accessible; many of the military buildings have been restored and are rented to businesses, services, and arts groups. Cockatoo Island now provides holiday homes for rent and houses a public campsite in the middle of Sydney Harbour. Numerous music and arts festivals are held on Cockatoo Island, including the Biennale of Sydney, an international arts festival.

While winning public access to space in these naturally and historically significant sites could be seen as a success story, maintaining public access depends upon governments continuing to fund the upkeep and operation of such sites for the public interest. Ongoing political and financial changes and the continued rise of corporate interests and restructured government finances means this issue of public access needs to be continually safeguarded. In a recent comparative analysis of how the Australian federal government and the state government of New South Wales dispose of public lands, Goh and Williams (2011:8) argue that financial pressures have led to various efforts to privatize public lands:

The work of individuals, community groups and government to have culturally or environmentally significant lands acquired or dedicated as public land in order to protect such sites, or to ensure public access to foreshore and scenic locations, is only half the task. Funding for ongoing maintenance must be allocated to these lands to ensure that privatisation is not the only management solution.

The story of various battles to maintain the historic, social, and environmental qualities of Sydney allows us to understand how the social informs the spatial and vice versa. Sydney is a beautiful natural landscape, and the people who live there have fought to preserve these qualities. In the process, strategic social connections are formed that transcend specific components of the "politics of identity". The Rocks protest linked enlightened labor leaders with social rights activists and historic preservationists. These people came from different social, economic, and ethnic backgrounds, but they all had a commitment to The Rocks, to this place and what it stood for, be it as a home, as a site of historical importance, or as a place where class-based resistance needed to be activated. Later, when the federal government considered selling decommissioned military sites to the private sector, some of those same activists who experienced success in The Rocks launched public protests and lobbied the government. This

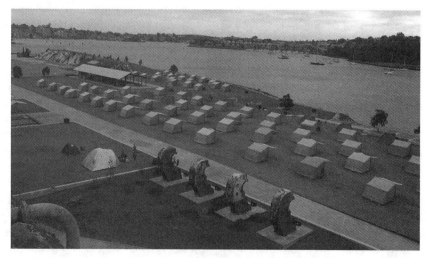

Cockatoo Island public camping site, Sydney Harbour Federation Trust Lands, 2009. This public camping area is on the site of an old naval shipyard. Today Cockatoo Island is a site promoting the arts and public recreation. (M. Walton-Roberts)

resulted in the formation of a government agency responsible for maintaining many of these sites to maximize public access (Mundey 1981). These social actions, then, have become embedded into these spatial landscapes, so that you can literally read them—for example, part of Argyle Street in The Rocks was named "Jack Mundey Place" in 2007—but if you know about the history of the area, you can also "read" the symbolic and material landscape and see how social activism creates and secures these public spaces.

Power and Exclusion in Place: Migrant Workers in UAE

Dominant societies do not just construct identities of others; they also construct the landscapes in which others might be obliged to live. In so doing, they construct landscapes that take certain characteristics for granted. For example, urban areas in the Western world have typically been built with the assumption that families are heterosexual and nuclear, that women are dependent on men, and that people are able-bodied. These assumptions, and the effects they have had on landscape formation, have reinforced the notion that those who do not conform to societal expectations are different. Difference can often translate into exclusion and disadvantage. The Rocks example mentioned previously was a case where mostly poor aboriginal tenants were the targets of state redevelopment schemes. It was only when this marginalized group was able to get the support of organized labor and heritage preservationists that it was able to resist displacement. If it were not for that strategic alliance, in all likelihood the tenants would have been easily displaced.

A distinctive feature of the contemporary world is the unsettling effect that expressions of "other" identities and their crossing of spatial boundaries have on dominant groups. Many collective identities that are protest identities can be interpreted as products of transgressions of normative spatial behavior. Such behaviors in certain places may be seen as a challenge to authority and an expression of resistance that results in the creation of contested landscapes. Consider, for example, the way young people use shopping malls. The film *Mallrats* (1995), directed by Kevin Smith, is a good example of this alternative use of space by a group of teenagers who antagonize a store manager because of their "lack of a shopping agenda". Or consider "take back the night" and "slut walks" that actively challenge normative assumptions about how women "should" behave and dress in public spaces at certain times.

Headland Park and Chowder Bay, Sydney Harbour Federation Trust Lands, 2009.
Local communities became involved in protecting this site from privatization when the Australian government decommissioned various military sites in Sydney Harbour. In their effort to maintain public access the community group called upon Jack Mundey and his experience in protesting similar forms of redevelopment. (M. Walton-Roberts)

People's identities become wrapped up in the spaces they occupy and use, for example, shopping malls, schools, churches, and public streets. If their use of these spaces is not in line with the general practice of others, it can be constructed as an act of defiance. Their identity might be read through the way they use and occupy space. Rowdy teenagers, graffiti artists, and homeless people are just some people who may occupy and use space in ways that run counter to the expected "norm".

One example of this conflict in use-space can be seen in countries that are home to large populations of migrant workers. In Singapore there has been conflict over the use of sidewalks for social gatherings on the weekend, when thousands of migrant domestic workers enjoy their day off. In gathering at popular sites to meet fellow workers, their presence has antagonized local Singaporeans who want to use the space to shop for groceries rather than socialize (Yeoh and Huang 1998). This clash represents the fundamentally different lifestyles enjoyed by domestic workers, whose "home" is effectively their workplace, and who only get to engage in socializing in public spaces outside of the homes where they work.

Another example is from the United Arab Emirates (UAE), where hundreds of thousands of mostly Asian migrant workers are employed. Dubai in the UAE is a city that depends upon migrant workers, yet their presence in the city space is highly constrained. Work visas demand employees complete a two-year work placement with their employer (who retains their passport), and only if the contract ends with "cordial relations" with the employer can the employee apply for another work visa and/or seek a new contract (Salama 2010). Conditions of work for many unskilled workers in the UAE are difficult; in fact the Indian consulate reports two suicides a week by Indian, mostly blue-collar, workers here (Malik 2011).

These restrictive conditions of living are evident across the city. Dubai is replete with intense luxury and extreme consumption, but it is also clearly dependent on armies of subservient migrant workers whose conditions of employment leave them little time, or space, for recreation. This privileging

of certain subjects is evident in the city, since Dubai is built for the automobile, and migrant workers have limited options when it comes to public transit (although a rapid transit system is being built by migrants). The vast networks of roads are constructed by migrant workers, but these roads and construction sites also become barriers to the mobility of workers around the city, since they often move around the city as pedestrians (Kendall 2012). Migrants have to walk alongside highways and overpasses that make little to no provision for pedestrians. On Fridays, the weekend holiday for domestic workers, the limited bus system is packed with Filipino domestic workers and other blue-collar laborers who are venturing to shopping malls and public beaches with their friends.

The photo right gives some sense of how constrained social activity is in public spaces. In this image we can see some green space, which is located on the Bur Dubai side of the Dubai Creek. This creek-side park has been developed within a tourist district that includes museums, the old souq (market), reconstructed heritage buildings, and a village at Shindagha that recounts traditional Arab architecture. What people are allowed to do in these spaces, however, is tightly controlled: no sports, no climbing trees, and no sitting on the grass.

Clearly such demarcation and restriction of access to space has been examined in previous chapters, especially Chapter 7 (consider the discussion about redlining, and the experiences of the Roma). In many countries overt discrimination to

Questions for Critical Thought

Think about how someone's identity is read geographically, for example, in terms of the place people may come from. One of the most obvious examples of this is how people tell jokes about regional identities. These jokes operate at several different levels: national, regional, and local. What do these jokes do? Are they just another form of social stereotyping, or are there other complexities at work that are both social and spatial?

Green space next to Dubai Creek, in Dubai, UAE, 2008. Notice that the sign is prohibiting exactly what the man in the background is doing—sitting on the grass. The sign is symbolic of the manner in which public space is regulated and controlled in this city. (M. Walton-Roberts)

materially exclude certain groups from public space is legally prohibited, but the process can still occur in informal ways. In certain cases people's identities might be conflated with how they use space, as well as where they live or come from.

Identity, Place, and Protest: Bhopal Social Movements

We have discussed how power and protest shape place and identity, and the following case explores an extreme and well-documented example of this. You may have heard of Bhopal in India, and if you have you are certainly aware that this is the site of one of the most infamous and tragic industrial disasters ever. The place and the name Bhopal are linked to the horror of this tragedy, but it is also now the site of well-developed and effective social movements, which have achieved various degrees of global and national success in light of the lessons learned from the Union Carbide disaster.

In the early hours of 3 December 1984, the city of Bhopal in Madhaya Pradesh, India, was the site of one of the world's worst industrial disasters. Union

Carbide India Ltd (UCIL), a US multinational (now owned by Dow Chemical), owned a chemical plant that experienced a poisonous gas leak. A failure in the cooling system allowed for a cloud of methyl isocyanate (MIC—a toxic chemical used in pesticide manufacturing that is fatal if inhaled) to spread into the neighboring community. While figures vary, it is suggested that over 500,000 people were exposed to the poisonous gas and estimates suggest that 8,000 people died that night. A 2006 government affidavit stated that the leak caused 558,125 injuries, including approximately 3,900 severely and permanently disabling injuries. In February 1989 Union Carbide Corporation and UCIL were ordered to pay a $470 million settlement; this amounted to US$370–$533 per victim. (For more information see the BBC documentary *One Night in Bhopal*, and the website http://www.bhopal.org/what-happened/.)

Several factors contributed to creating the situation that led up to this disaster. UCIL had been in India for some time, and in the 1960s expanded its operations to produce chemicals for "Green Revolution" crops. These were high-yielding crops that required a high amount of fertilizer input. In 1968 UCIL built a chemical and research and development plant in Bhopal on land two miles (three kilometers) from the city. The plant produced dangerous chemicals in the 1970s, even though state licenses described it as a general, not hazardous, industry. This license also informed the nature of planning surrounding the site. Once the plant was in situ, it was attractive to people—they could access jobs, electricity, and water. To the public it appeared as "vacant" government land. At first several people who arrived in the area "rented" huts from local landlords. Eventually the government granted rights to the squatters proximate to the plant, but did not improve water or sanitary services. In the 1980s, the UCIL plant faced losses due to local competition; it cut back on operations and cut staff. Various safeguards that should have prevented the gas leakage from occurring failed due to lack of maintenance and removal.

The town was clearly not planned, since any safeguards about locating near such industrial processing facilities were absent. There was very little in the way of warning systems, and even simple preventative methods of protection in the event of a leak had not been communicated to the nearby residents. Thousands of people were poisoned that night, and the toll inflicted ranged from death to long-term health complications, including respiratory and lymphatic illnesses. The story of the Bhopal gas victims and their fight for compensation is a long one that has not yet ended. There has been little in the way of effective compensation for many of the victims, with widespread charges of corruption and political influence having beset the numerous government compensation schemes and legal court cases, which continue to this day. The point of discussing the Bhopal case here is to explain how place frames identity politics and the culture of protest that emerged in the wake of the Bhopal tragedy.

Bhopal sits at the center of India, on the Malwa plateau. It has a hilly topography and is considered one of the greenest cities in India. As a city it is home to a number of religious groups, including a substantial Muslim population. After the Bhopal gas disaster several protest groups emerged, and in the years since many of them have worked tirelessly to campaign for the rights of the thousands of people left to suffer with horrendous health effects and little to no compensation to show for it. Many academics have looked at several of Bhopal's social movements as an example of groups that shifted from place-based local groups to transnational social movements. Scandrett and Mukherjee (2011) explore this issue in light of the idea that social movements can rise from "militant particularism" (based on the work of Raymond Williams and David Harvey) toward "divergent abstraction". Scandrett and Mukherjee (2011:199) turn to Harvey to explain how, when a protest movement goes global, there is a danger that in this process of abstraction (when you take an element of something to use in another context), something of the power of the original protest movement is lost. "The shift from one conceptual world, from one level of abstraction to another, can threaten the common purpose and values that ground the militant particularism achieved in particular places" (Harvey 1996:33). In the case of Bhopal, Scandrett and Mukherjee (2011) demonstrate that academics who have used Bhopal as an example of a social protest movement that effectively went global have greatly simplified the type of protest movements that have arisen in Bhopal since the disaster.

Bhopal is used as an example of a social movement that has gone global, but when you examine the case in more detail, as Scandrett and Mukherjee (2011) have, it is clear that this easy local-to-global trajectory is only relevant to some groups. The International Campaign for Justice in Bhopal (ICJB), which is a coalition of Bhopal-based groups and their international supporters, is the main group that has successfully linked to the global scale. This group sees global resonance in the theme of making a multinational corporation clean up its environmental pollution. In this regard the Bhopal disaster, when focused on the remnants of industrial pollution, finds resonance with other global sites of industrial pollution. The abstraction works because the places that are connected face the same problem of industrial waste and lack of corporate responsibility. However, in the case of Bhopal, in the wake of not only the actual industrial disaster but also the years of government, corporate, and legal stalemates following the search for social justice, numerous social movements, formed by the particular nature of social relations in this place, have emerged. For example, in addition to the ICJB, Scandrett and Mukherjee identify the Bhopal Gas Affected Women Workers' Campaign (or *Bhopal Gas Peedit Mahila Udyog Sangathan*). They argue that the abstractions that take place from various movements occur along different lines. In the case of the ICJB, the issue is environmental justice, but for other groups, class and gender analysis become the organizing framework for their militant particularism, and this does not necessarily abstract to the global level.

The environmental justice movement is the one strand of protest that has achieved the most global abstraction, through groups like ICJB. Bhopal is a global symbol of corporate environmental devastation; today even the groundwater surrounding the community is contaminated, and survivors of the poisonous gas continue to require medical care. This has encouraged ICJB to lobby on behalf of "pani peedit"—water-affected victims (residents who came to the community after 1984, but whose health has suffered due to consuming the groundwater), as well as "gas peedit"—gas-affected victims (Scandrett and Mukherjee 2011). The ICJB's global abstraction has led to several high-profile conference invitations and opportunities to align with other communities affected by industrial pollutants. The group has also received international recognition and awards for its environmental work. This abstraction, however, has caused personal rivalries and antagonism in terms of its relationship with other groups in Bhopal. As Scandrett and Mukherjee (2011) argue, the group's international recognition was not seen in Bhopal as a reflection of its hard work; rather, "[f]or many in this non-literate culture of personal loyalty, this ['courting' of international recognition] amounted to betrayal."

So in the case of Bhopal, the global abstraction from the militant particularism that accompanied the Bhopal gas victims' movement led to the loss of something. That loss was tied up with the particular local culture and place identity that accompanied this movement. As some groups found success internationally, they were derided by other groups who were tied to the local. The context of poverty, illiteracy, and immensely limited options created a sense of intense jealousy, and while leaders with ICJB gained international recognition, they were maligned at the local level for this detachment from the local.

The women workers, or *Bhopal Gas Peedit Mahila Udyog Sangathan*, on the other hand, emphasized the class dimension of their struggle. For this group it is the poor who have faced the most injustice, and it is only a struggle of the people that will correct it. The group has shunned or limited international support for fear it will undermine their integrity. Much of the work of the *Bhopal Gas Peedit* is in support of other Indian development groups, be it protests about tribal rights or large dam projects that cause widespread community displacement. For this group militant particularism is maintained along class lines and focused at the regional or national level, rather than in the courting of international connections. In some ways this approach is formed within the local context, where legitimacy close to the ground is most important, rather than international abstraction.

The third theme Scandrett and Mukherjee consider is gender. They explore how the *Bhopal Gas Peedit* is striking in its gender order. Indeed, the organization focuses on economic opportunities for women by offering sewing classes and promoting small-scale entrepreneurialism, and women have been central to many protests the group has been involved in, though they hold few top leadership positions. Scandrett and Mukherjee discuss the view that women are constrained by the actions of the local male leaders. A system of patriarchal patronage controls local positions of authority within the protest movements in Bhopal, and women are expected to abide by the norms and conditions men establish in terms of the nature and strategies of protest.

Before the disaster Bhopal had low levels of female literacy (about 20 per cent), and Muslim and Hindu women practiced different degrees of purdah (or separation). After the disaster, as women became active in protest movements, there has been increased freedom to exercise greater rights outside the control of male family members. Even though women have been active on the ground in terms of environmental and class struggle, when it comes to gender struggle, any emergent militant particularism is constrained by the patriarchal order, and so any global abstraction toward international feminist movements (the global equivalent of the women's movement at the local level) has been resisted by the women themselves:

> Whilst the daily gender regime in the movement, and consequently in the community, is certainly being renegotiated through the militant particularism of a highly gendered political practice, there seems to be no evidence that this is being generalised in any of the groups through commitments to any form of feminist universalism which might serve to critique the patriarchal gender order. (Scandrett and Mukherjee 2011:203–4).

Instead, women, even those at the heart of the movement, still support some of the central planks of the patriarchal system, including dowry, intra-caste marriage, and the right of the male head to determine what women in his household do.

The case of the Bhopal disaster and the development of different protest movements provides an important example of how place frames identities and protest movements. Groups that have successfully engaged in global abstraction in terms of environmental justice face the loss of local legitimacy. The *Bhopal Gas Peedit* group members seek to align their connectivity along the axis of class identity at the regional or national scale, preferring to limit international engagement for fear of how it might undermine their important status with local communities, who are often illiterate and below the poverty line. While women have come to hold important positions of leadership within many of the protest movements, abstracting the limited gains they have made in relation to universal feminist principles

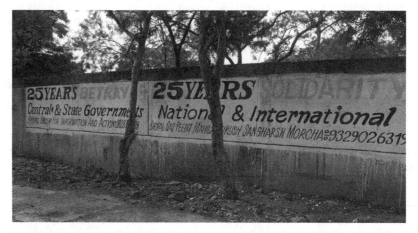

"Bhopal Gas Peedit Mahila Udyog Sangathan" message on the wall surrounding the decommissioned Union Carbide plant in Bhopal, 2010, sends a message to the community that local groups in the area represent their interests. (M.Walton-Roberts)

Protest mural on the Union Carbide plant, Bhopal, 2010. The image declares the important role of women in the "miltant particularism" of protest against the Bhopal gas disaster, but such activism has not evolved into "universal abstraction" to international feminism. (M. Walton-Roberts)

The view of the Union Carbide plant from one of the houses on the edge of the Union Carbide town, 2010. This image provides a sense of the proximity of houses to the factory site and, together with the lack of safety features and protections at the plant, explains why so many people were affected that night. (M.Walton-Roberts)

is constrained by local conditions and norms. The various improvements in gender rights women have experienced in light of the protest movement in Bhopal have been constrained by local context, and female identity remains demarcated by those local, rather than global, conditions.

In general, then, despite arguments about the power of global social movements, we can see the power of place and identity as still deeply relevant. Bhopal has become a site of powerful and meaningful

social protest movements that have undergone various degrees of global abstraction (the ICJB's link to environmental justice), but we also see how militant particularism exerts immense power—there has been a degree of gender empowerment, and women are active in various movements, but the most central aspects of patriarchy remain in place.

Concluding Comments

- This chapter examines the intersection of place and people in more detail using various examples that draw upon many of the issues we have explored in different parts of the book, such as landscape, identity, and intersectionality. Both people and place can be understood in material and symbolic terms, and the relationship between them is understood as a recursive one, which the term "socio-spatial" can be used to indicate.
  - › Building on questions of human and place identity, this chapter also stresses the symbolic character of landscape—landscapes as representations of cultural values and ideas and as outcomes of competing discourses. We considered this in how literature can

Questions for Critical Thought

———— ✥ ————

Did you know anything about Bhopal before you read about it in this chapter? If you had some exposure to the case before, how did it shape your reaction? Did you have a visceral or instinctual reaction based on any previous engagement with this story? Reflect on what you immediately felt when you read or heard the word "Bhopal", and consider how this might be reflective of "non-representational theory". Did you feel something before you consciously thought something? If so, can you critically express why that might be?

inform and reflect the nature of place and social relations in place. Landscapes are also materially produced and politically competed over, as all of the cases examined reveal, but particularly the issue of public land in Sydney Harbour. Landscapes of inequality can produce marginal identities, as the case of migrant workers in the UAE indicates, but we can also see how social protest as a response to experiences of exclusion challenge such processes—with varying social and spatial results—as the case of gas victim movements in Bhopal highlights.

- The cases explored in this chapter provide an example of how we can put all of the themes we have explored so far to work in our analysis of cultural and landscape change. Though not explicitly outlined, each case can be further examined from the various cultural geographic approaches we have covered in other chapters.

The approach to each case is informed by aspects of the cultural turn, in particular with regard to how the role of power is identified as playing a part in the construction of place and identity, whether that power is imposed (migrant workers in UAE), successfully challenged (protection of public space in Sydney Harbour), or continually reshaped (Bhopal gas victim movement). Each of these cases also encourages us to think more effectively about how we "read the landscape".

› In the case of the Angel of the North, we can see how, armed with some knowledge of local history of place and social relations, we can understand what is being communicated by landmark art monuments, but we can also recognize the nature of local contestation and begin to understand the complex social processes that frame such attempts at remaking landscapes.

Further Reading ➤➤➤

The following are useful sources for further reading on specific issues.

1. Various examples of this approach to cultural geography can be found in journals such as *Social and Cultural Geography, Cultural Geographies*, and *Space and Culture*, an interdisciplinary journal that focuses on social spaces such as homes, cities, workspaces, and virtual spaces.

2. Creswell's (2004) short book on place theory links everyday examples with the complex social theory that many cultural geographers now use in their work. Tuan (1977, 1989) provides seminal statements on the humanist concept of place. Smith, Light, and Roberts (1998) discuss philosophies and geographies of place.

3. Terkenli (1995) outlines processes through which places become homes. Dominy (2001) discusses home, place, and identity in the context of New Zealand.

4. Useful readings on non-representational theory include Thrift (2007), who first introduced the idea, and Anderson and Harrison (2010) who provide an overview of how the theory has been employed in different areas of geographic research. In terms of tracing the use of ideas of performativity in geography, see Nash's overview (2000).

5. Geographers, like other social scientists, are increasingly engaging with Foucault's work on governmentality and biopolitics. Duncan's (2007) examination of colonialism and the coffee plantation economy in nineteenth-century Ceylon explores the nature of power, control, and resistance. Crampton and Elden (2007) provide an edited collection that explores how a number of geographers are working with Foucault's ideas.

6. Geographers have examined socio-spatial manifestations of neoliberalism. For work on Toronto that engages with the everyday imagination of the city and neoliberalism and education, see Boudreau et al. (2009). David Harvey (2005) also focuses on this latest phase of capitalism.

7. Discussions of landscape as informed by humanism and the cultural turn include those by Olwig (1996b) and Kobayashi (1989). Walton (1995) discusses the idea of landscapes as texts. Granö (1997) outlines a traditional European interpretation of landscape. Hirsch and O'Hanlon (1995:22), in an apparent desire to establish an anthropological interest in landscape seen to be different from the interests of cultural geographers, propose the idea of landscape as process—"one which relates a 'foreground' everyday social life (us the way we are) to a 'background' social existence (us the way we might be)" (see also Bender 1993). Duncan (1990) examines the early nineteenth-century landscape in Sri Lanka; this is an influential landscape study in the newer tradition of landscape, arguing that accounts of the world and the world itself are both intertextual.

8. Both *Landscape* and the *Journal of Cultural Geography* are important outlets for discussions of ordinary landscapes. See also Zelinsky (especially 1994). An atlas of North American societies and cultures edited by Rooney, Zelinsky, and Loudon (1982) is a comprehensive description of one continental region. The wealth of interest in ordinary and sometimes not-quite-so-ordinary landscapes is reflected in the introductions to and some of the contributions included in the edited volumes *Landscape in America* (Thompson 1995), *Understanding Ordinary Landscapes* (Groth and Bressi 1997), and *Everyday America* (Wilson and Groth 2003). See also Jakle (1987) and Hart (1998).

9. Cosgrove and Daniels (1988) discuss iconographic analyses. See also Cosgrove (for example, 1984, 1997b) on the landscapes of Renaissance Venice and Vicenza within their larger cultural framework. Domosh (1989) discusses the example of a single newspaper building, the New York World Building constructed in 1890 in New York City, in terms of both personal egos and industrial capitalism (see also Domosh 1999). Eyles and Peace (1990) provide an iconographic analysis to help understand two prevailing but contradictory images of Hamilton, Ontario, namely those of smokestack city and cultured city. Also see Kong (1993) on religious buildings in Singapore and Cartier (1997) on the symbolism of bank buildings in Hong Kong.

10. Studies of geography and literature that emphasize a humanist perspective include those by Birch (1981), Barrell (1982), Herbert (1991), and Hudson (1992). See also the edited books by Pocock (1981) and Simpson-Housley and Mallory (1987). Studies of geography and literature influenced by the cultural turn include those by Sharp (1994, 1996, 2000), Kong and Tay (1998), Hugill (1999), and Kneale (2003). Statements about or examples of geographic studies of literature are provided by Salter and Lloyd (1977) and Noble and Dhussa (1990). Calder (1998) uncovers environmental determinist content in creative writing set in the Canadian Prairies. Donaldson (2001) shows how poetry can be used in the context of elucidating Pattison's four traditions of geography (spatial, area studies, human–environment, and earth science).

11. Osborne (1988) considers the iconography of national identity that was forged by painters, stressing the role played by the early twentieth-century "Group of Seven" and their concern with the Canadian North (see also Osborne 1992).

12. Feminist geographers have attended to issues of home and identity. McDowell (1999) examines home and identity as one element of gender, place, and identity. Blunt and Dowling (2006) examine the idea of home from multiple perspectives.

13. Slater (2003) outlines cultures of consumption. Goss (2004) reviews studies of geographies of consumption. Hopkins (1991), Goss (1993, 1999a, 1999b), and Gerlach and Janke (2001) describe malls as sites of consumption. See also the Special Issue of *Canadian Geographer* (1991, vol. 35, no. 3) on the subject of West Edmonton Mall and megamalls generally.

14. Ley and Olds (1988) and Squire (1994) provide examples of landscapes of consumption and spectacle. Yeoh and Teo (1996) discuss the example of Dragon World in Singapore.

15. Zelinsky (1988), Jeans (1988), and Johnson (1995) discuss monuments as a reflection of ethnic or national identity. Bell (1999) outlines the redefinition of national identity in Uzbekistan. Foote, Toth, and Arvay (2000) discuss the example of Hungary after 1989. Crampton (2001) considers one iconic South African monument, and Crampton (2003) reflects on political imagery at the South African National Gallery. Marshall (2004) explores the power of war memorials in everyday English landscapes. Analyses of memorials in landscape include those by Bender (2001), Crampton (2001), and Olsen and Timothy (2002). Leib (2002) discusses Monument Avenue in Richmond, Virginia; see also Monument Avenue (2004).

CHAPTER 10

Cultural Geography— Continuing and Unfolding

This book is a journey through a body of changing ideas and changing practices, and the aims of this final chapter are to assess where cultural geography stands today and where it may be heading. Both aims are difficult to achieve. Recall that cultural geography has been described as a "scholarly discourse that has shifted from the comfort of placid marginality toward the overheated vortex of ferment and creativity in today's human geography" (Zelinsky 1995:750). This chapter begins by commenting on ongoing contestations over cultural landscapes, and then suggests that it is necessary to focus more on global issues and to broaden analyses of difference and power. The chapter concludes with comments on the potential contributions contemporary cultural geography offers.

Within the subdiscipline of cultural geography, the debate about the relative merits of the Sauerian school and more socially and theoretically informed approaches continue to some degree today, though it is mostly directed at reprising Sauer's work in light of the torrents of criticism that have marked the development of cultural geography since the 1980s (for example, see Mathewson's (2009) defense of Sauer). To many cultural geographers whose studies are inspired by post-positivist approaches, the work of the late Allan Pred, another Berkeley School professor of cultural geography, is more emblematic of current trends in cultural geography today. Pred's work on supposed race, space, modernity, and urban change was influential in urban and historical as well as cultural geography. Pred was an outspoken critic of racism and state-sponsored violence, and his work was always passionate and animated in his efforts to communicate the importance of challenging violence and marginalization. Pred leaves a legacy of influential work on the American city and the rise of modernity in urban Sweden (for example, 1980, 2000). Pred was critical of Sauer's work, arguing that Sauer conflated racialized groups with the rural landscape, thereby diminishing their agency and subjectivity. In Sauer's defense, some have argued that he was concerned for all communities who were, he felt, being forced off the land due to various modernization processes (Mathewson 2009). Despite such turbulent disagreements, cultural geographers may still recognize some value in the early work of Sauer that may assist us in the current period. For example, questions about integrating physical and human geography have real implications for cultural geographers, especially those concerned with environmental issues. These debates center on understanding the role of culture, the meaning of landscape, the appropriateness of particular writing styles and research methods, as well as the analysis of power.

Not everything about cultural geography today need be seen in light of argumentation about the relative merits of the Sauerian framework. Times change and other concerns capture the attention of researchers. Another important intersection that is shaping cultural geography today is that of culture and economy. It is clear that throughout this book certain examples highlight the intersection of culture and economy; consider, for example, regionalization, globalization, marginalization, economic change, consumption behaviours, and intersections of identity. These matters are economically as well as culturally framed, and, increasingly, cultural and economic geographers have seen their interests and approaches merge. As a recent edited collection on the matter states: "In retrospect the long standing separation of economic and cultural geography seems ludicrous, even analytically disastrous" (Warf 2012).

In short, cultural geography is a complicated terrain, but, increasingly, we are witnessing strands of convergence, rather than contestation, between subdisciplinary approaches over issues of growing importance. This book is an attempt to reflect, incorporate, and, on a few occasions, reconcile or move beyond such differences.

 1

Cultural Landscapes

The study of cultural landscape has been a persistent concern of cultural geographers from the late nineteenth century to the present. Since it was first explicitly formulated by Ratzel, further developed by Vidal and Schlüter, decisively introduced into English-language geography by Sauer, and reinvigorated by Cosgrove, the idea of cultural landscape has been an almost endless source of conceptual inspiration and a rich vein for local and regional understanding. The impact of Sauer's methodological statements and the example set by the empirical studies conducted both by Sauer and by his students is difficult to exaggerate. For Sauer, at least in principle, there was a universal model that could be used to inform analyses of cultural landscapes, whereas contemporary interest in cultural landscape incorporates a wide variety of theoretically informed approaches that are more sensitive to the micro dynamics of power (for example, Pred 2000).

Of course, reading a landscape intelligently requires an awareness of authorship and context, and one of the earliest and most critical contributions associated with the cultural turn was the questions raised about the emphasis placed on cultures as groups and the apparent rejection of difference and human agency. Since about 1980, approaches to the study of cultural landscape have diversified even further, reflecting in part the proliferation of theoretical bases, and as a result there are today concerns with symbolic landscapes, with landscapes linked to prevailing power relations, with landscapes of production and consumption, emotion, violence, and exclusion, and with landscapes as texts.

Perhaps the most meaningful difference between the Sauerian approach to landscape and the more recently developed approaches to landscape lies in the transition from a concern with *culture* to a concern with *cultures*. Not all cultural geographers have welcomed this conceptual broadening of interest in cultural landscape. Hart (1995:23), for instance, writes: "I am quite content with the simple vernacular definition of landscape as 'the things we see', and I am saddened by the way the meaning of the word has been transmogrified. Some people have endowed the concept of landscape with magical, mystical, or symbolic significance, or have loaded it with metaphysical significance." In turn, Henderson (1997:95–6) criticizes Hart's point of view as one part of a more general criticism of those landscape studies that stress the richness and diversity of landscape at the expense of such considerations as human inequalities and related social struggles.

Although particular understandings of culture are central to most work in both traditions, two qualifications of this statement are worth noting. First, with reference to landscape school geographers through to the 1970s, Mikesell (1977:460) observed that "most cultural geographers have adopted a laissez-faire attitude towards the meaning of culture." Second, some cultural geographers have questioned the intent of the new cultural geography and have sought a major reorientation. These cultural geographers have argued that there is no such thing as culture, but rather only the idea of culture (Mitchell 1995). Further, they have claimed that there is no "need of a concept of culture as such", but only a need for "a much deeper, clearer, more operational conception of human behavior and development" (Wagner 1994:5).

Regardless of perspective, one key point emerges from this ongoing and diversifying interest in cultural landscape: landscapes, as the places we create, are important in any attempt to understand who we are—our human identity—and also how we live—our social, political, and economic systems of organization.

Global Cultural Geographies

As noted in Chapter 1, Wagner (1990:41) asserts that "a theoretically well-grounded, intellectually vigorous, and practically effective social and cultural geography might well assume, in time, a major role in guiding and guarding the evolution of humanity's environments." Are cultural geographers responding to this ambitious claim?

Given the concern with understanding people and place, cultural geography is certainly well positioned to contribute in various applied areas, and the ecological analyses and studies of cultural and regional inequalities discussed in this book attest to the contributions being made by cultural geographers. Cultural geography continues to embrace a diverse set of ideas and to address issues with multiple layers of relevance. But is this enough? Possibly not, as the following three observations suggest.

First, cultural geographers are saying little about questions of global environmental concern. Some studies within cultural ecology offer analyses of particular regional circumstances, but studying these issues at the global scale seems quite another matter. Not much is being written by cultural geographers about the all-too-evident gross inequalities of the human condition. Consider that in the early twenty-first century, about 1 billion people are malnourished, and that of these, about 650 million people live on the very edge of existence and lack a reliable supply of food, access to clean water, and sustained medical care. Facts such as these identify urgent global issues and are but suggestive of a long and depressing list of problems. While all scholarly work can yield valuable contributions, studying the lives and landscapes of the undernourished and the malnourished in the less developed world with the aim of identifying problems and proposing solutions is one aspect of the geography of consumption that merits much greater effort from cultural geographers. Certainly, major global differences of wealth and power and the forms of violence applied to environment and community are being critically addressed (Gregory and Pred 2007). As noted in Chapter 6, a great deal of the impetus for this is coming from political geographers: "We need to construct a new global geography that focuses on

the world map of global inequalities. We should very consciously proclaim this to be the most important map in the world and focus our Geography accordingly" (Taylor 1992:20). Cultural geographers might profitably pursue this line of argument, though concerning ourselves with such subdisciplinary divisions may be counterproductive as we promote this line of research. In some ways we may be witnessing a lack of analysis at the global scale, which may be the result of cultural geographers' reluctance to engage in grand theorizations that suppress too much of the difference and diversity we have been made aware of during the post–cultural turn. Nevertheless, the importance of environmental change and economic inequality at the global scale is becoming more and more apparent.

Second, cultural geographers have had relatively little to say about the formation of cultural regions at the global scale, despite the suggestion that this is, as Meinig (1976:35) put it, "the greatest topic in historical geography". Are there reasons to suggest that cultural geographers might pursue this topic further? One answer comes from an insightful and challenging commentary on Meinig's ambitious series of North American studies (discussed in Box 5.6) in which Harris (1999:10) notes that "there is awe from a dwindling group of historical geographers, but little response from the rest" because the work "is not tuned to theory currently in vogue". This comment must not be taken lightly. The key point made by Harris is that cultural geographers need to pursue a variety of conceptual inspirations, not simply those prompted by the cultural turn. More recently we can discern a rising interest in regional cultural geography based on the reemergence of the Asian economic powers China and India. As we examined in Chapter 6, current shifts in global economic power have encouraged revisions in our understanding of the historic formation of regional trading networks and power relations. Cultural geographers have a role to play not only in remapping the historical record but also in contributing to the work that critically reassesses how Western perceptions of the world were constructed through discourses of modernity and Eurocentrism.

Third, there is much discussion today about the possible emergence of a global culture and about the reassertion of some regional and local cultural identities in explicit opposition to globalization. In this context, cultural geographers might usefully evaluate, for example, the six possible global cultural scenarios outlined by Masini (1994:20–5):

1. Cultures, particularly living community cultures, might be reduced to just a relict, museum, or tourist role and be unable to challenge the dominant global identity that reflects Western technology and values.

2. There might be a combination of cultural continuity and change that allows the core elements of cultures to remain while involving some acceptance of global trends.

3. Many cultures might resist the dominant global trend, with any cultural change being internally generated.

4. There might be a widespread acceptance by most cultures that no one culture is complete in itself.

5. It might be that all cultures are influenced by information from other cultures, with Western culture as the most influential because of advances in information technologies.

6. Perhaps a non-Western culture, most likely an Asian culture, becomes globally dominant.

Certainly, cultural geographers today would do well to consider Masini's views on the future of both a global culture and of the myriad of local cultures. On the one hand, the seeming impossibility of isolation means that differences disappear, while on the other hand, local cultural differences seem able to resist many homogenizing forces.

Difference and Identity

It is clear from the material covered in this book that cultural geographers today are concerned with difference and identity. There are three closely related matters of concern regarding this current interest. First, there is more than one way to interpret claims about the emergence of alternative cultures, such as those based on sexual, regional, ethnic, or linguistic difference. In a discussion of the globalization of culture, Albrow (1997:150) states: "The multiplication of worlds means that individuals can inhabit several simultaneously, but secondly that each severally can only make a small selection from the many which coexist. The result of a plurality of individuals making their own selections is that each builds a different repertoire, and its total scope is obscure to everyone else." This is a challenging claim implying that the membership of individuals in particular groups is fluid and, more critically, not necessarily a meaningful marker of their identity. Our discussion of strategic essentialism and intersectionality (Chapter 8), and the examples of Sydney Harbour and Bhopal in Chapter 9, which demonstrate how movements can include social as well as spatial diversity in their composition, are illustrative of this fluidity.

Second, an emphasis on human difference rather than human unity often has the unfortunate consequence of reproducing not just difference but also inequality. Europeans during the period of colonial expansion saw numerous others as different, which became an important component of the practice of creating and reproducing inequalities. We have also seen how an obsession with fixed cultural difference can create global divides that appear intractable, such as the case with Huntington's "clash of civilizations", which we read about in Chapter 6. Huntington's reading of a homogeneous "Islamic culture" to explain political difference has arguably gained attention because we now understand culture and identity to be so significant. In this case, however, internal heterogeneity has been suppressed to create a singular "other". In this thesis cultural analysis and understanding was employed to create the other, but not taken far enough to demonstrate the internal instability at the core of that constructed other.

Third, when discussing questions of power and authority, as these relate to human differences, cultural geographers need to be more expansive about the values that underlie critical analyses of, for example, racist and sexist attitudes and practices. Indeed, it is possible that the current preoccupation with specific dimensions of difference might obscure the importance of intersections of identity, particularly where class and socio-economic axes of identity might provide lines of commonality between groups that we might otherwise overlook in our obsession with specific aspects of singular identities and difference.

The Past Is Prologue

From the perspective of this textbook, completely ignoring the Sauerian tradition is a mistake not only because it overlooks an extensive body of work in cultural geography but also because it fails to acknowledge substantive precedents to some current work. Admittedly, some cultural geographers may not wish either to lionize Sauer or to institutionalize a landscape school, but to largely omit these matters from meaningful discussion seems not just an inappropriate rewriting of the disciplinary past but also a rejection of much current work that continues to build on this tradition. There are important new critical approaches to contemporary problems that are evident in the field of cultural geography, but there are opportunities to continue to make contributions across the entire discipline and beyond that may find something meaningful from the earlier traditions of cultural geography, particularly the central importance of understanding environmental and social change.

An appraisal of current research, based on a survey of current books and journal articles, confirms that both the traditional and the newer cultural geography are with us today, and that both are changing. In 2009 a collection of writing by and on Carl Sauer was published (Denevan and Mathewson), revealing the breadth of influence his work has had on geography and related disciplines, not just cultural geography. Mathewson (2009), in particular, examines in detail the type of criticism Sauer's work has been subjected to. He accepts some, and rebuffs others, but in conclusion he argues that the current attention many scholars are directing at landscapes of destruction, such as those produced by global capitalism, imperialism, and past colonialism, were always central to Sauer's own interests (Mathewson 2009).

On the other hand, the new—much of which is no longer quite so new, of course—has certainly proven to be an enrichment, and most of the recent statements applauding the vitality of cultural geography reflect what can be seen as a "transformation" (McDowell 1994:146) or a "reinvention" (Price and Lewis 1993a) of the earlier tradition. As evident from the contents of this textbook, the use of both terms— *transformation* and *reinvention*—is understandable, as the new certainly reflects some significantly different conceptual content and includes a different set of empirical issues. In this sense, the new is not something that has evolved easily from the traditional; rather, as a significantly different set of ideas and practices, it has either been added on to or perhaps even replaced previous ideas and practices. Nevertheless, many links can be identified between the traditional and the new, including

- continuing interest in cultural landscapes and the powers that shape and destroy them;
- a mutual mistrust of scientific method;
- an interest in human-and-land relationships;
- a predilection for fieldwork;
- a concern for regional identity;
- a concern for group identity; and
- a willingness to conduct research at various spatial, temporal, and social scales.

To deny these continuing concerns seems unfortunate. →»» **2**

An Integrated Human Geography?

Clearly, a key question to ask about cultural geography—really a question about the larger discipline of geography, or indeed about the way in which knowledge is structured—concerns the viability of identifying a subdiscipline labeled "cultural geography". Described in Chapter 1 as a compositional subdiscipline—one concerned with some particular subject matter—cultural geography is always being questioned, as are all other compositional subdisciplines. Certainly, owing largely to the cultural turn, the links with social, political, and economic geographic subdisciplines are increasingly close. For example, in a discussion of economic geography, Crang (1997:4) considers five options for relating the economic and the cultural: a continued opposition of the economic and the cultural, effectively resisting the cultural turn; an export of the economic to the cultural; a view of the economic as embedded in a culturally constructed context; a view of the economic as represented through the cultural, acknowledging that places are cultural constructions; and an understanding of the cultural as materialized in the economic.

Similarly, one study of the discipline of geography was structured around themes rather than subject matter, using four principal headings as follows (Mabogunje 1997): conceptual fundamentals relating to space, time, theory, and methodology; environment and society, including human transformation of earth, climate change, and human response to hazards; spatial structures and social processes, including market and cultural forces, particular modes of production, transportation, and trade; and spatial organization and globalization processes, including the role played by the state, transnational corporations, and megacities. Such a framework for writing about geography effectively incorporates the cultural into a larger geographic context. This indeed may reflect the wider tendency toward interdisciplinarity that is evident across the social sciences.

Looking Forward

In light of the evident disagreements and uncertainties, what might be the future of cultural geography? As this textbook has often had occasion to stress, cultural geography is an evolving intellectual terrain. In most cases, textbook authors outline the way ahead as a continuation of current trends; at best, such predictions are but a partial truth. Certainly, it is reasonable to assume that the principal current trends will continue, but it is also reasonable to assume that new trends will appear.

It is worth repeating that this book has an ambitious and, for some, a questionable goal, namely that of integrating, or at least incorporating under some common headings, work that might be variously labeled traditional or new cultural geography. There is one principal reason—in addition, of course, to authorial limitations—why this goal is at best only partly attained: cultural geography is all too clearly a moving target, with new, often imported, concepts informing increasingly diverse analyses, and with ever-closer connections to such other traditional compositional subdivisions of human geography as social, political, and economic geography. These are certainly healthy signs, but just as certainly, it makes accommodating the contents of cultural geography within a few key themes—chapters in the context of this book—an unenviable task. Indeed, there is a tendency for reviewers of cultural geographic literature to focus on selections of material rather than on addressing some perceived totality.

For some cultural geographers the goal of integration, as it might appear to imply unity, is questionable because as Duncan (1994:402) argues, it "reflects the modernist will to order and discipline, to be governed under a master narrative". With these comments in mind, cultural geography might be described as a *heterotopia*—a site of incompatible discourses—and with this view in mind, Duncan (1994:402) writes approvingly of the fact that cultural geography "is no longer as much an intellectual site in the sense of sharing a common intellectual project as it is an institutional site containing significant epistemological differences".

Whatever the merits of the claim that integration is a questionable goal, this text accepts that there is value in discussing cultural geography—all cultural geography as broadly conceived including past, present, and possible future approaches—within a single organizational context, and this value is especially clear when the needs of student geographers are being considered. Organizational frameworks such as the one adopted in this book aid in understanding diverse ideas and analyses without in any sense imposing unwavering identities on ideas and analyses or suggesting that the framework used is in some sense correct and uncontestable, a point made in the opening chapter of this textbook.

To conclude, this text is sympathetic to the interpretation of cultural geography as "one of the most dynamic fields within geography today" and to the claim that the "great heterogeneity" of work being produced is "surely a sign of strength in a subdiscipline that was regarded by many as marginal as recently as the 1980s" (Murphy and Johnson 2000b:1). Indeed, it seems likely that cultural geographers will continue to move the subdiscipline forward in several directions, employing both traditional and new concepts to address issues of environments, landscapes, identities, and inequalities at both global and local scales.

Further Reading ➡️ ≫≫

The following are useful sources for further reading on specific issues.

1. Schein (2004) comments on the tradition of cultural geography, and Scott (2004) reviews recent developments. Gregory and Pred's (2007) edited collection on *Violent Geographies* gives an indication of some of the work at the interface of political and cultural geography.

2. Students interested in pursuing cultural geography further will benefit from appraising the contents of Murphy and Johnson (2000a), Anderson, Domosh, Pile, and Thrift (2003a), and Duncan, Johnson, and Schein (2004).

Glossary

ableism Discrimination on the basis of physical and/or mental abilities. Ableism typically involves the assumption that persons with disabilities are not "normal".

acculturation The process by which a minority or politically weaker cultural group undergoes gradual cultural change to become closer in character to a majority or dominant cultural group. Acculturation does not involve a complete cultural change.

actor–network theory (ANT) Approach developed by Bruno Latour that examines how humans and non-humans are integrated in networks. The network approach allows for non-human objects to have agency in various social processes. Relations are understood as occurring between both things—*materiality*—and ideas—*semiotics*.

adaptation Human adjustment to environmental conditions, both physical and cultural, in order to ensure that the needs of the group are met and that conflict is reduced. It is common to refer to the particular way in which a group meets those needs as an *adaptive strategy*. The term *cultural adaptation* refers to the idea that culture is an adaptive system.

affective geographies *Affect* refers to experience, feeling, or emotion. These are seen as non- or pre-cognitive responses to stimuli. (See **non-representational theory**, with which affect is linked.)

ageism Discrimination or prejudice on the basis of age. Ageism is often based on an inappropriate assumption that people of similar age have other characteristics in common. Unlike racism and sexism, it is likely that all people suffer from ageism, as identities based on age necessarily change through time.

anthropocentric Regarding humankind as the central or most important element of existence, especially as opposed to God or animals (compare **ecocentric**).

apartheid A system of policies governing the spatial separation of groups of people distinguished on the basis of a racial classification; also, the social distinctions made between those groups. The policies and practices of *apartheid*—an Afrikaans term meaning "apartness"—were applied in South Africa between 1948 and 1994.

assimilation The loss by members of a cultural group of all their previous cultural traits through the adoption of the traits of some other dominant group with which they are in contact. Assimilation involves a loss of identity.

authority The right, usually by mutual recognition, to require and receive submission by others.

Biopower term developed by Michel Foucault to indicate the regulation and governance of the body and society by the state. This regulation is often conducted through state technologies of biological-health-management and improvement.

body Especially in feminist geography, *body* is understood both as a personal space that contributes to identity formation and as a basis for distinguishing the self from others. In addition, the body is, literally, the basis for our interactions with physical and built environments. Increasingly, the body is one of the concerns of cultural geographers, being variously interpreted as a site of resistance (as in some feminist and postcolonial theory), as a producer of meanings, and as a declaration of subjectivity.

capitalism A social and economic system for the production of goods and services that is based on private enterprise, that involves a separation of the producer from the means of production, and that allows relatively few individuals to have access to resources.

Cartesianism A term that refers to the need for mechanical and mathematical explanations in physical science. Descartes was a key figure in the emergence of modern philosophical and scientific thinking. The Cartesian theory of the mind, *dualism*, maintained that the mind is entirely separate from the body, a view that was famously referred to by Gilbert Ryle as a "ghost in the machine".

catastrophism   The idea that the world has experienced and results from a series of sudden violent and unusual events (compare **uniformitarianism**).

civilization   A term with various contested meanings that is often used to refer to a culture with an agricultural surplus, a stratified social system, some specialization of labor, a form of central authority, and a system of keeping records. More generally, it is used to refer to any culture of global significance.

class   Generally, refers to a large group of individuals with similar social status, income, and culture. The presence of classes in a society explicitly reflects divisions based on the unequal distribution of economic goods, political power, and cultural status.

cognitive map   The model or mental representation of the world in which a person lives. Every individual's cognitive map is different.

colonialism   The process or policy, by a state or a people, of establishing and maintaining authority over another state or people.

conservation (or **constructivism**)   A general term referring to any form of environmental protection, including preservation.

constructionism   A position that claims that all of our conceptual underpinnings, such as ideas about identity, are necessarily contingent and dynamic, not given or absolute (compare **essentialism**; see **social construction**).

consumer culture   A term that, as discussed by Baudrillard, is understood with reference to institutions of consumption, such as shopping malls and advertisements. Consumption is a symbolic practice that needs to be interpreted.

contextualism   Broadly, the claim that it is necessary to take into account the specific discourse, or *context*, within which any account of ideas and facts takes place.

cornucopian thesis   The argument that advances in science and technology will continue to create resources sufficient to support the growing world population.

counterfactual   Used to designate a statement that asks what might have been the consequences if some detail of history had been different. Counterfactual statements may be used to analyze presumed cause-and-effect relationships through a comparison of the observed world and some hypothetical world derived from contemplating a different history.

creation myths   Stories, often invoking the supernatural, that explain both the origin of the world and the origin and early history of people in the world.

crisis of representation   Emerged alongside the *cultural turn* and the critical interrogation of the social construction of truth and meaning. In some disciplines, such as anthropology, this resulted in questioning the fundamental practice of research, which in some cases reproduced colonial power relations.

cultural ecology   The study of the interactions between humans and nature; more specifically, the analysis of culture as an adaptive system. The term is sometimes used to refer to the discipline of cultural anthropology.

cultural hearth (or **cultural core**)   The place of origin, the heartland, of a culture, from which it diffuses outwards.

cultural region   An area that is occupied by a cultural group and reflects that occupance both in the visible material and symbolic landscapes.

cultural studies   An area of academic interest concerned with various types of texts as these help us understand how values, meanings, and identities are produced. Cultural studies focus primarily on the political meaning of culture—on gender, sexuality, ethnicity, nationality, and class.

cultural trait   One element in the normal practice of a cultural group. A *cultural complex* is some related set of traits.

cultural turn A set of changes in social science, prompted by developments in philosophy and social theory, that recognize that culture is too important to be reduced to economics and politics.

culture Generally, the way of life of the members of a society as evident in their values, norms, and material goods. Note that there are many specific interpretations of this word.

cyborg A cyborg is a combination of human and non-human; for example, prosthesis limbs, reproductive fertility treatment, or information systems that combine technological and human capabilities. The term is shorthand for cybernetic organism, reflecting the integration of both the biological and artificial. *Cyborg theory* is linked to the work of Donna Haraway, who used it to critique the separation of nature and artificial organisms.

Darwinism The view that the development of a species results from competition among and within species, gradually eliminating the least fit and permitting the fittest to survive. The fittest are those having genetically based features that give them some competitive advantage over fellow members of their species. Darwinism is non-teleological and materialist.

deconstruction The principal method used in the postmodernist tradition to study texts. Deconstruction is based on the assumption that a text is not a self-contained entity whose meaning is determined by the intent of the author, but rather that a text exists in a context with both the production (writing) and receiving (reading) being affected by other texts. Deconstruction exposes the unquestioned, taken-for-granted, philosophical assumptions that underpin a text—assumptions that are inherited and typically contain uncritically accepted dualisms.

deep ecology An idea stressing the need for a new world view recognizing that we are all connected, that we need to work with rather than against nature, and that a central goal of human activity is the preservation of ecosystems.

dependency theory The idea that European overseas expansion produced a situation in which countries at the core of the world system could maintain the underdevelopment of countries at the periphery in order to maintain the dependence of these countries on the core. Dependency theory involves the belief that in order to become more developed, the more developed world had to create a less developed world.

determinism The general idea that everything that happens is an effect, that events are necessitated by earlier events. Philosophers such as Hobbes and Hume were determinists. There are many versions of determinism, mostly in physical science, with the principal example being the mechanistic clockwork universe of Newtonian physics. A dominant view in the nineteenth century, determinism was liberally applied in the social sciences in a series of varyingly effective versions. Philosophically, concerns with determinism include its application to the human world, with the implication that humans are unable to exercise free will, and moral responsibility.

development A term typically interpreted to mean a process of becoming larger, more mature, and better organized, often as measured by economic criteria. This is a term to be used cautiously, since it often involves an implicit general acceptance of a particular perspective (ethnocentrism).

dialectic The resolution of contradictions in the pursuit of truth. A dialectic is a method of reasoning that proceeds from thesis to antithesis to synthesis.

difference A term that recognizes the need to understand identities in relative terms. Studies of identity frequently acknowledge that identities are socially constructed in opposition to other identities, with power relations playing an important role.

diffusion The process of spread over space and growth through time of cultural traits, ideas, disease, or people, from a center or centers of origin.

discourse Language in which the meanings of words are specific to a community of users; for example, the technical vocabulary that distinguishes academic disciplines from each other. *Discourse* also refers more generally to the various social practices that enable the world to be made intelligible. Although sometimes used interchangeably with the term *text*, discourse is more appropriately seen as a way of connecting texts that share a common point of view, as in sexist discourse. Discourse is often interpreted in terms of the relationships between knowledge, language, and power.

disorganized capitalism   A new form of capitalism characterized by a process of disorganization and industrial restructuring (compare *organized capitalism*).

dualism   In the classic Cartesian formulation, based on earlier Greek ideas, the idea that mind and matter are separate. More generally, dualism refers to the separation of two things that might alternatively be seen as one—humans and nature, for example. It is often contrasted with *holism*.

ecocentric   Emphasizing the value of all parts of an ecosystem rather than, for example, placing humans at the center (compare *anthropocentric*).

ecosystem   An ecological system comprising interacting and interdependent organisms and their physical, chemical, and biological environment.

empiricism   An empiricist epistemology asserts that all factual knowledge is based on experience, with the human mind being a blank tablet (*tabula rasa*) before encountering the world. Empiricism is a fundamental component of the philosophy of positivism. It is also related to pragmatism.

Enlightenment   A European intellectual movement, the Age of Reason, of the seventeenth and eighteenth centuries, which is contrasted with the earlier period of irrationality and superstition. Generally, two assumptions reflect the philosophical ideas of the period: an empiricist epistemology and the idea that the aim of social science was to enhance social progress through revealing truths about ourselves.

environmental determinism   The argument that the physical environment is the principal cause of human behavior and human landscape creation.

episteme   The world views, or structures of thought, that a society holds at a particular time and that impose the same standards on all branches of knowledge. The concept is associated with Foucault.

epistemology   The branch of philosophy concerned with the nature, sources, and justification of knowledge. Epistemological questions are about how we know knowledge is knowledge (compare *ontology*).

essentialism   A belief in the existence of fixed, unchanging properties; the attribution of essential characteristics to a group instead of seeing such characteristics as being constructed socially (compare *constructionism* and *social construction*).

ethnic   Denoting a group whose members perceive themselves as different from others because of a common ancestry and shared culture.

ethnic cleansing   The forced removal of an ethnic group by another, more dominant group. Ethnic cleansing often involves extermination (see *genocide*).

ethnic religion   A religion that is associated with a particular group of people and that accordingly does not seek to convert others.

ethnocentric   Making judgments about a culture based on the values of one's own culture and, as a consequence, misrepresenting that culture. For example, Eurocentrism is an ethnocentric view that fails to acknowledge that other ethnic groups do not necessarily share the priorities of European discourse.

evolution   The idea that organisms have developed from primitive forms, through natural processes, to more complex forms. In Europe, belief in evolution was a product of Enlightenment thinking; prior to this, the prevailing view was that of creation. Lamarck was the first important evolutionist, but the critical contribution was from Darwin. There are many applications of an evolutionary epistemology to cultures, with Spencer as the major figure.

existentialism   A humanistic philosophy concerned with human existence, which often emphasizes human estrangement from the larger world with the aim of reunification.

feminism   Generally, the advocacy of the rights of women to equality with men, and recognizing that sexism prevails, is wrong, and needs to be eliminated; patriarchy is usually seen as the fundamental issue. There are numerous versions of feminism. Increasingly, feminism is concerned not only with the oppression or subordination of women but also with other oppressions, such as those based on social class, skin color, income, religion, age, culture, and geographic location.

feudalism A social and economic system, prevalent in Europe prior to the Industrial Revolution, that involved two principal groups: the land was controlled by lords, and peasants were bound to the land and subject to the lords' authority. The clergy were also a distinct group within the feudal system.

first effective settlement A phrase popularized by Zelinsky to refer to the group that first establishes a viable, self-perpetuating community in an area undergoing settlement from outside. It is similar to the concept of initial occupance noted by Kniffen.

flexible accumulation Industrial technologies, labor practices, relations between firms, and consumption patterns that are increasingly flexible.

forces of production A Marxist term that refers to the raw materials, tools, and workers that actually produce goods.

Fordism A group of industrial and broader social practices initiated by Henry Ford and dominant until recently in most industrial countries. Fordism is characterized by standardized products of mass production in large factories.

formal region A cultural region occupied by a relatively distinct cultural group and displaying a relatively uniform landscape.

frontier A zone of advance penetration by an incoming group. A frontier is often an area of conflict between an existing culture and an incoming culture.

frontier thesis An environmental determinist view, introduced by the American historian Turner, that sees the American frontier as the place where civilization encountered savagery and where civilization was continually, over a 300-year period, conditioned. The result of this extended experience was a new American, not European, culture.

functionalism A concern with the analysis of functions; a form of teleological philosophy that explains social situations through an account of roles.

functional region A cultural region that functions politically or economically as an integrated unit and may be related to some homogeneity in people and/or landscape.

gaze A particular way of looking at things that may influence the conduct and outcome of research. There is a particular need for researchers to recognize and then avoid a gaze that reflects power differences such as those associated with racism or sexism.

gender The social aspect of the relations between the sexes. The term does not refer to physical attributes but rather to the socially formed traits associated with masculine and feminine categories. Masculinity and femininity are not, therefore, naturally occurring; rather, they are the consequences of human history.

gender relations The idea that gender roles are explained in terms of power relations between women and men, especially that of patriarchy.

gender roles A set of behaviors traditionally assigned to women, characterized by passivity and relationship behavior (actions that facilitate human interaction)—femininity; and a set of behaviors traditionally associated with men, characterized by activity and instrumental behavior (goal-oriented actions)—masculinity.

genocide A form of one-sided mass killing in which a state or other authority intends to destroy a particular group, as that group and membership are defined by the perpetrator group.

ghetto A residential district in an urban area with a concentration of a particular ethnic group.

globalization A process whereby the population of the world is increasingly bonding into a single culture and economy. It is often related to the emergence of a world system.

governmentality An approach developed by Michel Foucault to indicate how governments act in a manner to control and improve the subjects of society, with the goal to make society governable in a manner desired by the state. It refers to the act of governing.

group   Perhaps the most general term used to refer to some collection of human beings. It usually implies that there is a set of relations existing between individual members of the group, and also that each member is conscious of the existence of the group.

hegemony   The ability of a group to exercise control over others without needing to rely on laws or the use of force. Hegemony involves an acceptance by others of fundamentally unequal circumstances.

hermeneutics   The study and interpretation of meaning; the interpretation of texts. Hermeneutics is the process of uncovering cultural meaning in everyday life by understanding the signs and symbols of one's own group in context with those of other groups. There are many versions; all assign a key role to the mental quality of humans (compare *naturalism*).

heteronormativity The dominant and powerful system of social organization that assumes the stability of heterosexual sex roles in society. Heteronormative systems assume conformity to certain sex roles based on biological gender, and deviation from these practices is marked as deviant behavior and marginalized.

heterosexuality   The orientation of emotions and/or sexual activity toward those of the opposite sex.

historical materialism   A method, associated with Marxism and centering on the material basis of society, of attempting to understand social change by reference to historical changes in social relations.

holism   The idea that the properties of individual elements in some complex arrangement are affected by relations and interactions with other elements. Holism emphasizes the value of studying groups of things together rather than apart; ecology is holistic. (Compare *dualism*).

homeland   A type of cultural region that involves interaction with the physical environment to evoke emotional attachment and bonding. It is usually associated with a particular group defined in ethnic terms.

homosexuality   The orientation of emotions and/or sexual activity toward those of the same sex.

humanism   An approach to the study of humans and human behavior that gives priority to the fact of being human. Although the term has many connotations, modern humanism flowered in the nineteenth century as part of the conflict between science, as exemplified by Darwinism, and fundamentalist interpretations of religion. Humanists emphasize our ability to make choices and our responsibility for the actions we take. The humanistic philosophical tradition includes such figures as Vico and Berkeley, both of whom questioned the idea that mathematics and mechanics are keys to understanding the world, and Hegel, whose work was an inspiration for various idealist perspectives.

hybridity Refers to the mixture of two distinct things. Rather than an assumption of "purity" or "essential" elements, hybridity allows theorists to reference mixing, be it racial, cultural, religious, or gender mixing. In some ways hybridity rejects colonial and imperialist notions of purity, essentialism, and separation, and, rather, promotes the fusion and amalgams that colonialism and now globalization actually generate.

iconography   The description and interpretation of visual images, including landscape, in order to uncover their symbolic meanings; the identity of a region as expressed through symbols.

idealism   **1.** A group of philosophies that suggest that what is real is related to and created by the contents of the human mind. Idealism is therefore opposed to both materialism and realism. According to Berkeley, nothing exists outside the mind. According to the transcendental idealism of Kant, the categories of terms that are used to describe the world are not objective characteristics of the world but are, rather, structures imposed by the mind. According to Hegel, history comprises a progressive realization of a single spirit (*Geist*). Generally, idealism is a metaphysical view that only minds and ideas exist. **2.** As introduced into cultural geography, a specific version of the larger idealist perspective, derived from the historical idealism proposed by the historian Collingwood, explicitly concerned with rethinking the thoughts behind human actions.

identity A term that refers to sameness and continuity, and that became popular with the rise of mass society and the related quest for understanding who we are. Contemporary accounts of identity derive from both sociological and psychoanalytic inspirations and focus on who people think they are, or their sense of self. It is recognized that identity is socially constructed and intimately associated with prevailing power relations. Most cultural geographic discussions of identity emphasize one or more of the myth of race, ethnicity, language, religion, class, gender, and sexuality, and typically discuss identity formation in terms of difference.

ideology A socially ordered system of cultural symbols; a body of ideas or a way of thinking. Ideology is closely related to the concept of power, as ideological systems serve to legitimize the differential power that groups possess.

idiographic Denoting a method that stresses the individuality and uniqueness of phenomena rather than the similarities between phenomena (compare **nomothetic**).

image The perception of reality held by an individual or group.

imperialism A relationship between states in which one is dominant over the other; the process of empire establishment that took place during the period of political colonialism.

Industrial Revolution A movement that converted a fundamentally rural society into an industrial society beginning in mid-eighteenth-century England. The Industrial Revolution was primarily a technological revolution associated with new energy sources.

infrastructure (or **base**) A Marxist term that refers to the economic structure of a society, especially as it gives rise to political, legal, and social systems.

innovation An idea that leads to change, often increasing individual and/or group productivity, or a cultural trait that is new to a group. An innovator is a person who leads change.

intersectionality References the way that multiple distinct identities interact and overlap in important ways to generate various hierarchies and structures of social organization. Intersecting identities may be important in positioning people as disadvantaged or advantaged in multiple ways depending upon the context.

intertextuality The idea that meaning is produced from one text to another rather than being produced between the world, including the author(s) and the historical context, and any given text.

knowledge production (see **production of knowledge**)

Lamarckianism An evolutionary process proposed by Lamarck, which was essentially abandoned after the acceptance of natural selection. Lamarckianism states that characteristics acquired by individual members of a species through their experiences can be inherited by their offspring. For some, it is appropriate to view cultural evolution, as opposed to biological evolution, in Lamarckian terms.

landscape For many cultural geographers, the basic object of study. Landscape has both material characteristics and symbolic identity.

landscape school The principal approach to studies in cultural geography, emphasizing landscape evolution, cultural regions, and ecology. Introduced by Sauer in the 1920s, the landscape school has been influential through to the present, although there have been challenges to it, beginning in the 1980s.

language family A group of related languages derived from a single common ancestral language.

lebensraum Literally, living space. As employed by Ratzel, it is the argument that a political state is similar to a living organism in that it might require space to grow.

limits-to-growth thesis The argument that, in the future, both world population and world economy may collapse because available world resources will be inadequate.

locale A setting or context for social interaction. This term, which may be applied at a range of scales, was introduced in structuration theory and has become popular in human geography as an alternative to place. The term may be employed at a range of scales.

logical positivism   A philosophical approach, initiated by the Vienna Circle group of scholars in the 1920s, that rejected all metaphysics in favor of a scientific approach.

material culture   The visible physical objects that are made or used by a group, including clothing, buildings, and tools.

materialism   The idea that humans are dependent on the natural world. Forms of materialism include the mechanistic materialism of early science and the dialectical materialism of Marxism. The term is sometimes used as a synonym for naturalism (compare *idealism*).

mechanistic   A term closely related to Cartesianism and determinism and more generally to the scientific approach emphasizing cause and effect. A mechanistic world is one that functions like clockwork: it exhibits regularity and predictability and is subject to the operation of laws.

metaphysics   The branch of philosophy that attempts to explore the world of the suprasensible, that is, the world beyond experience. Its proponents claim that metaphysics is able to deal with issues beyond the reach of science.

minority language   A language spoken by a minority group in a state in which the majority of the population speaks some other language. A minority language may or may not be an official language.

mode of production   A Marxist term that refers to the organized social relations through which a human society organizes productive activity. In Marxist theory, human societies are seen as passing through a series of these modes.

modernism   A view that assumes the existence of a reality characterized by structure, order, pattern, and causality.

multilinear evolution   A view of evolution claiming that there are various paths of development that cultures may follow.

nation   In general terms, a group of people sharing a common culture and an attachment to some territory. *Nation* is a difficult term to define objectively. It is now commonly understood to refer to imagined rather than real communities, and has unfortunate associations with the myth of race.

nationalism   The political expression of nationhood or aspiring nationhood. Nationalism reflects a consciousness of belonging to a nation.

nation-state   A political unit that contains one principal national group and that identifies itself and its territory with that group.

naturalism   The view that all things are natural; thus, the idea that human behavior can be explained by reference to mechanistic laws similar to those that apply in physical science. A fundamental decision that any social scientist must make is whether or not to adopt a naturalist philosophy. *Naturalism* also refers to the belief that what is studied by the physical and human sciences is all there is. It includes acceptance of natural selection. (Compare *hermeneutics*.)

natural selection   The process whereby organisms better adapted to their environment tend to survive and produce more offspring. This refers to the *survival of the fittest* (a term coined by Spencer in 1852) and is the principal contribution of Darwin to the question of evolution. It is a materialist explanation for evolutionary change.

nature   It is not helpful to suggest that there is a neat and clear definition of this term that is generally agreed to by cultural geographers and others. There are numerous specific meanings depending on the context in which the term is used. To suggest a contrast of nature with culture is a reasonable general understanding, given the cultural geographic tradition.

neocolonialism   Colonial economic and cultural circumstances in a different guise from the original political colonialism.

new cultural geography   An agenda for cultural geography, first proposed in the 1980s, that stresses the need to understand rather than merely describe cultures and places. Proponents of new cultural geography reject the Sauerian concept of culture as cause, emphasizing instead the diversity of cultures, the role of power relations, and markers of human identity such as "race", gender, and sexuality. In addition, this approach to cultural geography stresses the constructive power of language, reflecting a shift in focus from modern to postmodern.

niche An ecological address; the space occupied by an organism and the activities that allow it to survive.

nomothetic Denoting a method that stresses the similarities between phenomena. It is concerned with seeking laws and is associated with positivism (compare *idiographic*).

non-material culture The oral components of culture, such as beliefs, customs, songs, paintings, and poetry.

non-representational theory Builds on the importance of affect. It focuses on practices and performance rather than representation and meaning as cognitively processed and produced. This approach requires registering sensory, emotional, performative, and visceral forms of being in and interacting with the world (see *affective geographies*).

norms Rules of conduct that identify appropriate and inappropriate behavior for members of a group. Norms are reinforced by sanctions of some form.

Occidentalism The reverse idea of Orientalism, whereby the "East" constructs an image of the "West" that reflects assumptions and essentialisms that characterize the West as militaristic and Islamophobic.

occupance The inhabiting and modification of an area by humans.

ordinary landscape The landscapes of everyday urban space—rooms, buildings, backyards, streets, and neighborhoods—and of everyday rural space—fields, fences, barns, and farmhouses.

organized capitalism A form of capitalism that developed after the Second World War, characterized by increased growth of major (often multinational) corporations and increased involvement by the state (often in the form of public ownership) in the economy.

Orientalism The particular perspective that Western scholars of the Orient have of that area. The term implies a view of the periphery from the center. Although Orientalism is now associated with postcolonial theory, primarily with the work of Said, it is clear that the idea of the Orient as a foil for the West dates back to classical Greece.

other Philosophically, an ambiguous term that derives from the writings of Hegel and that has been used especially by Lacan; it is a prominent concept in feminist and postcolonial theory. *Other* is usually defined in opposition to the same or to self and is based on the assumption that identities are defined not autonomously but by reference to something that can be either excluded or contradicted. The term may be capitalized when used in this specific context. In cultural geography, it is often used to refer to subordinate groups as these are viewed by and contrasted to dominant groups. Thus, it implies both difference and inferiority; for example, masculine identity is defined in terms of the exclusion of the feminine other.

palimpsest A writing material on which writing has been removed or partly removed to allow for subsequent writings. The term has been adopted by historical geographers to refer to cultural landscapes that contain features from a series of occupations. There is a suggestive association with the idea of a landscape as a text than can be read.

paradigm A term used to describe the stable pattern of scientific activity prevailing within a discipline.

patriarchy Literally, the rule of the fathers. The term refers to a social system in which men dominate, oppress, and exploit women. The system is often seen as comprising six factors: the family household, employment, sexuality, violence, culture, and the state. Most feminist movements seek to combat patriarchy.

perception The process by which humans acquire information about physical and social environments.

performativity A feminist-inspired term referring to the idea that bodily performance creates both identities and places.

phenomenology   The science of phenomena as distinct from that of the nature of being. *Phenomenology* refers to various philosophies that provide non-empirical descriptions of phenomena. The modern version, based on the work of Husserl, tries to reveal phenomena as intuited essences through direct awareness.

place   Location. In humanistic geography, this term has acquired a particular meaning as a context for human action that is rich in human significance and meaning. Use of this term usually implies a rejection of various scientific approaches, including positivism.

placelessness   Homogeneous and standardized landscapes that lack local variety and character. Placelessness is sometimes the result of the spread of popular culture at the expense of local cultures. Globalization is an extreme version of this.

plural society   A culture comprising several ethnic groups, each living in a community largely separate from the others.

political ecology   An approach to the study of human-and-nature relationships that employs Marxist concepts to stress political dimensions of the relationship and that typically focuses on rural and agrarian issues in the less developed world.

positivism   A movement introduced by Comte, related to both empiricism and naturalism, that organized knowledge and technology into a consistent whole. It posits that the history of human thought evolved through three stages—religious, metaphysical, and scientific—and that the sciences form a natural hierarchy ranging from mathematics through to the human science of sociology. The philosophy of logical positivism is often equated with a scientific approach.

postcolonialism   An aspect of the cultural studies perspective, more specifically of postmodernism, that explicitly opposes the ethnocentrism seen as a fundamental component of the European cultural tradition. Postcolonialism sees national cultures of previously colonial areas as defined by the tensions related to the history of colonial domination.

post-Fordism   A group of industrial and broader social practices evident in industrial countries since about 1970. Post-Fordism involves more flexible production methods than those associated with Fordism. It involves the decentralized use of information technologies.

postmodernism   A movement in philosophy, social science, and the arts, arguing that reality cannot be studied objectively and stressing that multiple interpretations are both possible and legitimate.

poststructuralism   A complex body of philosophical thought that is sometimes loosely equated with postmodernism or even with deconstruction. It is, however, better regarded as a set of ideas that expand upon the logic of structuralism to insist that meaning is produced within language and that human actions are constrained by structures.

power   In general, the capacity to affect outcomes; more specifically, the capacity to dominate others by means of violence, force, manipulation, or authority. It is characteristic of an oligarchic society in which power is in the hands of an elite group. Interpreting culture in terms of power enables us to consider what are often called cultural politics and the politics of identity.

preadaptation   The condition of being adapted to life in conditions not previously encountered. A *preadapted* culture already possesses the necessary cultural traits to allow successful occupation of a new environment prior to movement to that environment; groups with these characteristics have a competitive advantage.

prejudice   Preconceived ideas about an individual or group, which foster resistance to change even in the light of contrary evidence. The term means, literally, "prejudging".

production of knowledge   Critical awareness of how meaning is produced through institutional rules and discourses that frame specific epistemes (in Foucault's terms). The *production of knowledge* is therefore bound-up systems of power, but such systems are themselves discontinuous and subject to transformation.

queer theory A group of ideas developed in gay and lesbian studies and concerned with oppressed sexualities in terms of both social rights and cultural politics.

race Subspecies; a physically distinguishable population within a species. The term is not applicable to the human species, which comprises one and only one species.

racialization While "race" is not scientifically valid as a category of difference, racialization is a real process with real outcomes. Racialization refers to discursive constructions of difference that create negative material outcomes for specific subpopulations. Complex and multiple social, ethnic, regional, and religious differences are collapsed and essentialized and used to negatively categorize people using predetermined lines of assumption. Sometimes this includes differences formed by racist perspectives, but the process can also apply to religious and other differences.

racism A particular form of prejudice attributing characteristics of superiority or inferiority to a group of people who share some physically inherited characteristics.

radical geography A strand of geography that emerged in the 1960s and 1970s that was explicitly directed at challenging systems of oppression and inequality. The approach encouraged the practical and political objectives of change inspired by socialist and feminist activism.

realism A philosophical view holding that material objects exist independently of sense experiences. Realism aims to reveal the causal mechanisms through which events are situated within underlying structures (compare *idealism*).

redlining A term used to indicate the institutional practice of linking ethnicity to higher financial risk in the spatialization of variable rates of mortgage repayment risk.

reductionism Any doctrine that claims to be able to make seemingly complex matters comprehensible in more simple and limited terms.

reflexivity The need for researcher self-reflection during the research process in order to ensure that the positionality of the researcher is not unduly affecting any aspect of the research activity.

relations of production A Marxist term that refers to the ways in which the production process is organized, specifically to the relationships of ownership and control.

representation Traditionally, the claim that there is a real world that can be mirrored by cultural geographers in their writings. This assumes a neutrality on the part of the author. It is now widely acknowledged that such a mode of representation is not possible; rather, cultural geographers interpret the world through particular lenses, especially in the context of prevailing power relations and discourses.

research The process of systematic study and investigation in order to gather evidence, data, and information to create and add to knowledge.

resource Something material or abstract that can be used by humans to satisfy a need or perceived deficiency.

restructuring In a capitalist economy, changes in or between the various component parts of an economic system, resulting from economic change.

satisficing behavior A model of human behavior that rejects the rationality assumptions of the economic operator, assuming instead that the objective is to reach a level of satisfaction that is acceptable.

semiotics The study of signs. Semiotics assumes that the meaning of cultural materials is to be understood through an analysis of signs in the context of a particular discourse. It is possible to study many different visual materials in this way, including the built environment.

sense of place The deep attachments that humans have to specific locations, such as home, and also to particularly distinctive locations.

sequent occupance An approach to evolutionary landscape analysis that recognizes a series of stages during which the cultural landscape is essentially unchanging, with periods of rapid and profound change occurring between stages.

sexism   Attitudes or beliefs that serve to justify sexual inequalities by incorrectly attributing or denying certain capacities either to women or to men.

sexuality   In some feminist and psychoanalytic theory, a cultural construct rather than a biological given that is aligned with power and control.

simplification   A process of cultural change involving the creation of a less complex culture. Simplification is often experienced by groups moving to a different environment.

simulation   A method of representing a real process in an abstract form for the purposes of experimentation.

situated knowledge   The idea that knowledge is not neutral and cannot be acquired in some detached and disembodied manner, but all knowledge is partial and located somewhere. Work that emphasizes the situated character of knowledge often highlights the positionality of the researcher and the subjects of research. This is an idea that rejects notions of researcher authority and impartiality.

slavery   Labor that is controlled through compulsion and does not involve remuneration. In Marxist terminology, slavery is one example of a mode of production.

social construction   The recognition that all knowledge reflects the fact that we are born into an existing society that precedes any individual development, with that social knowledge becoming a part of our world view and ideology (compare *essentialism*). The idea of social construction is often applied in the context of identity and has long been a fundamental part of much sociological thinking.

social Darwinism   An interpretation of Darwinian concepts that is applied to human societies. Initially proposed by Spencer, social Darwinism represents an evolutionary and naturalist conception of society.

socialism   A social and economic system that involves common ownership of the means of production and distribution.

society   The system of interrelationships that connect individuals as members of a culture.

sociobiology   A modern growth area within what some call neo-Darwinism. Sociobiology is a concern with the evolutionary interpretation of species behavior, especially concerning social interaction. Behavior is interpreted in terms of strategies that have selective advantage in that they increase the chances of survival. Applied to humans, sociobiology has proven controversial because of the implication that human behavior is genetically determined.

spectacle   Any place or event that is an example of carefully created mass leisure and consumption.

stereotype   A collection of expectations and beliefs about a particular group of people that effectively serves as a blueprint affecting how they are perceived and understood by others.

Stoicism   A Greek philosophical tradition that was materialist, viewing the earth as designed and fit for human life.

strategic essentialism   The decision to emphasize gender at the expense of other markers of identity on the grounds that such a position is justified for practical and/or political reasons.

structuralism   A range of philosophies sharing the view that the empirical world of observable phenomena results from underlying structures.

structuration theory   A social theory, developed by Anthony Giddens, that aims to integrate knowledgeable human agents and the social structures of which they are a part.

subaltern (studies)   A term used in the cultural studies perspective to refer to groups that are considered, on the basis of class, caste, gender, race, or culture, to be socially inferior to other groups. The term is associated with the concept of hegemony, with its focus on relationships between dominant and subordinate groups.

superorganic   Denoting an interpretation of culture that sees it as being above both nature and individuals and therefore as the principal cause of the human world. The superorganic view of culture represents a form of cultural determinism.

superstructure   A Marxist term that refers to the political, legal, and social systems of a society.

surrogate   Substitute data used to represent a variable when precise data pertaining to the desired variable are not available.

taken-for-granted world   The world of everyday living and thinking, sometimes called the lifeworld and most closely associated with phenomenology; the intersubjective world of lived experience and shared meanings.

teleology   The doctrine that everything in the world has been designed by God. The term also refers to the study of purposiveness in the world, the idea that some phenomena are best explained in terms of ends (what they have become or what they achieve) rather than in terms of causes. Teleology is sometimes used to refer to a recurring theme in history, such as progress or class conflict.

text   A term that originally referred to the written page but that has broadened to include all human activities, products, and representations that can be read, for example, maps and landscape. Texts can be regarded as indicators of deeper cultural realities. Postmodernists recognize that there are any number of realities depending on how a text is read. The term is sometimes seen as synonymous with discourse.

topophilia   The affective ties that humans have with particular places; literally, love of place.

tourist gaze   A concept introduced by sociologist John Urry to refer to how peoples and places are viewed and consumed by tourists, and to how viewings and consumptive preferences and practices are directed by those responsible for constructing and marketing the tourist site.

transnational corporations (or transnational companies)   Large business organizations that operate in two or more countries.

uniformitarianism   The idea that those physical processes that affected the earth in the past continue to operate today, and vice versa (compare *catastrophism*).

unilinear evolution   A view of cultural evolution that claims that all cultures pass through the same sequence of stages.

universalizing religion   A religion that does not have a restricted domain because of its claim that its beliefs are appropriate for all people.

values   Ideas held by an individual or a group concerning what is good, bad, appropriate, and inappropriate. Different cultures possess different values.

vernacular region   A vernacular cultural region is identified as such on the basis of the perceptions held by those both inside and outside the region. A vernacular region usually has a generally accepted name.

verstehen   A research method, associated primarily with phenomenology, that involves the researcher adopting the perspective of the individual or group under investigation. Verstehen is a German term that is best translated as "sympathetic or empathetic understanding".

world system   A cultural system of global dimensions linking different cultures in some key respects.

References

Abler, R., J.S. Adams, and P. Gould. 1971. *Spatial Organization: A Geographer's View of the World*. Englewood Cliffs, NJ: Prentice Hall.

Abu-Lughod, J.L. 1989. *Before European Hegemony: The World System A.D. 1250–1350*. New York: Oxford University Press.

Acemoglu, D., S. Johnson, and J. Robinson. 2001. "The Colonial Origins of Comparative Development: An Empirical Observation". *The American Economic Review* 91, no. 1, 5: 1369–1401.

Adger, W.N. 2003. "Social Capital, Collective Action, and Adaptation to Climate Change". *Economic Geography* 79, no. 4:387–404.

Adriansen, H.K. 2008. "Understanding Pastoral Mobility: The Case of Senegalese Fulani". *Geographical Journal* 174:207–22.

Agnew, J.A. 1989. "The Devaluation of Place in Social Science". In *The Power of Place: Bringing Together Sociological and Geographical Imaginations*, edited by J.A. Agnew and J.S. Duncan, 9–29. Boston: Unwin Hyman.

———. 1997a. "Geographies of Nationalism and Ethnic Conflict". In *Political Geography: A Reader*, edited by J.A. Agnew, 317–24. New York: Arnold.

———. 1997b. "Places and the Politics of Identities". In *Political Geography: A Reader*, edited by J.A. Agnew, 249–55. New York: Arnold.

———, and J.S. Duncan. 1989. "Introduction". In *The Power of Place: Bringing Together Sociological and Geographical Imaginations*, edited by J.A. Agnew and J.S. Duncan, 1–8. Boston: Unwin Hyman.

Aitchison, J., and H. Carter. 1990. "Battle for a Language". *Geographical Magazine* 42, no. 3:44–6.

Aitken, S.C. 2001. *Geographies of Young People: The Morally Contested Spaces of Modernity*. New York: Routledge.

Alboim, N., and The Maytree Foundation. 2009. "Adjusting the Balance: Fixing Canada's Economic Immigration Policies". <http://www.maytree.com/wp-content/uploads/2009/07/adjustingthebalance-final.pdf>.

Albrow, M. 1997. *The Global Age: State and Society Beyond Modernity*. Stanford, CA: Stanford University Press.

Althusser, L. 1969. *For Marx*, translated by B. Brewster. New York: Penguin.

Alvarez, A. 2001. *Governments, Citizens and Genocide: A Comparative and Interdisciplinary Approach*. Indianapolis: Indiana University Press.

Amnesty International. 2009. "Europe's Roma Community Still Facing Massive Discrimination". <http://www.amnesty.org/en/news-and-updates/feature-stories/europes-roma-community-still-facing-massive-discrimination-20090408>.

Anderson, B. 1983. *Imagined Communities: Reflections on the Origins and Spread of Nationalism*. London: Verso.

Anderson, Ben. 2009. "Non-representational Theory". In *The Dictionary of Human Geography*, 5th edn, edited by D. Gregory, R. Johnston, G. Pratt, M. Watts, and S. Whatmore. Padstow, UK: Wiley-Blackwell, 135–8.

———. 2010. "Morale and the Affective Geographies of the 'War on Terror'". *Cultural Geographies* 17, no. 2:219–36.

Anderson, B., and P. Harrison. 2010. *Taking-Place: Non-representational Theories and Geography*. Ashgate Publishing Company.

Anderson, K.J. 1991. *Vancouver's Chinatown: Racial Discourse in Canada, 1875–1980*. Montreal: MQUP.

——— .1995. "Culture and Nature at the Adelaide Zoo: At the Frontiers of 'Human' Geography". *Transactions of the Institute of British Geographers* NS 20:275–94.

———. 1997. "A Walk on the Wild Side: A Critical Geography of Domestication". *Progress in Human Geography* 21:463–85.

———. 2000. "'The Beast Within': Race, Humanity, and Animality". *Environment and Planning D: Society and Space* 18:301–20.

———, M. Domosh, S. Pile, and N. Thrift, eds. 2003a. *Handbook of Cultural Geography*. Thousand Oaks, CA: Sage.

———, M. Domosh, S. Pile, and N. Thrift. 2003b. "Preface". In *Handbook of Cultural Geography*, edited by K. Anderson, M. Domosh, S. Pile, and N. Thrift, xviii–xix. Thousand Oaks, CA: Sage.

———, and F. Gale, eds. 1999. *Cultural Geographies*, 2nd edn. Harlow, UK: Longman.

———. 2007. *Race and the Crisis of Humanism*. New York: Routledge.

———. 2008. "'Race' in Post-universalist Perspective". *Cultural Geographies* 15:155–71.

Anderson, W. 2001. *Lost Diners and Roadside Restaurants of New England and New York: A Tribute to over 100 of the Northeast's Old Time Best*. Utica, NY: North Country Books.

Andrews, H.F. 1984. "The Durkheimians and Human Geography: Some Contextual Problems in the Sociology of Knowledge". *Transactions of the Institute of British Geographers* NS 9:315–36.

Angel Symposium, The. 2008 Conference Report. <http://www.publicartonline.org.uk/resources/reports/documents/AngelSymposium.pdf>.

APCOM (Asian Pacific Coalition on Male Sexual Health). 2012. "South Asia Legal Environments for Men Who Have Sex with Men and Transgender People". <http://www.apcom.org/tl_files/2012_resources/12_05_resources/PolicyBrief-SouthAsia.pdf>.

Arai, M., and P.S. Thoursie. 2009. "Renouncing Personal Names: An Empirical Examination of Surname Change and Earnings". *Journal of Labor Economics*, *27*(1), 127–47.

Arendt, H. 1951. *The Origins of Totalitarianism*. New York: Harcourt Brace.

Arkell, T. 1991. "Geography on Record". *Geographical Magazine* 63, no. 7:30–4.

Arnold, D. 1996. *The Problem of Nature: Environment, Culture and European Expansion*. Cambridge, MA: Blackwell.

Arreola, D.D. 2006. "Spencer, J.E. and Thomas, W.L. Jr 1969: *Cultural Geography: An Evolutionary Introduction to Our Humanized Earth*". *Progress in Human Geography* 30: 667–71.

Aschmann, H. 1965. "Athapaskan Expansion in the Southwest". *Association of Pacific Coast Geographers Yearbook* 32:79–97.

———. 1987. "Carl Sauer, A Self Directed Career". In *Carl O. Sauer: A Tribute*, edited by M.S. Kenzer, 137–43. Corvallis: Oregon State University Press.

Atkinson, D., P. Jackson, D. Sibley, and N. Washbourne. 2005. *Cultural Geography: A Critical Dictionary of Key Concepts*. London: I.B. Tauris.

Augelli, J.P. 1958. "Cultural and Economic Changes of Bastos, a Japanese Colony on Brazil's Paulista Frontier". *Annals of the Association of American Geographers* 48:3–19.

Axford, B. 1995. *The Global System: Economics, Politics and Culture*. New York: St Martin's Press.

Azaryahu, M., and A. Kellerman. 1999. "Symbolic Places of National History and Revival: A Study in Zionist Mythical Geography". *Transactions of the Institute of British Geographers* NS 24:109–23.

Bacevich, A. 2008. "He Told Us to Go Shopping. Now the Bill Is Due". *Washington Post* (5 October). <http://www.washingtonpost.com/wp-dyn/content/article/2008/10/03/AR2008100301977.html>.

Badcock, B. 1996. "'Looking-Glass' Views of the City". *Progress in Human Geography* 20:91–9.

Baker, A.R.H. 2003. *Geography and History: Bridging the Divide*. New York: Cambridge University Press.

———. 2007. "Classifying Geographical History". *Professional Geographer* 59:344–56.

———, and R.A. Butlin, eds. 1973. *Studies of Field Systems in the British Isles*. New York: Cambridge University Press.

Barber, B.R. 1995. *Jihad vs McWorld: How Globalism and Tribalism Are Reshaping the Modern World*. New York: Ballantine Books.

Barcus, H.R. 2007. "The Emergence of New Hispanic Settlement Patterns in Appalachia". *Professional Geographer* 59:298–315.

Barker, R.G. 1968. *Ecological Psychology: Concepts and Methods for Studying the Environment and Behavior*. Stanford, CA: Stanford University Press.

Barnes, T.J. 1996a. *Logics of Dislocation: Models, Metaphors, and Meanings of Economic Space*. New York and London: Guilford Press.

———. 1996b. "Political Economy II: Compliments of the Year". *Progress in Human Geography* 20:521–8.

———. 2003. "Introduction: 'Never Mind the Economy. Here's Culture'". In *Handbook of Cultural Geography*, edited by K. Anderson, M. Domosh, S. Pile, and N. Thrift, 89–97. Thousand Oaks, CA: Sage.

———, and M. Curry. 1983. "Towards a Contextualist Approach to Geographical Knowledge". *Transactions of the Institute of British Geographers* NS 8:467–82.

———, and J.S. Duncan. 1992. "Introduction: Writing Worlds". In *Writing Worlds: Discourse, Text and Metaphor in the Representation of Landscape*, edited by T.J. Barnes and J.S. Duncan, 1–17. New York: Routledge.

Barnett, C. 2009. "Culture". In *The Dictionary of Human Geography*, 5th edn, edited by D. Gregory, R. Johnston, G. Pratt, M. Watts, and S. Whatmore, 135–8. Padstow, UK: Wiley Blackwell.

Barrell, J. 1982. "Geographies of Hardy's Wessex". *Journal of Historical Geography* 8:347–61.

Barrows, H.H. 1923. "Geography as Human Ecology". *Annals of the Association of American Geographers* 13:1–14.

Barth, F. 1956. "Ecologic Relationships of Ethnic Groups in Swat, North Pakistan". *American Anthropologist* 58:1079–89.

———, ed. 1969. *Ethnic Groups and Boundaries*. Boston: Little Brown.

Bassett, T.J. 1988. "The Political Ecology of Peasant–Herder Conflicts in the Northern Ivory Coast". *Annals of the Association of American Geographers* 78:453–72.

Batterbury, S. 2001. "Landscapes of Diversity: A Local Political Ecology of Livelihood Diversification in South-Western Nigeria". *Ecumene* 8:437–64.

Bauer, R. 2003. *The Cultural Geography of Colonial American Literatures*. New York: Cambridge University Press.

Bayly, C.A. 2003. *The Birth of the Modern World, 1780–1914: Global Connections and Comparisons*. New York: Blackwell.

BBC News. 2008. "Angel Delight" (15 February). <http://news.bbc.co.uk/2/hi/uk_news/magazine/7245477.stm>.

Bebbington, A.J., and S. Batterbury. 2001. "Transnational Livelihoods and Landscapes: Political Ecologies of Globalization". *Ecumene* 8:369–80.

Bell, D., and G. Valentine. 1995. "Introduction: Orientations". In *Mapping Desire: Geographies of Sexualities*, edited by D. Bell and G. Valentine, 1–27. New York: Routledge.

———, and ———. 1999. "Geographies of Sexuality—A Review of Progress". *Progress in Human Geography* 23:175–87.

Bell, J. 1999. "Redefining National Identity in Uzbekistan: Symbolic Tensions in Tashkent's Official Public Landscape". *Ecumene* 6:183–211.

Bell-Fialkoff, A. 1993. "'A Brief History of Ethnic Cleansing'". *Foreign Affairs* 72, no. 3:110–21.

Bender, B. 1993. *Landscape: Politics and Perspectives.* Oxford: Berg.

———. 2001. "Introduction". In *Contested Landscapes: Movement, Exile and Place*, edited by Barbara Bender and Margot Winer, 1–17. New York: Berg.

Benediktsson, K. 2007. "'Schenophobias'. Geography and the Aesthetic Politics of Landscape". *Geografiska Annaler* 89B:203–17.

Benjaminsen, T.A., and B. Ba. 2009. "Farmer-Herder Conflicts, Pastoral Marginalisation and Corruption: A Case Study from the Inland Niger Delta of Mali". *Geographical Journal* 175:71–81.

Benko, G., and U. Strohmayer. 1997. "Preface". In *Space and Social Theory: Interpreting Modernity and Postmodernity*, edited by G. Benko and U. Strohmayer, xiii–xvi. Malden, MA: Blackwell.

Bennett, J.W. 1976. *The Ecological Transition: Cultural Anthropology and Human Adaptation.* New York: Pergamon.

———. 1993. *Human Ecology as Human Behavior: Essays in Environmental and Developmental Anthropology.* New Brunswick, NJ: Transaction Publishers.

Bennion, L.C. 2001. "Mormondom's Desert Homeland". In *Homelands: A Geography of Culture and Place across America*, edited by R.L. Nostrand and L.E. Estaville, 184–209. Baltimore: Johns Hopkins University Press.

Berdoulay, V. 1978. "The Vidal–Durkheim Debate". In *Humanistic Geography*, edited by D. Ley and M. Samuels, 77–90. Chicago: Maaroufa Press.

Beresford, M.W. 1957. *History on the Ground: Six Studies in Maps and Landscapes.* London: Lutterworth.

Berg, L.D. 1993. "Between Modernism and Postmodernism". *Progress in Human Geography* 17:490–507.

———, and R.A. Kearns. 1996. "Naming as Norming: 'Race', Gender, and the Identity Politics of Naming Places in Aotearoa/New Zealand". *Environment and Planning D: Society and Space* 14:99–122.

———, and R.A. Kearns. 1997. "Constructing Cultural Geographies of Aotearoa". *New Zealand Geographer* 53, no. 2:1–2.

Berger, P.L., and T. Luckman. 1966. *The Social Construction of Reality: A Treatise in the Sociology of Knowledge.* Garden City, NY: Doubleday.

———, and S.P. Huntington, eds. 2002. *Many Globalizations: Cultural Diversity in the Contemporary World.* New York: Oxford University Press.

Berry, B.J.L. 1995. "Editorial: The Postmodernist Pursuit of Pragna". *Urban Geography* 16:95–7.

———, E.C. Conkling, and D.M. Ray. 1997. *The Global Economy in Transition*, 2nd edn. Upper Saddle River, NJ: Prentice Hall.

Berry, J.W. 1984. "Cultural Ecology and Individual Behavior". In *Human Behavior and Environment, Advances in Theory and Research, Volume 4, Environment and Culture*, edited by I. Altman, A. Rapaport, and J.F. Wohlwill, 83–106. New York: Plenum Press.

———. 1997. "Immigration, Acculturation, and Adaptation". *Applied Psychology: An International Review* 46:5–68.

Bhaskar, R. 1989. *The Possibility of Naturalism: A Philosophical Critique of the Contemporary Human Sciences*, 2nd edn. Brighton, England: Harvester Press.

Bhattacharyya, Sambit. 2009. "Root Causes of African Underdevelopment". *Journal of African Economies* 18, no. 5:745–80. doi:10.1093/jae/ejp009.

Binnie, J. 1997. "Coming Out of Geography: Towards a Queer Epistemology". *Environment and Planning D: Society and Space* 15:223–37.

———, and G. Valentine. 1999. "Geographies of Sexuality—A Review of Progress". *Progress in Human Geography* 23:175–87.

Birch, B.P. 1981. "Wessex, Hardy and the Nature Novelists". *Transactions of the Institute of British Geographers* NS 6:348–58.

Blaikie, P. 1978. "The Theory of Spatial Diffusion of Innovations: A Spacious Cul-de-Sac". *Progress in Human Geography* 2:270–95.

———. 1985. *The Political Economy of Soil Erosion in Developing Countries.* London: Longman.

———, and H. Brookfield. 1987. *Land Degradation and Society.* London: Methuen.

Blainey, G. 1983. *A Land Half Won.* Melbourne: Sun Books.

Blau, P.M., and J.W. Moore. 1970. "Sociology". In *A Reader's Guide to the Social Sciences*, 2nd edn, edited by B.F. Hoselitz, 1–40. New York: Free Press.

Blaut, J.M. 1977. "Two Views of Diffusion". *Annals of the Association of American Geographers* 67:343–9.

———. 1980. "A Radical Critique of Cultural Geography". *Antipode* 12:25–9.

———. 1984. "Commentary on Nostrand's 'Hispanos' and Their 'Homeland'". *Annals of the Association of American Geographers* 74:157–64.

———. 1993a. "Mind and Matter in Cultural Geography". In *Culture, Form, and Place: Essays in Cultural and Historical Geography*, edited by K. Mathewson. *Geoscience and Man* 32:345–56. Baton Rouge: Louisiana State University, Department of Geography and Anthropology.

———. 1993b. *The Colonizer's Model of the World: Geographical Diffusionism and Eurocentric History.* New York: Guilford Press.

Blumler, M.A. 1996. "Ecology, Evolutionary Theory and Agricultural Origins". In *The Origins and Spread of Agriculture and Pastoralism in Eurasia*, edited by D.R. Harris, 25–50. Washington, DC: Smithsonian Institution Press.

Blunt, A., and R. Dowling. 2006. *Home.* New York: Routledge.

Blunt, A., and C. McEwan, eds. 2002. *Postcolonial Geographies.* New York: Continuum.

———, and J. Wills. 2000. *Dissident Geographies: An Introduction to Radical Ideas and Practice.* New York: Prentice Hall.

———, et al. 2003. *Cultural Geography in Practice.* New York: Arnold.

Boas, F. 1928. *Anthropology and Modern Life.* New York: Norton.

Bondi, L. 1997. "In Whose Words? On Gender Identities, Knowledge and Writing Practices". *Transactions of the Institute of British Geographers* NS 22:245–58.

———, and J. Davidson. 2003. "Troubling the Place of Gender". In *Handbook of Cultural Geography*, edited by K. Anderson, M. Domosh, S. Pile, and N. Thrift, 325–43. Thousand Oaks, CA: Sage.

———, and M. Domosh. 1992. "Other Figures in Other Landscapes: On Feminism, Postmodernism and Geography". *Environment and Planning D: Society and Space* 10:199–213.

Bonnett, A. 2000. *White Identities.* New York: Prentice Hall.

———, and A. Nayak. 2003. "Cultural Geographies of Racialization—The Territory of Race". In *Handbook of Cultural Geography*, edited by K. Anderson, M. Domosh, S. Pile, and N. Thrift, 300–12. Thousand Oaks, CA: Sage.

Boone, J.A. 1996. "Queer Sites in Modernism: Harlem/ The Left Bank/Greenwich Village". In *The Geography of Identity*, edited by P. Yaeger, 243–72. Ann Arbor: University of Michigan Press.

Botkin, D. 1990. *Discordant Harmonies: A New Ecology for the Twenty-First Century.* New York: Oxford University Press.

Boudreau, J.A., R. Keil, and D. Young. 2009. *Changing Toronto: Governing Urban Neoliberalism.* University of Toronto Press.

Boulding, K.E. 1950. *A Reconstruction of Economics.* New York: Wiley.

———. 1956. *The Image.* Ann Arbor: University of Michigan Press.

Bourdieu, P. 1977. *Outline of a Theory of Practice*, translated by R. Nice. New York: Cambridge University Press.

———. 1994. *Distinction: A Social Critique of the Judgement of Taste.* New York: Routledge.

Bowden, M.J. 1976. "The Great American Desert in the American Mind: The Historiography of a Geographical Notion". In *Geographies of the Mind: Essays in Historical Geosophy in Honor of John Kirtland Wright*, edited by D. Lowenthal and M.J. Bowden, 119–47. New York: Oxford University Press.

Bowen, D.S. 1996. "Carl Sauer, Field Exploration, and the Development of American Geographical Thought". *Southeastern Geographer* 36:176–91.

Bowen, E. 1981. *Empiricism and Geographical Thought.* New York: Cambridge University Press.

Bowler, P.J. 1992. *The Fontana History of the Environmental Sciences.* London: Fontana Press.

Bowman, I. 1934. *Geography in Relation to the Social Sciences.* New York: Charles Scribner.

Boyd, R. 1999. *The Coming of the Spirit of Pestilence.* Vancouver: UBC Press.

Boyd, R.T., G. Guilmet, D.L. Whited, and N. Thompson. 1991. "The Legacy of Introduced Disease: The Southern Coast Salish". *American Indian Culture and Research Journal* 15, 4, no. 7–8:11.

Brace, C. 1999. "Finding England Everywhere: Regional Identity and the Construction of National Identity, 1840–1940". *Ecumene* 6:90–109.

Brack, D., and I. Dale, eds. 2003. *Prime Minister Portillo— And Other Things That Never Happened.* London: Politico's.

Breitbart, M.M. 1981. "Peter Kropotkin: The Anarchist Geographer". In *Geography, Ideology and Social Concern*, edited by D.R. Stoddart, 134–53. Totowa, NJ: Barnes and Noble.

Broek, J.O.M. 1932. *The Santa Clara Valley, California: A Study in Landscape Change.* Utrecht: Oosthoek.

———. 1965. *Compass of Geography.* Columbus: Merrill.

———, and J.W. Webb. 1978. *A Geography of Mankind*, 2nd edn. New York: McGraw-Hill.

Brook, T. 2008. *Vermeer's Hat: The Seventeenth Century and the Dawn of the Global World.* London: Profile Books.

Brookfield, H.C. 1964. "Questions on the Human Frontiers of Geography". *Economic Geography* 40:283–303.

Browett, J. 1980. "Development, the Diffusionist Paradigm, and Geography". *Progress in Human Geography* 4:57–79.

Brown, M., and L. Knopp. 2003. "Queer Cultural Geographies—We're Here! We're Queer! We're Over There Too". In *Handbook of Cultural Geography*, edited by K. Anderson, M. Domosh, S. Pile, and N. Thrift, 313–24. Thousand Oaks, CA: Sage.

Brunn, S.D. 1963. "A Cultural Plant Geography of the Quince". *Professional Geographer* 15:16–18.

Bryant, R., 1982. "Rent Strike in the Gorbals". *Community Development Journal*, 17, 1: 41–6.

Bryant, R.L. 1997. "Beyond the Impasse: The Power of Political Ecology in Third World Environmental Research". *Area* 29:5–19.

———. 2001. "Political Ecology: A Critical Agenda for Change". In *Social Nature: Theory, Practice, and Politics*, edited by N. Castree and B. Brown, 151–69. Malden, MA: Blackwell.

Buchanan, R.H., E. Jones, and D. McCourt, eds. 1971. *Man and His Habitat: Essays Presented to Emyr Estyn Evans*. London: Routledge and Kegan Paul.

Buckingham-Hatfield, S. 2000. *Gender and Environment*. New York: Routledge.

Bunkse, E.V. 1996. "Humanism: Wisdom of the Heart and Mind". In *Concepts in Human Geography*, edited by C. Earle, K. Mathewson, and M.S. Kenzer, 355–81. Lanham, MD: Rowman and Littlefield.

Buruma, I., and A. Margalit. 2004. *Occidentalism: The West in the Eyes of Its Enemies*. New York: Penguin.

Butler, J. 1990. *Gender Trouble: Feminism and the Subversion of Identity*. New York: Routledge.

———. 1993. *Bodies That Matter: On the Discursive Limits of "Sex"*. New York: Routledge.

Butler, R., and H. Parr, eds. 1999. *Mind and Body Spaces: Geographies of Disability, Illness and Impairment*. New York: Routledge.

Butler, W.F. 1872. *The Great Lone Land*. London: Sampson Low.

Butlin, R.A. 1993. *Historical Geography: Through the Gates of Space and Time*. New York: Arnold.

Buttimer, A. 2003. "Cultural Geographers Forum: What are the Five Most Important Principles that Should be Covered in an Introductory Course in Cultural Geography and Why?" *Place and Culture, The Newsletter of the Cultural Geography Specialty Group of the Association of American Geographers* Spring, 3–4.

Butzer, K.W. 1980. "Civilizations: Organisms or Systems?" *American Scientist* 68:517–23.

———. 1989a. "Hartshorne, Hettner, and *The Nature of Geography*". In *Reflections on Richard Hartshorne's* The Nature of Geography, edited by J.N. Entrikin and S.D. Brunn, 35–52. Washington, DC: Association of American Geographers, Occasional Publication.

———. 1989b. "Cultural Ecology". In *Geography in America*, edited by G.L. Gaile and C.J. Willmott, 192–208. Columbus, OH: Merrill.

Calder, A. 1998. "'The Nearest Approach to a Desert': Implications of Environmental Determinism in the Criticism of Canadian Prairie Writing". *Prairie Forum* 23:171–82.

Callard, F. 1998. "The Body in Theory". *Environment and Planning D: Society and Space* 16:387–400.

Cameron, J.M.R. 1974. "Information Distortion in Colonial Promotion: The Case of Swan River Colony". *Australian Geographical Studies* 12:57–76.

———. 1977. *Coming to Terms: The Development of Agriculture in Pre-convict Western Australia*. Geowest 11. Perth: University of Western Australia, Department of Geography.

Carlson, A.W. 1990. *The Spanish-American Homeland: Four Centuries of Change in New Mexico's Río Arriba*. Baltimore: Johns Hopkins University Press.

Carlstein, T. 1982. *Time, Resources, Society and Ecology: On the Capacity for Human Interaction in Space and Time. Volume 1: Preindustrial Societies*. London: Allen and Unwin.

Carney, George, ed. 1994. *The Sounds of People and Places: A Geography of American Folk and Popular Music*, 3rd edn. Lanham, MD: Rowman and Littlefield.

———. 1996. "Western North Carolina: Culture Hearth of Bluegrass Music". *Journal of Cultural Geography* 16, no. 1:65–87.

Carson, R. 1962. *Silent Spring*. Boston: Houghton Mifflin.

Carter, E., J. Donald, and J. Squires, eds. 1995. *Cultural Remix: Theories of Politics and the Popular*. London: Lawrence and Wishart.

Carter, G.F. 1948. "Clark Wissler: 1870–1947". *Annals of the Association of American Geographers* 38:145–6.

———. 1968. *Man and the Land: A Cultural Geography*, 2nd edn. New York: Holt, Rinehart and Winston.

———. 1977. "A Hypothesis Suggesting a Single Origin of Agriculture". In *Origins of Agriculture*, edited by C.A. Reed, 99–109. The Hague: Mouton.

———. 1978. "Context as Methodology". In *Diffusion and Migration: Their Roles in Cultural Development*, edited by P.G. Duke et al., 55–64. Calgary: University of Calgary, Archaeological Association.

———. 1980. *Earlier Than You Think: A Personal View of Man in America*. College Station: Texas A&M University Press.

———. 1988. "Cultural Historical Diffusion". In *The Transfer and Transformation of Ideas and Material Culture*, edited by P.J. Hugill and D.B. Dickson, 3–22. College Station: Texas A&M University Press.

Carter, H., and J.G. Thomas. 1969. "The Referendum on the Sunday Opening of Licensed Premises in Wales as a Criterion of a Cultural Region". *Regional Studies* 3:61–71.

Cartier, C. 1997. "Symbolic Landscape in High Rise Hong Kong". *Focus* 44, 3:13–21.

Carvel, J. 1997. "Global Study Finds the World Speaking in 10,000 Languages". *The Guardian* (22 July).

Castells, M. 2004. *The Power of Identity*, 2nd edn. Malden, MA: Blackwell.

Castree, N. 2003. "Strange Natures: Geography and the Study of Human–Environment Relationships". In *The Student's Companion to Geography*, 2nd edn, edited by A Rogers, H. Viles, and A Goudie, 82–7. Cambridge, MA: Blackwell.

———. 2005. *Nature*. New York: Routledge.

Castree, N., and B. Braun, eds. 2001. *Social Nature: Theory, Practice and Politics*. Malden, MA and Oxford: Blackwell Publishers.

Cater, J., and T. Jones. 1989. *Social Geography: An Introduction to Contemporary Issues*. New York: Arnold.

Cavalli-Sforza, L.L., and M.W. Feldman. 1981. *Cultural Transmission and Evolution: A Quantitative Approach.* Princeton, NJ: Princeton University Press.

Chaalo Gujarat. 2012. <http://www.chaalogujarat.com/>.

Chagnon, Napoleon A. *The Yanomamo*, 6th edn. New York: Wadsworth, 2012.

Chalk, F., and K. Jonassohn. 1990. *The History and Sociology of Genocide.* New Haven, CT: Yale University Press.

Chambers, I. 1994. *Migrancy, Culture, Identity.* New York: Routledge.

Chapman, G.P. 1977. *Human and Environmental Systems: A Geographer's Appraisal.* New York: Academic Press.

Chapple, C.K., and M.E. Tucker. 2000. *Hinduism and Ecology: The Intersection of Earth, Sky, and Water.* Cambridge, MA: Harvard University Press.

Chari, S. 2009. "Subaltern Studies" entry in *The Dictionary of Human Geography*, by D. Gregory, R. Johnson, G. Pratt, M. Watts, and S. Whatmore. Wiley-Blackwell, 727.

Charlesworth, A. 1992. "Towards a Geography of the Shoah". *Journal of Historical Geography* 18:464–9.

———. 1994. "Contesting Places of Memory: The Case of Auschwitz". *Environment and Planning D: Society and Space* 12:579–93.

———. 2003. "Landscapes of the Holocaust: Schindler, Authentic History and the Lie of Landscape". In *Studying Cultural Landscapes*, edited by I. Robertson and P. Richards, 93–107. New York: Arnold.

Childe, V.G. 1936. *Man Makes Himself.* London: Watts and Co.

Chorley, R.J. 1973. "Geography as Human Ecology". In *Directions in Geography*, edited by R.J. Chorley, 155–69. London: Methuen.

Chouinard, V. 1997. "Guest Editorial. Making Space for Disabling Differences: Challenging Ableist Geographies". *Environment and Planning D: Society and Space* 15:379–87.

Christopher, A.J. 1973. "Environmental Perception in Southern Africa". *South African Geographical Journal* 55:14–22.

———. 1982. "Towards a Definition of the Nineteenth Century South African Frontier". *South African Geographical Journal* 64:97–113.

———. 1984. *Colonial Africa.* Totowa, NJ: Barnes and Noble.

———. 1994. *The Atlas of Apartheid.* New York: Routledge.

Church News. 1979. *Special Edition: The Era of Mormon Colonization.* Salt Lake City: Deseret News (26 May).

Clark, A.H. 1954. "Historical Geography". In *American Geography: Inventory and Prospect*, edited by P.E. James and C.F. Jones, 70–105. Syracuse: Syracuse University Press.

———. 1959. *Three Centuries and the Island.* Toronto: University of Toronto Press.

———. 1975. "The Conceptions of 'Empires' of the St Lawrence and the Mississippi: An Historico-Geographical View with Some Quizzical Comments on Environmental Determinism". *American Review of Canadian Studies* 5:4–27.

Clarkson, J.D. 1970. "Ecology and Spatial Analysis". *Annals of the Association of American Geographers* 60:700–16.

Clay, G. 1980. *Close Up: How to Read the American City.* Chicago: University of Chicago Press.

———. 1987. *Right Before Your Eyes.* Washington, DC: APA Planners Press.

———. 1994. *Real Places: An Unconventional Guide to America's Generic Landscape.* Chicago: University of Chicago Press.

Clayton, D. 2003. "Critical Imperial and Colonial Geographies". In *Handbook of Cultural Geography*, edited by K. Anderson, M. Domosh, S. Pile, and N. Thrift, 354–68. Thousand Oaks, CA: Sage.

Clements, F.E. 1905. *Research Methods in Ecology.* Lincoln: University of Nebraska Publishing Company.

Clifford, N., and G. Valentine, eds. 2003. *Key Methods in Geography.* Thousand Oaks, CA: Sage.

Cloke, P., C. Philo, and D. Sadler. 1991. *Approaching Human Geography: An Introduction to Contemporary Theoretical Debates.* New York: Guilford Press.

Coates, P. 1998. *Nature: Western Attitudes since Ancient Times.* Berkeley: University of California Press.

Cobb, J.C. 1992. *The Most Southern Place on Earth: The Mississippi Delta and the Roots of Regional Identity.* New York: Oxford University Press.

Cohen, S.B. 2003. *Geopolitics of the World System.* New York: Rowman and Littlefield.

Cole, T., and G. Smith. 1995. "Ghettoization and the Holocaust: Budapest 1944". *Journal of Historical Geography* 21:300–16.

Conzen, M.P. 1990. "Ethnicity on the Land". In *The Making of the American Landscape*, edited by M.P. Conzen, 221–48. Boston: Unwin Hyman.

———. 1993. "The Historical Impulse in Geographical Writing About the United States, 1850–1900". In *A Scholar's Guide to Geographical Writing on the American and Canadian Past*, by M.P. Conzen, T.A. Rumney, and G. Wynn, 3–90. Chicago: University of Chicago Press.

———. 2001. "American Homelands: A Dissenting View". In *Homelands: A Geography of Culture and Place across America*, edited by R.L. Nostrand and L.E. Estaville, 238–71. Baltimore: Johns Hopkins University Press.

———, ed. 2010. *The Making of the American Landscape*, 2nd edn. New York: Routledge.

———, T.A. Rumney, and G. Wynn. 1993. *A Scholar's Guide to Geographical Writing on the American and Canadian Past.* Chicago: University of Chicago Press.

Cook, I., D. Crouch, S. Naylor, and J. Ryan. 2000. *Cultural Turns/Geographical Turns: Perspectives on Cultural Geography.* New York: Prentice Hall.

Cook, N.D., and W.G. Lovell, eds. 1992. *"Secret Judgments of God": Old World Disease in Colonial Spanish America*. Norman: University of Oklahoma Press.

Coon, C. 1962. *The Origin of Races*, 2nd edn. New York: Knopf.

Coones, P., and J. Patten. 1986. *The Penguin Guide to the Landscape of England and Wales*. New York: Penguin.

Cooper, R. 2003. *The Breaking of Nations: Order and Chaos in the Twenty-First Century*. London: Atlantic Books.

Cosgrove, D. 1978. "Place, Landscape and the Dialectics of Cultural Geography". *Canadian Geographer* 22:66–72.

———. 1983. 'Towards a Radical Cultural Geography: Problems of Theory". *Antipode* 15:1–11.

———. 1984. *Social Formation and Symbolic Landscape*. London: Croom Helm.

———. 1985. "Prospect, Perspective and the Evolution of the Landscape Idea". *Transactions of the Institute of British Geographers* NS 10:45–62.

———. 1997a. "Spectacle and Society: Landscape as Theater in Premodern and Postmodern Cities". In *Understanding Ordinary Landscapes*, edited by P. Groth and T.W. Bressi, 99–110. New Haven, CT: Yale University Press.

———. 1997b. *Social Formation and Symbolic Landscape*, 2nd edn. Madison: University of Wisconsin Press.

———. 2003. "Landscape and the European Sense of Sight—Eyeing Nature". In *Handbook of Cultural Geography*, edited by K. Anderson, M. Domosh, S. Pile, and N. Thrift, 249–68. Thousand Oaks, CA: Sage.

———, and S. Daniels, eds. 1988. *The Iconography of Landscape*. New York: Cambridge University Press.

———, and P. Jackson. 1987. "New Directions in Cultural Geography". *Area* 19:95–101.

Cowen, T. 2000. *Creative Destruction: How Globalization is Changing the World's Cultures*. Princeton, NJ: Princeton University Press.

Crampton, A. 2001. "The Voortrekker Monument, the Birth of Apartheid, and Beyond". *Political Geography* 20:221–46.

———. 2003. "The Art of Nation-building: (Re)present-ing Political Transition at the South African National Gallery". *Cultural Geographies* 10:218–42.

Crampton, J.W., and S. Elden. 2007. *Space, Knowledge and Power: Foucault and Geography*. Aldershot, UK, and Burlington, VT: Ashgate.

Crang, M. 1998. *Cultural Geography*. New York: Routledge.

Crang, P. 1994. "It's Showtime: On the Workplace Geographies of Display in a Restaurant in Southeast England". *Environment and Planning D: Society and Space* 12:675–704.

———. 1997. "Cultural Turns and the (Re)constitution of Economic Geography: Introduction to Section One". In *Geographies of Economies*, edited by R. Lee and J. Wills, 3–15. New York: Arnold.

Cream, J. 1995. "Re-solving Riddles: The Sexed Body". In *Mapping Desire: Geographies of Sexuality*, edited by D. Bell and G. Valentine, 31–40. New York: Routledge.

Cresswell, T. 2004. *Place: A Short Introduction*. Oxford: Blackwell.

Cronk, L., N. Chagnon, and W. Irons. 2000. *Adaptation and Human Behavior: An Anthropological Perspective*. New York: Aldine de Gruyter.

Cronon, W. 1983. *Changes in the Land: Indians, Colonists, and the Ecology of New England*. New York: Hill and Wang.

———. 1992. "A Place for Stories: Nature, History, and Narrative". *Journal of American History* 78:1347–76.

———. 1995. "Introduction: In Search of Nature". In *Uncommon Ground: Toward Reinventing Nature*, edited by W. Cronon, 23–56. New York: W.W. Norton.

Crosby, A.W. 1978. "Ecological Imperialism: The Overseas Migration of Western Europeans as Biological Phenomenon". *Texas Quarterly* 21:10–22.

———. 1986. *Ecological Imperialism: The Biological Expansion of Europe, 900–1900*. New York: Cambridge University Press.

———. 1995. "The Past and Present of Environmental History". *American Historical Review* 100:1177–89.

Cross, J.A 2007. "The Expanding Role of the Amish in America's Dairy Industry". *Focus* 50, 3:7–16.

Croucher, S.L. 2004. *Globalization and Belonging: The Politics of Identity in a Changing World*. Lanham, MD: Rowman and Littlefield.

Crowley, W.K. 1978. "Old Order Amish Settlement: Diffusion and Growth". *Annals of the Association of American Geographers* 68:249–64.

Crystal, D. 1997. *English as a Global Language*. New York: Cambridge University Press.

Cumberland, K.B. 1949. "Aotearoa Maori: New Zealand about 1780". *Geographical Review* 39:401–24.

Cunfer, G. 2005. *On the Great Plains: Agriculture and Environment*. College Station: Texas A&M University Press.

Dalby, S. 1991. "Critical Geopolitics: Discourse, Difference, and Dissent". *Environment and Planning D: Society and Space* 9:261–83.

———. 1996. "Reading Robert Kaplan's 'Coming Anarchy'". *Ecumene* 3:472–96.

Daniels, S. 1989. "Marxism, Culture and the Duplicity of Landscape". In *New Models in Geography, 2*, edited by R. Peet and N. Thrift, 196–220. Boston: Unwin Hyman.

Darby, H.C. 1940. *The Draining of the Fens*. New York: Cambridge University Press.

———. 1956. "The Clearing of the Woodland in Europe". In *Man's Role in Changing the Face of the Earth*, edited by W.L. Thomas Jr, C.O. Sauer, M. Bates, and L. Mumford, 183–216. Chicago: University of Chicago Press.

———. 1973. *The New Historical Geography of England*. New York: Cambridge University Press.

———. 1977. *Domesday England*. New York: Cambridge University Press.

Darby, W.J. 2000. *Landscape and Identity: Geographies of Nation and Class in England*. New York: Berg.

Darden, J., and E. Wyly. 2010. "Cartographic Editorial: Mapping the Racial/Ethnic Topography of Subprime Inequality in Urban America". *Urban Geography* 31, 4:425–33. doi:10.2747/0272-3638.31.4.425

Davidson, W.V. 1974. *Historical Geography of the Bay Islands, Honduras*. Birmingham, AL: Southern University Press.

Dawson, G. 1994. *Soldier Heroes: British Adventure, Empire and the Imagining of Masculinities*. New York: Routledge.

Day, K. 1999. "Embassies and Sanctuaries: Women's Experiences of Race and Fear in Public Space". *Environment and Planning D: Society and Space* 17:307–28.

Dear, M. 1988. "The Postmodern Challenge: Reconstructing Human Geography". *Transactions of the Institute of British Geographers* NS 13:262–74.

———, and S. Flusty, eds. 2000. *The Spaces of Postmodernity: Readings in Human Geography*. Malden, MA: Blackwell.

———, and J.R. Wolch. 1987. *Landscapes of Despair*. Princeton, NJ: Princeton University Press.

Dei, G.S. 1996. *Antiracism Education: Theory and Practice*. Halifax: Fernwood Publications.

de Lauretis, Teresa. 1991. "Queer Theory: Lesbian and Gay Sexualities". *Differences: A Journal of Feminist Cultural Studies* 3, 2:iii-xviii.

De Longueville, B., N. Ostländer, and C. Keskitalo. 2009. "Addressing Vagueness in Volunteered Geographic Information (VGI)—A case study". Article under review for the *International Journal of Spatial Data Infrastructures Research*, Special Issue on GSDI-11.

DeLyser, D., S. Herbert, M. Crang, S. Aitken, and L. McDowell, eds. 2009. The SAGE Handbook of Qualitative Geography. Thousand Oaks, CA: Sage.

Demeritt, D. 1994. "The Nature of Metaphors in Cultural Geography and Environmental History". *Progress in Human Geography* 18:163–85.

Denevan, W.M. 1983. "Adaptation, Variation and Cultural Geography". *Professional Geographer* 35:399–406.

———, ed. 1992. *The Native Population of the Americas in 1492*. Madison: University of Wisconsin Press.

———. 2001. *Cultivated Landscapes of Native Amazonia and the Andes*. New York: Oxford University Press.

———, and K. Mathewson, eds. 2009. *Carl Sauer on Culture and Landscape: Readings and Commentaries*. Baton Rouge: Louisiana State University Press.

Dennis, R.J., and H. Clout. 1980. *A Social Geography of England and Wales*. New York: Pergamon.

Des Jardins, J.R. 2001. *Environmental Ethics: An Introduction to Environmental Philosophy*. Belmont, CA: Wadsworth.

de Steiguer, J.E. 1997. *The Age of Environmentalism*, 3rd edn. New York: McGraw-Hill.

Devereaux, R. 2012. "Occupy Wall Street Joining Communities in Call for NYPD Commissioner to Quit". *The Guardian* (20 March).

Diamond, J. 1997. *Guns, Germs, and Steel: A Short History of Everybody for the Last 13,000 Years*. New York: W.W. Norton.

Dicken, P. 1992. *Global Shift: The Internationalization of Economic Activity*, 2nd edn. New York: Guilford Press.

Dicken, S., and E. Dicken. 1979. *The Making of Oregon: A Study in Historical Geography*. Portland: Oregon Historical Society.

Dickens, P. 1996. *Reconstructing Nature: Alienation, Emancipation and the Division of Labour*. New York: Routledge.

Dickinson, R.E. 1969. *The Makers of Modern Geography*. London: Routledge and Kegan Paul.

Dijkink, G. 1996. *National Identity and Geopolitical Visions: Maps of Pride and Pain*. New York: Routledge.

Dittmer, J., and S. Larsen. 2007. "*Captain Canuck*, Audience Response and the Project of Canadian Nationalism". *Social and Cultural Geography* 8:735–53.

Dodgshon, R.A. 2008. "Geography's Place in Time". *Geografiska Annaler B* 90:1–15.

Dodson, B. 2000. "Dismantling Dystopia: New Cultural Geography for a New South Africa". In *The Geography of South Africa in a Changing World*, edited by R. Fox and K. Rowntree, 138–57. New York: Oxford University Press.

Doel, M. 1999. *Poststructuralist Geographies: The Diabolical Art of Spatial Science*. Edinburgh: Edinburgh University Press.

———. 2004. "Poststructuralist Geographies: The Essential Selection". In *Envisioning Human Geographies*, edited by P. Cloke, P. Crang, and M. Goodwin, 146–71. New York: Arnold.

———, and D.B. Clarke. 1998. "Figuring the Holocaust: Singularity and the Purification of Space". In *Rethinking Geopolitics*, edited by G. Ó Tuathail and S. Dalby, 39–61. New York: Routledge.

Dohrs, F.E., and L.M. Sommers, eds. 1967. *Cultural Geography: Selected Readings*. New York: Crowell.

Dollar, D., and A. Kraay. 2002. "Spreading the Wealth". *Foreign Affairs* 81, no. 1:120–33.

Dominy, M.D. 2001. *Calling the Station Home: Place and Identity in New Zealand's High Country*. Lanham, MD: Rowman and Littlefield.

Domosh, M. 1989. "A Method for Interpreting Landscape: A Case Study of the New York World Building". *Area* 21:347–55.

———. 1996. "Feminism and Human Geography". In *Concepts in Human Geography*, edited by C. Earle, K. Mathewson, and M.S. Kenzer, 411–27. Lanham, MD: Rowman and Littlefield.

———. 1999. "Corporate Cultures and the Modern Landscape of New York City". In *Cultural Geographies*, 2nd edn, edited by K. Anderson and F. Gale, 95–111. Harlow, UK: Longman.

———, and J. Seager. 2001. *Putting Women in Place: Feminist Geographers Make Sense of the World*. New York: Guilford Press.

Donaldson, D.P. 2001. "Teaching Geography's Four Traditions With Poetry". *Journal of Geography* 100:24–31.

Donkin, R.A. 1997. "A 'Servant of Two Masters'?". *Journal of Historical Geography* 23:247–66.

Doob, L.W. 1996. "Foreword". In *Nations without States: A Historical Dictionary of Contemporary National Movements*, edited by J. Minahan. Westport, CT: Greenwood Press.

Doolittle, W.E. 2001. *Cultivated Landscapes of Native North America*. New York: Oxford University Press.

Douglas, I., R. Huggett, and M. Robinson. 1996. "Preface". In *Companion Encyclopedia of Geography: The Environment and Humankind*, edited by I. Douglas, R. Huggett, and M. Robinson, ix–xi. New York: Routledge.

Dowling, R., and P.M. McGuirk. 1998. "Gendered Geographies in Australia, Aotearoa/New Zealand and the Asia-Pacific". *Australian Geographer* 29:279–91.

Dunbar, G.S. 1974. "Geographic Personality". In *Man and Cultural Heritage: Papers in Honor of Fred B. Kniffen*, edited by H.J. Walker and W.G. Haag. *Geoscience and Man* 5:25–33. Baton Rouge: Louisiana State University, Department of Geography and Anthropology.

———. 1977. "Some Early Occurrences of the Term 'Social Geography'". *Scottish Geographical Magazine* 93:15–20.

Duncan, J.S. 1978. "The Social Construction of Unreality: An Interactionist Approach to the Tourist's Cognition of Environment". In *Humanistic Geography*, edited by D. Ley and M.S. Samuels, 269–82. Chicago: Maaroufa Press.

———. 1980. "The Superorganic in American Cultural Geography". *Annals of the Association of American Geographers* 7:181–98.

———. 1990. *The City as Text: The Politics of Interpretation in the Kandyan Kingdom*. New York: Cambridge University Press.

———. 1993. "Commentary on 'The Reinvention of Cultural Geography'". *Annals of the Association of American Geographers* 83:517–19.

———. 1994. "After the Civil War: Reconstructing Cultural Geography as Heterotopia". In *Re-reading Cultural Geography*, edited by K.E. Foote, P.J. Hugill, K. Mathewson, and J.M. Smith, 401–8. Austin: University of Texas Press.

———. 1998. "Author's Response". *Progress in Human Geography* 22:571–3.

———. 2007. *In the Shadows of the Tropics: Climate, Race and Biopower in Nineteenth Century Ceylon*. Aldershot, UK: Ashgate.

———, and N. Duncan. 1988. "(Re)reading the Landscape". *Environment and Planning D: Society and Space* 6:117–26.

———, N.C. Johnson, and R.H. Schein, eds. 2004. *A Companion to Cultural Geography*. Malden, MA: Blackwell.

———, and D. Ley. 1993. "Introduction: Representing the Place of Culture". In *Place/Culture/Representation*, edited by J.S. Duncan and D. Ley, 1–21. New York: Routledge.

———, and J.P. Sharp. 1993. "Confronting Representation(s)". *Environment and Planning D: Society and Space* 11:473–86.

Duncan, N. 1996. "Postmodernism in Human Geography". In *Concepts in Human Geography*, edited by C. Earle, K. Mathewson, and M.S. Kenzer, 429–58. Lanham, MD: Rowman and Littlefield.

Dunn, K.M. 1997. "Cultural Geography and Cultural Policy". *Australian Geographical Studies* 35:1–11.

———, and A. McDonald. 2001. "The Geographies of Racisms in NSW: A Theoretical Exploration and Some Preliminary Findings from the mid 1990s". *Australian Geographer* 32:29–44.

Durham, W.H. 1976. "The Adaptive Significance of Cultural Behavior". *Human Ecology* 4:89–121.

Dwivedi, O.P. 1990. "Satyagraha for Conservation: Awakening the Spirit of Hinduism". In *Ethics of Environment and Development: Global Challenge, International Response*, edited by J.R. Engel and J.G. Engel, 201–12. Tucson: University of Arizona Press.

Dwyer, C. and C. Bressey, eds. 2008. *New Geographies of Race and Racism*. Aldershot, UK: Ashgate.

Earle, C. 1995. "Review Article: Historical Geography in Extremis? Splitting Personalities on the Postmodern Turn". *Journal of Historical Geography* 21:455–9.

Earley, J. 1997. *Transforming Human Culture: Social Evolution and the Planetary Crisis*. Albany: State University of New York Press.

Eden, S. 2001. "Environmental Issues: Nature versus the Environment?" *Progress in Human Geography* 25:79–85.

Eder, K. 1996. *The Social Construction of Nature*. London: Sage.

Edwards, G. 1996. "Alternative Speculations on Geographical Futures: Towards a Postmodern Perspective". *Geography* 81:217–24.

Ehrenreich, B., and A.R. Hochschild, eds. 2003. *Global Woman: Nannies, Maids and Sex Workers in the New Economy*. New York: Metropolitan Books.

Eichengreen, B. 1998. "Geography as Destiny: A Brief History of Economic Growth". *Foreign Affairs* 77, no. 2:128–33.

Eigenheer, R.A. 1973–4. "The Frontier Hypothesis and Related Spatial Concepts". *California Geographer* 14:55–69.

Elazar, D.J. 1984. *American Federalism: A View From the States*. New York: Harper and Row.

——. 1994. *The American Mosaic: The Impact of Space, Time and Culture on American Politics*. Boulder, CO: Westview Press.

Eliades, D.K. 1987. "Two Worlds Collide: The European Advance into North America". In *A Cultural Geography of North American Indians*, edited by T.E. Ross and T.G. Moore, 33–44. Boulder, CO: Westview Press.

Ellen, R. 1982. *Environment, Subsistence, and System: The Ecology of Small-Scale Social Formations*. New York: Cambridge University Press.

——. 1988. "Persistence and Change in the Relationship Between Anthropology and Human Geography". *Progress in Human Geography* 12:229–61.

Eller, J.D. 1999. *From Culture to Ethnicity to Conflict: An Anthropological Perspective on International Ethnic Conflict*. Ann Arbor: University of Michigan Press.

England, K.V.L. 1994. "Getting Personal: Reflexivity, Positionality, and Feminist Research". *Professional Geographer* 46:80–9.

English, P.W., and R.C. Mayfield. 1972. "Ecological Perspectives". In *Man, Space and Environment*, edited by P.W. English and R.C. Mayfield, 115–20. New York: Oxford University Press.

Ennals, P., and D. Holdsworth. 1981. "Vernacular Architecture and the Cultural Landscape of the Maritime Provinces: A Reconnaissance". *Acadiensis* 10:86–105.

Entrikin, J.N. 1976. "Contemporary Humanism in Geography". *Annals of the Association of American Geographers* 66:615–32.

——. 1980. "Robert Park's Human Ecology and Human Geography". *Annals of the Association of American Geographers* 70:43–58.

——. 1984. "Carl Sauer: Philosopher in Spite of Himself". *Geographical Review* 74:387–407.

——. 1988. "Diffusion Research in the Context of the Naturalism Debate in Twentieth-Century Social Thought". In *The Transfer and Transformation of Ideas and Material Culture*, edited by P.J. Hugill and D.B. Dickson, 165–78. College Station: Texas A&M University Press.

Ericksen, E.G. 1980. *The Territorial Experience: Human Ecology as Symbolic Interaction*. Austin: University of Texas Press.

Estaville, L.E. 1993. "The Louisiana-French Homeland". *Journal of Cultural Geography* 13, no. 2:31–45.

——. 2001. "*Nouvelle Acadie*: The Cajun Homeland". In *Homelands: A Geography of Culture and Place across America*, edited by R.L. Nostrand and L.E. Estaville, 83–100. Baltimore: Johns Hopkins University Press.

European Commission. 2012. Justice: EU and Roma. <http://ec.europa.eu/justice/discrimination/roma/index_en.htm>.

Evans, E.E. 1973. *The Personality of Ireland: Habitat, Heritage, and History*. New York: Cambridge University Press.

——. 1996. *Ireland and the Atlantic Heritage: Selected Writings*. Dublin: Lilliput Press.

Evernden, N. 1992. *The Social Creation of Nature*. Baltimore: Johns Hopkins University Press.

Eyles, J., and W. Peace. 1990. "Signs and Symbols in Hamilton: An Iconology of Steeltown". *Geografiska Annaler B* 72:73–88.

——, and D.M. Smith. 1978. "Social Geography". *American Behavioral Scientist* 22:41–58.

Fairhead, J., and M. Peach. 1996. *Misreading the African Landscape: Society and Ecology in a Forest-Savanna Mosaic*. New York: Cambridge University Press.

Farish, M. 2010. *The Contours of America's Cold War*. Minneapolis: University of Minnesota Press.

Febvre, L. 1925. *A Geographical Introduction to History*. New York: Knopf.

Feder, K.L., and M.A. Park. 1997. *Human Antiquity: An Introduction to Physical Anthropology and Archaeology*, 3rd edn. Mountain View, CA: Mayfield.

Ferguson, N. 1997. *Virtual History: Alternatives and Counterfactuals*. London: Picador.

Fernández-Armesto, F. 2003. *The Americas: The History of a Hemisphere*. London: Weidenfeld and Nicolson.

Finlay, R. 2004. "How Not to (Re)Write World History: Gavin Menzies and the Chinese Discovery of America". *Journal of World History* 15, 2:229–42.

Firey, W. 1945. "Sentiment and Symbolism as Ecological Variables". *American Sociological Review* 10:140–8.

——. 1947. *Land Use in Central Boston*. Cambridge, MA: MIT Press.

Fitzsimmons, M. 1989. "The Matter of Nature". *Antipode* 21:106–20.

——. 2004. "Emerging Ecologies". In *Envisioning Human Geographies*, edited by P. Cloke, P. Crang, and M. Goodwin, 30–47. New York: Arnold.

Fleure, H.J. 1919. "Human Regions". *Scottish Geographical Magazine* 35:94–105.

——. 1951. *Natural History of Man in Britain*. London: Collins.

Flew, A. 1978. *A Rational Animal and Other Philosophical Essays on the Nature of Man*. Oxford: Clarendon Press.

Fogel, R.W. 1964. *Railroads and American Economic Growth: Essays in Econometric History*. Baltimore: Johns Hopkins University Press.

Foley, M.A. 1976. "Culture Area". In *Encyclopedia of Anthropology*, edited by D.E. Hunter and P. Whitten, 104. New York: Harper and Row.

Foote, D.C., and B. Greer-Wootten. 1968. "An Approach to Systems Analysis in Cultural Geography". *Professional Geographer* 20:86–91.

Foote, K.E., A. Toth, and A. Arvay. 2000. "Hungary After 1989: Inscribing a New Past on Place". *Geographical Review* 90:301–34.

Forrest, J., and K. Dunn. 2006. "Racism and Intolerance in Eastern Australia: A Geographic Perspective". *Australian Geographer* 37:167–86.

Forsyth, T. 2003. *Critical Political Ecology: The Politics of Environmental Science*. New York: Routledge.

Foucault, M. 1970. *The Order of Things: An Archaeology of the Human Sciences*, translated by A. Sheridan. New York: Random House.

———. 1977. *Discipline and Punish: The Birth of the Prison*, translated by A. Sheridan. New York: Pantheon Books.

———. 1978. *The History of Sexuality, Volume 1: An Introduction*, translated by R. Hurley. New York: Pantheon Books.

———. 1980. *Power/Knowledge*. New York: Pantheon Books.

———. 1991. "Politics and the Study of Discourse". In *The Foucault Effect: Studies in Governmentality*, edited by G. Burchell, C. Gordon, and P. Miller. Chicago: University of Chicago Press.

Fox, E.W. 1991. *The Emergence of the Modern European World: From the Seventeenth to the Twentieth Century*. Cambridge, MA: Blackwell.

Francaviglia, R.V. 1978. *The Mormon Landscape*. New York: AMS Press.

———. 1991. *Hard Places: Reading the Landscape of America's Historic Mining Districts*. Iowa City: University of Iowa Press.

———. 1994. "Elusive Land: Changing Geographic Images of the Southwest". In *Essays on the Changing Images of the Southwest*, edited by R.V. Francaviglia and D. Narrett, 8–39. College Station: Texas A&M University Press.

Francis, R.D. 1988. "The Ideal and the Real: The Image of the Canadian West in the Settlement Period". In: *Rupert's Land: A Cultural Tapestry*, edited by R.C. Davis, 253–73. Waterloo, ON: Wilfrid Laurier Press.

Frank, A. 1998. *ReOrient: Global Economy in the Asian Age*. Berkeley: University of California Press.

Franklin, A. 2002. *Nature and Social Theory*. Thousand Oaks, CA: Sage.

Fredrickson, G.M. 1981. *White Supremacy: A Comparative Study in American and South African History*. New York: Oxford University Press.

———. 2003. *Racism: A Short History*. Princeton, NJ: Princeton University Press.

Freeman, D.B. 1985. "The Importance of Being First: Preemption by Early Adopters of Farming Innovations in Kenya". *Annals of the Association of American Geographers* 75:17–28.

Friesen, G. 1987. *The Canadian Prairies: A History*. Toronto: University of Toronto Press.

Fudge, J., and F. MacPhail. 2009. "The Temporary Foreign Worker Program in Canada: Low-Skilled Workers as an Extreme Form of Flexible Labour". *Comparative Labor Law and Policy Journal* 5: 5–45.

Gade, D.W. 1997. "'Germanic Towns in Southern Brazil". *Focus* 44, no. 1:1–6.

———. 1999. *Nature and Culture in the Andes*. Madison: University of Wisconsin Press.

Gagen, E.A. 2004. "Landscapes of Childhood and Youth". In *A Companion to Cultural Geography*, edited by J.S. Duncan, N.C. Johnson, and R.H. Schein, 404–19. Malden, MA: Blackwell.

Garrison, J.R. 1990. "Review of *The American Backwoods Frontier: An Ethnic and Ecological Interpretation* by T.G. Jordan and M. Kaups". *Annals of the Association of American Geographers* 80:639–41.

Garst, R.D. 1974. "Innovation Diffusion among the Gusii of Kenya". *Economic Geography* 50:300–12.

Gastil, R.D. 1975. *Cultural Regions of the United States*. Seattle: University of Washington Press.

Gates, H.L. Jr, ed. 1986. *"Race", Writing, and Difference*. Chicago: University of Chicago Press.

Gateshead Council. 2008. "Angel of the North: Background and History". <http://www.gateshead.gov.uk/ Leisure%20and%20Culture/attractions/Angel/ Background/Background2.aspx>.

Gay, J. 1971. *Geography of Religion in England*. London: Duckworth.

Geertz, C. 1963. *Agricultural Involution: The Processes of Ecological Change in Indonesia*. Berkeley: University of California Press.

———. 1973. *The Interpretation of Cultures*. New York: Basic Books.

———. 1996. "Afterword". In *Senses of Place*, edited by S. Feld and K.H. Basso, 259–62. Santa Fe: School of American Research Press.

———. 2003. "Off the Menu: A Review of T. Cowen, *Creative Destruction: How Globalization is Changing the World's Cultures*". *The New Republic* (17 February) 228:27–30.

Gellner, E. 1997. "Knowledge of Nature and Society". In *Nature and Society in Historical Context*, edited by M. Teich, R. Porter, and B. Gustafsson, 9–17. New York: Cambridge University Press.

Gerber, J. 1997. "Beyond Dualism—The Social Construction of Nature and the Natural and Social Construction of Human Beings". *Progress in Human Geography* 21:1–17.

Gerlach, J., and J. Janke. 2001. "The Mall of America as a Tourist Attraction". *Focus* 46, no. 3:32–6.

Gerlach, R.L. 1976. *Immigrants in the Ozarks: A Study in Ethnic Geography*. Columbia: University of Missouri Press.

Gertler, M.S. 2003. "A Cultural Economic Geography of Production". In *Handbook of Cultural Geography*, edited by K. Anderson, M. Domosh, S. Pile, and N. Thrift, 131–46. Thousand Oaks, CA: Sage.

Giddens, A. 1984. *The Constitution of Society: Outline of the Theory of Structuration*. Cambridge: Polity Press.

———. 1987. *Sociology: A Brief but Critical Introduction*, 2nd edn. New York: Harcourt Brace Jovanovich.

———. 1991. *Introduction to Sociology*. New York: W.W. Norton.

———, and J.H. Turner. 1987. "Introduction". In *Social Theory Today*, edited by A. Giddens and J.H. Turner, 1–10. Stanford, CA: Stanford University Press.

Gilbert, A. 1988. "The New Regional Geography in English and French-Speaking Countries". *Progress in Human Geography* 12:208–28.

Giliomee, H. 2003. *The Afrikaners: Biography of a People*. Charlottesville: University of Virginia Press.

Gills, B.K. 1995. "Capital and Power in the Processes of World History". In *Civilizations and World Systems: Studying World-Historical Change*, edited by S.K. Sanderson, 136–62. Walnut Creek, CA: Altamira Press.

Gil-White, F.J. 2001. "Are Ethnic Groups Biological 'Species' in the Human Brain?" *Current Anthropology* 42:515–54.

Ginsburg, N. 1970. Geography. In *A Reader's Guide to the Social Sciences*, 2nd edn, edited by B.F. Hoselitz, 293–318. New York: Free Press.

Glacken, C. 1967. *Traces on the Rhodian Shore: Nature and Culture in Western Thought from Ancient Times to the End of the Eighteenth Century*. Berkeley: University of California Press.

———. 1985. "Culture and Environment in Western Civilization during the Nineteenth Century". In *Environmental History*, edited by K.E. Bailes, 46–57. New York: University Press of America.

Glasscock, R.E. 1991. "Obituary: E. Estyn Evans, 1905–1989". *Journal of Historical Geography* 17:87–91.

Gleeson, B. 1999. *Geographies of Disability*. New York: Routledge.

Goh, L., and P. Williams. 2011. "Neoliberalism and the Role of the State in Surplus Public Land Management: Protecting Sydney Harbour's Open Space Legacy". <http://soac.fbe.unsw.edu.au/2011/papers/SOAC2011_0114_final.pdf>.

Gökariksel, B., and A. Secor. 2012. "'Even I Was Tempted': The Moral Ambivalence and Ethical Practice of Veiling Fashion in Turkey". *Annals of the Association of American Geographers* 102, no. 4:847–62.

Gold, J.R. 1980. *An Introduction to Behavioral Geography*. New York: Oxford University Press.

Goldhagen, D.J. 1996. *Hitler's Willing Executioners: Ordinary Germans and the Holocaust*. New York: Knopf.

Goldschmidt, W. 1965. "Theory and Strategy in the Theory of Cultural Adaptability". *American Anthropologist* 67:402–8.

Golledge, R.G. 1969. "The Geographical Relevance of Some Learning Theories". In *Behavioral Problems in Geography: A Symposium*, edited by K.R. Cox and R.G. Golledge, 101–45. Evanston, IL: Northwestern University Press, Studies in Geography, no. 17.

———. 1987. "Environmental Cognition". In *Handbook of Environmental Psychology, Volume 1*, edited by D. Stokols and I. Altman, 131–74. New York: Wiley.

———. 1993. "Geography and the Disabled: A Survey with Special Reference to Vision Impaired and Blind Populations". *Transactions of the Institute of British Geographers* NS 18:63–85.

———. 2003. "Reflections on Recent Cognitive Behavioral Research with an Emphasis on Research in the United States of America". *Australian Geographical Studies* 41:117–30.

Goode, P. 1926. *The Geographic Background of Chicago*. Chicago: University of Chicago Press.

Gore, A. 1992. *Earth in the Balance: Ecology and the Human Spirit*. Boston: Houghton Mifflin.

Goss, J. 1993. "The 'Magic of the Mall': An Analysis of Form, Function, and Meaning in the Contemporary Built Environment". *Annals of the Association of American Geographers* 83:18–47.

———. 1999a. "Modernity and Postmodernity in the Retail Landscape". In *Cultural Geographies*, 2nd edn, edited by K. Anderson and F. Gale, 199–219. Harlow, UK: Longman.

———. 1999b. "Once-upon-a-Time in the Commodity World: An Unofficial Guide to the Mall of America". *Annals of the Association of American Geographers* 89:45–75.

———. 2004. "Geography of Consumption I". *Progress in Human Geography* 28:369–80.

Gottlieb, R.S. 1996. *This Sacred Earth: Religion, Nature, Environment*. New York: Routledge.

Gould, P. 1969. *Spatial Diffusion*. Washington, DC: Association of American Geographers, Resource Paper, no. 4.

Gould, S.J. 1990. "The Golden Rule—A Proper Scale for Our Environmental Crisis". *Natural History* 99, no. 9:24–30.

Graham, B.J., ed. 1997. *In Search of Ireland: A Cultural Geography*. New York: Routledge.

———, and L.J. Proudfoot. 1993. "A Perspective on the Nature of Irish Historical Geography". In *An Historical Geography of Ireland*, edited by B.J. Graham and L.J. Proudfoot, 1–18. New York: Academic Press.

Graham, E. 1997. "Philosophies Underlying Human Geographic Research". In *Methods in Human Geography: A Guide for Students Doing a Research Project*, edited by R. Flowerdew and D. Martin, 6–30. Harlow, UK: Longman.

Granö, J.G. [1929] 1997. *Pure Geography*, edited by O. Granö and A. Paasi, translated by M. Hicks. Baltimore: Johns Hopkins University Press.

Gregory, D. 1981. "Human Agency and Human Geography". *Transactions of the Institute of British Geographers* NS 6:1–18.

———. 1994. *Geographical Imaginations*. Cambridge, MA: Blackwell.

———. 2004. *The Colonial Present: Afghanistan, Palestine, Iraq*. Oxford: Blackwell.

——. 2009. "Orientalism". In *The Dictionary of Human Geography*, 5th edn., edited by D. Gregory, R. Johnston, G. Pratt, M. Watts, and S. Whatmore, 513–15. Padstow, UK: Wiley-Blackwell.

——, and A. Pred. 2007. *Violent Geographies: Fear, Terror and Political Violence*. New York and London: Routledge.

Gregson, N. 1987. "Structuration Theory: Some Thoughts on the Possibilities for Empirical Research". *Environment and Planning D: Society and Space* 5:73–91.

——. 1993. "'The Initiative': Delimiting or Deconstructing Social Geography". *Progress in Human Geography* 17:525–30.

——. 1995. "And Now It's All Consumption?" *Progress in Human Geography* 19:135–41.

——, and G. Rose. 2000. "Taking Butler Elsewhere: Performativities, Spatialities and Subjectivities". *Environment and Planning D: Society and Space* 18:433–52.

——, G. Rose, J. Cream, and N. Laurie. 1997. "Conclusions". In *Feminist Geographies: Explorations in Diversity and Difference*, edited by Women and Geography Study Group of the Royal Geographical Society with the Institute of British Geographers, 191–200. Harlow, UK: Longman.

——, et al. 1997. "Gender in Feminist Geography". In *Feminist Geographies: Explorations in Diversity and Difference*, edited by the Women and Geography Study Group of the Royal Geographical Society with the Institute of British Geographers, 49–85. Harlow, UK: Longman.

Grey, A. 1994. *Aotearoa and New Zealand: A Historical Geography*. Christchurch: Canterbury University Press.

Griffiths, T. 1997. "Introduction. Ecology and Empire: Towards an Australian History of the World". In *Ecology and Empire: Environmental History of Settler Societies*, edited by T. Griffiths and L. Robin, 1–16. Seattle: University of Washington Press.

Gritzner, C.F. 1966. "The Scope of Cultural Geography". *Journal of Geography* 65:4–11.

——. 2002. "What is Where, Why There, and Why Care?" *Journal of Geography* 101:38–40.

Grossman, L. 1977. "Man–Environment Relationships in Anthropology and Geography". *Annals of the Association of American Geographers* 67:126–44.

——. 1984. *Peasants, Subsistence, Ecology, and Development in the Highlands of Papua New Guinea*. Princeton, NJ: Princeton University Press.

——. 1993. "The Political Ecology of Banana Exports and Local Food Production in St Vincent, Eastern Caribbean". *Annals of the Association of American Geographers* 83:347–67.

Groth, P. 1997. "Frameworks for Cultural Landscape Study". In *Understanding Ordinary Landscapes*, edited by P. Groth and T.W. Bressi, 1–21. New Haven, CT: Yale University Press.

——, and T.W. Bressi, eds. 1997. *Understanding Ordinary Landscapes*. New Haven, CT: Yale University Press.

Grove, R.H. 1992. "Origins of Western Environmentalism". *Scientific American* 267:42–7.

Guelke, L. 1971. "Problems of Scientific Explanation in Geography". *Canadian Geographer* 15:38–53.

——. 1974. "An Idealist Alternative in Human Geography". *Annals of the Association of American Geographers* 64:193–202.

——. 1975. "On Rethinking Historical Geography". *Area* 7:135–8.

——. 1982. *Historical Understanding in Geography*. New York: Cambridge University Press.

——. 1987. "Frontier Settlement and Human Values: A Comparative Look at North America and South Africa". In *Abstract Thoughts: Concrete Solutions. Essays in Honour of Peter Nash*, edited by L. Guelke and R. Preston, 181–99. Waterloo: University of Waterloo, Department of Geography, Publication Series, no. 29.

Habermas, J. 1972. *Knowledge and Human Interests*. London: Heinemann.

Hage, G. 1996. "The Spatial Imagery of National Practices: Dwelling–Domesticating/Being–Exterminating". *Environment and Planning D: Society and Space* 14:463–85.

Hagen, J. 2004. "The Most German of Towns: Creating an Ideal Nazi Community in Rothenburg ob der Tauber". *Annals of the Association of American Geographers* 94:207–27.

Hägerstrand, T. 1951. *Migration and the Growth of Culture Regions*. Lund: Gleerup, Lund Studies in Geography, Series B, no. 3.

——. 1952. *The Propagation of Innovation Waves*. Lund: Gleerup, Lund Studies in Geography, Series B, no. 4.

——. 1967. *Innovation Diffusion as a Spatial Process*, translated by A. Pred. Chicago: University of Chicago Press.

Haggett, P. 2000. *The Geographical Structure of Epidemics*. Oxford: Clarendon Press.

——. 2001. *Geography: A Global Synthesis*. New York: Prentice Hall.

Hall, C.S., and G. Lindzey. 1978. *Theories of Personality*, 3rd edn. New York: Wiley.

Hall, S. 1996. "Introduction: Who Needs Identity?" In *Questions of Cultural Identity*, edited by S. Hall and P. duGay, 1–17. London: Sage.

Hamnett, C., ed. 1996. *Social Geography: A Reader*. New York: Arnold.

——. 2003. "Editorial. Contemporary Human Geography: Fiddling while Rome Burns". *Geoforum* 34:1–3.

Hannaford, I. 1996. *Race: The History of an Idea in the West*. Baltimore: Johns Hopkins University Press.

Hanson, S. 1992. "Geography and Feminism: Worlds in Collision". *Annals of the Association of American Geographers* 82:569–86.

———, and G. Pratt. 1995. *Gender, Work, and Space*. New York: Routledge.

Haraway, D.J. 1985. "Manifesto for Cyborgs: Science, Technology, and Socialist Feminism in the 1980s". *Socialist Review* 80:65–108.

———. 1988. "Situated Knowledges: The Science Question in Feminism and the Privilege of Partial Perspective". *Feminist Studies* 14:575–99.

———. 1989. *Primate Visions: Gender, Race, and Nature in the World of Modern Science*. London and New York: Routledge.

———. 1991. *Simians, Cyborgs, and Women: The Reinvention of Nature*. New York: Routledge.

Hardesty, D.L. 1977. *Ecological Anthropology*. New York: Wiley.

Hardill, I., D.T. Graham, and E. Kofman. 2001. *Human Geography of the UK*. New York: Routledge.

Harley, J. Brian. 1989. "Deconstructing the Map". *Cartographica: The International Journal for Geographic Information and Geovisualization* 26, no. 2:1–20.

———. 1992. "Rereading the Maps of the Columbian Encounter". *Annals of the Association of American Geographers* 82, 3:522–42.

Harris, L. 2004. *Civilization and Its Enemies*. New York: Free Press.

Harris, M. 1968. *The Rise of Anthropological Theory*. London: Routledge and Kegan Paul.

———. 1979. *Cultural Materialism: The Struggle for a Science of Culture*. New York: Random House.

Harris, R.C. 1971. "Theory and Synthesis in Historical Geography". *The Canadian Geographer* 15, no. 3:157–72.

———. 1977. "The Simplification of Europe Overseas". *Annals of the Association of American Geographers* 67:469–83.

———. 1978. "The Historical Geography of North American Regions". *American Behavioral Scientist* 22:115–30.

———. 1999. "Comments on 'The Shaping of America, 1850–1915'". *Journal of Historical Geography* 25:9–11.

Harrison, L.E. 2000. "Why Culture Matters". In *Culture Matters: How Values Shape Human Progress*, edited by L.E. Harrison and S.P. Huntington, xvii–xxxiv. New York: Basic Books.

Harrison, R., and D.N. Livingstone. 1979. "There and Back Again: Towards a Critique of Idealist Human Geography". *Area* 11:75–9.

Hart, J.F. 1975. *The Look of the Land*. Englewood Cliffs, NJ: Prentice Hall.

———. 1982. "The Highest Form of the Geographer's Art". *Annals of the Association of American Geographers* 72:1–29.

———. 1995. "Reading the Landscape". In *Landscape in America*, edited by G.F. Thompson, 23–42. Austin: University of Texas Press.

———. 1998. *The Rural Landscape*. Baltimore: Johns Hopkins University Press.

Hartmann, H. 1979. "The Unhappy Marriage of Marxism and Feminism: Towards a More Progressive Union". *Capital & Class* 3:1–33.

Hartshorne, R. 1939. *The Nature of Geography*. Lancaster, PA: Association of American Geographers.

———. 1959. *Perspective on the Nature of Geography*. Chicago: Rand McNally.

Harvey, D.W. 1969. *Explanation in Geography*. New York: Arnold.

———. 1973. *Social Justice and the City*. Baltimore: Johns Hopkins University Press.

———. 1989. *The Condition of Postmodernity: An Enquiry into the Origins of Cultural Change*. Cambridge, MA: Blackwell.

———. 1993. "Class Relations, Social Justice and Politics of Difference". In *Place and the Politics of Identity*, edited by M. Keith and S. Pile, 41–66. New York: Routledge.

———. 2005. *A Brief History of Neoliberalism*. Oxford: Oxford University Press.

Hauptman, L.M., and R.G. Knapp. 1977. "Dutch–Aboriginal Interaction in New Netherland and Formosa: An Historical Geography of Empire". *Proceedings of the American Philosophical Society* 121:166–82.

Hausladen, G.J., ed. 2003. *Western Places, American Myths: How We Think About the West*. Reno: University of Nevada Press.

Hawley, A.H. 1950. *Human Ecology: A Theory of Community Structure*. New York: Ronald Press Co.

———. 1968. "Human Ecology". In *International Encyclopedia of the Social Sciences, Volume 4*, edited by D.L. Sills, 328–37. New York: Free Press.

———. 1998. "Human Ecology, Population, and Development". In *Continuities in Sociological Human Ecology*, edited by M. Micklin and D.L. Poston Jr, 11–25. New York: Plenum Press.

Hay, I., ed. 2010. *Qualitative Research Methods in Human Geography*. Oxford: Oxford University Press.

Haynes, R.M. 1980. *Geographical Images and Mental Maps*. New York: Macmillan.

Head, L. 2007. "Cultural Ecology: The Problematic Human and the Terms of Engagement". *Progress in Human Geography* 31:837–46.

Headland, T. 1997. "Revisionism in Ecological Anthropology". *Current Anthropology* 38:605–9.

Heathcote, R.L. 1972. "The Visions of Australia, 1770–1970". In *Australia as Human Setting*, edited by A. Rapoport, 77–98. Sydney: Angus and Robertson.

Heffernan, M. 1998. *The Meaning of Europe: Geography and Geopolitics*. New York: Arnold.

Henderson, G. 1997. "'Landscape is Dead, Long Live Landscape': A Handbook for Sceptics". *Journal of Historical Geography* 24:94–100.

Herb, G.H. 2004. "Double Vision: Territorial Strategies in the Construction of National Identities in Germany, 1949–1979". *Annals of the Association of American Geographers* 94:140–64.

Herbert, D. 1991. "Place and Society in Jane Austen's England". *Geography* 76:193–208.

Heyer, P. 1982. *Nature, Human Nature, and Society: Marx, Darwin, Biology and the Human Sciences*. Westport, CT: Greenwood.

Hilliard, S.B. 1972. *Hog Meat and Hoecake: Food Supply in the Old South, 1840–1860*. Carbondale: Southern Illinois University Press.

Hirsch, E., and M. O'Hanlon, eds. 1995. *The Anthropology of Landscape: Perspectives on Space and Place*. Oxford: Clarendon.

Hobsbawm, E. 1993. "The New Threat to History". *The New York Review of Books* (16 December):62–4.

Hodge, S. 1996. "Disadvantage and 'Otherness' in Western Sydney". *Australian Geographical Studies* 34:32–44.

Hoggart, R. 1957. *The Uses of Literacy: Changing Patterns in English Mass Culture*. Fair Lawn, NJ: Essential Books.

Hoke, G.W. 1907. "The Study of Social Geography". *Geographical Journal* 29:64–7.

Holloway, L., and P. Hubbard. 2001. *People and Place: The Extraordinary Geographies of Everyday Life*. New York: Prentice Hall.

Holloway, S.L. 2003. "Outsiders in Rural Society? Constructions of Rurality and Nature–Society Relations in the Racialisation of English Gypsy-Travellers, 1869–1934". *Environment and Planning D: Society and Space* 21:695–715.

——, and G. Valentine, eds. 2002. *Children's Geographies: Living, Playing, Learning*. New York: Routledge.

Homans, G. 1987. "Behaviorism and After". In *Social Theory Today*, edited by A. Giddens and J.H. Turner, 58–81. Stanford, CA: Stanford University Press.

Honderich, T. 1995. *The Oxford Companion to Philosophy*. New York: Oxford University Press.

Hooson, D.J.M. 1981. "Carl O. Sauer". In *The Origins of Academic Geography in the United States*, edited by B.W. Blouet, 165–74. Hamden, CT: Archon Books.

Hopkins, J.S.P. 1990. "West Edmonton Mall: Landscape of Myths and Elsewhereness". *Canadian Geographer* 34:2–17.

——. 1991. "West Edmonton Mall as a Centre for Social Interaction". *Canadian Geographer* 35:261–7.

Hopkins, T.K., and I. Wallerstein. 1996. "The World-System: Is There a Crisis?" In *The Age of Transition: Trajectory of the World System, 1945–2025*, coordinated by T.K. Hopkins and I. Wallerstein, 1–10. Atlantic Highlands, NJ: Zed Books.

Hornbeck, D., C. Earle, and C.M. Rodrigue. 1996. "The Way We Were: Deployments (and Redeployments) of Time in Human Geography". In *Concepts in Human Geography*, edited by C. Earle, K. Mathewson,

and M.S. Kenzer, 33–61. Lanham, MD: Rowman and Littlefield.

Hoskins, W.G. 1955. *The Making of the English Landscape*. London: Hodder and Stoughton.

Hough, M. 1990. *Out of Place: Restoring Identity to the Regional Landscape*. New Haven, CT: Yale University Press.

Hubbard, P. 2000. "Desire/Disgust: Mapping the Moral Contours of Heterosexuality". *Progress in Human Geography* 24:191–217.

——, R. Kitchin, B. Bartley, and D. Fuller. 2002. *Thinking Geographically: Space, Theory and Contemporary Human Geography*. New York: Continuum.

Huckle, J. 2002. "Reconstructing Nature: Towards a Geographical Education for Sustainable Development". *Geography* 87:64–72.

Hudson, B.J. 1992. "Arnold Bennett's Five Towns: 'Internal and External Vision'". *Journal of Cultural Geography* 11, no. 2:21–30.

Hudson, J.C. 1977. "Theory and Methodology in Comparative Frontier Studies". In *The Frontier*, edited by D.H. Miller and J.O. Steffen, 11–32. Norman: University of Oklahoma Press.

Hudson, W., and G. Bolton. 1997. "Creating Australia". In *Creating Australia: Changing Australian History*, edited by W. Hudson and G. Bolton, 1–11. St Leonards, NSW: Allen and Unwin.

Hufferd, J. 1980. "Toward a Transcendental Human Geography of Places". *Antipode* 12, 3:18–23.

Huggett, R. 1980. *Systems Analysis in Geography*. Oxford: Clarendon Press.

Hugill, P.J. 1997. "Review Article: World-System Theory: Where's the Theory?" *Journal of Historical Geography* 23:344–9.

——. 1999. "Imperialism and Manliness in Edwardian Boys' Novels". *Ecumene* 6:318–40.

——, and D.B. Dickson, eds. 1988. *The Transfer and Transformation of Ideas and Material Culture*. College Station: Texas A&M University Press.

Hunter, L.M., and M.B. Toney. 2005. "Religion and Attitudes toward the Environment: A Comparison of Mormons and the General US Population". *The Social Science Journal* 42(1): 25–38.

Huntington, E. 1915. Civilization and Climate. New Haven, CT: Yale University Press.

——. 1920. *Principles of Human Geography*, 5th edn. New York: Wiley.

——. 1945. *Mainsprings of Civilization*. New York: Wiley.

Huntington, S.P. 1993. "The Clash of Civilizations". *Foreign Affairs* 73, no. 2:22–49.

——. 1996. *The Clash of Civilizations and the Remaking of World Order*. New York: Simon and Schuster.

——. 2000. "Culture Counts". In *Culture Matters: How Values Shape Human Progress*, edited by L.E. Harrison and S.P. Huntington, xiii–xvi. New York: Basic Books.

Hurt, D.A. 2003. "Defining American Homelands: A Creek Nation Example, 1828–1907". *Journal of Cultural Geography* 21, no. 1:19–43.

Hutcheon, P.D. 1996. *Leaving the Cave: Evolutionary Naturalism in Social-Scientific Thought*. Waterloo, ON: Wilfrid Laurier University Press.

Hutchinson, J., and A.D. Smith, eds. 1996. *Ethnicity*. New York: Oxford University Press.

Huxley, J.S., and A.C. Haddon. 1936. *We Europeans*. London: Jonathan Cape.

Imrie, R. 1996. *Disability and the City: International Perspectives*. London: Paul Chapman.

Inglehart, R. 1990. *Culture Shift in Advanced Industrial Society*. Princeton, NJ: Princeton University Press.

Innis, H.A. 1930. *The Fur Trade in Canada*. New Haven, CT: Yale University Press.

Isajiw, W. 1974. "Definitions of Ethnicity". *Ethnicity* 1:111–24.

Ittelson, W.H., H.M. Proshansky, L.G. Rivlin, and G.H. Winkel. 1974. *An Introduction to Environmental Psychology*. New York: Holt, Rinehart and Winston.

Iveson, K. 2010. "Seeking Spatial Justice: Some Reflections from Sydney". *City: Analysis of Urban Trends, Culture, Theory, Policy, Action* 14:6, 607–11.

Jackdaw, The. 2011. "Eh up Arup! Laura Gascoigne Looks into the Real Brains behind the Public Sculptures We Don't Need". *The Jackdaw*. <http://www.thejackdaw .co.uk/?p=92>.

Jackson, J.B. 1974. *A Sense of Place, A Sense of Time*. New Haven, CT: Yale University Press.

———. 1984. *Discovering the Vernacular Landscape*. New Haven, CT: Yale University Press.

———. 1997a. *Landscape in Sight: Looking at America*, edited by H.L. Horowitz. New Haven, CT: Yale University Press.

———. 1997b. "Foreword". In *The Evolving Landscape: Homer Aschmann's Geography*, edited by M.J. Pasqualetti, vii–viii. Baltimore: Johns Hopkins University Press.

Jackson, P. 1980. "A Plea for Cultural Geography". *Area* 12:110–13.

———. 1987. "The Idea of 'Race' and the Geography of Racism". In *Race and Racism: Essays in Social Geography*, edited by P. Jackson, 3–21. Boston: Allen and Unwin.

———. 1989. *Maps of Meaning: An Introduction to Cultural Geography*. Boston: Unwin Hyman.

———. 1990. "The Cultural Politics of Masculinity: Towards a Social Geography". *Transactions of the Institute of British Geographers* NS 16:199–213.

———. 2000. "Rematerializing Social and Cultural Geography". *Social and Cultural Geography* 1:9–14.

———, and J. Penrose, eds. 1994. *Constructions of Race, Place, and Nation*. Minneapolis: University of Minnesota Press.

———, and S.J. Smith. 1984. *Exploring Social Geography*. London: Allen and Unwin.

Jackson, R.H. 1978. "Mormon Perception and Settlement". *Annals of the Association of American Geographers* 68:317–34.

———, and R. Henrie. 1983. "Perception of Sacred Space". *Journal of Cultural Geography* 3, no. 2:94–107.

———, and R.L. Layton. 1976. "The Mormon Village: Analysis of a Settlement Type". *Professional Geographer* 28:136–41.

Jacobs, W.R. 1997. "Perspectives on Scarcity, Population, Ecology, and the American West". *Journal of the West* 36, no. 3:3–5.

Jakle, J.A. 1987. *The Visual Elements of Landscape*. Amherst: University of Massachusetts Press.

———. 1994. *The Gas Station in America*. Baltimore: Johns Hopkins University Press.

———. 2001. *City Lights: Illuminating the American Night*. Baltimore: Johns Hopkins University Press.

———, S.D. Brunn, and C.C. Roseman. 1976. *Human Spatial Behavior: A Social Geography*. North Scituate, MA: Duxbury.

———, and K.A. Sculle. 1999. *Fast Food: Roadside Restaurants in the Automobile Age*. Baltimore: Johns Hopkins University Press.

———, and ———. 2004. *Lots of Parking: Land Use in a Car Culture*. Charlottesville: University of Virginia Press.

———, and ———. 2008. *Motoring: The Highway Experience in America*. Athens: University of Georgia Press.

———, and ———. 2009. *America's Main Street Hotels: Transiency and Community in the Early Auto Age*. Knoxville: University of Tennessee Press.

James, P.E. 1964. *One World Divided: A Geographer Looks at the Modern World*. New York: Blaisdell Publishing Co.

Jameson, F. 1991. *Postmodernism, or The Cultural Logic of Late Capitalism*. Durham, NC: Duke University Press.

Jamieson, D. 1985. "Against Zoos". In *In Defense of Animals*, edited by P. Singer, 108–17. Cambridge, MA: Blackwell.

———. 1997. "Zoos Revisited". In *The Philosophy of the Environment*, edited by T.D.J. Chappell, 180–92. Edinburgh: Edinburgh University Press.

———, ed. 2001. *A Companion to Environmental Philosophy*. Malden, MA: Blackwell.

Jeans, D.N. 1981. "Mapping the Regional Patterns of Australian Society: Some Preliminary Thoughts". *Australian Historical Geography Bulletin* 2:1–6.

———. 1988. "The First World War Memorials in New South Wales: Centres of Meaning in the Landscape". *Australian Geographer* 19:259–67.

Jeffrey, C., P. Jeffrey, and R. Jeffrey. 2008. *Degrees without Freedom? Education, Masculinities, and Unemployment in North India*. Stanford, CA: Stanford University Press.

Johnson, N. 1995. "Cast in Stone: Monuments, Geography, and Nationalism". *Environment and Planning D: Society and Space* 13:51–65.

Johnson, R. 1996. "Revising Culture Through 'Women and Nature': An Introduction to the Special Issue". *Women's Studies* 25:v–xi.

Johnston, J.A. 1979. "Image and Reality: Initial Assessments of Soil Fertility in New Zealand, 1839–55". *Australian Geographer* 14:160–5.

———. 1981. "The New Zealand Bush: Early Assessments of Vegetation". *New Zealand Geographer* 37, no. 1:19–24.

Johnston, R.J. 1985. "Introduction: Exploring the Future of Geography". In *The Future of Geography*, edited by R.J. Johnston, 3–26. New York: Methuen.

———. 1990. "The Challenge for Regional Geography: Some Proposals for Research Frontiers". In *Regional Geography: Current Developments and Future Prospects*, edited by R.J. Johnston, J. Hauer, and G.A. Hoekveld, 122–39. New York: Routledge.

———. 1991. *A Question of Place: Exploring the Practice of Human Geography*. Cambridge, MA: Blackwell.

———. 1997. *Geography and Geographers: Anglo-American Geography since 1945*, 5th edn. New York: Wiley.

Jones, E.L. 1987. *The European Miracle: Environments, Economics and Geopolitics in the History of Europe and Asia*, 2nd edn. New York: Cambridge University Press.

Jones, R. 2004. "What Time Human Geography?" *Progress in Human Geography* 28:287–304.

Jones, R. 2008. "Relocating Nationalism: On the Geographies of Reproducing Nations". *Transactions of the Institute of British Geographers* NS 33:319–34.

Jordan, T.G. 1966. *German Seed in Texas Soil*. Austin: University of Texas Press.

———. 1983. "A Reappraisal of Fenno-Scandian Antecedents for Midland American Log Construction". *Geographical Review* 73:58–94.

———. 1985. *American Log Buildings: An Old World Heritage*. Chapel Hill: University of North Carolina Press.

———, M. Domosh, and L. Rowntree. 1997. *The Human Mosaic: A Thematic Introduction to Cultural Geography*, 7th edn. New York: Harper and Row.

———, and M. Kaups. 1989. *The American Backwoods Frontier: An Ethnic and Ecological Interpretation*. Baltimore: Johns Hopkins University Press.

———, J.T. Kilpinen, and C.F. Gritzner. 1997. *The Mountain West: Interpreting the Folk Landscape*. Baltimore: Johns Hopkins University Press.

Jordan-Bychkov, T.G. 2003. *The Upland South: The Making of an American Folk Region and Landscape*. Santa Fe: Center for American Places.

———, and B. Bychkova Jordan. 2002. *The European Culture Area: A Systematic Geography*, 4th edn. Lanham, MD: Rowman and Littlefield.

Judd, C.M., and A.J. Ray, eds. 1980. *Old Trails and New Directions: Papers of the Third North American Fur Trade Conference*. Toronto: University of Toronto Press.

Judkins, G., M. Smith, and E. Keys. 2008. "Determinism within Human-Environmental Research and the Rediscovery of Environmental Causation". *Geographical Journal* 174:17–29.

Juergensmeyer, M. 1995. "Religious Nationalism: A Global Threat?" *Current History* 95, 604:372–6.

Kamin, B. 2005. "Lincoln Land". *Chicago Tribune* (10 April). <http://www.chicagotribune.com/news/specials/chi-0504100474apr10,1,2472954.story?coll=chi-homepagenews2-utl&ctrack=1&cset=true>.

Kaplan, R.D. 1994. "The Coming Anarchy". *Atlantic Monthly* 273, 2:44–76.

———. 1996. *The Ends of the Earth: A Journey at the Dawn of the 21st Century*. New York: Random House.

———. 2000. *The Coming Anarchy: Shattering the Dreams of the Post-Cold War*. New York: Random House.

Karskens, G. 2008. "The Rocks". *The Dictionary of Sydney*. <http://www.dictionaryofsydney.org/entry/the_rocks>.

Kates, R.W. 1971. "Natural Hazard in Human Ecological Perspective: Hypotheses and Models". *Economic Geography* 47:438–51.

Katz, C. 2001. "On the Grounds of Globalization: A Topography for Feminist Political Engagement". *Signs: A Journal of Women in Culture and Society* 26 1214–34.

Kearns, K.C. 1974. "Resuscitation of the Irish Gaeltacht". *Geographical Review* 64:82–110.

Keil, R., D. Bell, P. Penz, and L. Fawcett, eds. 1998. *Political Ecology: Global and Local*. New York: Routledge.

Keith, M., and S. Pile, eds. 1993. *Place and the Politics of Identity*. New York: Routledge.

Keith, W.J. 1980. *The Poetry of Nature: Rural Perspectives in Poetry from Wordsworth to the Present*. Toronto: University of Toronto Press.

Kelly, K. 1974. "The Changing Attitudes of Farmers to Forest in Nineteenth Century Ontario". *Ontario Geography* 8:67–77.

Kendall, D. 2012. "Always Let the Road Decide: South Asian Labourers along the Highways of Dubai, UAE: A photographic Essay". *South Asian Diaspora* 4, 1:45–55.

Khan, S. 2009. *Of Hockey and Hijab: Reflections of a Canadian Muslim Woman*. Toronto: TSAR Books.

Kingsbury, P. 2004. "Psychoanalytic Approaches". In *A Companion to Cultural Geography*, edited by J.S. Duncan, N.C. Johnson, and R.H. Schein, 108–20. Malden, MA: Blackwell.

Kinsley, D. 1994. *Ecology and Religion*. Englewood Cliffs, NJ: Prentice Hall.

Kirk, W. 1951. "Historical Geography and the Concept of the Behavioral Environment". *Indian Geographical Journal* 25:152–60.

———. 1963. "Problems in Geography". *Geography* 48:357–71.

Kitchin, R.M. 1996. "Increasing the Integrity of Cognitive Mapping Research: Appraising Conceptual Schemata of Environment–Behavior Interaction". *Progress in Human Geography* 20:56–84.

———. 1998. "'Out of Place'. 'Knowing One's Place': Space, Power and the Exclusion of Disabled People". *Disability and Society* 13:343–56.

———, M. Blades, and R.G. Golledge. 1997. "Relations between Geography and Psychology". *Environment and Behavior* 29:554–73.

Kjekshus, H. 1977. *Ecology Control and Economic Development in East Africa: The Case of Tanganyika*. Berkeley: University of California Press.

Klare, M.T. 1996. "Redefining Security: The New Global Schisms". *Current History* 95:353–8.

Knapp, G.W. 1991. *Andean Ecology: Adaptive Dynamics in Ecuador*. Boulder, CO: Westview Press.

Kneale, J. 2003. "Secondary Worlds: Reading Novels as Geographic Research". In *Cultural Geography in Practice*, edited by A. Blunt et al., 39–54. New York: Arnold.

Kniffen, F. 1932. "Lower California Studies III: The Primitive Cultural Landscape of the Colorado Delta". *University of California Publications in Geography* 5:43–66.

———. 1936. "Louisiana House Types". *Annals of the Association of American Geographers* 26:179–93.

———. 1949. "The American Agricultural Fair: The Pattern". *Annals of the Association of American Geographers* 39:264–82.

———. 1951a. "The American Agricultural Fair: Time and Place". *Annals of the Association of American Geographers* 41:42–57.

———. 1951b. "The American Covered Bridge". *Geographical Review* 41:114–23.

———. 1965. "Folk Housing: Key to Diffusion". *Annals of the Association of American Geographers* 55:549–57.

———. 1974. "Material Culture in the Geographic Interpretation of the Landscape". In *The Human Mirror*, edited by M. Richardson, 252–67. Baton Rouge: Louisiana State University Press.

———, and H. Glassie. 1966. "Building in Wood in the Eastern United States: A Time–Place Perspective". *Geographical Review* 56:40–65.

Knight, D. 1982. "Identity and Territory: Geographical Perspectives on Nationalism and Regionalism". *Annals of the Association of American Geographers* 72:514–31.

Knox, P., and J. Agnew. 1994. *The Geography of the World Economy*, 2nd edn. New York: Arnold.

Kobayashi, A. 1989. "A Critique of Dialectical Landscape". In *Remaking Human Geography*, edited by A. Kobayashi and S. Mackenzie, 164–83. Boston: Unwin Hyman.

Koelsch, W.A. 1969. "The Historical Geography of Harlan H. Barrows". *Annals of the Association of American Geographers* 59:632–51.

Kofman, E., and G. Young, eds. 1996. *Globalization: Theory and Practice*. New York: Pinter.

Kollmorgen, W.M. 1941. "A Reconnaissance of Some Cultural-Agricultural Islands in the South". *Economic Geography* 17:409–30.

———. 1943. "Agricultural–Cultural Islands in the South, Part II". *Economic Geography* 19:109–17.

Kong, L. 1993. "Ideological Hegemony and the Political Symbolism of Religious Buildings in Singapore". *Environment and Planning D: Society and Space* 11:23–45.

———. 1997. "A 'New' Cultural Geography: Debates about Invention and Reinvention". *Scottish Geographical Magazine* 113:177–85.

———, and L. Tay. 1998. "Exalting the Past: Nostalgia and the Construction of Heritage in Children's Literature". *Area* 30:133–43.

Kroeber, A.L. 1917. "The Superorganic". *American Anthropologist* 19:163–213.

———. 1928. "Native Cultures of the Southwest". *University of California Publications in American Archaeology and Ethnology* 23, no. 9:375–98.

———. 1939. *Cultural and Natural Areas of Native North America*. Berkeley: University of California Press, Publications in American Archaeology and Ethnology, 38.

———, and C. Kluckhohn. 1952. *Culture: A Critical Review of Concepts and Definitions*. Cambridge, MA: Papers of the Peabody Museum of American Archaeology and Ethnology, Harvard University, 47, 1.

———, and T. Parsons. 1958. "The Concepts of Culture and of Social System". *American Sociological Review* 23:582–3.

Kuper, A. 1999. *Culture: The Anthropologists' Account*. Cambridge, MA: Harvard University Press.

Kuznar, L.A. 1997. *Reclaiming a Scientific Anthropology*. Walnut Creek, CA: Altamira Press.

Lakshman, S.Y., and R.C. Mayfield. 1978. "Non-adoption of Innovations: Evidence from Discriminant Analysis". *Economic Geography* 54, 2: 145–56.

Lamal, P.A. 1991. "Preface". In *Behavioral Analysis of Societies and Cultural Practices*, edited by P.A. Lamal, xiii–xiv. New York: Hemisphere Publishing Corporation.

Lambert, A.M. 1985. *The Making of the Dutch Landscape: An Historical Geography of the Netherlands*, 2nd edn. New York: Academic Press.

Lambert, D. 2002. "Geography, 'Race' and Education". *Geography* 87:297–304.

Landes, D.S. 1998. *The Wealth and Poverty of Nations: Why Some Are So Rich and Some So Poor*. New York: W.W. Norton.

Landzelius, M. 2004. "The Body". In *A Companion to Cultural Geography*, edited by J.S. Duncan, N.C. Johnson, and R.H. Schein, 279–97. Malden, MA: Blackwell.

Langton, J. 1979. "Darwinism and the Behavioral Theory of Sociocultural Evolution: An Analysis". *American Journal of Sociology* 85:288–309.

Latour, B. 1993. *We Have Never Been Modern*. Cambridge, MA: Harvard University Press.

Latour, B., and S. Woolgar. 1979. *Laboratory Life: The Social Construction of Scientific Facts*. Beverly Hills: Sage.

Leach, S., M. Lewis, C. Chenery, G. Müldner, and H. Eckardt. 2009. "Migration and Diversity in Roman Britain: A Multidisciplinary Approach to the Identification of Immigrants in Roman York, England". *American Journal of Physical Anthropology* 140, 546–61.

Leaf, M.J. 1979. *Man, Mind and Science: A History of Anthropology*. New York: Columbia University Press.

Leary, R. 1999. "Youth Cultures of the Hobart Rivulet". In *Australian Cultural Geographies*, edited by E. Stratford, 153–69. New York: Oxford University Press.

Lee, R. 1990. "Regional Geography: Between Scientific Geography, Ideology, and Practice (or What Use is Regional Geography?)". In *Regional Geography: Current Developments and Future Prospects*, edited by R.J. Johnston, J. Hauer, and G.A. Hoekveld, 103–21. New York: Routledge.

———, and J. Wills, eds. 1997. *Geographies of Economies*. New York: Arnold.

Lefebvre, H. 1991. *The Production of Space*. Cambridge, MA: Blackwell.

Legrain, P. 2002. *Open World: The Truth About Globalisation*. London: Abacus.

Lehr, J.C. 1973. "Ukrainian Houses in Alberta". *Alberta Historical Review* 21:9–15.

Leib, J.I. 2002. "Separate Times, Shared Spaces: Arthur Ashe, Monument Avenue and the Politics of Richmond, Virginia's Symbolic Landscape". *Cultural Geographies* 9:286–312.

Leighly, J. 1954. "Innovation and Area". *Geographical Review* 44:439–41.

———, ed. 1963. *Land and Life: A Selection from the Writings of Carl Ortwin Sauer*. Berkeley: University of California Press.

———. 1976. "Carl Ortwin Sauer, 1889–1975". *Annals of the Association of American Geographers* 66:337–48.

———. 1978. "Town Names of Colonial New England in the West". *Annals of the Association of American Geographers* 68:233–48.

———. 1987. "Ecology as Metaphor: Carl Sauer and Human Ecology". *Professional Geographer* 39:405–12.

Lemon, J.T. 1972. *The Best Poor Man's Country: A Geographical Study of Early Southeastern Pennsylvania*. Baltimore: Johns Hopkins University Press.

Leopold, A. 1949. *A Sand County Almanac: And Sketches Here and There*. New York: Oxford University Press.

Lester, A. 1996. *From Colonization to Democracy: A New Historical Geography of South Africa*. New York: I.B. Tauris Publishers.

———. 2000. "Historical Geography". In *The Geography of South Africa in a Changing World*, edited by R. Fox and K. Rowntree, 60–85. New York: Oxford University Press.

Levison, M., R.G. Ward, and J.W. Webb. 1973. *The Settlement of Polynesia. A Computer Simulation*. New York: Oxford University Press.

Lewin, K. 1944. "Constructs in Psychology and Ecological Psychology". In *Authority and Frustration*, edited by K. Lewin et al., 17–23. Iowa City: University of Iowa Press.

———. 1951. *Field Theory in Social Science: Selected Theoretical Papers*, edited by D. Cartwright. New York: Harper.

Lewis, K.E. 2002. *West to Far Michigan: Settling the Lower Peninsula, 1815–1860*. East Lansing: Michigan State University Press.

Lewis, M.W. 1991. "Elusive Societies: A Regional-Cartographical Approach to the Study of Human Relatedness". *Annals of the Association of American Geographers* 81:605–26.

———, and K.E. Wigen. 1997. *The Myth of Continents: A Critique of Metageography*. Berkeley: University of California Press.

Lewis, P. 1975. "Common Houses, Cultural Spoor". *Landscape* 19, no. 2:1–22.

———. 1979. "Axioms for Reading the Landscape". In *The Interpretation of Ordinary Landscapes: Geographical Essays*, edited by D.W. Meinig, 11–32. New York: Oxford University Press.

———. 1983. "Learning from Looking: Geographic and Other Writing about the American Cultural Landscape". *American Quarterly* 35:242–61.

Ley, D. 1977. "Social Geography and the Taken-for-Granted World". *Transactions of the Institute of British Geographers* NS 2:498–512.

———. 1981. "Behavioral Geography and the Philosophies of Meaning". In *Behavioral Problems in Geography Revisited*, edited by K.R. Cox and R.G. Golledge, 209–30. New York: Methuen.

———. 1998. "Classics in Human Geography Revisited: Author's Response". *Progress in Human Geography* 22:78–80.

——— 2008. "Are There Limits to Gentrification? The Contexts of Impeded Gentrification in Vancouver". *Urban Studies* 45, 12:2471–98.

———, and K. Olds. 1988. "Landscape as 'Spectacle': World's Fairs and the Culture of Heroic Consumption". *Environment and Planning D: Society and Space* 6:191–212.

———, and M.S. Samuels. 1978. "Contexts of Modern Humanism in Geography". In *Humanistic Geography*, edited by D. Ley and M.S. Samuels, 1–18. Chicago: Maaroufa Press.

Llosa, M.V. 2001. "The Culture of Liberty". *Foreign Policy* 122:66–71.

Lloyd J., and L. Johnson. 2004. "Dream Stuff: The Postwar Home and the Australian Housewife, 1940–60". *Environment and Planning D: Society and Space* 22, no. 2:251–72.

Lomborg, B. 2001. *The Skeptical Environmentalist: Measuring the Real State of the World*. New York: Cambridge University Press.

Longhurst, R. 1994. "Reflections on and a Vision for Feminist Geography". *New Zealand Geographer* 50, no.1:14–19.

———. 1997. "(Dis)embodied Geographies". *Progress in Human Geography* 21:486–501.

Louder, D.R., C. Morissonneau, and E. Waddell. 1983. "Picking up the Pieces of a Shattered Dream: Quebec and French America". *Journal of Cultural Geography* 4, no. 1:44–56.

Lovell, W.G. 1985. *Conquest and Survival in Colonial Guatemala*. Kingston, ON: McGill-Queen's University Press.

———. 1992. "'Heavy Shadows and Black Night': Disease and Depopulation in Colonial Spanish America". *Annals of the Association of American Geographers* 82:426–43.

Lovelock, J. 1982. *Gaia: A New Look at Life on Earth*. New York: Oxford University Press.

Low, N., and B. Gleeson. 1998. *Justice, Society and Nature: An Exploration of Political Ecology*. New York: Routledge.

Lowenthal, D. 1968. "The American Scene". *Geographical Review* 58:61–88.

———. 2000. *George Perkins Marsh: Prophet of Conservation*. Seattle: University of Washington Press.

———, and H.C. Prince. 1964. "The English Landscape". *Geographical Review* 54:309–46.

———, and ———. 1965. "English Landscape Tastes". *Geographical Review* 55:186–222.

Lowie, R.H. 1937. *History of Ethnological Theory*. New York: Farrar and Rinehart.

Luebke, F.C. 1984. "Regionalism and the Great Plains: Problems of Concept and Method". *Western Historical Quarterly* 15:19–38.

Mabogunje, A. 1997. *State of the Earth: Contemporary Geographic Perspectives*. Malden, MA: Blackwell.

McBride, B. 1999. "The (Post)colonial Landscape of Cathedral Square: Urban Redevelopment and Representation in the 'Cathedral City'". *New Zealand Geographer* 55, no. 1:3–11.

McCall, L. 2005. "The Complexity of Intersectionality". *Signs* 30, 3 (Spring), 1771–800.

McCann, E.J. 1997. "Where Do You Draw the Line? Landscape, Texts and the Politics of Planning". *Environment and Planning D: Society and Space* 15:641–61.

McClintock, A. 1995. *Imperial Leather: Race, Gender and Sexuality in the Colonial Contest*. London: Routledge.

MacDonald, R.H. 1994. *The Language of Empire: Myths and Metaphors of Popular Imperialism, 1880–1918*. Manchester: Manchester University Press.

McDowell, L. 1994. "The Transformation of Cultural Geography". In *Human Geography: Society, Space and Social Science*, edited by D. Gregory, R. Martin, and G. Smith, 146–73. New York: Macmillan.

———. 1995. "Body Work: Heterosexual Gender Performances in City Workplaces". In *Mapping Desire*, edited by D. Bell and G. Valentine, 75–95. New York: Routledge.

———. 1997a. "Women/Gender/Feminisms: Doing Feminist Geography". *Journal of Geography in Higher Education* 21:381–400.

———. 1997b. *Capital Culture: Gender at Work in the City*. Oxford: Basil Blackwell.

———. 1999. *Gender, Identity and Place: Understanding Feminist Geographies*. Minneapolis: University of Minnesota Press.

McEwan, C. 2003. "The West and Other Feminisms". In *Handbook of Cultural Geography*, edited by K. Anderson, M. Domosh, S. Pile, and N. Thrift, 405–19. Thousand Oaks, CA: Sage.

McIlwraith, T.F. 1997. *Looking for Old Ontario: Two Centuries of Landscape Change*. Toronto: University of Toronto Press.

McLeay, C. 1997. "Inventing Australia: A Critique of Recent Cultural Policy Rhetoric". *Australian Geographical Studies* 35:40–6.

McLynn, F. 2004. *1759: The Year Britain Became Master of the World*. New York: Atlantic Monthly Press.

McNeill, W.H. 1995. "The Changing Shape of World History". *History and Theory* 34:8–26.

Macpherson, A. 1987. "Preparing for the National Stage: Carl Sauer's First Ten Years at Berkeley". In *Carl O. Sauer: A Tribute*, edited by M.S. Kenzer, 69–89. Corvallis: Oregon State University Press.

MacPherson, Alan D., J.E. McConnell, A. Vance, V. Vanchan. 2006. "The Impact of US Government Antiterrorism Policies on Canada–US Cross-Border Commerce: An Exploratory Study from Western New York and Southern Ontario". *Professional Geographer* 58, no. 3:266–77.

McPherson, B.D. 1998. *Aging as a Social Process: An Introduction to Individual and Population Aging*, 3rd edn. New York: Harcourt Brace.

McQuillan, D.A. 1978. "Territory and Ethnic Identity: Some New Measures of an Old Theme in the Cultural Geography of the United States". In *European Settlement and North American Development: Essays on Geographical Change in Honour and Memory of Andrew Hill Clark*, edited by J.R. Gibson, 3–24. Toronto: University of Toronto Press.

———. 1990. *Prevailing Over Time: Ethnic Adjustment on the Kansas Prairies, 1875–1925*. Urbana: University of Illinois Press.

———. 1993. "Historical Geography and Ethnic Communities in North America". *Progress in Human Geography* 17:355–66.

Malik, N. 2011. "Dubai's Skyscrapers, Stained by the Blood of Migrant Workers". *The Guardian* (27 May). <http://www.guardian.co.uk/commentisfree/2011/may/27/dubai-migrant-worker-deaths>.

Malin, J.C. 1947. *The Grasslands of North America: Prolegomena to its History*. Lawrence, KS: J.C. Malin.

Mann, C.C. 2002. "1491". *Atlantic Monthly* 289, no. 3:41–53.

Mannion, J. 1974. *Irish Settlements in Eastern Canada: A Study of Culture Transfer and Adoption*. Toronto: University of Toronto, Department of Geography, Research Publication no. 12.

Marger, M.N. 2003. *Race and Ethnic Relations: American and Global Perspectives*. Belmont, CA: Wadsworth.

Marsh, G.P. [1864] 1965. *Man and Nature, or Physical Geography as Modified by Human Action*, edited by D. Lowenthal. Cambridge, MA: Harvard University Press.

Marshall, D. 2004. "Making Sense of Remembrance". *Social and Cultural Geography* 5:37–54.

Marshall, H.W. 1995. *Paradise Valley, Nevada: The People and Buildings of an American Place*. Tucson: University of Arizona Press.

Martin, C. 1978. *Keepers of the Game: Indian–Animal Relationships and the Fur Trade*. Berkeley: University of California Press.

Martin, D.G. 1991. *Psychology: Principles and Applications*. Englewood Cliffs, NJ: Prentice Hall.

———. 2007. Bureaucratising Ethics: Institutional Review Boards and Participatory Research. *ACME: An International E-Journal for Critical Geographies* 6:3, 319–28.

Martin, G.J. 1987a. "Foreword". In *Carl O. Sauer: A Tribute*, edited by M.S. Kenzer, ix–xvi. Corvallis: Oregon State University Press.

———. 1987b. "The Ecologic Tradition in American Geography". *Canadian Geographer* 31:74–7.

———, and P.E. James. 1993. *All Possible Worlds: A History of Geographical Ideas,* 3rd edn. New York: Wiley.

Masini, E. 1994. "The Futures of Cultures: An Overview". In *The Futures of Cultures*, compiled by Division of Studies and Programming, UNESCO, 9–28. Paris: UNESCO.

Mason, O. 1895. "Influence of Environment upon Human Industries and Arts". *Annual Report of the Smithsonian Institution* 639–65.

Massey, D.B. 1985. "New Directions in Space". In *Social Relations and Spatial Structures*, edited by D. Gregory and J. Urry, 9–19. New York: Macmillan.

Mathewson, K. 1996. "High/Low, Back/Center: Culture's Stages in Human Geography". In *Concepts in Human Geography*, edited by C. Earle, K. Mathewson, and M.S. Kenzer, 97–125. Lanham, MD: Rowman and Littlefield.

———. 1999. "Cultural Landscape and Ecology II: Regions, Retrospects, Revivals". *Progress in Human Geography* 23:267–81.

———, and M.S. Kenzer, eds. 2003. *Culture, Land, and Legacy: Perspectives on Carl O. Sauer and the Berkeley School. Geoscience and Man 37*. Baton Rouge: Louisiana State University, Department of Geography and Anthropology.

———. 2009. "Carl Sauer and His Critics". In *Carl Sauer on Culture and Landscape: Readings and Commentaries*, edited by W. Denevan and K. Mathewson. Baton Rouge: Louisiana State University Press, 9–28.

Matthews, H., and M. Limb. 1999. "Defining an Agenda for the Geography of Children: Review and Prospect". *Progress in Human Geography* 23:61–90.

Mayer, J.D. 1996. "The Political Ecology of Disease". *Progress in Human Geography* 20:441–56.

Mazrui, A.A. 1997. "Islamic and Western Values". *Foreign Affairs* 76, no. 5:118–32.

Mead, W.R. 1981. *An Historical Geography of Scandinavia*. London and New York: Academic Press.

Meggers, B.J. 1954. "Environmental Limitation on the Development of Culture". *American Anthropologist* 56:801–24.

Meigs, P. 1935. "The Dominican Mission Frontier of Lower California". *University of California Publications in Geography*, no. 7.

Meinig, D.W. 1965. "The Mormon Culture Region: Strategies and Patterns in the Geography of the American West, 1847–1964". *Annals of the Association of American Geographers* 55:191–220.

———. 1969. *Imperial Texas: An Interpretive Essay in Cultural Geography*. Austin: University of Texas Press.

———. 1971. *Southwest: Three Peoples in Geographical Change*. New York: Oxford University Press.

———. 1972. "American Wests: Preface to a Geographical Introduction". *Annals of the Association of American Geographers* 62:159–84.

———. 1976. "Spatial Models of a Sequence of Transatlantic Interactions". In *International Geography* 76, 9 (*Historical Geography*):30–5. Toronto: Pergamon.

———, ed. 1979a. *The Interpretation of Ordinary Landscapes: Geographical Essays*. New York: Oxford University Press.

———. 1979b. "Reading the Landscape: An Appreciation of W.G. Hoskins and J.B. Jackson". In *The Interpretation of Ordinary Landscapes: Geographical Essays*, edited by D.W. Meinig, 195–244. New York: Oxford University Press.

———. 1979c. "Symbolic Landscapes: Some Idealizations of American Communities". In *The Interpretation of Ordinary Landscapes: Geographical Essays*, edited by D.W. Meinig, 164–92. New York: Oxford University Press.

———. 1986. *The Shaping of America: A Geographical Perspective on 500 Years of History. Volume 1: Atlantic America, 1492–1800*. New Haven, CT: Yale University Press.

———. 1993. *The Shaping of America: A Geographical Perspective on 500 Years of History. Volume 2: Continental America, 1800–1867*. New Haven, CT: Yale University Press.

Melko, M. 1969. *The Nature of Civilizations*. Boston: Porter Sargent.

Mellor, M. 1997. *Feminism and Ecology*. New York: New York University Press.

Mels, T., and G. Setten. 2007. "Romance, Practice and Substantiveness: What Do Landscapes Do?". *Geografiska Annaler B* 89:197–202.

Menzies, G. 2002. *1421: The Year China Discovered the World*. London: Bantam Press.

———. 2008. *1434: The Year a Magnificent Chinese Fleet Sailed to Italy and Ignited the Renaissance*. New York: Harper Collins Publishers.

Merchant, C. 1980. *The Death of Nature: Women, Ecology, and the Scientific Revolution*. San Francisco: Harper and Row.

———. 1989. *Ecological Revolutions: Nature, Gender, and Science in New England*. Chapel Hill, NC: University of North Carolina Press.

———. 1990. "Gender and Environmental History". *Journal of American History* 76:1117–21.

———. 1995. *Earthcare: Women and the Environment*. New York: Routledge.

Merrens, H.R. 1969. "The Physical Environment of Early America: Images and Image Makers in Colonial South Carolina". *Geographical Review* 59:530–56.

Meyer, D.K. 2000. *Making the Heartland Quilt: A Geographical History of Settlement and Migration in Early Nineteenth-Century Illinois*. Carbondale: Southern Illinois University Press.

Michaels, A. 1996. *Fugitive Pieces*. New York: Vintage.

Mikesell, M.W. 1960. "Comparative Studies in Frontier History". *Annals of the Association of American Geographers* 50:62–74.

———. 1961. "Northern Morocco: A Cultural Geography". *University of California Publications in Geography* 14:1–136.

———. 1967. "Geographic Perspectives in Anthropology". *Annals of the Association of American Geographers* 57:617–34.

———. 1968. "Landscape". In *International Encyclopedia of the Social Sciences, Volume 8*, 249–64. New York: Macmillan and Free Press.

———. 1969. "The Borderlands of Geography as a Social Science". In *Interdisciplinary Relationships in the Social Sciences*, edited by M. Sherif and C.W. Sherif, 227–48. Chicago: Aldine.

———. 1976. "The Rise and Decline of Sequent Occupance: A Chapter in the History of American Geography". In *Geographies of the Mind: Essays in Historical Geosophy in Honor of John Kirtland Wright*, edited by D. Lowenthal and M.J. Bowden, 149–69. New York: Oxford University Press.

———. 1977. "Cultural Geography". *Progress in Human Geography* 1:460–4.

———. 1978. "Tradition and Innovation in Cultural Geography". *Annals of the Association of American Geographers* 68:1–16.

Miller, V.P., Jr. 1971. "Some Observations on the Science of Cultural Geography". *Journal of Geography* 70:27–35.

Mills, S.F. 1997. *The American Landscape*. Edinburgh: Keele University Press.

Minca, C., ed. 2001. *Postmodern Geography: Theory and Praxis*. Malden, MA: Blackwell.

———. 2007. "Humboldt's Compromise, or the Forgotten Geographies of Landscape". *Progress in Human Geography* 31:179–93.

Ministry of Overseas Indian Affairs (MOIA). 2009. "India and its Diaspora". <http://moia.gov.in/accessories.aspx?aid=10>.

Mitchell, D. 1995. "There's No Such Thing as Culture: Towards a Reconceptualization of the Idea of Culture in Cultural Geography". *Transactions of the Institute of British Geographers* NS 20:102–16.

———. 1996. "Sticks and Stones: The Work of Landscape (A Reply to Judy Walton's 'How Real(ist) Can You Get?')". *Professional Geographer* 48:94–6.

———. 2000. *Cultural Geography: A Critical Introduction*. Malden, MA: Blackwell.

———. 2003. "Cultural Geographers Forum: What are the Five Most Important Principles That Should be Covered in an Introductory Course in Cultural Geography and Why?" *Place and Culture, The Newsletter of the Cultural Geography Specialty Group of the Association of American Geographers*, Spring, 2–3.

———. 2004. "Historical Materialism and Marxism". In *A Companion to Cultural Geography*, edited by J.S. Duncan, N.C. Johnson, and R.H. Schein, 51–65. Malden, MA: Blackwell.

Mitchell, R.D. 1977. *Commercialism and Frontier: Perspectives on the Early Shenandoah Valley*. Charlottesville: University Press of Virginia.

———. 1978. "The Formation of Early American Cultural Regions". In *European Settlement and Development in North America*, edited by J.R. Gibson, 66–90. Toronto: University of Toronto Press.

Mitra, S.K., and R.A. Lewis. 1996. *Subnational Movements in South Asia*. Boulder, CO: Westview Press.

Mogey, J. 1971. "Society, Man and Environment". In *Man and His Habitat*, edited by R.H. Buchanan, E. Jones, and D. McCourt, 79–92. New York: Barnes and Noble.

Monk, J. 1996. "Challenging the Boundaries: Survival and Change in a Gendered World". In *Companion Encyclopedia of Geography: The Environment and Humankind*, edited by I. Douglas, R. Huggett, and M. Robinson, 888–905. New York: Routledge.

———. 2004. "Women, Gender, and the Histories of American Geography". *Annals of the Association of American Geographers* 94:1–22.

Montagu, A. 1942. *Man's Most Dangerous Myth: The Fallacy of Race*. New York: Harper.

———. 1997. *Man's Most Dangerous Myth: The Fallacy of Race*, 6th edn. Walnut Creek, CA: Altamira Press.

Monument Avenue. 2004. http://xroads.virginia.edu/~UG97/monument/begin.html

Morgan, L.H. [1877] 1974. *Ancient Society, or Researches in the Lines of Human Progress from Savagery through Barbarism to Civilization*. Gloucester, MA: Peter Smith.

Morgan, W.B., and R.P. Moss. 1965. "Geography and Ecology: The Concept of the Community and Its Relationship to Environment". *Annals of the Association of American Geographers* 55:339–50.

Morin, K. 2003. "Landscape and Environment: Representing and Interpreting the World". In *Key Concepts in Geography*, edited by S.L. Holloway, S.P. Rice, and G. Valentine, 319–34. London: Sage.

Morley, D., and K. Robins. 2001. *British Cultural Studies: Geography, Nationality, and Identity*. New York: Oxford University Press.

Morrill, R.L. 1965. *Migration and the Spread and Growth of Urban Settlement*. Lund: Gleerup, Lund Studies in Geography, Series B, no. 26.

———. 2002. "Pausing for Breath". In *Geographical Voices: Fourteen Autobiographical Essays*, edited by P. Gould and F.R. Pitts, 211–36. Syracuse, NY: Syracuse University Press.

Mosely, C., and R.E. Asher. 1994. *Atlas of the World's Languages*. New York: Routledge.

Moser, S. 2008. "Personality: A New Positionality". *Area* 40:383–92.

Moss, P., ed. 2002. *Feminist Geography in Practice: Research and Methods*. Malden, MA: Blackwell.

———, and I. Dyck. 2003. "Embodying Social Geography". In *Handbook of Cultural Geography*, edited by K. Anderson, M. Domosh, S. Pile, and N. Thrift, 58–73. Thousand Oaks, CA: Sage.

———, and K. Falconer Al-Hindi, eds. 2008. *Feminisms in Geography: Rethinking Space, Place and Knowledges*. Lanham, MD: Rowman and Littlefield.

Mundey, J. 1981. *Green Bans and Beyond*. London: Angus and Robertson.

Murdoch, J. 1997. "Inhuman/Nonhuman/Human: Actor-Network Theory and the Prospects for a Nondualistic and Symmetrical Perspective on Nature and Society". *Environment and Planning D: Society and Space* 15:731–56.

Murphy, A.B., and D.L. Johnson, eds. 2000a. *Cultural Encounters with the Environment: Enduring and Evolving Geographic Themes*. Lanham, MD: Rowman and Littlefield.

———, and D.L. Johnson. 2000b. "Introduction: Encounters with Environment and Place". In *Cultural Encounters with the Environment: Enduring and Evolving Geographic Themes*, edited by A.B. Murphy

and D.L. Johnson, 1–13. Lanham, MD: Rowman and Littlefield.

Myrdal, G. 1944. *An American Dilemma: The Negro Problem and Modern Democracy*. New York: Harper.

Naess, A. 1989. *Ecology, Community and Lifestyle*, translated and edited by D. Rothenberg. New York: Cambridge University Press.

Nash, C. 2000. "Performativity in Practice: Some Recent Work in Cultural Geography". *Progress in Human Geography* 24:653–64.

———. 2002. "Cultural Geography: Postcolonial Cultural Geographies". *Progress in Human Geography* 26:219–30.

———. 2003. "Cultural Geography: Anti-racist Geographies". *Progress in Human Geography* 27:637–48.

———. 2004. "Postcolonial Geographies: Spatial Narratives of Inequality and Interconnection". In *Envisioning Human Geographies*, edited by P. Cloke, P. Crang, and M. Goodwin, 104–27. New York: Arnold.

Nathanson, P., and K.K. Young. 2001. *Spreading Misandry: The Teaching of Contempt for Men in Popular Culture*. Montreal: McGill-Queen's University Press.

Neisser, U. 1967. *Cognitive Psychology*. New York: Appleton Century Crofts.

———. 1976. *Cognition and Reality: Principles and Implications of Cognitive Psychology*. New York: Freeman.

Netting, R.M. 1977. *Cultural Ecology*. Menlo Park, CA: Cummings.

Neuman, R.P. 2005. *Human Geography in the Making: Political Ecology*. London: Hodder Arnold.

Newcomb, R.M. 1969. "Twelve Working Approaches to Historical Geography". *Yearbook of the Association of Pacific Coast Geographers* 31:27–50.

Newson, L.A. 1976. "Cultural Evolution: A Basic Concept for Human and Historical Geography". *Journal of Historical Geography* 2:239–55.

Newton, M. 1974. "Cultural Preadaptation and the Upland South". In *Man and Cultural Heritage: Papers in Honor of Fred B. Kniffen*, edited by H.J. Walker and W.G. Haag. *Geoscience and Man* 5:143–54. Baton Rouge: Louisiana State University, Department of Geography and Anthropology.

Nietschmann, B.Q. 1973. *Between Land and Water: The Subsistence Ecology of the Miskito Indians, Eastern Nicaragua*. New York: Seminar Press.

Nisbett, J. 2003. *The Geography of Thought: How Asians and Westerners Think Differently . . . and Why*. New York: Free Press.

Noble, A.G. 1984. *Wood, Brick, and Stone: The North American Settlement Landscape*, 2 vols. Amherst: University of Massachusetts Press.

———, and R. Dhussa. 1990. "Image and Substance: A Review of Literary Geography". *Journal of Cultural Geography* 10, no. 2:49–65.

Nolin, C., and J. Stephens. 2010. "We Have to Protect the Investors: 'Development' and Canadian Mining Companies in Guatemala". *Journal of Rural and Community Development*, 5, no. 2:37–70.

Norberg, J. 2001. *In Defence of Global Capitalism*. Stockholm: Timbro.

North, D.C. 1977. "The New Economic History After Twenty Years". *American Behavioral Scientist* 21:187–200.

Norton, W. 1984. *Historical Analysis in Geography*. London: Longman.

———. 1987. "Humans, Land and Landscape: A Proposal for Cultural Geography". *Canadian Geographer* 31:21–30.

———. 1988. "Abstract Cultural Landscapes". *Journal of Cultural Geography* 8, no. 2:67–80.

———. 1989. *Explorations in the Understanding of Landscape: A Cultural Geography*. Westport, CT: Greenwood.

———. 1995. "State Boundaries and Agricultural Change in the South Eastern Australian Wheat Belt: Counterfactual Analyses, 1891–1911". *Australian Geographical Studies* 33:228–41.

———. 2003. "'After-the-Fact Causality': A Different Direction for Cultural Geography". *Area* 35:418–26.

Norwine, J., and T.D. Anderson. 1980. *Geography as Human Ecology*. Lanham, MD: University Press of America.

Nostrand, R.L. 1992. *The Hispano Homeland*. Norman: University of Oklahoma Press.

———. 2001. "The Highland Hispano Homeland". In *Homelands: A Geography of Culture and Place across America*, edited by R.L. Nostrand and L.E. Estaville, 155–67. Baltimore: Johns Hopkins University Press.

———. 2003. *El Cerrito: Eight Generations in a Spanish Village*. Norman: University of Oklahoma Press.

———, and L.E. Estaville. 1993a. "Introduction: The Homeland Concept". *Journal of Cultural Geography* 13, no. 2:1–4.

———, and L.E. Estaville, eds. 1993b. *Special Issue. Journal of Cultural Geography* 13, no. 2.

———, and L.E. Estaville. 2001a. "Introduction: Free Land, Dry Land, Homeland". In *Homelands: A Geography of Culture and Place across America*, edited by R.L. Nostrand and L.E. Estaville, xiii–xxiii. Baltimore: Johns Hopkins University Press.

———, and L.E. Estaville, eds. 2001b. *Homelands: A Geography of Culture and Place across America*. Baltimore: Johns Hopkins University Press.

———, and S.B. Hilliard, eds. 1988. *The American South. Geoscience and Man* 25. Baton Rouge: Louisiana State University, Department of Geography and Anthropology.

Oakes, J., R. Riewe, K. Kinew, and E. Maloney. 1998. *Sacred Lands: Aboriginal World Views, Claims, and Conflicts*. Occasional Publication no. 43. Edmonton: Canadian Circumpolar Institute Press.

O'Brien, P.K., ed. 1999. *Oxford Atlas of World History*. New York: Oxford University Press.

O'Connor, J. 1998. *Natural Causes: Essays in Ecological Marxism*. New York: Guilford Press.

O'Dwyer, B. 1997. "Pathways to Homelessness: A Comparison of Gender and Schizophrenia in Inner-Sydney". *Australian Geographical Studies* 35:294–307.

Oelschlaeger, M. 1991. *The Idea of Wilderness*. New Haven, CT: Yale University Press.

Ohmae, K. 1993. "The Rise of the Region State". *Foreign Affairs* 76, no 2:78–87.

Olds, K. 1995. "Globalization and the Production of New Urban Spaces: Pacific Rim Mega-projects in the Late 20th Century". *Environment and Planning A* 27:1713–43.

Olsen, D.H., and D.J. Timothy. 2002. "Contested Religious Heritage: Differing Views of Mormon Heritage". *Tourism Recreation Research* 27, no. 2:7–15.

Olwig, K. 1980. "Historical Geography and the Society/Nature 'Problematic': The Perspective of J.F. Schouw, G.P. Marsh and E. Reclus". *Journal of Historical Geography* 6:29–45.

———. 1996a. "Nature—Mapping the Ghostly Traces of a Concept". In *Concepts in Human Geography*, edited by C. Earle, K. Mathewson, and M.S. Kenzer, 63–96. Lanham, MD: Rowman and Littlefield.

———. 1996b. "Recovering the Substantive Nature of Landscape". *Annals of the Association of American Geographers* 86:630–53.

———. 2002. *Landscape, Nature, and the Body Politic: From Britain's Renaissance to America's New World*. Madison: University of Wisconsin Press.

———. 2003. "Landscape: The Lowenthal Legacy". *Annals of the Association of American Geographers* 93:871–7.

Ortner, S.B. 1972. "Is Female to Male as Nature Is to Culture?" *Feminist Studies* 1:5–31.

———, and H. Whitehead. 1981. "Introduction: Accounting for Sexual Meaning". In *Sexual Meanings: The Cultural Construction of Gender and Sexuality*, edited by S.B. Ortner and H. Whitehead, 1–27. New York: Cambridge University Press.

Osborne, B.S. 1988. "The Iconography of Nationhood in Canadian Art". In *The Iconography of Landscape*, edited by D. Cosgrove and S. Daniels, 162–78. New York: Cambridge University Press.

———. 1992. "Interpreting a Nation's Identity: Artists as Creators of National Consciousness". In *Ideology and Landscape in Historical Perspective*, edited by A.R.H. Baker and G. Biger, 230–54. New York: Cambridge University Press.

Osella, C., and F. Osella. 2008. "Nuancing the Migrant Experience: Perspectives from Kerala, South India". In *Transnational South Asians: The Making of a Neo-Diaspora*, edited by S. Koshy and R. Radhakrishnan, 146–78. New Delhi and New York: Oxford University Press.

Ostergren, R.C. 1988. *A Community Transplanted: The Trans-Atlantic Experience of a Swedish Immigrant Settlement in the Upper Midwest, 1835–1915*. Madison: University of Wisconsin Press.

———, and J.G. Rice. 2004. *The Europeans: A Geography of People, Culture, and Environment*. New York: Guilford Press.

Ó Tuathail, G. 1995. "Political Geography I: Theorizing History, Gender and World Order amidst Crises of Global Governance". *Progress in Human Geography* 19:260–72.

———. 1996. *Critical Geopolitics: The Politics of Writing Global Space*. Minneapolis: University of Minnesota Press.

Paehlke, R., ed. 1995. *Conservation and Environmentalism: An Encyclopedia*. New York: Garland.

Pahl, R.E. 1965. "Trends in Social Geography". In *Frontiers in Geographical Teaching*, edited by R.J. Chorley and P. Haggett, 81–100. London: Methuen.

Pain, R.H. 1997. "Social Geographies of Women's Fear of Crime". *Transactions of the Institute of British Geographers* NS 22:231–44.

———. 1999. "The Geography of Fear". *Geography Review* 12, no. 5:22–5.

———, and C. Bailey. 2004. "British Social and Cultural Geography: Beyond Turns and Dualisms?" *Social and Cultural Geography* 5:319–29.

———, et al. 2001. *Introducing Social Geographies*. New York: Oxford University Press.

Panelli, R. 2004. *Social Geographies: From Difference to Action*. Thousand Oaks, CA: Sage.

Park, C.C. 1994. *Sacred Worlds: An Introduction to Geography and Religion*. New York: Routledge.

Park, R.E. 1915. "The City: Suggestions for the Investigation of Human Behavior in the City Environment". *American Journal of Sociology* 20:577–612.

———. 1952. *Human Communities: The City and Human Ecology. The Collected Papers of Robert Ezra Park*, vol. 2. Glencoe, IL: Free Press.

———, and E.W. Burgess. 1921. *Introduction to the Science of Sociology*. Chicago: University of Chicago Press.

Parr, H., and C. Philo. 1995. "Mapping 'Mad' Identities". In *Mapping the Subject: Geographies of Cultural Transformation*, edited by S. Pile and N. Thrift, 199–225. New York: Routledge.

Parsons, J.J. 1979. "The Later Sauer Years". *Annals of the Association of American Geographers* 69:9–15.

Paulson, S., and L.L. Gezon, eds. 2005. *Political Ecology across Spatial Scales and Social Groups*. New Brunswick, NJ: Rutgers University Press.

Pawson, E. 1999. "Postcolonial New Zealand". In *Cultural Geographies*, 2nd edn, edited by K. Anderson and F. Gale, 25–50. Harlow, UK: Longman.

Peach, C. 1999. "Social Geography". *Progress in Human Geography* 23:282–8.

———. 2002. "Social Geography: New Religions and Ethnoburbs—Contrasts with Cultural Geography". *Progress in Human Geography* 26:252–60.

Peet, R. 1977a. *Radical Geography*. Chicago: Maaroufa Press.

———. 1977b. "The Development of Radical Geography in the United States". *Progress in Human Geography* 1:240–63.

———. 1985. "The Social Origins of Environmental Determinism". *Annals of the Association of American Geographers* 75:309–33.

———. 1996a. "Structural Themes in Geographical Discourse". In *Companion Encyclopedia of Geography: The Environment and Humankind*, edited by I. Douglas, R. Huggett, and M. Robinson, 860–87. New York: Routledge.

———. 1996b. "Discursive Idealism in the 'Landscape-as-Text' School". *Professional Geographer* 48:96–8.

———, and N. Thrift. 1989. "Political Economy and Human Geography". In *New Models in Geography: Volume 1*, edited by R. Peet and N. Thrift, 3–29. Boston: Unwin Hyman.

———, and M. Watts. 1996. "Liberation Ecology: Development, Sustainability and Environment in an Age of Market Triumphalism". In *Liberation Ecologies: Environment, Development, Social Movements*, edited by R. Peet and M. Watts, 1–45. New York: Routledge.

Pelto, P.J. 1966. *The Nature of Anthropology*. Columbus, OH: Merrill.

Peluso, N.L., and M. Watts, eds. 2001. *Violent Environments*. Ithaca, NY: Cornell University Press.

Penn, M., and F. Lukermann. 2003. "Chorology and Landscape: An Internalist Reading of 'The Morphology of Landscape'". In *Culture, Land, and Legacy: Perspectives on Carl O. Sauer and the Berkeley School*, edited by K. Mathewson and M.S. Kenzer. *Geoscience and Man* 37:233–60. Baton Rouge: Louisiana State University, Department of Geography and Anthropology.

Penrose, J. 1997. "Construction, De(con)struction and Reconstruction. The Impact of Globalization and Fragmentation on the Canadian Nation-State". *International Journal of Canadian Studies* 16:15–49.

Perramond, E.P. 2007. "Tactics and Strategies in Political Ecology Research". *Area* 39:499–507.

Peters, B.C. 1972. "Oak Openings or Barrens: Landscape Evaluation on the Michigan Frontier". *Proceedings of the Association of American Geographers* 4:84–6.

Peterson, J., and J. Anfison. 1985. "The Indian and the Fur Trade: A Review of Recent Literature". *Manitoba History* 10 (Autumn):10–18.

Phillips, R. 2004. "Sexuality". In *A Companion to Cultural Geography*, edited by J.S. Duncan, N.C. Johnson, and R.H. Schein, 265–78. Malden, MA: Blackwell.

Philo, C. 1988. "Conference Report: New Directions in Cultural Geography". *Journal of Historical Geography* 14:178–81.

————. 1989. "Enough to Drive One Mad: The Organization of Space in 19th Century Lunatic Asylums". In *The Power of Geography*, edited by J. Wolch and M. Dear, 258–90. Boston: Unwin Hyman.

————, compiler. 1991. *New Words, New Worlds: Reconceptualising Social and Cultural Geography*. Aberystwyth, Wales: Cambrian Printers.

————, and H. Parr. 2003. "Introducing Psychoanalytic Geographies". *Social and Cultural Geography* 4:283–93.

————, and C. Wilbert, eds. 2000. *Animal Spaces, Beastly Places: New Geographies of Human–Animal Relations*. New York: Routledge.

Pickles, J. 1985. *Phenomenology, Science, and Geography*. New York: Cambridge University Press.

Pile, S. 1993. "Human Agency and Human Geography Revisited: A Critique of "New Models" of the Self". *Transactions of the Institute of British Geographers* NS 18:122–39.

————. 1996. *The Body and the City: Psychoanalysis, Space and Subjectivity*. New York: Routledge.

————, and N. Thrift, eds. 1995. *Mapping the Subject: Geographies of Cultural Transformation*. New York: Routledge.

Pillsbury, R. 1970. "The Urban Street Pattern as a Cultural Indicator". *Annals of the Association of American Geographers* 60:428–46.

Pinkerton, J.P. 1997. "Enviromanticism: The Poetry of Nature as Political Force". *Foreign Affairs* 76, no. 3:2–7.

Pittock, M.G.H. 1999. *Celtic Identity and the British Image*. New York: Manchester University Press.

Platt, R.S. 1962. "The Rise of Cultural Geography in America". In *Readings in Cultural Geography*, edited by P.L. Wagner and M.W. Mikesell, 35–43. Chicago: University of Chicago Press.

Plummer, V. 1993. *Feminism and the Mastery of Nature*. New York: Routledge.

Pocock, D.C.D., ed. 1981. *Humanistic Geography and Literature*. London: Croom Helm.

Pomeranz, K. 2000. *The Great Divergence: China, Europe, and the Making of the Modern World Economy*. Princeton, NJ: Princeton University Press.

Porter, P.W. 1965. "Environmental Potentials and Economic Opportunities: A Background for Cultural Adaptation". *American Anthropologist* 67:409–20.

————. 1978. "Geography as Human Ecology". *American Behavioral Scientist* 22:15–39.

————. 1979. *Food and Development in the Semi-Arid Zone of East Africa*. Syracuse, NY: Syracuse University, Maxwell School of Citizenship and Public Affairs.

————. 1987. "Ecology as Metaphor: Sauer and Human Ecology". *Professional Geographer* 39:414.

Powell, J.M. 1977. *Mirrors of the New World: Images and Image Makers in the Settlement Process*. Folkestone, England: Dawson.

————. 1980. "Taylor, Stefansson and the Arid Centre: An Historic Encounter of 'Environmentalism' and Possibilism". *Journal of the Royal Australian Historical Society* 66:163–83.

Pratt, G. 1992. "Spatial Metaphors and Speaking Positions". *Environment and Planning D: Society and Space* 10, 241–4.

————. 2004. "Feminist Geographies: Spatialising Feminist Politics". In *Envisioning Human Geographies*, edited by P. Cloke, P. Crang, and M. Goodwin, 128–45. New York: Arnold.

————. 2008. "Complexity and Connection". In *Feminisms in Geography: Rethinking Space, Place and Knowledges*, edited by P. Moss and K. Falconer Al-Hindi. Lanham, MD: Rowman and Littlefield.

Pred, A. 1980. *Urban Growth and City-Systems in the United States, 1840–1860*. Cambridge, MA: Harvard University Press.

————. 1984. "Place as Historically Contingent Process: Structuration and the Time Geography of Becoming Places". *Annals of the Association of American Geographers* 74:279–97.

————. 1985. "The Social Becomes the Spatial, the Spatial Becomes the Social: Enclosures, Social Change and the Becoming of Place in the Swedish Province of Skåne". In *Social Relations and Spatial Structures*, edited by D. Gregory and J. Urry, 336–75. London: Macmillan.

————. 2000. *Even in Sweden: Racisms, Racialized Spaces, and the Popular Geographical Imagination*. Berkeley and Los Angeles: University of California Press.

Price, M., and M. Lewis 1993a. "The Reinvention of Cultural Geography". *Annals of the Association of American Geographers* 83:1–17.

————. 1993b. "Reply: On Reading Cultural Geography". *Annals of the Association of American Geographers* 83:520–2.

Prince, H.C. 1971. "Real, Imagined and Abstract Worlds of the Past". In *Progress in Geography, Volume 3*, edited by C. Board et al., 1–86. New York: Arnold.

Proctor, J.D. 1998. "The Social Construction of Nature: Relativist Accusations, Pragmatist and Critical Responses". *Annals of the Association of American Geographers* 88:352–76.

Pryce, W.T.R. 1975. "Migration and the Evolution of Culture Areas: Cultural and Linguistic Frontiers in North-East Wales, 1750 and 1851". *Transactions of the Institute of British Geographers* NS 65:79–107.

Pudup, M.B. 1988. "Arguments within Regional Geography". *Progress in Human Geography* 12:369–90.

Pyle, G.F. 1969. "The Diffusion of Cholera in the United States in the Nineteenth Century". *Geographical Analysis* 1:59–75.

Quani, M. 1982. *Geography and Marxism*. Cambridge, MA: Blackwell.

Quinn, J.A. 1950. *Human Ecology*. Englewood Cliffs, NJ: Prentice Hall.

Qvistrom, M. 2007. "Landscapes Out of Order: Studying the Inner Urban Fringe beyond the Rural-Urban Divide". *Geografiska Annaler B* 89:269–82.

Raby, S. 1973. "Indian Land Surrenders in Southern Saskatchewan". *Canadian Geographer* 17:36–52.

Radcliffe, S. 1994. "(Representing) Post-colonial Women: Authority, Difference and Feminisms". *Area* 26:25–32.

Radding, C. 1997. *Wandering Peoples: Colonialism, Ethnic Spaces, and Ecological Frontiers in Northwestern Mexico, 1700–1850*. Durham, NC: Duke University Press.

Raento, P., and C.J. Watson. 2000. "Gernika, Guernica, Guernica? Contested Meanings of a Basque Place". *Political Geography* 19:707–36.

Raitz, K.B. 1973. "Theology on the Landscape: A Comparison of Mormon and Amish-Mennonite Land Use". *Utah Historical Quarterly* 41:23–34.

———. 1979. "Themes in the Cultural Geography of European Ethnic Groups in the United States". *Geographical Review* 69:79–94.

Rankin, K.N. 2003. "Anthropologies and Geographies of Globalization". *Progress in Human Geography* 27:708–34.

Rapoport, A. 1969. *House Form and Culture*. Englewood Cliffs, NJ: Prentice Hall.

Rappaport, R.A. 1963. "Aspects of Man's Influence on Island Ecosystems: Alteration and Control". In *Man's Place in the Island Ecosystem*, edited by F.R. Fosberg, 155–74. Honolulu: Bishop Museum Press.

Ray, A.J. 1974. *Indians in the Fur Trade: Their Role as Hunters, Trappers and Middlemen in the Lands Southwest of Hudson Bay, 1660–1870*. Toronto: University of Toronto Press.

———. 1996. *I Have Lived Here Since the World Began: An Illustrated History of Canada's Native People*. Toronto: Lester Publishing, Key Porter Books.

Redekop, C. 2000. *Creation & the Environment: An Anabaptist Perspective on a Sustainable World*. Baltimore: Johns Hopkins University Press.

Rediscovering Geography Committee, National Research Council. 1997. *Rediscovering Geography: New Relevance for Science and Society*. Washington: National Academies Press.

Rees, R. 1988. *New and Naked Land: Making the Prairies Home*. Saskatoon: Western Producer Prairie Books.

Rehder, J.B. 2004. *Appalachian Folkways*. Baltimore: Johns Hopkins University Press.

Relph, E. 1970. "An Inquiry into the Relations Between Phenomenology and Geography". *Canadian Geographer* 14:193–201.

———. 1976. *Place and Placelessness*. London: Pion.

———. 1981. *Rational Landscapes and Humanistic Geography*. Totowa, NJ: Barnes and Noble.

———. 1985. "Geographical Experiences and Being-in-the-World: The Phenomenological Origins of Geography". In *Dwelling, Place and Environment: Towards a Phenomenology of Person and World*, edited by D. Seamon and R. Mugerauer, 15–31. New York: Columbia University Press.

———. 1997. "Sense of Place". In *Ten Geographic Ideas That Changed the World*, edited by S. Hanson, 205–26. New Brunswick, NJ: Rutgers University Press.

Renfrew, C. 1988. *Archaeology and Language: The Puzzle of Indo-European*. New York: Cambridge University Press.

Renner, M. 1996. *Fighting for Survival: Environmental Decline, Social Conflict, and the New Age of Insecurity*. New York: Norton.

Reynolds, H. 1982. *The Other Side of the Frontier: Aboriginal Resistance to the European Invasion of Australia*. New York: Penguin.

Rice, J.G. 1977. "The Role of Culture and Community in Frontier Prairie Farming". *Journal of Historical Geography* 3:155–75.

Robbins, P. 2004. "Cultural Ecology". In *A Companion to Cultural Geography*, edited by J.S. Duncan, N.C. Johnson, and R.H. Schein, 180–93. Malden, MA: Blackwell.

Robinson, G.M. 1998. *Methods and Techniques in Human Geography*. New York: Wiley.

Rogers, E.M. 1962. *Diffusion of Innovations*. New York: Free Press of Glencoe.

Rogerson, R.J., and A. Gloyer. 1995. "Gaelic Cultural Revival or Language Decline". *Scottish Geographical Magazine* 111:46–53.

Roinila, M. 2006. The *Finnish Immigrant Experience in North America, 1880–2000: Studies in Cultural Geography*. New York: Edward Mellen Press.

Rolston, H., III. 1997. "Nature for Real: Is Nature a Social Construct?" In *The Philosophy of the Environment*, edited by T.D.J. Chappell, 38–64. Edinburgh: Edinburgh University Press.

Rooney, J.R, W. Zelinsky, and D.R. Loudon, eds. 1982. *This Remarkable Continent: An Atlas of United States and Canadian Society and Cultures*. College Station: Texas A&M University Press.

Rose, A.J. 1972. "Australia as a Cultural Landscape". In *Australia as Human Setting*, edited by A. Rapoport, 58–74. Sydney: Angus and Robertson.

Rose, C. 1981. "William Dilthey's Philosophy of Historical Understanding: A Neglected Heritage of Contemporary Humanistic Geography". In *Geography, Ideology and Social Concern*, edited by D.R. Stoddart, 99–133. Totowa, NJ: Barnes and Noble.

Rose, G. 1993. *Feminism and Geography: The Limits of Geographical Knowledge*. Minneapolis: University of Minnesota Press.

———. 1997. "Situating Knowledges: Positionality, Reflexivities and Other Tactics". *Progress in Human Geography* 21:305–20.

———. 2001. *Visual Methodologies*. Thousand Oaks, CA: Sage.

———, V. Kinnaird, M. Morris, and C. Nash. 1997. "Feminist Geographies of Environment, Nature and Landscape". In *Feminist Geographies: Explorations in Diversity and Difference*, edited by the Women and Geography Study Group of the Royal Geographical Society with the Institute of British Geographers, 146–90. Harlow, UK: Longman.

———, et al. 1997. "Introduction". In *Feminist Geographies: Explorations in Diversity and Difference*, edited by the Women and Geography Study Group of the Royal Geographical Society with the Institute of British Geographers, 1–12. Harlow, UK: Longman.

Rose, N. 2006. *The Politics of Life Itself: Biomedicine, Power and Subjectivity in the Twenty-First Century*. Princeton, NJ: Princeton University Press.

Rosenau, P.M. 1992. *Post-Modernism and the Social Sciences: Insights, Inroads, and Intrusions*. Princeton, NJ: Princeton University Press.

Rosenberg, M. 2004. "Ethnic and Race Relations". In *Sociology: A Canadian Perspective*, edited by L. Tepperman and J. Curtis, 402–32. Toronto: Oxford University Press.

Rossi, I., and E. O'Higgins. 1980. "Unit 1: Theories of Culture and Anthropological Methods". In *People in Culture: A Survey of Cultural Inquiry*, edited by I. Rossi, 31–78. New York: Praeger.

Rostlund, E. 1956. "Twentieth-Century Magic". *Landscape* 5, no. 3:3–6.

Rowles, G. 1978. *Prisoners of Space: Exploring the Geographical Experience of Older People*. Boulder, CO: Westview Press.

Rowntree, L.B. 1996. "The Cultural Landscape Concept in American Cultural Geography". In *Concepts in Human Geography*, edited by C. Earle, K. Mathewson, and M.S. Kenzer, 127–59. Lanham, MD: Rowman and Littlefield.

Roxby, P.M. 1930. "The Scope and Aims of Human Geography". *Scottish Geographical Magazine* 46:276–99.

Ruggie, J.G. 1998. *Constructing the World Polity: Essays on International Institutionalization*. New York: Routledge.

Rupesinghe, K. 1996. "Governance and Conflict Resolution in Multi-ethnic Societies". In *Ethnicity and Power in the Contemporary World*, edited by K. Rupesinghe and V.A. Tishkov, 10–31. New York: United Nations University Press.

Russell, R.J., and F.B. Kniffen. 1951. *Culture Worlds*. New York: Macmillan.

Ryan, J.R. 2004. "Postcolonial Geographies". In *A Companion to Cultural Geography*, edited by J.S. Duncan, N.C. Johnson, and R.H. Schein, 469–84. Malden, MA: Blackwell.

Saarinen, T.F. 1974. "Environmental Perception". In *Perspectives on Environment*, edited by I.R. Manners and M.W. Mikesell, 252–89. Washington: Association of American Geographers, Commission on College Geography Publication, no. 13.

Sachs, J.B., A.D. Mellinger, and J.L. Gallup. 2001. "The Geography of Poverty and Wealth". *Scientific American* 284: 70–5.

Sack, R.D. 2003. *A Geographical Guide to the Real and the Good*. New York: Routledge.

Said, E. 1978. *Orientalism*. New York: Vintage.

———. 1993. *Culture and Imperialism*. London: Chatto and Windus.

Salama, S. 2010. "New Work Permit Rules in UAE from 2011 Will Ease Job Switch". <http://gulfnews.com/news/gulf/uae/employment/new-work-permit-rules-in-uae-from-2011-will-ease-job-switch-1.732116>.

Salerno, R.A. 2003. *Landscapes of Abandonment: Capitalism, Modernity, and Estrangement*. Albany: State University of New York Press.

Salter, C.L. 1971a. "The Mobility of Man". In *The Cultural Landscape*, edited by C.L. Salter, 1–4. Belmont, CA: Duxbury.

———. 1971b. "Introductory Comments to 'Some Curious Analogies in Explorer's Preconceptions of Virginia' by G. Dunbar". In *The Cultural Landscape*, edited by C.L. Salter, 18. Belmont, CA: Duxbury.

———. 1972. "A Speculative Cultural Geography: A Free-for-All Approach to Introductory Cultural Geography". *Journal of Geography* 71:533–40.

———. 1981. "John Steinbeck's *The Grapes of Wrath* as a Primer for Cultural Geography". In *Humanistic Geography and Literature*, edited by D.C.D. Pocock, 142–58. Totowa, NJ: Barnes and Noble.

———, and W.J. Lloyd. 1977. *Landscape in Literature*. Resource Papers for College Geography, no. 76-3. Washington DC: Association of American Geographers.

Samuels, M.S. 1978. "Existentialism and Human Geography". In *Humanistic Geography: Prospects and Problems*, edited by D. Ley and M.S. Samuels, 22–40. London: Croom Helm.

———. 1979. "The Biography of Landscape: Cause and Culpability". In *The Interpretation of Ordinary Landscapes*, edited by D.W. Meinig, 51–88. New York: Oxford University Press.

———. 1981. "An Existential Geography". In *Themes in Geographic Thought*, edited by M.E. Harvey and B.P. Holly, 115–32. London: Croom Helm.

Sanderson, S.K. 1990. *Social Evolutionism: A Critical History*. Cambridge, MA: Blackwell.

———, ed. 1995. *Civilizations and World Systems: Studying World–Historical Change*. Walnut Creek, CA: Altamira Press.

———, and T.D. Hall. 1995. "World System Approaches to World–Historical Change". In *Civilizations and*

World Systems: Studying World-Historical Change, edited by S.K. Sanderson, 95–108. Walnut Creek, CA: Altamira Press.

Sauer, C.O. 1924. "The Survey Method in Geography and Its Objectives". *Annals of the Association of American Geographers* 14:17–33.

———. 1925. "The Morphology of Landscape". *University of California Publications in Geography* 2:19–53.

———. 1927. "Recent Developments in Cultural Geography". In *Recent Developments in the Social Sciences*, edited by E.C. Hayes, 154–212. Philadelphia: J.B. Lippincott.

———. 1931. "Cultural Geography". In *Encyclopedia of the Social Sciences, Volume 6*, 621–4. New York: Macmillan.

———. 1935. Aboriginal Populations of North-Western Mexico. *Ibero-Americana* 10. Berkeley: University of California Press.

———. 1940. *Culture Regions of The World: Outline of Lectures*. Berkeley: University of California at Berkeley, Department of Geography.

———. 1941a. "Foreword to Historical Geography". *Annals of the Association of American Geographers* 31:1–24.

———. 1941b. "The Personality of Mexico". *Geographical Review* 31:353–64.

———. 1952. *Agricultural Origins and Dispersals*. New York: American Geographical Society.

———. 1963. "Historical Geography and the Western Frontier". In *Land and Life: A Selection from the Writings of Carl Sauer*, edited by J. Leighly, 45–52. Berkeley: University of California Press.

———. 1968. "Human Ecology and Population". In *Population and Economics*, edited by P. DePrez, 207–14. Winnipeg: University of Manitoba Press.

———. 1969. *Seeds, Spades, Hearths, and Herds: The Domestication of Animals and Foodstuffs*. Cambridge, MA: MIT Press.

———. 1970. "Plants, Animals and Man". In *Man and His Habitat*, edited by R.H. Buchanan, E. Jones, and D. McCourt, 34–61. London: Routledge and Kegan Paul.

———. 1971. *Sixteenth Century North America*. Berkeley: University of California Press.

———. 1974. "The Fourth Dimension of Geography". *Annals of the Association of American Geographers* 64:189–92.

———. 1981. *Selected Essays, 1963–1975*. Berkeley, CA: Turtle Island Foundation.

———. 1985. *North America: Notes on Lectures by Professor Carl O. Sauer at the University of California at Berkeley, 1936*. Northridge, CA: California State University, Northridge, Department of Geography, Occasional Paper no. 1.

Sayare, S. 2012. "Hopes Raised, Roma in France Still Face a Date with the Bulldozer". *New York Times* (30 August).

Sayer A. 1992. *Method in Social Science: A Realist Approach*, 2nd edn. New York: Routledge.

Scandrett, E., and S. Mukherjee. 2011. "Globalisation and Abstraction in the Bhopal Survivors' Movement". *Interface* 3, no. 1:195–209 (May).

Schaefer, F. 1953. "Exceptionalism in Geography: A Methodological Examination". *Annals of the Association of American Geographers* 43:226–49.

Schein, R.H. 1997. "The Place of Landscape: A Conceptual Framework for Interpreting an American Scene". *Annals of the Association of American Geographers* 87:660–80.

———. 2004. "Cultural Traditions". In *A Companion to Cultural Geography*, edited by J.S. Duncan, N.C.5Johnson, and R.H. Schein, 11–23. Malden, MA: Blackwell.

Schnore, L.F. 1961. "Geography and Human Ecology". *Economic Geography* 37:207–17.

Scholte, J.A. 2000. *Globalization: A Critical Introduction*. New York: St Martin's Press.

Schroeder, W.A. 2002. *Opening the Ozarks: A Historical Geography of Missouri's Ste Genevieve District, 1760–1830*. Columbia: University of Missouri Press.

Scott, H. 2004. "Cultural Turns". In *A Companion to Cultural Geography*, edited by J.S. Duncan, N.C. Johnson, and R.H. Schein, 24–37. Malden, MA: Blackwell.

Seamon, D. 1979. *A Geography of the Lifeworld*. London: Croom Helm.

Seig, L. 1963. "The Spread of Tobacco: A Study in Cultural Diffusion". *Professional Geographer* 15:17–21.

Semple, E. 1911. *Influences of Geographic Environment*. New York: Henry Holt.

Senn, P.R. 1971. *Social Science and Its Methods*. Boston: Holbrook Press Inc.

Shamir, R. 2008. "The Age of Responsibilization: On Market-Embedded Morality". *Economy and Society* 37:1–19.

Sharp, J. 2004. "Feminisms". In *A Companion to Cultural Geography*, edited by J.S. Duncan, N.C. Johnson, and R.H. Schein, 66–78. Malden, MA: Blackwell.

Sharp, J.P. 1994. "A Topology of 'Post' Nationality: (Re)mapping Identity in the Satanic Verses". *Ecumene* 1:65–76.

———. 1996. "Locating Imaginary Homelands: Literature, Geography and Salman Rushdie". *Geojournal* 38:119–27.

———. 2000. "Towards a Critical Analysis of Fictive Geographies". *Area* 32:327–34.

Shelley-Clark, M. 2004. "Cockatoo Island—The Fight for Australia's Heritage". <http://www.afloat.com.au/images/magazine-articles/MAGAZINE/2004/0804/0804_MC_Cockatoo.pdf>.

Shiva, V. 1988. *Staying Alive: Women, Ecology and Development*. London: Zed Books.

Short, J.R. 2001. *Global Dimensions: Space, Place and the Contemporary World*. London: Reaktion Books.

Shortridge, J.R. 1995. *Peopling the Plains: Who Settled Where in Frontier Kansas.* Lawrence: University Press of Kansas.

Shurmer-Smith, P., ed. 2002. *Doing Cultural Geography.* Thousand Oaks, CA: Sage.

———, and K. Hannam. 1994. *Worlds of Desire, Realms of Power: A Cultural Geography.* New York: Arnold.

Sibley, D. 1995. *Geographies of Exclusion: Society and Difference in the West.* New York: Routledge.

———. 1999a. "Outsiders in Society and Space". In *Cultural Geographies*, 2nd edn, edited by K. Anderson and F. Gale, 135–51. Harlow, UK: Longman.

———. 1999b. "Creating Geographies of Difference". In *Human Geography Today*, edited by D. Massey, J. Allen, and P. Sarre, 115–28. Cambridge: Polity Press.

Sidaway, J.D. 2000. "Postcolonial Geographies: An Exploratory Essay". *Progress in Human Geography* 24:591–612.

Silverstone, R. and Hirsch, E., eds. 1992. *Consuming Technologies: Media and Information in Domestic Spaces.* London: Routledge.

Simmons, C.W. 2004. "The Political Economy of Land Conflict in the Eastern Brazilian Amazon". *Annals of the Association of American Geographers* 94:183–206.

Simon, J., and H. Kahn, eds. 1984. *The Resourceful Earth.* Cambridge, MA: Blackwell.

Simpson-Housley, P., and A.H. Mallory, eds. 1987. *Geography and Literature: A Meeting of Disciplines.* Syracuse, NY: Syracuse University Press.

Singer, P. 1993. *Practical Ethics*, 2nd edn. New York: Cambridge University Press.

Skelton, T., and G. Valentine, eds. 1998. *Cool Places: Geographies of Youth Cultures.* New York: Routledge.

Skinner, B.F. 1969. *Contingencies of Reinforcement: A Theoretical Analysis.* New York: Appleton Century Crofts.

Slater, D. 2003. "Cultures of Consumption". In *Handbook of Cultural Geography*, edited by K. Anderson, M. Domosh, S. Pile, and N. Thrift, 147–63. Thousand Oaks, CA: Sage.

Slater, T., and P. Jarvis, eds. 1982. *Field and Forest: An Historical Geography of Warwickshire and Worcestershire.* Norwich, England: Geobooks.

Smil, V. 1993. *Global Ecology: Environmental Changes and Social Flexibility.* New York: Routledge.

Smith, A.D. 1986. *The Ethnic Origins of Nations.* Cambridge, MA: Blackwell.

Smith, C.T. 1965. "Historical Geography: Current Trends and Prospects". In *Frontiers in Geographical Teaching*, edited by R.J. Chorley, 118–43. London: Methuen.

Smith, D.M. 2000. *Moral Geographies: Ethics in a World of Difference.* Edinburgh: Edinburgh University Press.

Smith, J.M., A. Light, and D. Roberts. 1998. "Introduction: Philosophies and Geographies of Place". In *Philosophy and Geography III: Philosophies of Place*, edited by A. Light and J.M. Smith, 1–19. Lanham, MD: Rowman and Littlefield.

———. 2004. "Ethics and the Human Environment". In *A Companion to Cultural Geography*, edited by J.S. Duncan, N.C. Johnson, and R.H. Schein, 209–20. Malden, MA: Blackwell.

Smith, J.S. 1999. "Anglo Intrusion on the Old Sangre de Cristo Land Grant". *Professional Geographer* 51:170–83.

———. 2003. Cultural Geography: A Survey of Perceptions Held by Cultural Geography Specialty Group Members. *Professional Geographer* 55:18–30.

Smith, N. 1984. *Uneven Development: Nature, Capital and the Production of Space.* Oxford: Basil Blackwell.

Smith, R. 1997. *The Norton History of the Human Sciences.* New York: W.W. Norton.

———, and B.N. White. 2004. "Detached from Their Homeland: The Latter-day Saints of Chihuahua, Mexico. *Journal of Cultural Geography* 21 (2):57–76.

Smith, T.R., J.W. Pellegrino, and R.G. Golledge. 1982. "Computational Process Modeling of Spatial Cognition and Behavior". *Geographical Analysis* 14:305–25.

Soja, E.W. 1989. *Postmodern Geographies.* London: Verso.

Solot, M. 1986. "Carl Sauer and Cultural Evolution". *Annals of the Association of American Geographers* 76:508–20.

Sopher, D.E. 1967. *Geography of Religions.* Englewood Cliffs, NJ: Prentice Hall.

———. 1972. "Place and Location: Notes on the Spatial Patterning of Culture". *Social Science Quarterly* 53:321–37.

———. 1981. "Geography and Religions". *Progress in Human Geography* 5:510–24.

Southwick, C.H. 1996. *Global Ecology in Human Perspective.* New York: Oxford University Press.

Sowell, T. 1994. *Race and Culture: A World View.* New York: Basic Books.

———. 1996. *Migrations and Cultures: A World View.* New York: Basic Books.

Sparke, M. 2000. "Excavating the Future in Cascadia: Geoeconomics and the Imagined Geographies of a Cross-Border Region". *BC Studies* 127: 5–44.

Spate, O.H.K. 1952. "Toynbee and Huntington: A Study in Determinism". *Geographical Journal* 118:406–28.

Spencer, C., and M. Blades. 1986. "Pattern and Process: A Review Essay on the Relationship Between Behavioral Geography and Environmental Psychology". *Progress in Human Geography* 10:230–48.

Spencer, J.E. 1960. "The Cultural Factor in Underdevelopment". In *Essays on Geography and Economic Development*, edited by N. Ginsburg. Department of Geography Research Paper no. 62. Chicago: University of Chicago.

———. 1978. "The Growth of Cultural Geography". *American Behavioral Scientist* 22:79–92.

———, and R.J. Horvath. 1963. "How Does an Agricultural Region Originate?" *Annals of the Association of American Geographers* 53:74–92.

———, and W.L. Thomas Jr. 1973. *Introducing Cultural Geography*. New York: Wiley.

Speth, W.W. 1967. "Environment, Culture and the Mormon in Early Utah". *Yearbook of the Association of Pacific Coast Geographers* 29:53–67.

———. 1987. "Historicism: The Disciplinary World View of Carl O. Sauer". In *Carl O. Sauer: A Tribute*, edited by M.S. Kenzer, 11–39. Corvallis: Oregon State University Press.

Spring, D., and R. Spring. 1974. *Ecology and Religion in History*. New York: Harper and Row.

Squire, S.J. 1994. "Accounting for Cultural Meanings: The Interface Between Geography and Tourism Studies Re-Examined". *Progress in Human Geography* 18:1–16.

Stacey, J. 1998. *Brave New Families: Stories of Domestic Upheaval in Late-Twentieth-Century America*. Berkeley: University of California Press.

Stanfield, J.H. 1997. "Preface to the Sixth Edition". In *Man's Most Dangerous Myth: The Fallacy of Race*, 6th edn, by A. Montagu, 25–9. Walnut Creek, CA: Altamira Press.

Stanislawski, D. 1946. "The Origin and Spread of the Grid Pattern Town". *Geographical Review* 36:105–20.

Starrs, P.F. 2009. "Meetinghouses in the Mormon Mind: Ideology, Architecture, and Turbulent Streams of an Expanding Church". *Geographical Review* 99:323-355.

Steiman, L.B. 1998. *Paths to Genocide: Antisemitism in Western History*. New York: St Martin's Press.

Stein, B. 2006. "In Class Warfare, Guess Which Class Is Winning". *New York Times* (26 November). <http://www.nytimes.com/2006/11/26/business/yourmoney/26every.html>.

Steward, J. 1936. "The Economic and Social Basis of Primitive Bands". In *Essays in Anthropology Presented to A.L. Kroeber*, edited by R.H. Lowie, 331–45. Berkeley: University of California Press.

———. 1938. *Basin-Plateau Aboriginal Sociopolitical Groups*. Washington, DC: Smithsonian Institution.

———. 1955. *Theory of Culture Change: The Methodology of Multilinear Evolution*. Urbana: University of Illinois Press.

Stewart, K. 2007. *Ordinary Affects*. Durham, NC: Duke University Press.

Stilgoe, J.R. 1982. *Common Landscape of America: 1580–1845*. New Haven, CT: Yale University Press.

Stoddart, D.R. 1965. "Geography and the Ecological Approach". *Geography* 50:242–51.

———. 1966. "Darwin's Impact on Geography". *Annals of the Association of American Geographers* 56:683–98.

———. 1967. "Organism and Ecosystem as Geographical Models". In *Models in Geography*, edited by R.J. Chorley and P. Haggett, 511–48. London: Methuen.

Straussfogel, D. 1997. "World Systems Theory: Toward a Heuristic and Pedagogic Conceptual Tool". *Economic Geography* 73:118–30.

Straw, R.A., and H.T. Blethen, eds. 2004. *High Mountains Rising: Appalachia in Time and Place*. Urbana: University of Illinois Press.

Strohmayer, U. 2003. "Cultural Geographers Forum: What are the Five Most Important Principles That Should be Covered in an Introductory Course in Cultural Geography and Why?" *Place and Culture, The Newsletter of the Cultural Geography Specialty Group of the Association of American Geographers*, Spring, 5–6.

Stump, R.W. 1986. "Introduction". *Journal of Cultural Geography* 7, no. 1:1–3.

Sumartojo, R. 2004. "Contesting Place: Anti-gay and Lesbian Hate Crime in Columbus, Ohio". In *Spaces of Hate: Geographies of Discrimination and Intolerance in the USA*, edited by C. Flint, 87–108. New York: Routledge.

Sutter, R.E. 1973. *The Next Place You Come To: A Historical Introduction to Communities in North America*. Englewood Cliffs, NJ: Prentice Hall.

Tarrow, S. 1992. "Mentalities, Political Cultures, and Collective Action Frames". In *Frontiers in Social Movement Theory*, edited by A. Morris and C. Mueller, 174–202. New Haven, CT: Yale University Press.

Tatham, G. 1951. "Environmentalism and Possibilism". In *Geography in the Twentieth Century*, edited by G. Taylor, 128–62. New York: Philosophical Library.

Taussig, M. 1987. *Shamanism, Colonialism, and the Wild Man: A Study in Terror and Healing*. Chicago and London: University of Chicago Press.

Taylor, P.J. 1991. "The English and Their Englishness: 'A Curiously Mysterious, Elusive and Little Understood People'". *Scottish Geographical Magazine* 107:146–61.

———. 1992. "Understanding Global Inequalities: A World-Systems Approach". *Geography* 77:10–21.

———. 1994. "The State as Container: Territoriality in the Modern World-System". *Progress in Human Geography* 18:151–62.

———. 1996. *The Way the Modern World Works: World Hegemony to World Impasse*. New York: Wiley.

———, and C. Flint. 2000. *Political Geography: World-Economy, Nation-State, and Locality*, 4th edn. New York: Prentice Hall.

Taylor, T.G. 1928. *Australia in Its Physiographic and Economic Aspects*, 5th edn. Oxford: Clarendon.

———. 1937. *Environment, Race, and Migration*. Toronto: University of Toronto Press.

———. 1951. *Australia: A Study of Warm Environments and Their Effect on British Settlement*, 6th enlarged edn. New York: Dutton.

Teather, E.K., ed. 1999. *Embodied Geographies: Spaces, Bodies and Rites of Passage*. New York: Routledge.

Teich, M., R. Porter, and B. Gustafsson, eds. 1997. *Nature and Society in Historical Context*. New York: Cambridge University Press.

Terkenli, T.S. 1995. "Home as a Region". *Geographical Review* 85:324–34.

Tetlock, P.E., and A. Belkin, eds. 1996. *Counterfactual Thought Experiments in World Politics: Logical, Methodological, and Psychological Perspectives.* Princeton, NJ: Princeton University Press.

Tharoor, S. 2007. *The Elephant, the Tiger, and the Cellphone: Reflections on India in the Twenty-first Century.* New Delhi: Penguin Books India.

Thomas, W.L., Jr. 1956. "Introductory". In *Man's Role in Changing the Face of the Earth*, edited by W.L. Thomas Jr., C.O. Sauer, M. Bates, and L. Mumford, xxi–xxxviii. Chicago: University of Chicago Press.

———, C.O. Sauer, M. Bates, and L. Mumford, eds. 1956. *Man's Role in Changing the Face of the Earth.* Chicago: University of Chicago Press.

Thompson, E.P. 1968. *The Making of the English Working Class.* London: Gollancz.

Thompson, G.F., ed. 1995. *Landscape in America.* Austin: University of Texas Press.

Thomson, G.M. 1975. *The North-West Passage.* London: Secker and Warburg.

Thornthwaite, C.W. 1940. "The Relation of Geography to Human Ecology". *Ecological Monograph* 10:343–8.

Thrift, N. 1994. "Taking Aim at the Heart of the Region". In *Human Geography: Society, Space and Social Science*, edited by D. Gregory, R. Martin, and G. Smith, 200–31. New York: Macmillan.

———. 2007. *Non-representational Theory: Space, Politics, Affect.* London: Routledge.

Tickell, A., and Peck, J. 1996. "The Return of the Manchester Men: Men's Words and Men's Deeds in the Remaking of the Local State". *Transactions of the Institute of British Geographers* 595–616.

Toney, M.B., C. Keller, and L.M. Hunter. 2003. "Regional Cultures, Persistence and Change: A Case Study of the Mormon Culture Region". *Social Science Journal* 40:431–45.

Toynbee, A. 1934–61. *A Study of History*, 12 vols. New York: Oxford University Press.

Trépanier, C. 1991. "The Cajunization of French Louisiana: Forging a Regional Identity". *Geographical Journal* 157:161–71.

Trigger, B.G. 1982. "Response of Native Peoples to European Contact". In *Early European Settlement and Exploitation in Atlantic Canada*, edited by G.M. Story, 139–55. St John's: Memorial University of Newfoundland.

———. 1985. *Natives and Newcomers: Canada's "Heroic Age" Reconsidered.* Kingston: McGill–Queen's University Press.

———. 1998. *Sociocultural Evolution: Calculation and Contingency.* Malden, MA: Blackwell.

Tuan, Y.-F. 1971. "Geography, Phenomenology and the Study of Human Nature". *Canadian Geographer* 15:181–92.

———. 1972. "Structuralism, Existentialism and Environmental Perception". *Environment and Behavior* 4:319–42.

———. 1974. *Topophilia: A Study of Environmental Perception, Attitudes, and Values.* Englewood Cliffs, NJ: Prentice Hall.

———. 1975. "Images and Mental Maps". *Annals of the Association of American Geographers* 65:205–13.

———. 1977. *Space and Place: The Perspective of Experience.* Minneapolis: University of Minnesota Press.

———. 1989. *Morality and Imagination: Paradoxes of Progress.* Madison: University of Wisconsin Press.

———. 1997. "Alexander von Humboldt and His Brother: Portrait of an Ideal Geographer in Our Time". Los Angeles: University of California at Los Angeles, Department of Geography, Alexander von Humboldt Lecture.

Turner, B.S. 1994. *Orientalism, Postmodernism and Globalism.* New York: Routledge.

Turner, F.J. 1961. *Frontier and Section*, edited by R.A. Billington. Englewood Cliffs, NJ: Prentice Hall.

Turner, T. 1993. "Anthropology and Multiculturalism: What Is Anthropology That Multiculturalists Should Be Mindful of It?" *Cultural Anthropology* 8:411–29.

Tylor, E.B. 1916. *Anthropology: An Introduction to the Study of Man and Civilization.* New York: Appleton.

———. 1924. *Primitive Culture: Researches into the Development of Mythology, Philosophy, Religion, Language, Art, and Custom.* New York: Brentano's.

United Nations Development Programme. 2002. *Arab Human Development Report, 2002: Creating Opportunities for Future Generations.* New York: United Nations Development Programme.

Unstead, J.F. 1922. "Geography and Historical Geography". *Geographical Journal* 41:55–9.

Unwin, T. 1992. *The Place of Geography.* New York: Longman.

Valentine, G. 1989. "The Geography of Women's Fear". *Area* 21:385–90.

———. 1993. "(Hetero)sexing Space: Lesbian Perceptions and Experiences of Everyday Spaces". *Environment and Planning D: Society and Space* 11:395–413.

———. 1998. "'Sticks and Stones May Break My Bones': A Personal Geography of Harrassment". *Antipode* 30:305–32.

———. 1999. "A Corporeal Geography of Consumption". *Environment and Planning D: Society and Space* 17:329–51.

———. 2001. *Social Geographies: Space and Society.* New York: Prentice Hall.

———, D. Sporton, and K.B. Nielsen. 2008. "Language Use on the Move: Sites of Encounter, Identities, and Belonging". *Transactions of the Institute of British Geographers* NS 33:376–87.

Vayda, A.P., and R.A. Rappaport. 1968. "Ecology, Cultural and Noncultural". In *Introduction to Cultural Anthropology*, edited by J.A. Clifton, 477–97. Boston: Houghton Mifflin.

Veblen, T.T. 1977. "Native Population Decline in Totoncapán, Guatemala". *Annals of the Association of American Geographers* 67:484–99.

Vertovec, S. 2007. "Super-Diversity and Its Implications". *Ethnic and Racial Studies* 30, no. 6:1024–54.

Vibert, E. 1997. *Trader's Tales: Narratives of Cultural Encounters in the Columbia Plateau, 1807–1846.* Norman: University of Oklahoma Press.

Vidal de la Blache, P. 1926. *Principles of Human Geography,* edited by E.D. Martonne, translated by M.T. Bingham. London: Constable.

von Maltzahn, K.E. 1994. *Nature as Landscape: Dwelling and Understanding.* Kingston: McGill–Queen's University Press.

Wacker, P.O. 1968. *The Musconetcong Valley of New Jersey: A Historical Geography.* New Brunswick, NJ: Rutgers University Press.

——. 1975. *Land and People: A Cultural Geography of Pre-Industrial New Jersey, Origins and Settlement Patterns.* New Brunswick, NJ: Rutgers University Press.

——, and P.G.E. Clemens. 1995. *Land Use in Early New Jersey: A Historical Geography.* Newark: New Jersey Historical Society.

Wagner, P.L. 1958a. "Nicoya: A Cultural Geography". *University of California Publications in Geography* 12:195–250.

——. 1958b. "Remarks on the Geography of Language". *Geographical Review* 48:86–97.

——. 1972. *Environments and Peoples.* Englewood Cliffs, NJ: Prentice Hall.

——. 1974. "Cultural Landscapes and Regions: Aspects of Communication". In *Man and Cultural Heritage: Papers in Honor of Fred B. Kniffen,* edited by H.J. Walker and W.G. Haag. *Geoscience and Man* 5:133–42. Baton Rouge: Louisiana State University, Department of Geography and Anthropology.

——. 1975. "The Themes of Cultural Geography Rethought". *Yearbook of the Association of Pacific Coast Geographers* 37:7–14.

——. 1990. "Review of *Explorations in the Understanding of Landscape: A Cultural Geography* by W. Norton". *Journal of Geography* 89:40–1.

——. 1994. "Foreword: Culture and Geography: Thirty Years of Advance". In *Re-reading Cultural Geography,* edited by K.E. Foote, P.J. Hugill, K. Mathewson, and J.M. Smith, 3–8. Austin: University of Texas Press.

——. 1996. *Showing Off: The Geltung Hypothesis.* Austin: University of Texas Press.

——, and M.W. Mikesell. 1962. "The Themes of Cultural Geography". In *Readings in Cultural Geography,* edited by P.L. Wagner and M.W. Mikesell, 1–24. Chicago: University of Chicago Press.

Walker, H.J., and R.A. Detro, eds. 1990. *Cultural Diffusion and Landscapes: Selections by Fred B. Kniffen. Geoscience and Man* 27. Baton Rouge: Louisiana State University, Department of Geography and Anthropology.

Walker, P.A. 2003. "Reconsidering 'Regional' Political Ecologies: Toward a Political Ecology of the Rural American West". *Progress in Human Geography* 27:7–24.

——. 2005. "Political Ecology: Where is the Ecology?" *Progress in Human Geography* 29:73–82.

——. 2007. "Political Ecology: Where is the Politics?" *Progress in Human Geography* 31:363-9.

Walker, R. 1997. "Unseen and Disbelieved: A Political Economist among Cultural Geographers". In *Understanding Ordinary Landscapes,* edited by P. Groth and T.W. Bressi, 162–73. New Haven, CT: Yale University Press.

Wallerstein, I. 1974a. *The Modern World-System: Capitalist Agriculture and the Origins of the European World-Economy in the Sixteenth Century.* New York: Academic Press.

——. 1974b. "The Rise and Future Demise of the World-Capitalist System: Concepts for Comparative Analysis". *Comparative Studies in Society and History* 16:387–415.

——. 1980. *The Modern World-System II: Mercantilism and the Consolidation of the European World-Economy, 1600–1750.* New York: Academic Press.

——. 1989. *The Modern World-System III: The Second Era of Great Expansion of the Capitalist World-Economy, 1730–1840s.* New York: Academic Press.

——. 1996. "The Global Picture". In *The Age of Transition: Trajectory of the World System, 1945–2025,* coordinated by T.K. Hopkins and I. Wallerstein, 209–25. Atlantic Highlands, NJ: Zed Books.

Walmsley, D.J., and G.J. Lewis. 1984. *Human Geography: Behavioral Approaches.* London: Longman.

Walton, J. 1995. "How Real(ist) Can You Get?" *Professional Geographer* 47:61–5.

Walton-Roberts, M. 2004. "Globalization, National Autonomy and Non-Resident Indians". *Contemporary South Asia* 13, no. 1:53–69.

Ward, K.G., and M. Jones. 1999. "Researching Local Elites: Reflexivity, 'Situatedness' and Political-Temporal Contingency". *Geoforum* 30, 4:301–12.

Warf, B., ed. 2012. *Encounters and Engagements Between Economic and Cultural Geography.* Dordrecht: Springer.

——, and S. Arias, eds. 2008. *The Spatial Turn: Interdisciplinary Perspectives.* Abingdon, UK, and New York: Routledge.

Warren, K.J. 1997. "Taking Empirical Data Seriously: An Ecofeminist Philosophical Perspective". In *Ecofeminism: Women, Culture, Nature,* edited by K.J. Warren, 3–20. Bloomington: Indiana University Press.

——. 2000. *Ecofeminist Philosophy: A Western Perspective on What It Is and Why It Matters.* Lanham, MD: Rowman and Littlefield.

Watson, J.B. 1913. "Psychology as the Behaviorist Views It". *Psychological Review* 20:158–77.

Watson, J.W. 1951. "The Sociological Aspects of Geography". In *Geography in the Twentieth Century*, edited by G. Taylor, 463–99. New York: Philosophical Library.

Watts, M. 2007. "Petro-Insurgency or Criminal Syndicate? Conflict and Violence in the Niger Delta". *Review of African Political Economy* 34, no. 114, 637–60.

———. 1983. *Silent Violence: Food, Famine, and Peasantry in Northern Nigeria*. Berkeley: University of California Press.

Watts, S.J., and S.J. Watts. 1978. "On the Idealist Alternative in Geography and History". *Professional Geographer* 30:123–7.

Webb, N.L. 1990–1. "Deconstruction and Human Geography: Exploring Four Basic Themes". *South African Geographer* 18:123–33.

Webb, W.P. 1931. *The Great Plains*. New York: Grosset and Dunlap.

———. 1964. *The Great Frontier*. Austin: University of Texas Press.

Webber, M.J., and D.L. Rigby. 1996. *The Golden Age Illusion: Rethinking Postwar Capitalism*. New York: Guilford Press.

Weitz, E.D. 2003. *A Century of Genocide: Utopias of Race and Nation*. Princeton, NJ: Princeton University Press.

Wekerle, G.R., and C. Whitzman. 1995. *Safe Cities: Guidelines for Planning*. New York: Van Nostrand Reinhold.

Wertz, F.J. 1998. "The Role of the Humanistic Movement in the History of Psychology". *Journal of Humanistic Psychology* 38:42–70.

West, R.C., and J.P. Augelli. 1966. *Middle America: Its Lands and Peoples*. Englewood Cliffs, NJ: Prentice Hall.

West Edmonton Mall. n.d. "Map and Directory: The Wonder of It All".

Western, J. 1981. *Outcast Cape Town*. Minneapolis: University of Minnesota Press.

Whatmore, S. 1999. "Hybrid Geographies: Rethinking the 'Human' in Human Geography". In *Human Geography Today*, edited by J. Allen and P. Sarre, 22–39. Cambridge: Polity Press.

———. 2002. *Hybrid Geographies: Natures, Cultures, Spaces*. Thousand Oaks, CA: Sage.

———. 2009. "Nature". In *The Dictionary of Human Geography*, 5th edn, edited by D. Gregory, R. Johnston, G. Pratt, M. Watts, and S. Whatmore. Padstow, UK: Wiley-Blackwell, 492–4.

———, and L. Thorne. 1998. "Wild(er)ness: Reconfiguring the Geographies of Wildlife". *Transactions of the Institute of British Geographers* NS 23:435–54.

Wheeler, J.O. 2002. "Editorial: From Urban Economic to Social/Cultural Urban Geography, 1980–2001". *Urban Geography* 23:97–102.

White, C.L., and G.T. Renner. 1936. *Geography: An Introduction to Human Ecology*. New York: Appleton Century.

———, and ———. 1948. *Human Geography: An Ecological Study of Society*. New York: Appleton Century Crofts.

White, G.W. 2000. *Nationalism and Territory: Constructing Group Identity in Southeastern Europe*. Lanham, MD: Rowman and Littlefield.

White, L., Jr. 1967. "The Historical Roots of our Ecological Crisis". *Science* 155:1203–7.

White, L.A. 1949. *The Science of Culture: A Study of Man and Civilization*. New York: Grove Press.

White, R. 1990. "Environmental History, Ecology, and Meaning". *Journal of American History* 76:1111–16.

White, D.F., and C. Wilbert, eds. 2009. *Technonatures: Environments, Technologies, Spaces, and Places in the Twenty-First Century*. Waterloo, ON: Wilfrid Laurier University Press.

Whittlesey, D. 1929. "Sequent Occupance". *Annals of the Association of American Geographers* 19:162–5.

Widdis, R. 1993. "Saskatchewan: The Present Cultural Landscape". In *Three Hundred Prairie Years*, edited by H. Epp, 142–57. Regina: Canadian Plains Research Centre.

———. 2010. "Crossing an Intellectual and Geographic Border: The Importance of Migration in Shaping the Canadian–American Borderland at the Turn of the Twentieth Century". *Social Science History* 34, no. 4:445–97.

Wilkinson, D. 1987. "Central Civilization". *Comparative Civilizations Review* 17:31–59.

Wilkinson, H.R. 1951. *Maps and Politics: A Review of the Ethnographic Cartography of Macedonia*. Liverpool: Liverpool University Press.

Williams, M. 1974. *The Making of the South Australian Landscape*. New York: Academic Press.

———. 1983. "The Apple of My Eye: Carl Sauer and Historical Geography". *Journal of Historical Geography* 9:1–28.

———. 1987. "Historical Geography and the Concept of Landscape". *Journal of Historical Geography* 15:92–104.

Williams, R. 1958. *Culture and Society, 1780–1950*. New York: Columbia University Press.

———. 1976. *Keywords: A Vocabulary of Culture and Society*. New York: Oxford University Press.

———. 1981. *Culture*. London: Fontana.

Willsher, K. 2010. "France's Deportation of Roma Shown to be Illegal in Leaked Memo, Say Critics". *The Guardian* (13 September). <http://www.guardian.co.uk/world/2010/sep/13/france-deportation-roma-illegal-memo>.

Wilmer, F. 1997. "Identity, Culture, and Historicity". *World Affairs* 160, no. 1:3–16.

Wilson, A. 1992. *The Culture of Nature*. Cambridge, MA: Blackwell.

Wilson, A.G. 1981. *Geography and the Environment: Systems Analytical Methods*. New York: Wiley.

Wilson, C., and P. Groth, eds. 2003. *Everyday America: Cultural Landscape Studies After J.B. Jackson*. Berkeley: University of California Press.

Wilson, J.Q. 1963. "Planning and Politics: Citizen Participation in Urban Renewal". *Journal of the American Institute of Planners*, 29(4), 242–9.

Wilson, R.K. 1999. "'Placing Nature': The Politics of Collaboration and Representation in the Struggle for La Sierra in San Luis, Colorado". *Ecumene* 6:1–28.

Wilton, R.D. 1998. "The Constitution of Difference: Space and Psyche in Landscapes of Exclusion". *Geoforum* 29:173–85.

Wishart, D. 2004. "Period and Region". *Progress in Human Geography* 28:305–19.

———, A. Warren, and R.H. Stoddard. 1969. "An Attempted Definition of a Frontier Using a Wave Analogy". *Rocky Mountain Social Science Journal* 6:73–81.

Wisner, B. 1978. "Does Radical Geography Lack an Approach to Environmental Relations?" *Antipode* 10:84–95.

Wissler, C. 1917. *The American Indian: An Introduction to the Anthropology of the New World*. New York: McMurtrie.

———. 1923. *Man and Culture*. New York: Crowell.

Withers, C.W.J. 1988. *Gaelic Scotland: The Transformation of a Culture Region*. London: Routledge.

———. 1995. "How Scotland Came to Know Itself: Geography, National Identity and the Making of a Nation". *Journal of Historical Geography* 21:371–97.

Wolch, J.R., and J. Emel. 1995. "Bringing the Animals Back In". *Environment and Planning D: Society and Space* 13:632–6.

———, J. Emel, and C. Wilbert. 2003. "Reanimating Cultural Geography". In *Handbook of Cultural Geography*, edited by K. Anderson, M. Domosh, S. Pile, and N. Thrift, 184–206. Thousand Oaks, CA: Sage.

Wolf, E. 1982. *Europe and the People without History*. Berkeley: University of California Press.

Wolpert, J. 1964. "The Decision Process in Spatial Context". *Annals of the Association of American Geographers* 54:537–58.

Women and Geography Study Group of the Royal Geographical Society with the Institute of British Geographers. 1997. *Feminist Geographies: Explorations in Diversity and Difference*. Harlow, UK: Longman.

Wood, D. 2002. "The Map as a Kind of Talk: Brian Harley and the Confabulation of the Inner and Outer Voice". *Visual Communication* 1, no. 2:139–61.

Wood, W.B. 2001. "Geographic Aspects of Genocide: A Comparison of Bosnia and Rwanda". *Transactions of the Institute of British Geographers* NS 26:57–75.

Worster, D. 1984. "History as Natural History: An Essay on Theory and Method". *Pacific Historical Review* 53:1–19.

———. 1988. "Doing Environmental History". In *The Ends of the Earth: Perspectives on Modern Environmental History*, edited by D. Worster, 289–307. New York: Cambridge University Press.

———. 1990. "Transformations of the Earth: Toward an Agroecological Perspective in History". *Journal of American History* 76:1087–106.

Wright, E.O. 1983. "Gidden's Critique of Marxism". *New Left Review* 138:11–35.

Wright, J.B., and C.L. Campbell. 2008. "Landscape Change in Hispano and Chicano Villages of New Mexico". *Geographical Review* 98:551–65.

Wright, J.K. 1947. "Terra Incognitae: The Place of Imagination in Geography". *Annals of the Association of American Geographers* 37:1–15.

Wyckoff, W. 1999. *Creating Colorado: The Making of a Western American Landscape, 1860–1940*. New Haven, CT: Yale University Press.

Yapa, L.S. 1977. "The Green Revolution: A Diffusion Model". *Annals of the Association of American Geographers* 67:350–9.

———. 1996. "Innovation Diffusion and Paradigms of Development". In *Concepts in Human Geography*, edited by C. Earle, K. Mathewson, and M.S. Kenzer, 231–70. Lanham, MD: Rowman and Littlefield.

Yeoh, B.S.A., and S. Huang. 1998. "Negotiating Public Space: Strategies and Styles of Migrant Female Domestic Workers in Singapore". *Urban Studies* 35, no. 3:583–602.

———, and P. Teo. 1996. "From Tiger Balm Gardens to Dragon World: Philanthropy and Profit in the Making of Singapore's First Cultural Theme Park". *Geografiska Annaler B* 76:27–42.

Yorgason, E.R. 2003. *Transformation of the Mormon Culture Region*. Urbana: University of Illinois Press.

Young, G.L. 1974. "Human Ecology as an Interdisciplinary Concept: A Critical Inquiry". *Advances in Ecological Research* 8:1–105.

———. 1983. "Introduction". In *Origins of Human Ecology*, edited by G.L. Young, 1–9. Stroudsburg, PA: Hutchinson Press.

Zelinsky, W. 1961. "An Approach to the Religious Geography of the United States: Patterns of Church Membership in 1952". *Annals of the Association of American Geographers* 51:139–93.

———. 1973. *The Cultural Geography of the United States*. Englewood Cliffs, NJ: Prentice Hall.

———. 1980. "North America's Vernacular Regions". *Annals of the Association of American Geographers* 70:1–16.

———. 1988. *Nation into State: The Shifting Symbolic Foundations of American Nationalism*. Chapel Hill: University of North Carolina Press.

———. 1992. *The Cultural Geography of the United States*, revised edn. Englewood Cliffs, NJ: Prentice Hall.

———. 1994. *Exploring the Beloved Country: Geographic Forays into American Society and Culture*. Iowa City: University of Iowa Press.

———. 1995. "Review of *Re-Reading Cultural Geography*, edited by K.E. Foote, P.J. Hugill, K. Mathewson, and

J.M. Smith". *Annals of the Association of American Geographers* 85:750–3.

———. 1997. "Seeing beyond the Dominant Culture". In *Understanding Ordinary Landscapes*, edited by P. Groth and T.W. Bressi, 157–61. New Haven, CT: Yale University Press.

Zimmerer, K.S. 1994. "'Human Ecology and the "New Ecology"': The Prospect and Promise of Integration". *Annals of the Association of American Geographers* 84:108–25.

———. 1996a. "Ecology as Cornerstone and Chimera in Human Geography". In *Concepts in Human Geography*, edited by C. Earle, K. Mathewson, and M.S. Kenzer, 161–88. Lanham, MD: Rowman and Littlefield.

———. 1996b. *Changing Fortunes: Biodiversity and Peasant Livelihood in the Peruvian Andes*. Berkeley: University of California Press.

———, and T.J. Bassett. 2003. *Political Ecology: An Integrative Approach to Geography and Environment-Development Studies*. New York: Guilford Press.

———, and K.R. Young, eds. 1998. *Nature's Geography: New Lessons for Conservation in Developing Countries*. Madison: University of Wisconsin Press.

Index